This volume contains the proceedings of the international workshop on the 'Evolution of Interstellar Matter and Dynamics of Galaxies' held at the Center for Theoretical Study in Prague in May 1991. All of the invited lectures are included, covering a broad area embracing star formation inside molecular clouds, HI supershells in interstellar space, and the dynamical evolution of the interstellar medium in galaxies. Contributed and poster papers are also included.

Evolution of Interstellar Matter and Dynamics of Galaxies

Evolution of Interstellar Matter and Dynamics of Galaxies

Proceedings of the International Conference, Prague, 21–25 May 1991

Center for Theoretical Study, Workshop No. 1

Edited by
Jan Palouš,
Center for Theoretical Study, Prague
W. Butler Burton,
Leiden Observatory
Per Olof Lindblad,
Stockholm Observatory

CAMBRIDGE
UNIVERSITY PRESS

Published by the Press Syndicate of the University of Cambridge
The Pitt Building, Trumpington Street, Cambridge CB2 1RP
40 West 20th Street, New York, NY 10011–4211, USA
10 Stamford Road, Oakleigh, Victoria 3166, Australia

First published 1992

Printed in Great Britain at the University Press, Cambridge

A catalogue record for this book is available from the British Library

Library of Congress cataloguing in publication data available

ISBN 0 521 41984 0 hardback

CONTENTS

Z.Lojda: Welcome Address 1
I.M.Havel: Introduction 3
L.Perek: Going a Long Way 5

Part I: SHELLS, SUPERSHELLS, WORMS, CHIMNEYS AND SUPERNOVA REMNANTS **11**

C.Heiles: Shells, Supershells, Worms, Chimneys—and the 'Worm–Ionized Medium' 12
W.B. Burton, T.M. Bania, Dap Hartmann, and Tang Yuan: Anomalous–Velocity HI Structures Associated with Infrared Cirrus 25
N.V. Bystrova: Neutral Hydrogen Observations of Extended Galactic Structures 58
D.J. Bomans, U. Hopp: Supergiant Shells in the Irregular Galaxy NGC 2366 63
Jan Palouš: 3D Simulations of Expanding Bubbles 65
S.A. Silich: 3D Supershells 72
R. Kunze, H.W. Yorke, R. Spurzem: Numerical Simulations of Multiple Supernova Remnants in a Galactic Disk with a Clumpy Interstellar Medium ... 77
J. Franco, P. Bodenheimer, G. Tenorio–Tagle, and M. Różyczka: Supernovae in a Porous Interstellar Medium: Single and Multi–Supernova Remnants 83
M.–M. Mac Low, R.I. Klein, C.F. McKee, M.L. Norman, and J.M. Stone: MHD Shock Interactions with Interstellar Clouds 93
N. Junkes, E. Fürst: Supernova Remnants as Sites of Star Formation 95
A. Ferrara, G. Einaudi: Galactic Fountains and Instabilities 101
A.I. Gómez de Castro, R.E. Pudritz, P. Bastien: Magnetic Field Geometry Towards the High Velocity Clouds (HVCs) 108
F. Comerón: The Collision of Large High Velocity Clouds with the Galactic Disk 114
A.M. Bykov: The Effect of Cosmic Rays on Superbubble Dynamics 116
E.A. Dorfi: Evolution of SNR's with Cosmic Rays 122
P. Polechová, P. Mayer: Infrared Properties of Supernova Parent Galaxies 128
J. Franco, A. Ferrara: Dynamical Evolution of Dust Grains: Dusty Flows in Galaxies 130
C. Heiles: Possible Detection of the D I 327 MHz Line 136
V. Vanýsek: Models of the Deuterium Enrichment in ISMC Dust Grains 138

J. Köppen, A. Acker, M. Samland, & B. Stenholm: Abundances from Planetary Nebulae: The Past ISM ... 141

J.A. Lopez: Evolutionary Effects in Bipolar Planetary Nebulae ... 143

E.B. Kostyakova: The Study of Planetary Nebula Variability ... 145

M. Wilmes, J. Köppen: The N–O Relation in the Interstellar Medium ... 147

J. Guarinos: Visual Interstellar Extinction in the Galactic Disk ... 149

D. Proga, R. Szczerba: Peak Position and Width of the 2175 ÅBump in the MRN Model ... 151

M. Šolc, H. Danielczyk: SiC and Iron Rich Grains in Cometary Dust ... 153

Part II: INTERSTELLAR CLOUDS AND STAR FORMATION **157**

Leo Blitz: Large Scale Properties of Giant Molecular Clouds ... 158

B.G. Elmegreen: Propagating Star Formation ... 178

J.M. Dickey, E. Brinks: Is the Cloud Population of M31 Different from That of the Milky Way? ... 197

H. Pongracic, S.J. Chapman, J.R. Davies, M.J. Disney, A.H. Nelson, and A.P. Whitworth: Numerical Simulations of Star Formation ... 210

H.W. Yorke, R. Spurzem, R. Kunze: The Evolution of Clumpy Gas in Galaxies . 222

L.S. Marochnik, A.A. Suchkov: State of the Interstellar Medium, Phase Transitions, and Galactic Shocks ... 234

I.G. Kolesnik: Gravitationally Induced Giant Molecular Cloud Formation Inside Superclouds ... 249

I. Kovalenko, Yu. Shchekinov: Heat and Radiation Driven Implosion of Interstellar Clouds ... 251

T. Zimmermann, J. Stutzki, G. Winnewisser: The Fractal Appearance of the Closest Molecular Cloud L1457 ... 254

C. Kramer, J. Stutzki, G. Winnewisser: Multiline Studies in Orion ... 257

C. Feldt, H.J.Wendker: HI Self-Absorption in a Molecular Cloud ... 260

M. Tapia, M. Roth, J.A. Lopez: High Resolution IR Images of the Young Cluster GM–24 ... 263

A. Just, B.M. Deiss: The Interaction by Dynamical Friction of Embedded Young Stellar Clusters and the ISM ... 264

S.A. Kutuzov, L.P. Ossipkov, E.M. Nezhinskij: The Dynamics of the Subsystem of Open Clusters in the Two–Component Model of our Galaxy ... 266

P. Kroupa, Ch.A. Tout: The Mass Function For Faint Stars ... 269

N. Schneider, J. Stutzki, G. Winnewisser: The KOSMA 3m Radiotelescope ... 272

PART III: EVOLUTION OF GALAXIES **275**

D.M. Elmegreen, B.G. Elmegreen: Density Wave Resonances in Galaxies 276

Chi Yuan: Formation of the Inner Ring in the Grand Design Galaxies 286

F. Combes, S. Garcia–Burillo, M. Gerin: Gas Cloud Dynamics in Spiral Galaxies 291

F. Ferrini: Multi–Phase Models for the Evolution of Galaxies 304

G. Gilmore: The Gravitational Field Near the Sun 318

D. Pfenniger: Gravitational Dynamics of Galaxies 328

J.A. Sellwood: Gas Flows in Spirals and Bars 343

A.B. Romeo: The Cold Interstellar Gas and the Stability of Thick Galactic Discs .355

I. Dvořák: Dynamical Systems: Transdisciplinary Mathematical Tool or Fiction? .360

B. Koribalski, M. Dahlem: Dynamics of HI–Gas in the Peculiar Galaxy NGC 1808 372

F. Matteucci, M. Stiavelli: Abundance Gradients in Ellipticals 374

I.G. Kolesnik: Polar–Ring Galaxies and the Possibility of their Formation in Triaxial Potential 376

C. Hunter, P.T. de Zeeuw: Triaxial Galaxy Model with Stäckel Potentials 381

P. Berczik, I.G. Kolesnik: Hoag's Object as a Polar–Ring Galaxy 383

P. Berczik, I.G. Kolesnik: IC 2006 as an Elliptical Galaxy with a Polar Ring 385

Part IV: INTERACTING GALAXIES AND STARBURSTS **387**

I.F. Mirabel: Interstellar Gas in Colliding Galaxies 388

A.A. Suchkov, V.G. Berman: "Hot" Model of Galaxy Formation: Evolution of a Cloudy Protogalaxy Interacting with a Galactic Superwind 392

Yu. Shchekinov: Starbursts Stimulated by Thermal Conductivity and Heating Radiation 401

N. Brouillet, P. Notni, W. Bronkalla, A. Baudry, C. Henkel: Star Formation Activity in the Interacting Galaxy NGC 2976 406

Gastão B. Lima Neto, F. Combes: Merging in Groups of Galaxies 408

Part V: CONCLUDING REMARKS **411**

I.R. King: Star Formation in Galaxies—Some Thoughts on This Workshop 412

Foreword

The Center for Theoretical Study was founded in Prague in 1990 as an institution devoted to advanced theoretical interdisciplinary research on subjects ranging from mathematical and natural sciences to the humanities. The CTS intends to create opportunities to overcome barriers of scientific specialization, to stimulate close international cooperation, and to influence in a positive way the existing universities and scientific institutions in Czechoslovakia. The Center is affiliated with Charles University in Prague.

Sponsoring international Workshops is one of the principal activities of the Center. CTS Workshop No. 1, held in Prague from May 21 to May 25, 1991, was organized by the working group in astronomy. Such a Workshop seemed appropriate since astronomy is a subject central between mathematics and physics and its substance is inherently interdisciplinary. CTS Workshop No. 2 will be held in 1992, and will be devoted to Phase Transitions: Physics, Mathematics, Biology, Society.

The CTS Workshop No. 1 on the "Evolution of Interstellar Matter and Dynamics of Galaxies" was organized in cooperation with the Astronomical Institute of the Czechoslovak Academy of Sciences. The organizers wished to make it clear that cooperation between the Academy and Charles University is not only possible but also very fruitful.

The backbone of the scientific program of the Workshop was centered on invited lectures. In the present volume, we publish all of them. They cover a broad area embracing star formation inside molecular clouds, HI supershells in interstellar space, and dynamical evolution of the interstellar medium in galaxies. Contributed and poster papers are also included in these proceedings.

The Workshop was ceremonially opened in the old Aula Magna of Charles University; all lectures took place in the Blue and Green halls of the University. The non-scientific program included astronomical sight–seeing in Prague, a concert of the music festival Prague Spring, and a trip to the Karlštejn Castle.

We wish to express our thanks to the members of the organizing committees, particularly to Jitka Bošková, Jiřina Dočkálková, Dáša Hanáková, Bruno Jungwiert, Jiři Kopecký, and Jindřiška Příhodová. We are especially grateful to Marion Blonk, in Leiden, and Jiři Kopecký, in Prague, for their invaluable help in editing the proceedings.

Jan Palouš *W. Butler Burton* *Per Olof Lindblad*

Preface

For the participants of the workshop *Evolution of Interstellar Matter and Dynamics of Galaxies*, 21 – 25 May 1991, it was felt as a privilege to attend this Workshop No. 1 of the Center for Theoretical Study in Prague. The creation of and the plans for this new Center characterize the ambitions of the scientists of the country and of the new directorship of the Charles University as spurred by the recent political development in Czechoslovakia.

The ceremonial opening in the Aula Magna, where the Vice–Rector Professor Zdeněk Lojda addressed the participants in fluent latin, recalled the old history of this famous university. The friendly reception given by the Rector, Professor Radim Palouš, was marked by the hopes and challenges of the future.

The evolution of interstellar matter and the dynamics of galaxies are both important fields of modern astronomy. The combination of these fields is the area where the Center for Theoretical Study plans to concentrate its efforts in astronomy over the near future. Thus, it was highly appropriate to assemble in Prague a number of researchers active in these fields for discussions within the frame of this the Center's first workshop. These discussions, including posters, are documented in the present proceedings.

The charming city of Prague in the center of Europe, with so much of old culture, is the city where Johannes Kepler met with Tycho Brahe, a meeting that was crucial to the modern development of Science. This is indeed an appropriate place for astronomers to meet.

Per Olof Lindblad

LIST OF PARTICIPANTS:

Andrle P., Astronomical Institute, Academy of Sciences, Budečská 6, 120 23, Prague 2, Czechoslovakia

Blitz L., Astronomy. Program, University of Maryland, College Park, MD 20742, U.S.A.

Bomans D., Universität Sternwarte, Auf dem Hügel 71, D-5300 Bonn 1, Germany

Brouillet N., M.P.I. für Radioastronomie, Auf dem Hügel 69, D-5300 Bonn 1, Germany

Burton W.B., Sterrewacht Leiden, Leiden University, Postbus 9513, 2300 RA Leiden, The Netherlands

Bykov A., A.F.Ioffe Institute of Physics and Technology, 194021 St Petersburg, USSR

Bystrova N.V., SAO, 196140 St Petersburg, Pulkovo, USSR

Canto J., Instituto de Astronomia, UNAM Ap.Postal 70-264, MX-04510 Mexico D.F., Mexico

Combes F., Observatoire de Meudon Demirm, Observatoire de Meudon, F-92190 Meudon, France

Comeron F., Dep. d'Astronomia & Meteor. Univ. of Barcelona, E-08028 Barcelona, Spain

Dickey J., Dept. of Astronomy, Univ. of Minnesota, 116 Church Street S.E., Minneapolis, Minnesota 55455, U.S.A.

Dorfi E., Institut für Astronomie, Türkenschanzstr.17, A-1180 Wien, Austria

Ducati J.R., Dept. of Astr., Inst. de Fisica, Avenue Bento Goncalves, c.p. 1505, 91500-Porto Alegre, RS-Brasil, Brasil

Dvořák I., Proslunĕná 15, 152 00, Prague 5, Czechoslovakia

Egret D., Observatoire de Strasbourg, 11 rue de l'Université, F-67000 Strasbourg, France

Elmegreen B.G., IBM Watson Research Center, P.O.Box 218, Yorktown Heights, NY 10598, U.S.A.

Elmegreen D.M., Dept. Phys. and Astronomy, Vassar College, Poughkeepsie, NY 12601, U.S.A.

Feldt Ch., Univ. Hamburg, Hamburger Sternwarte, Gojenbergsweg 112, D-2050 Hamburg 80, Germany

Ferrara A., Univ. di Firenze, Dip. di Astronomia, Largo Fermi 5, 50125 Firenze, Italy

Ferrini F., Dipartimento di Fisica, Univ. di Pisa, Piazza Torricelli 2, Pisa, Italy

Fiedler R., NRL Code 4210, Washington D.C., 20375-5000, U.S.A.

Franco J., Instituto de Astronomia, UNAM Apartado postal 70-264, MX-04510 Mexico D.F., Mexico

Mexico D.F., Mexico
Gilmore G., Institute of Astronomy, Madingley Road, Cambridge CB3 OHA, U.K.
Gómez de Castro A.I., IUE Observat.Villa Franca d.Castillo, P.O. Box 50727, 2808-Madrid, Spain
Grygar J., Institute of Physics, Řež, Czechoslovakia
Guarinos J., Observatoire Astronomique, 11 rue de l'Universite, F-67000 Strasbourg, France
Havel I.M., CTS, Charles University, Celetná 20, 111 00, Prague 1, Czechoslovakia
Heiles C., University of California, Astronomy, Joan Road, Berkeley CA 94720, U.S.A.
Hermann F., MPI für Extraterr. Physik, Karl-Schwarzschild-Strasse 1, Garching bei München, Germany
Hunter Ch., Math. Dept., Florida State University, Tallahassee, FL 32306-3027, U.S.A.
Jungwiert B., Astronomical Institute, Charles University, Švédská 8, 150 00, Prague 5, Czechoslovakia
Junkes N., Inst. für. Theor. Physic, Univ. Sternwarte, Olshausenstr. 40, D-2300 Kiel 1, Germany
Just A., Inst. für Theor. Physik, Univ. Frankfurt, Robert-Mayer-str.8-10, D-6000 Frankfurt 11, Germany
King I.R., Astronomy Dept., Univ.of California, Berkeley, CA 94720, U.S.A.
Ko Ch.-M., MPI für Kernphysik, Garching bei München, Germany
Kolesnik I.G., Main Astronomical Observatory, Ukrainian Academy of Science, 252127 Kiev, Goloseevo, Ukraina
Kopecký J., Astronomical Institute, Academy of Sciences, Budečská 6, 120 23, Prague 2, Czechoslovakia
Köppen J., Inst. für Theor. Astrophysik, Im Neuenheimer Feld 561, D-6900 Heidelberg, Germany
Koribalski B., M.P.I. für Radioastronomie, Auf dem Hugel 69, D-5300 Bonn 1, Germany
Kostyakova E.B., Sternberg State Astronomical Institute, 117234 Moscow, USSR
Kramer C., Physikal. Inst. Universität Köln, Zulpicher Strasse 77, D-5000 Köln 41, Germany
Kroupa P., Institute of Astronomy, Madingley Rd., Cambridge CB3 OHA, U.K.
Kunze R., Inst.für Theoretische Physik, Oslhausenstr.40, D-2300 Kiel 1, Germany
Kutuzov S.A., Leningrad State University, Bibliotechnaya ploshchad 2, 198904 St Petersburg, USSR
Lindblad P.O., Stockholm Observatory, S-13336 Saltsjöbaden, Sweden
Lojda Z., Charles University, Prague, Czechoslovakia
Lopez J.A., Instituto de Astronomia, UNAM-OAN, P.O.Box 439027, San Diego,

Ca 92143-9027, Mexico
Mac Low M.-M., Ames Research Center, Mail Stop 245-3, Moffet Field, CA 94035, U.S.A.
Malkan M., Department of Astronomy, UCLA Los Angeles, Calif. 90024-1562, U.S.A.
Marochnik L., Space Inst. of USSR Acad. of Sci., Profsojuznaja 84/32, 117810 Moscow, USSR
Matteucci F., MPI für Astrophysics, D-4086 Garching bei München, Germany
Mirabel I.F., Cen. Saclay, 91191 Gif-Sur-Yvette, France
Nelson A.H., Dept. of Physics, Univ.of Wales, P.O. Box 913, Cardiff CF1 3TH, U.K.
Nezhinsky E.M., Institute for Theoretical Astronomy, Naberezhnaja Kutuzova 10, 191187 St Petersburg, USSR
Ollongren A., Dep. of Computer Sc., Leiden Univ., Niels Bohr Weg 1, Nl-2333 CA Leiden, The Netherlands
Ossipkov Z.P., Inst.of Num.Maths, Leningrad St.Univ., Bibliotechnaya ploshchad 2, St Petersburg, USSR
Palouš J., Astronomical Institute, Academy of Sciences, Budečská 6, 120 23, Prague 2, Czechoslovakia
Papayannopoulos Th., Panepistimiopolis, GR-157 83, Zografos, Greece
Perek L., Astronomical Institute, Academy of Sciences, Budečská 6, 120 23, Prague 2, Czechoslovakia
Pfenniger D., Observatoire de Geneve, CH-1290, Sauverny, Switzerland
Polechová P., Štefánikova Hvězdárna, Praha Petřín, Prague, Czechoslovakia
Příhoda P., Prague Observatory, Prague, Czechoslovakia
Příhodová J., Astronomical Institute, Academy of Sciences, Budečská 6, 120 23, Prague 2, Czechoslovakia
Proga D., Universitet Kopernika, ul. Chopina 12/18, Pl-87100 Torun, Polska
Ruprecht J., Astronomical Institute, Academy of Sciences, Budečská 6, 120 23, Prague 2, Czechoslovakia
Sehnal L., Astronomical Institute, Academy of Sciences, Ondřejov Observatory, 251 65, Czechoslovakia
Sellwood J.A., Space Telescope Science Institute, 3700 San Martin Drive, Baltimore, MD 21218, U.S.A.
Shchekinov Y., Inst.of Physics, University of Rostov, Stachki 194, 344 104 Rostov on Don, USSR
Silich S., Main Astr. Observ., Ukrainian Acad. Sci., Kiev, USSR
Šolc M., Astronomical Institute, Charles University, Švédská 8, 150 00, Prague 5, Czechoslovakia
Šolcová A., Technical Museum, Prague, Czechoslovakia

Spurzem R., Inst. für Theor. Phys. & Observ., Olshausenstr. 40, D-2300 Kiel 1, Germany

Stařecký T., Astronomical Institute, Academy of Sciences, Budečská 6, 120 23, Prague 2, Czechoslovakia

Štork R., Astronomical Institute, Charles University, Švédská 8, 150 00, Prague 5, Czechoslovakia

Suchkov A., Astrophys. Dep., Rostov St. University, Zorge 5, Rostov-on-Don, 344104 USSR

Torra J., Dept. D'Astronomia. Univ. de Barcelona, Av. Diagonal 647, E-08028 Barcelona, Spain

Trnovská E., CTS, Charles University, Celetná 20, 111 00, Prague 1, Czechoslovakia

Vanýsek V., Astronomical Institute, Charles University, Švédská 8, 150 00, Prague 5, Czechoslovakia

Wolf M., Astronomical Institute, Charles University, Švédská 8, 150 00, Prague 5, Czechoslovakia

Yorke H.W., Inst. f. Astron. & Astrophys. Am Hubland, D-8700 Würzburg, Germany

Yuan Chi, Dep of Physics, Tsing Hua University, Taiwan

Zimmermann T., KOSMA, Phys. Inst., Univ. zu Köln, Zulpricher Str. 77, D-5000 Köln 41, Germany

Welcome address

Z. LOJDA

Vice-President of the Charles University

DOMINI CLARISSIMI ATQUE DOCTISSIMI IN PRAESIDIO, PARTICIPANTES ILLUSTRISSIMAE ET ILLUSTRISSIMI IN SYMPOSIO INTERNATIONALI DE ASTRONOMIA, HOSPITES HONORATI, AMICI CARISSIMI:

Honorem mihi duco vos in aula magna universitatis Carolinae Pragensis in nomine rectoris magnifici salutare. Universitas nostra antiquissima in Europa centrali est et anno millesimo trecentesimo quadragesimo octavo a rege bohemico et imperatore romano Carolo quarto in lingua latina condita est. Lingua ista in universitate intra quadringentos annos resonabat et nunc intra promotiones et celebrationes dicitur quam ob rem vobis salutationen in lingua ista feci. Conventum prosperum efficacemque et mansionem suavem Pragae in urbe pulcherrima vobis opto.

Ladies and gentlemen:

It is a great pleasure for me to welcome you in the great hall of the Charles university, the oldest university in the central and east Europe. It was founded by the Bohemian king and Roman emperor Charles IV in the Latin language in 1348. This language was used for 400 years as teaching language and you can hear it during graduation ceremonies also nowadays. This is the reason for my Latin greeting address. This hall has witnessed many important cultural and scientific events. I have no doubt that your meeting will also belong to them. The hall is in the university's oldest building founded in the 12th century and has even after its reconstruction a very nice atmosphere reminding ancient as well as recent festive meetings and ceremonies. I would like to draw your attention to the tapestry on the frontal wall. In its center you see the enlarged seal of the university which has been used since the 14th century. The kneeing university's founder Charles IV with the royal Bohemian crown adorned with three lillies places the founding charter of the university under the protection of the standing Saint Wenceslas, patron of Bohemia. On the tapistery there are also symbols of lands belonging to the Bohemian kingdom and at the top symbols of the four faculties belonging to the university since its founding:

the pelican - faculty of medicine; the earth, moon and star - faculty of arts (philosophy); the balance - faculty of law; and the eagle - faculty of theology. Notice, please, that the faculty of arts is demonstrated with astronomic pictures. Of the seven "free arts" (elementary trivium: grammar, rhetoric and logic and subsequent quadrivium : mathematics, geometry, astronomy, and music) the least attention was given to grammar and rhetoric. Instructions was based on Aristotle, supplemented by the learning and commentaries of medieval authors. The basis of all social-political teachings was the course of Aristotle's Metaphysics, Ethics and Politics. Euclid's work was the basis in geometry and astronomy. The later achieved a significant level. Almagestum Ptolemai, the famous astronomical work, was also studied. Beside the University the astronomy was actually astrology and the interest in events in the sky was dictated by the idea of connection between them and the human fate. The astronomical science in the sixteenth century had in Prague several famous names. It is first of all the physician at the court of Rudolph II Tadeáš Hájek of Hájek, known as Nemicus. His work "Tractatus de sphaera" was used at the University besides Ptolemy's Almagest in studies of astronomy. Hájek studied comets. He described a new one in his treaties "De investigatione loci novae stellae" and "Dialexis de nova et prius incognita stella"(1574). The merits of Hájek were recognized by Tycho Brahe, who came to Prague in 1599 and contributed very much to the foundation of modern astronomy particularly by his observations of the planet Mars. The name of the Danish astronomer Tycho Brahe is well known in Prague although he worked here 2 years only. His tomb is in the church of St. Mary of Týn, which is very close to the university and you certainly will pay a visit there. The last famous astronomer working in Prague in this time was Johannes Kepler, who spent in Prague 12 years, used Brahe's study and continued an excellent astronomical work. He condemned astrology, formulated two of his rules and wrote contributions to the optics. After the death of Rudolph II in 1612 he leaved Prague for Linz.

Ladies and gentlemen, from these examples you can see that the astronomy has had a very high level at Charles university so that it is legitime to be the host of your meeting. I wish you a prosperous proceeding and an enjoyable stay in Prague.

Introduction

I.M.HAVEL

Center for Theoretical Study, Charles University, Prague

One of the most profound, but not yet completely appreciated, intellectual outcomes of scientific endeavor in the twentieth century has been a new vision of the relationship between the Large and the Small. Some old beliefs were shaken, among them the reductionist attitude that the Small is essentially simpler than the Large and the fragmentary practice of scientific specializations restricting the scope of study to entities and events at a limited spacial and temporal scale.

We have learned from fractal geometry that the Small may be precisely as complex as the Large and we know from high energy physics that the Small and the Large may obey quite different laws. But the lesson of astrophysics is that the Small and the Large form a unity: there is an intimate relationship between the behavior of large-scale objects on the one hand and events occuring at the smallest scale on the other hand. Thus, by building a bridge between the Large and the Small - as well as between the Slow and the Fast, the Dense and the Sparse, and the Bright and the Dark - astrophysics plays a dominant role among physical sciences, similarly as physics plays a dominant role among natural sciences.

This volume contains overviews contributed papers and posters presented at the workshop on Evolution of Interstellar Matter and Dynamics of Galaxies. It is the first workshop of our Center for Theoretical Study, a young and small international research institution affiliated with Charles University in Prague. The Center has a difficult but challenging aim: to facilitate at least temporary cooperation of scientists of different fields and from different countries, to create a stimulating environment for their interaction, and to contribute to the flow of knowledge across the borders of traditional disciplines (including exact sciences, natural sciences, social sciences, and humanities).

We believe that the new topics, concepts, and theories with a salient transdisciplinary character, which appeared in the last two or three decades, are symptomatic for a revival of an integrative trend in science. A typical and well-known example of such a topic are the ubiquitous dynamical systems and the associated nonlinear and chaotic phenomena. Dynamical system theory provides a common language for processes in as different domains as galactic astronomy and brain research. There are several other cross-disciplinary topics, some of them less developed or lacking a proper common language, some others yet to be discovered. We hope to be able to form special ad hoc working groups in our Center for such topics, with various size, duration, and professional structure. There are reasons to believe that mathematicians will offer an integrative language and that computers will provide a magical tool for simulations under arbitrary shifts of scale of space and time. Cognitive scientists may try to indicate the physical ingredient of mind and philosophers the humanly ingredient of nature.

The subject area of our first workshop and of this volume fits rather well into the scheme of our Center and I would like to thank to all scientists who contributed in whatever way, but specifically to Jan Palouš who conceived the idea of the meeting, who was its main *spiritus agens*, and who lead the team of enthusiasts who undertook the organizational burden.

Prague, August 1991

GOING A LONG WAY

LUBOŠ PEREK

Astronomical Institute, Czechoslovak Academy of Sciences, Prague

The atmosphere of this historic hall requires that a few words be said on what happened in the past, at least selecting a few moments of interest.

Astronomy was taught at the Charles University since its foundation in 1348. The purpose of teaching astronomy was, of course, not the quest for new knowledge but to meet practical requirements of the calendar, in particular determining the dates of moving feasts. The level of teaching was that of antique science as represented by Ptolemy's Almagest and its Arabic translations and elaborations. It is reported that the name of the first professor of astronomy was Havel of Strahov, a personal physician of Charles IV.

Original thinking was slow to appear. It can be found in writings of Tadeáš Hájek z Hájku, known in Latin as Thaddeus Hagecius, who became deeply interested in astronomy after the appearance of Nova Cassiopeiae 1572. Finding no parallax of the Nova, Hájek concluded that it had to be more distant than the Moon and that, consequently, the sphere of stars was not as fixed as traditional astronomy required.

Of paramount importance was the proposal by Hájek to the emperor Rudolph II to invite Tycho Brahe from Denmark to his court and to establish an observatory. Brahe found favorable conditions in Prague. There were artisans capable of manufacturing precise sextants and quadrants which added to Brahe's own collection of astronomical instruments. Brahe had contacts with the scientific community in Prague, among them rabbi David Gans, an author of a textbook on astronomy in hebraeic. They must have worked rather closely together because Gans reported details about Tycho's observing methods.

Brahe, in his turn, proposed to the emperor the name of another visiting professor, Johannes Kepler. Thus originated a team of excellent instrument makers, of an exceptional observer and of a theoretician who was much ahead of his time. Kepler, a strong supporter of Copernicus, spent 12 years in Prague and during that time wrote many of his books on astronomy and formulated his law on elliptical orbits and the law of areas. Being no historian I cannot be quite sure but it is my opinion that Kepler was the first to use the idea of a mathematical model with a highly critical eye on the reliability of his observing material while searching for a simple and elegant interpretation. He had many followers in times to come.

The thirty years war destroyed the entire scientific life in central Europe. While Newton was formulating his theories and while large observatories with new instruments were established in France and England, astronomy in this part of the world was in the hands of the jesuit order. Astronomical methods were still needed for geographic orientation in new countries visited by jesuits and therefore taught at the Prague University but the heliocentric explanation of planetary motions was forbidden. It is one of the paradoxes which often visit this country, that at a time when no scientific developments were possible, the Prague observatory in the Clementinum was founded by a jesuit, Joseph Stepling. And, as happened many times since, the main obstacle in a meaningful development of the observatory was a lack of funds. Also the strong competition of the most advanced astronomical establishment in the Austrian empire, the Vienna Observatory, was felt in the provincial town of Prague.

By the middle of the nineteenth century Prague Observatory was a very old-fashioned institution located at an unfavorable site at the bottom of the Vltava valley on a tower where the instruments had to be carried to the balcony for observations. In 1882, when Prague university split into a Czech and a German university, the number of posts in astronomy increased, the first private observatories were established and the first astrophysical observations were carried out. These were the observations of variable stars by a professor of chemistry, Vojtěch Šafařik. One of the private observatories was established in 1898 in Ondřejov, the present site of the Astronomical Institute. The location at 500m altitude on a hill 40km South-East of Prague was excellent at the time of foundation and is still possible today. The end of the nineteenth century saw also the first textbook on theoretical astronomy and a high-level popular book by Grus. They were much read for decades to come.

At the conclusion of the First World War, astronomy in our area had a brief glimpse of hope for a bright future. A student from Slovakia, Milan R. Štefánik, after his graduation in 1904 went to Paris. He happened to live in the same building as Camille Flammarion whose books he translated. Flammarion introduced Štefánik to Janssen. Already the following year Štefánik became a member of an expedition to MtBlanc to observe the Sun in infrared up to 1 micron and in ultraviolet down to 383nm. He observed also planets, found indications for a fast rotation of Venus and, among others, volunteered for exploring the bottom of a crevasse of the glacier under the observing site. He went six times to MtBlanc, observed solar eclipses in central Asia, in Spain, and in 1910 departed for Tahiti to observe Halley's Comet, to set up a catalogue of southern nebulae and to establish a radio station for ships en route to the Panama canal which was in an advanced state of construction at that time. He also visited another site for a radio station on an island in the Galapagos to which France had some claims. The beginning of the First World War reached Štefánik in Morocco and he immediately volunteered for the French Air Force. His career was brilliant and fast. By 1917 he was a general and after the creation of Czechoslovakia became a minister in the first government. In 1919 he was returning home piloting a plane but in a dramatic turn which can happen only in real life, he crashed shortly before landing and lost his life at the age of thirty-eight.

Can any one imagine how astronomy and all other sciences would have developed in this country had this whirlwind of energy and courage returned home safely and could have initiated or at least supported astrophysical observations with his world-wide contacts, ideas for a southern observatory, for global radio networks and for anything else that might occur to a young well versed scientist.

Life wrote a different scenario. Two university professors were in Prague, one theoretician working on the three- body-problem, the other interested in the construction of a circumzenital for precise position measurements. None of those two tried to established better working conditions and more modern instruments for their successors than they had themselves. The next generation - fortunately there always is a new generation - had to seek specialized education abroad. B. Šternberk studied in Babelsberg, operated the then largest telescope of 60cm diameter in southern Slovakia and established the chronometrical service in Prague. A. Bečvář, whose name may be familiar from his Atlas Coeli, founded the observatory at Skalnaté Pleso. E. Buchar worked at the Algiers observatory in the 1920's and later became professor of geodetical astronomy. J. M. Mohr worked for some time in Algiers, studied in France and Holland and introduced stellar astronomy into this country. V. Guth founded a school on interplanetary matter and the last of the six brave men, F. Link worked at Pic du Midi where he started his extensive work on upper atmosphere studied by astronomical methods. Each of them did something for the generations to follow, either establishing or directing an institution or providing new instruments.

The Second World War interrupted the continuous development of all sciences in Czechoslovakia. Universities were closed down for six years but, as it happens with living organisms, the punishment did not kill sciences. On the contrary, the adversities and frustrations of the war brought forward a new hardened generation wanting to work even under difficult conditions. Thus after the end of the war a highly motivated group of young people joined astronomical observatories and institutes. Fortunately, it was possible to increase gradually the staff of most institutions. Also instruments, as far as they were available on this side of the iron curtain, could be provided, among them the two-meter telescope, mainly used for spectrography, located at the Ondřejov Observatory, and the home-made 60cm telescope for photoelectric photometry of the Masaryk University in Brno. Briefly stated, astronomy flourished. If the number of Presidents and Vice-Presidents of IAU Commissions is any measure, we held four posts between 1964-1967, six posts between 1967-1970 and again four posts between 1970-1973, indicating a strong maximum of activity. In 1967, the IAU in a friendly gesture held its General Assembly in Prague and two Symposia, one on planetary nebulae and another one on the physics and dynamics of meteors in the Tatra mountains.

There was, of course, the other side of the coin. The beginning of the war prevented Z. Kopal, of the close double-stars fame, from returning home. He stayed at the Harvard University after the expiration of his fellowship. Political restrictions, enacted after 1948, in travelling abroad and other consequences of a rather tough regime induced, and sometimes forced, people to emigrate to countries with a more quiet climate.

In 1949, Gus Bakoš stayed in Leiden. In the 1960's and 1970's, M. Blaha went to the Maryland University, Z. Švestka and his wife, L. Fritzová, to ESTEC in the Netherlands, M. Plavec and his wife, Z. Plavcová, to the University of California, F. Link to the Institut d'Astrophysique in Paris, Z. Sekanina to the JPL in Pasadena, L. Kohoutek to the Hamburg Observatory, Z. Kviz to the Broken Hill University in Australia, V. Matas to the Astronomisches Rechen-Institut in Heidelberg, I. Hubený to Boulder, Colorado, and others to other places. I cannot help imagining that if all those who had left Czechoslovakia were the staff of a single institute, it would be quite a nice institute, indeed, worthy of attention.

We are slowly coming to the present time, when the only limits to travel abroad and to participation in international teams are personal qualification and financial funds, when the institutes can decide their own affairs and when, in spite of what the ecologists tell us, the atmosphere is much more agreeable to breathe than it was before the 17 November 1989. Let us hope that the next maximum in astronomical activities in this country is not far from coming. The first swallow was the European Regional Astronomy Meeting in 1987, the second swallow - which may make the summer - evidently is this Workshop of the Center for Theoretical Study of the Charles University.

May I conclude this walk along a long way by wishing full success to your sessions and a nice place in history to all of you. Thank you.

Part I

SHELLS, SUPERSHELLS, WORMS, CHIMNEYS AND SUPERNOVA REMNANTS

SHELLS, SUPERSHELLS, WORMS, CHIMNEYS—AND THE 'WORM-IONIZED MEDIUM'

CARL HEILES

University of California, Berkeley, U.S.A.

1. INTRODUCTION

For nearly two decades the importance of *individual* supernovae on the global properties of the interstellar medium has been widely appreciated. However, supernova that arise from massive stars occur in *clusters*. These inject enormous amounts of energy, concentrated in space and time, into the interstellar medium. This affects the interstellar medium by producing enormous structures and causing mass interchange between the gaseous disk and halo. IAU Symposium number 144 (Bloemen 1991) concentrated on the disk/halo interaction and remains an excellent contemporary review of the topic.

Here I concentrate on the directly observable structures produced by the clustered supernovae. Observations of both our Galaxy and external galaxies are being produced and published at a rapid rate. It is both difficult and relatively pointless to produce a complete list of references because the rate of current publication in this area is so large. Therefore, I will concentrate on my own specialty, the Galaxy, and mention some of the extragalactic studies only briefly.

The terminology is important. A 'shell' is the observable dense, usually cool shell of material located just behind the shock produced by one or a few supernovae. The theory of shell dynamics is well-understood for the case of a smooth interstellar medium (Cioffi, McKee, and Bertschinger 1988). If interstellar clouds are dissipated by photoevaporation in the H II region that precedes the supernovae explosions, then the appropriate volume density at $z = 0$ is $n_0 \approx 0.4$ cm^{-3} (Heiles 1990). At this density, the shell produced by a single supernova with energy 10^{51} erg merges with the interstellar medium at a radius $R \approx 50$ pc after $\sim$ 2.8 Myr. Old remnants of interstellar shells, or portions thereof, are common in the interstellar medium (Kulkarni and Heiles 1987 [KH]).

A 'supershell' is the multi-supernova case of a shell—a very large shell caused by dozens or more supernovae. These, too, are common in the interstellar medium, both in the Galaxy (Heiles 1984) and in external galaxies. The theory and many observations are reviewed by Tenorio-Tagle and Bodenheimer (1988). Recent theoretical developments concentrate on the effects of the extended, low-density 'Lockman' H I and 'Reynolds' H II extended layers and many are being presented and reviewed at this meeting. There is a caution: some extremely large interstellar structures are probably formed not by aggregates of supernovae but instead by collisions of high-velocity clouds with the galactic disk. The premier example is in M 101 (van der Hulst and Sancisi 1988).

A 'worm' is a wiggly-looking interstellar structure that lies roughly perpendicular to the Galactic plane ('crawls out of the Galactic plane' [Heiles 1984]). The gas in a worm is thought to be the gas that originally resided in the volume enclosed by a very large supershell whose size exceeds the scale height of the gaseous disk. At high z the interstellar gas density becomes smaller and the supershell shock moves faster. If there is enough energy, the supershell blows a hole in the gaseous disk. Even if it does not, the vertical portions of the supershell wall contain most of the gas that originally resided in the hole and should appear as 'worms'.

A 'chimney' is the most extreme version of a supershell. If the energy is high enough and if the gaseous disk does not have too much of a z-extended, low-density component, then a supershell can break out of the disk, forming a pipeline for direct injection of both the hot interior gas and cold clouds into the halo. Its walls should also appear as worms.

There is a crucial distinction between worms and chimneys. A 'worm' is defined *empirically*: the term arises because of its observational appearance, and there is no associated astrophysical interpretation. A 'chimney' is a *theoretical* construct: a supershell that breaks out of the gaseous disk has dense walls and acts as a pipeline for hot gas to escape vertically from the gaseous disk—just as a chimney acts as a pipeline for the hot gas from a fire to escape into the Earth's atmosphere. The analogy goes further: the hot escaping interstellar gas is polluted by the heavy elements produced by the supernovae, just as the terrestrial gas is polluted by the products of combustion. This very clever term was invented by Norman and Ikeuchi (1989).

The point is that not all worms must be chimneys (Heiles 1991). We would like to know the fraction of worms that are chimneys, because this has ramifications for gaseous galactic halos and overall galactic evolution. One can try to estimate the fraction from the properties of interstellar gas, the distribution of supernovae in clusters, and complicated numerical simulations of superexplosive events, but the estimates depend on the details and are correspondingly uncertain.

Finally, the 'Warm Ionized Medium' (WIM) is a diffusely-distributed ionized gas

($T \sim 8000$ K in the Solar neighborhood). It is one of the four major constituents of the diffuse interstellar medium (KH). In the Galactic interior it becomes observable in low-frequency radio recombination lines, which means it becomes much denser than it is in the Solar neighborhood. Its existence is puzzling because an enormous energy input is required to keep it ionized. Below we show that this puzzle may evaporate upon considering the relation between the WIM and clustered supernovae.

2. SHELLS

Until recently, our list of observable supernova shells was limited to those well-known objects seen optically, such as the Cygnus Loop, plus some other objects. However, the IRAS satellite has provided images of a number of fairly young shells (a survey by Arendt 1989; CTB 80 by Fesen, Shull, and Saken 1988; RCW 86 by Greidanus and Strom 1990b [also see references therein for other remnants]); and new 21-cm line observations reveal rapidly expanding shells around a number of cataloged supernova remnants. With H I, the ability to measure both velocities and column densities affords a fairly detailed comparison with theoretical models of shells.

The first fast shell seen in H I is CTB 80 (Koo *et al.* 1990). This object is particularly edifying (a 'Rosetta Stone') because there is a pulsar associated with CTB 80, and the age of the pulsar must be the same as the age of the expanding shell. The expansion velocity is 72 km s^{-1}, the radius is $19 d_{2kpc}$ pc, and the swept-up mass $1200 d_{2kpc}^2$ $M_\odot$. The age of the shell, inferred solely from its observed properties and a comparison with standard theory, is $77000 d_{2kpc}$ yr, only $\sim 20\%$ smaller than the pulsar age. The energy of the supernova explosion is inferred to be $0.44 \times 10^{51} d_{2kpc}^2$ ergs or greater. These results are nicely consistent with expectation for a supernova-produced shell (but not with a stellar wind-produced shell) and are correspondingly satisfying. An aspect that is *not* expected is that the shell contains *too much* slow-moving mass, which Koo *et al.* ascribe to clumping. The theoretical treatment of clumping is neither well understood nor extensive, and observations of more objects are desirable to investigate the effects of clumping. There is also a very impressive optical study of CTB 80 in the Hα line by Greidanus and Strom (1990a), who find velocities ranging from -130 to $+94$ km s^{-1} and extensive clumping. However, the optical shell is much smaller than the H I shell and has a correspondingly short expansion time scale of ~ 1100 years. The optical shell is interpreted as resulting not from the shock associated with the much older expanding H I shell but instead from excitation by the pulsar itself.

Spurred on by CTB 80—and in particular the fact that the first observational indication of the fast H I shell was obtained with the lowly 85-foot Hat Creek telescope, on which observing time is easily accessible to us—we went on to survey all 103 known (from radio-continuum emission: Green 1988) supernova remnants north of $\delta \approx -40°$ (Koo and Heiles 1991). The result: 15 definite detections with expansion velocities ranging from 70 to 160 km s^{-1} and 12 possible detections with smaller expansion velocities. For many shells the parameters are consistent with our overall expectations. However, as with CTB 80, the shells seem to exhibit too much slow-moving mass; again, Koo and Heiles ascribe this to clumping. This set of $\sim$ 25 rapidly-expanding supernova shells deserves to be studied intensively: the result should be definitive verification of theoretical predictions and a greatly increased understanding of the effects of clumping on the evolution of expanding shells. With sufficient attention, this set might also provide observational information on thermal instability in the shell, cosmic ray production, and magnetic fields.

As shells age they slow down and get larger. Remnants of old shells abound in interstellar space and are probably what we call 'diffuse interstellar clouds' (KH). Because these old shell remnants abound, some are nearby and can be studied in detail.

However, not all of these objects necessarily came from supernova-produced expanding shells. One of the most easily-studied is a $\sim$ 30° diameter shell that runs through the North Celestial Pole, the 'NCP Loop' (Meyerdierks, Heithausen, and Reif 1991). It is easy to study because it is the only interstellar object along the line of sight so there is no possibility for confusing it with unrelated foreground or background gas. The H I data do not mimic what is expected from an expanding shell, and Meyerdierks *et al.* model it instead as a cylindrical cavity, probably produced by infall of a high-velocity cloud from the Galactic halo.

No matter what the originating mechanism, the loop undoubtedly outlines an interstellar shock. The loop has a strong magnetic field, $\gtrsim 10\ \mu$G (Heiles 1989), and if the gas in the shell were not clumped then the magnetic forces would dominate all others. However, the gas *is* clumped: the IRAS image exhibits a delicately interconnected network of filaments whose angular widths are $\lesssim 4'$ (Meyerdierks *et al.* 1991). If the H I that exhibits Zeeman splitting is concentrated in these filaments, then the magnetic forces are *not* very important. The H I may not be so concentrated, because many of the filaments contain observable CO (Heithausen, Mebold, and de Vries 1987; de Vries, Heithausen, and Thaddeus 1987; Heithausen and Thaddeus 1990). Unfortunately, the angular resolution of existing H I observations is insufficient to pinpoint the H I. What is needed to resolve the importance of magnetic forces are high-resolution observations of the filaments over a reasonably large fraction of the total area occupied by the shell remnant (which is many square degrees).

At this point I want to mention a theoretical prediction on shells that needs to be observationally verified. Ferrière, Mac Low, and Zweibel (1991) considered the case of a shell expanding in the presence of a uniform magnetic field. The field lines are, of course, swept up by the expansion, and they cause motions of the gas *tangential* to the field lines (*nonradial* motions). The velocity should be $\sim$ the Alfvén velocity in the ambient medium. This should be an observable effect.

A quick summary of this section: we are beginning to obtain detailed kinematic and structural studies of both young and old shells. There is now a catalog of young shells, which cry out for detailed observation. Further studies should reveal detailed observational information on the effects of clumping and, perhaps, magnetic fields, cosmic rays, and thermal instability in the shells.

3. SUPERSHELLS, WORMS, AND CHIMNEYS

Supershells reach enormous sizes, $\sim$ 2 kpc in diameter. They are easily observed in the Galaxy with modest telescopes in the H I line, less easily in external galaxies with optical astronomy in the Hα, N II, and S II lines, and with difficulty using synthesis techniques in the H I line. In the Galaxy, Heiles (1979, 1984) provides incomplete lists derived from H I observations. Detailed maps sometimes exhibit open tops, providing a channel for disk gas to escape into the halo. Cold clouds are occasionally tossed into the halo; one is visible in Figure 1 of Heiles (1979) [$(\ell, b, v) \approx (71°, -5°, -87 \text{ km s}^{-1})$].

In external galaxies, H I supershells are directly observed in NGC 3079 (Irwin and Seaquist 1990) and are indirectly observed, through the existence of large holes in the gaseous disk, in M31 (Brinks and Bajaja 1986), M33 (Deul and den Harton 1990), and IC 10 (Shostak and Skillman 1989). A most impressive H I hole in M 101 was observed by Kamphuis, Sancisi, and van der Hulst (1991). This hole is $\sim$ 1.5 kpc in diameter and, most importantly, the 'top' and 'bottom' of the hole exhibit almost symmetrical vertical motions amounting to $\sim$ 50 km s^{-1}. The associated kinetic energy is $\sim 3 \times 10^{53}$ erg; production of this object would require $\sim$ 1000 clustered supernovae.

In external galaxies, optical observations are easier than radio observations. Ionized 'froth', interpreted in terms of supershells, is observed by Hunter and Gallagher (1990). Rand, Kulkarni, and Hester (1990) observe worms extending to high $|z|$ in the edge-on galaxy NGC 891 and interpret them as chimneys. Direct evidence for gas expanding through chimneys into the halo is given for NGC 4258 by Rubin and Graham (1990) and for NGC 2363 by Roy *et al.* (1991). An inclination effect in supernova rates for Sc and Sd galaxies is ascribed to the existence of clustered O stars in the H I holes produced by the preceding supernovae in the same clusters by van den Bergh and McClure (1990). The classic observations of supershells ('supergiant shells') in the Large Magellanic Cloud by Meaburn (1980) are very impressive; expansion velocities of $\sim$ 40 km s^{-1} are measured for one of the smaller shells by Meaburn and Laspias (1991), and we can look forward to further measurements for the supershells. Comparable velocities for supershells, together with evidence for self-propagating star formation, are seen in the H I line (Dopita, Mathewson, and Ford 1985). Bruhweiler, Fitzurka, and Gull (1991) find that SN1987a in the LMC is located inside a large supershell that may have triggered more recent star formation.

X-ray observations should reveal the hot gas inside supershells and chimneys directly, and with the impending storm of data from the ROSAT satellite we can hope to learn a great deal. Some *Einstein* satellite observations already exhibit X-rays from the associated hot gas in LMC supershells (Chu and Mac Low 1990; Wang and Helfand 1991).

4. WORMS, DIFFUSE RECOMBINATION LINES, AND THE 'WORM-IONIZED MEDIUM'

4.1. The Warm Ionized Medium and Diffuse Radio Recombination Lines

Koo and Heiles (1992) have reason to believe that the inner-Galaxy version of the Warm Ionized Medium is intimately related to worms and chimneys and in fact can equally well be described as the 'Worm-Ionized Medium'. The WIM is one of the four major components of the diffuse interstellar medium (KH). It amounts to $\sim$ 25% of H I mass in the Solar neighborhood. In the Solar neighborhood it has $\langle n_e^2 \rangle^{1/2} \approx 0.09$ cm^{-3} at $z = 0$ and a scale height for $\langle n_e \rangle$ of $\sim$ 1000 pc; in the Perseus arm the scale height for $\langle n_e^2 \rangle \sim 300$ pc.

In the Galactic interior ($|\ell| \lesssim 50^\circ$), a much more powerful version of the WIM becomes observable as the emitter of 'diffuse recombination lines' (diffuse RRL's) observed at radio frequencies $\sim$ 1 GHz (usually the H166α line). The diffuse RRL emission exhibits considerable angular structure and is somewhat correlated with H II regions, but is observable even in 'empty' regions; thus, it is truly distributed over large volumes of space, but at the same time it is clumpy. Surveys of this line in the north by Lockman (1980) and in the south by Cersosimo (1989, 1990) reveal an H166α line emissivity $\sim$ 2 K-kHz per kpc (more accurate values are considered below). This is about 100 times larger than would be produced by the WIM in the Solar vicinity! The RRL intensity $\propto T^{-3/2}$, and if the gas temperature T is smaller in the interior then the emissivity per unit emission measure (EM) is larger. Nevertheless, the r.m.s. electron density $\langle n_e^2 \rangle^{1/2}$ in the Galactic interior must be much larger than the local value. Presumably, this is a result of the larger flux of ionizing photons available from the hot, young stars in the Galactic interior.

The WIM in the Galactic interior was previously considered by Mezger (1978), who called it the 'Extended Low-Density Ionized' (ELD) gas. He estimated its global requirement, as we also do below (§4.3). Galacticwide, an enormous energy input, comparable to the total energy released to the ISM by supernovae, is required to keep the WIM ionized. The most likely mechanism, given the observed line strengths, is photoionization by distant O stars (Mathis 1986)—which is also the only known source that is sufficiently powerful (Reynolds 1984). Mezger (1978) and Anantharamaiah (1985, 1986) consider this gas to be the low-density outer halo of ordinary H II regions. However, there is a problem with this idea: H II regions should have sharp boundaries, and it is not clear why they should have extended, low-density haloes. In more generality, the problem with O stars as the ionizing agent for the WIM is that the stars are '...rare and hence it is not clear how their Lyman continuum radiation can reach every nook and corner of the Galaxy to produce the widespread WIM' (KH).

4.2. The WIM and supershells

In theory, worms and chimneys can provide unobstracted pathways for the photons. The cylindrical cavities in the gaseous disk, which are produced by clustered supernovae, contain mainly the very diffuse Hot Ionized Medium, which does not absorb the ionizing photons. During the beginning stages of the cavity, only a fraction of the clusered O stars have exploded; the rest are still shining. Their photons escape stars in straight lines, both ionizing the cylinder walls and streaming up to gas at high z in a cone with the apex located at the original star cluster and the cone angle defined by the diameter of the cylindrical cavity (Norman 1989).

There are now some observational indications that this actually happens. First, Koo, Heiles, and Reach (1991) have prepared a catalog of 118 Galactic worms. The worms are visible as extended vertical structures not only in the H I line (whose velocity provides the distance) but also in the IRAS maps and in the 408 MHz radio continuum data of Haslam *et al.* (1982). Typical z-heights are 160 pc, widths are 70 pc, and N_{HI}'s are 2×10^{20} cm^{-2}. 56 of the worms are associated with large H II regions in the Galactic plane, which is expected if the worms are produced by clustered supernovae and some of the massive stars have not yet exploded. For these 56 worms there is a good correlation between the 408 MHz and 60 μm IRAS brightnesses. This correlation is consistent with much of the 408 MHz emission being *thermal* (surprising for classical radio astronomers!), with all the Lyα photons being absorbed by dust grains and converted to IR photons. (A significant contribution from nonthermal radiation may also exist; the North Polar Spur provides a local example of this extreme). The similarities between IRAS and radio continuum morphologies suggest that the walls of the cylindrical hole made in the gaseous disk by the clustered supernovae are in fact partially ionized by the remaining O stars in the cluster.

Second, the thermal radio emission from two worms conforms to a simple theoretical model for a cylindrical cavity whose walls are ionized by photons from a cluster located at the center. One worm is GW 31.6–5.9, located near $\ell = 30°$ and extending several degrees towards negative latitudes; it is associated with the H II region W 43. Koo and I have observed this worm for H166α emission, with positive results. This worm exhibits integrated line intensities ranging up to 70 K kHz, with the line intensity decreasing rapidly with $|b|$ and remaining detectable for $b \gtrsim -2.6°$. The other worm is GW 17.8+3.0, located near $\ell = 18°$ and extending towards positive latitudes; it is associated with the H II region S 54. This worm was studied in the radio continuum and the RRL emission by Müller, Reif, and Reich (1987). Again, the continuum and line intensities decrease rapidly with $|b|$.

The simple theoretical model consists of a cylindrical cavity of radius R with a cluster of OB stars at its center that produces L_{51} ionizing photons per second (units: 10^{51} photons s^{-1}). Suppose that the ionizing photons travel in straight lines to the cylinder walls, where they ionize the gas. Then the emission measure (EM; units cm^{-6} pc) of the cylinder wall as a function of height z above the cluster is

$$EM = 159 \; L_{51} T_3^{0.7} R_{100}^{-2} \left(1 + \frac{z^2}{R^2}\right)^{-3/2} \tag{1}$$

where the subscript 100 implies units of 100 pc. For the above two worms, the $|b|$-dependence of the observed intensity fits this z-dependence within a factor of two.

But for GW 31.6–5.9 we can say more. The total radio emission from the worm walls, integrated over angle, requires ~ 20 times more ionizing photons than does the W43 H II region itself (which is located at the center of the worm). Thus 95% of the ionizing photons from the W 43 star cluster escape the H II region and ionize the distant chimney walls. This is precisely what is expected for a cluster in which many supernovae have already exploded, because the explosions should have dissipated most of the interstellar gas that originally resided in the immediate vicinity of the cluster.

4.3. Global Energy and Ionization Requirements

In this light we now consider the global energy and ionization requirements for the diffuse Galactic RRL emission. The recombination rate, and thus the ionization rate, are $\propto$ the RRL emissivity. Thus, by suitably integrating the RRL line intensity over solid angle we can estimate quite accurately the total ionization rate in the inner Galaxy. This, in turn, tells us the total UV ionizing photon luminosity, which can be related to the total Galactic supernova rate.

The physics of the elementary processes give for the ratio of recombination rate per cm^2 to the frequency-integrated RRL brightness temperature $\int T_B\,d\nu$ (units: K kHz), assuming that the RRL emission is LTE,

$$\frac{rec.\ rate\ per\ cm^2}{\int T_B\,d\nu} \approx 8.1 \times 10^7\ T_3^{0.8}\ , \tag{2}$$

where T_3 is the temperature of the ionized gas in units of 1000 K.

Lockman (1980) argues $T_3 \lesssim 3$; we adopt $T_3 = 3$ (more accurately, $\langle T_3^{-x}\rangle^{-1/x} = 3$, where $x \sim 2$) on the basis of the following argument. The reader may wish to skip this paragraph, which is nothing more than a rather involved justification for taking $T_3 = 3$, and proceed to the next paragraph. Lockman's Figure 5 uses limits on the free-free opacity (which is $\propto T^{-2.3}$) and the H166α brightness (which is $\propto T^{-1.5}$) to constrain the allowed region of the EM-T_3 plane. The region must lie close to the line $\log EM = (1.4 \log T) - 1.85$ and must have $T_3 \lesssim 3.7$. We suppose that the pressure in the emitting region is a factor χ larger than the standard interstellar pressure $p/k = 4000\ \mathrm{cm}^{-3}$ K; this provides $n_e = 2\chi/T_3$. With $EM = n_e^2 L$, these provide the relation $L = 56\ T_3^{3.4}\chi^{-2}$ pc. In addition, the emitting region contributes a fraction d of the total pulsar-determined dispersion measure DM, which is about 250 cm^{-3} pc; thus $n_e L = 250d$, or $L = 125 T_3 d/\chi$. The supershell shock expands into the ambient interstellar medium, which we take as having $n_H = 0.5\ \mathrm{cm}^{-3}$, with velocity $v \sim 20$ km s^{-1}; behind this shock we have $p/k \approx 3.4 \times 10^4\ \mathrm{cm}^{-3}$ pc, or $\chi = 8.6$. Thus $T_3 \approx 3.4 d^{0.42}$. d is likely to be close to unity; $d = 0.74$ gives $T_3 = 3$, and is a quite reasonable value. This choice also gives $L = 48.6 d^{1.4} = 32$ pc, which also seems like a reasonable value. As described above, this temperature is not a true average but instead is biased to low values because it is derived from two types of observations, both of which are sensitive to gas at lower temperatures and provide $x \sim 2$ above.

The RRL observations give $\langle \int T_B\,d\nu \rangle \approx 10$ K kHz for $|\ell| \lesssim 50°$ at $b = 0°$. We assume that the scale height of $\langle n_e^2 \rangle \sim 300$ pc, as it is in the Perseus arm (KH), and that the average distance of the emitting gas is the Solar Galactocentric radius, 8 kpc. Multiplying all this out in equation (2), and setting the ionization rate equal to the recombination rate, gives:

$$Total\ Galactic\ ionization\ rate \sim 6.5 \times 10^{52}\ T_3^{0.8}\ \mathrm{s}^{-1}\ . \tag{3}$$

Mezger (1978) performed a more sophisticated analysis based on less extensive data and, with his assumed value of $T_3 = 7$ obtained an almost identical result. Heiles (1990) used a Scalo (1986) stellar mass function for clusters to relate the ionizing photon luminosity to the supernova rate for clusters; the conversion factor is 4.8×10^{62} ionizing photons per supernova. Thus, for $T_3 = 3$ the above ionization rate corresponds to a Galactic Type II and Ib supernova rate of one every 97 years.

This rate is just about right. It is about half the total rate (one every 45 years; Ratnatunga and van den Bergh 1989). Thus about half of the OB stars can put their ionizing photons into ordinary H II regions, which do not produce diffuse RRL emission. When these explode and create a supershell, worm, or chimney, the remaining stars have little interstellar gas left in the immediate vicinity, so their ionizing photons are used to ionize the chimney walls and produce the diffuse RRL emission.

If this picture is correct, then the diffuse RRL emission should be correlated in a loose sense with the discrete H II regions because some of these H II regions, such as W43 discussed above, are associated with the clusters that also produce ionization in the fairly distant chimney walls. Lockman's (1980) refreshingly clear discussion of the observations reveals that such a 'loose correlation' does, in fact, exist, as follows: 85% of the small, discrete H II regions seen in the higher-frequency (*non*-diffuse) H110α observations have counterparts in the diffuse (H166α) RRL profiles for angular displacements $\lesssim 1°$. And the H166α profile at any position resembles the area-summed 'composite' profile of nearby (within $\sim 1°$) discrete H II regions observed at H110α. However, while the line components are similar in velocity, the relative intensities differ 'a great deal', and the diffuse H166α RRL profiles are broader and more complex than the composite H110α profiles. Extra velocity width would be expected from expanding chimney walls! This argues that the compact H II regions capture only a fraction of the photons from the exciting stars and the rest get out to the diffuse gas. Furthermore, the physical properties expected from this picture match reasonably well those that are deduced from careful consideration of RRL data taken at lower frequencies (Anantharamaiah 1985) and from free-free absorption at low radio frequencies (Kassim 1989).

We now go further and consider the local WIM observed by Reynolds. The local ionization requirements are about two orders of magnitude smaller than that of the diffuse RRL-emitting gas; correspondingly, the local OB star formation rate is much smaller. It seems reasonable to assume that the local WIM is produced by the same process that produces the diffuse RRL-emitting gas in the Galactic interior. However, there are some problems with this simple extension of reason. One, the local WIM has $T_3 \approx 8$ (KH), much larger than the $T_3 \approx 3$ that characterizes the diffuse RRL emitting gas; but various factors determine the RRL gas temperature, such as the Oxygen abundance, and in addition our value of $T_3 = 3$ is biased toward a low value. Two, the filling factor for the local WIM was derived by KH to be large, $\sim 11\%$ at $z = 0$, but that for the diffuse RRL gas should be considerably smaller in our picture; but the KH derivation depends on assuming that the properties of the WIM in the Perseus arm are the same as those of the local WIM. Clearly, it is premature to worry too strongly about these problems, but they definitely need to be investigated.

In summary, our proposed scenario is that the WIM throughout the Galaxy is the ionized walls of supershells, worms, and chimneys. The ionization is produced by distant OB stars that reside in the same cluster that provided the multiple supernovae that produced the supershell in the first place. This scenario is supported by the detailed observational results on two worms/chimneys (GW17.8+3.0 and GW31.6–5.9), by the overall observational characteristics of diffuse RRL emission, and by the fact that the global Galactic ionization rate of supershell walls is reasonably close to that expected from the scenario.

There are several types of observational and theoretical attacks that should be applied to this picture. Some of the crucial observational parameters that might do this, and in addition distinguish the diffuse RRL from the local WIM, are gas temperature, kinematics, filling factor, and diagnostic line ratios such as [S II]/Hα. If these attacks confirm our proposed picture, then the deeply puzzling questions regarding the origin of the WIM will have been answered. Given the importance of the WIM from the standpoints of energy and mass, these questions definitely deserve attention.

5. SUMMARY

The summary can be quick: clustered supernovae produce huge structures in the interstellar medium. Some of these are chimneys, providing a direct connection between the disk gas and the halo. The supernovae also may be responsible for producing one of the major constituents of the interstellar medium, the WIM—which then becomes, perhaps, the 'Worm-Ionized Medium'.

It is a real pleasure to acknowledge collaboration and conversation with Dr. Bon-Chul Koo regarding general problems of shells/supershells, their thermal radio and RRL emission, comments on an earlier version of this paper, and the suggestion (claimed to be an innocent typographical error in an e-mail message, but his former thesis advisor knows better) for the title of this paper. Another real pleasure is to thank Shri Kulkarni for his usual penetrating, insightful, and embarrassing comments. This research was supported by NSF grant AST-8818544 to the author.

6. REFERENCES

Anantharamaiah, K.R. 1985, *J. Ap. Astron.*, **6**, 203.
Anantharamaiah, K.R. 1986, *J. Ap. Astron.*, **7**, 131.
Arendt, R.G. 1989, *Ap. J. Suppl.*, **70**, 181.
Bloemen, J.B.G.M. 1991, *I.A.U. Symposium No. 144, The Interstellar Disk-Halo Connection in Galaxies*, Kluwer Academic Publ.
Brinks, E. and Bajaja, E. 1986, *Astron. Ap.*, **169**, 14.
Bruhweiler, F.C., Fitzurka, M.A., and Gull, T.R. 1991, *Ap. J.*, **370**, 551.
Cersosimo, J.C. 1989 *Ap. J.*, **349**, 67.
Cersosimo, J.C. 1990, in *Radio Recombination Lines: 25 Years of Investigation*, ed. M.A. Gordon and R.L. Sorochenko, Kluwer Academic Publishers, p. 237.
Chu, Y-H. and Mac Low, M-M. 1990, *Ap. J.*, **365**, 510.
Cioffi, D.F., McKee, C.F., and Bertschinger, E. 1988, *Ap. J.*, **334**, 252.
Deul, E.R. and den Hartog, R.H. 1990, *Astron. Ap.*, **229**, 362.
de Vries, H.W., Heithausen, A., and Thaddeus, P. 1987, *Ap. J.*, **319**, 723.
Dopita, M.A., Mathewson, D.S., and Ford, V.L. 1985, *Ap. J.*, **297**, 599.
Ferrière, K.M., Mac Low, M., and Zweibel, E.G. 1991, *Ap. J.*, **375**, 239.
Fesen, R.A., Shull, J.M., and Saken, J.M. 1988, *Nature,* **334**, *229.*
Green, D.A. 1988, *Ap. Space Sci.*, **148**, 3.
Greidanus, H. and Strom, R.G. 1990a, *Astron. Ap.*, **240**, 376.
Greidanus, H. and Strom, R.G. 1990b, *Astron. Ap.*, **240**, 385.
Haslam, C.G.T., Salter, C.J., Stofell,H., and Wilson, W.E. 1982, *Astron. Ap. Suppl.*, **47**, 1.
Heiles, C. 1979, *Ap. J.*, **229**, 533.
Heiles, C. 1984, *Ap. J. Suppl.*, **55**, 585.
Heiles, C. 1989, *Ap. J.*, **336**, 808.
Heiles, C. 1990, *Ap. J.*, **354**, 483.
Heiles, C. 1991, in *I.A.U. Symposium No. 144, The Interstellar Disk-Halo Connection in Galaxies*, ed. J.G.B.M. Bloemen, Kluwer Academic Publ.
Heithausen,A., Mebold, U., and de Vries, H.W. 1987, *Astron. Ap.*, **179**, 263.
Heithausen, A. and Thaddeus, P. 1990, *Ap. J.*, **353**, L49.
Hunter, D. A. and Gallagher, J.S. 1990, *Ap. J.*, **362**, 480.
Irwin, J.A. and Seaquist, E.R. 1990, *Ap. J.*, **353**, 469.
Kamphuis, J., Sancisi, R., and van der Hulst, T. 1991, *Astron. Ap.*, **244**, L29.
Kassim, N. 1989, *Ap. J.*, **347**, 915.
Koo, B-C. and Heiles 1991, *Ap. J.*, submitted.
Koo, B-C. and Heiles 1992, *Ap. J.*, in preparation.
Koo, B-C., Heiles, C., and Reach, W.T. 1991, *Ap. J.*, **000**, 000.
Koo, B-C., Reach, W.T., Heiles, C., Fesen, R.A., and Shull, J.M. 1990, *Ap. J.*, **364**, 178.

Kulkarni, S.R. and Heiles, C. 1987, in *Interstellar Processes*, ed. D. J. Hollenbach and H.A. Thronson, Jr., D. Reidel (Dordrecht), p. 87 (KH).
Lockman, F.J. 1980, in *Radio Recombination Lines*, ed. P.A. Shaver, Reidel Publishing Company, p. 185.
Mathis, J.S. 1986, *Ap. J.*, **301**, 423.
Meaburn, J. 1980, *Mon. Not. Roy. Astron. Soc.*, **192**, 365.
Meaburn, J. and Laspias, V.N. 1991, *Astron. Ap.*, **245**, 635.
Meyerdierks, H., Heithausen, A., and Reif, K. 1991, *Astron. Ap.*, **245**, 247.
Mezger, P.G. 1978, *Astron. Ap.*, **70**, 565.
Müller, P., Reif, K., and Reich, W. 1987, *Astron. Ap.*, **183**, 327.
Norman, C.A. 1991, in *I.A.U. Symposium No. 144, The Interstellar Disk-Halo Connection in Galaxies*, ed. J.G.B.M. Bloemen, Kluwer Academic Publ.
Norman, C.A. and Ikeuchi, S. 1989, *Ap. J.*, **345**, 372.
Rand, R.J., Kulkarni, S.R., and Hester, J.J. 1990, *Ap. J.*, **352**, L1.
Ratnatunga, K.U. and van den Bergh, S. 1989, *Ap. J.*, **343**, 713.
Reynolds, R.J. 1984, *Ap. J.*, **282**, 191.
Roy, J-R., Boulesteix, J., Joncas, G., and Grundseth, B. 1991, *Ap. J.*, **367**, 141.
Rubin, V.C. and Graham, J.A. 1990, *Ap. J.*, **362**, L5.
Scalo, J.M. 1986, *Fund, Cosmic Phys.*, **11**, 1.
Shostak, G.S. and Skillman, E.D. 1989, *Astron. Ap.*, **214**, 33.
Tenorio-Tagle, G. and Bodenheimer, P. 1988, *Ann. Rev. Astron. Ap.*, **27**, 145.
van der Hulst, T. and Sancisi, R. 1988, *Astron. J.*, **95**, 1354.
van den Bergh, S. and McClure, R.D. 1990, *Ap. J.*, **359**, 277.
Wang, Q. and Helfand, D.J. 1991, *Ap. J.*, **370**, 541.

Anomalous-Velocity HI Structures Associated with Infrared Cirrus

W.B. BURTON[1], T.M. BANIA[2], DAP HARTMANN[1], AND TANG YUAN[1]

[1] Sterrewacht Leiden, P.O. Box 9513, 2300 RA Leiden, The Netherlands
[2] Boston University, Department of Astronomy, Boston, MA 02215, USA

ABSTRACT
We examine the kinematic aspects of atomic hydrogen gas correlated tightly with infrared cirrus dust. Cirrus clouds commonly show highly anomalous velocities, in several cases embracing the kinematic regime of the intermediate-velocity-cloud phenomenon. Evidently many cirrus clouds are not quiescent wisps, but involve substantial energetics.

1 ASSOCIATIONS OF GAS AND DUST IN INTERSTELLAR CIRRUS CLOUDS

Interstellar cirrus clouds are probably the principal carrier of the diffuse, cool interstellar medium in the Milky Way. At low latitudes, cirrus clouds are heavily blended along long lines of sight, so that individual clouds are usually not resolved. Nevertheless, the tight general correlation of radio observations at 21 cm of emission from atomic hydrogen with IRAS observations of emission at 100 μm from dust has revealed a number of properties of the gas and dust at low latitudes: the tight correlation shows, for example, that the HI optical depth of the blended cirrus is typically low, the dust temperature is rather constant, the molecular content is low, and that the cirrus are not important stellar nurseries (see e.g. Burton and Deul, 1987).

Contrary to the situation at low latitudes, however, the gas/dust correlation can be studied at high latitudes in individual, resolved cirrus clouds. The kinematic structure of unblended dust clouds can be unravelled by examining the gas/dust correlation as a function of velocity across the 21-cm spectra of the associated gas. Deul and Burton (1990) investigated 7 cirrus fields in these terms, and found, for all of the fields considered, that the cirrus were line-of-sight superpositions of kinematically different structures, and found, in addition, at least for several of the fields, that the dust emission was correlated with gas emitting at distinctly anomalous velocities, extending to the regime of the intermediate-velocity-cloud phenomenon. Here, we consider the matter of kinematic unravelling of cirrus further, using high-resolution Arecibo HI data toward several of the Deul and Burton features, and HI data from the new Leiden-Dwingeloo survey in several additional fields.

Our point of departure is the question: what are the velocity characteristics of cirrus clouds? It appears that this question leads, at least for some cirrus, to the answer: cirrus traced by HI show highly anomalous velocities. The consequences of this answer are puzzling in several regards. It suggests that the energetics of cirrus are substantially greater than anticipated for quiescent wills-o'-the-wisp; that the role of turbulence in the interstellar medium at high latitudes might be currently underestimated; that the cirrus material might contain a significant fraction of the kinetic energy of the local interstellar medium; and that the concept of an ensemble of distinct features, with volume, and even area, filling factors less than unity, might be relevant to the material contributing at least part of the emission in the high-dispersion wings of high-latitude HI spectra. The association of some cirrus with HI features belonging to the category of intermediate-velocity-clouds furthermore renews interest in some of the questions posed, and never fully answered, by the early IVC investigations by Wesselius (1972), Wesselius and Fejes (1973), and others.

At low latitudes, HI emission characteristics vary slowly, both in position and in velocity: little fine-scale emission structure (other than that from localized self-absorption features), either in angle or in position, is revealed by observing with, say, the 4-arcminute beam of the Arecibo 1000-foot telescope rather than with the 36-arcminute beams of the 25-m Hat Creek or Dwingeloo antennae. (This point is illustrated by the comparison made in Figure 3 of Bania and Lockman (1984) of four 21-cm spectra observed in the galactic plane at l=36°87 at 36', 21', 11', and 4' resolution.) The total-velocity HI column depth measured at low $|b|$ varies by only about 10% or less over angular scales of several degrees. This same evidently slow variation of N(HI) pertains at high $|b|$, if the data are effectively smoothed by a large sampling interval; but observations of HI associated with cirrus, on the other hand, show that observations made with high angular (and kinematic) resolution reward the effort with a wealth of small-scale structure.

2 HI OBSERVATIONS

The Arecibo data used here were taken during October and November, 1990, by Bania and Burton. The half-power beamwidth at 21 cm of the 1000-foot telescope equipped with the dual circular feeds is 4.0 arcminutes. The spectra were taken at grid spacings in b and l cos(b) of 6 arcminutes; the kinematic resolution is 0.5 km/s per channel. The spectra have not been corrected for stray-radiation contamination. (Although such contamination is no doubt present at intensity levels less than a few tenths K, it does not invalidate the uses of the data made here.)

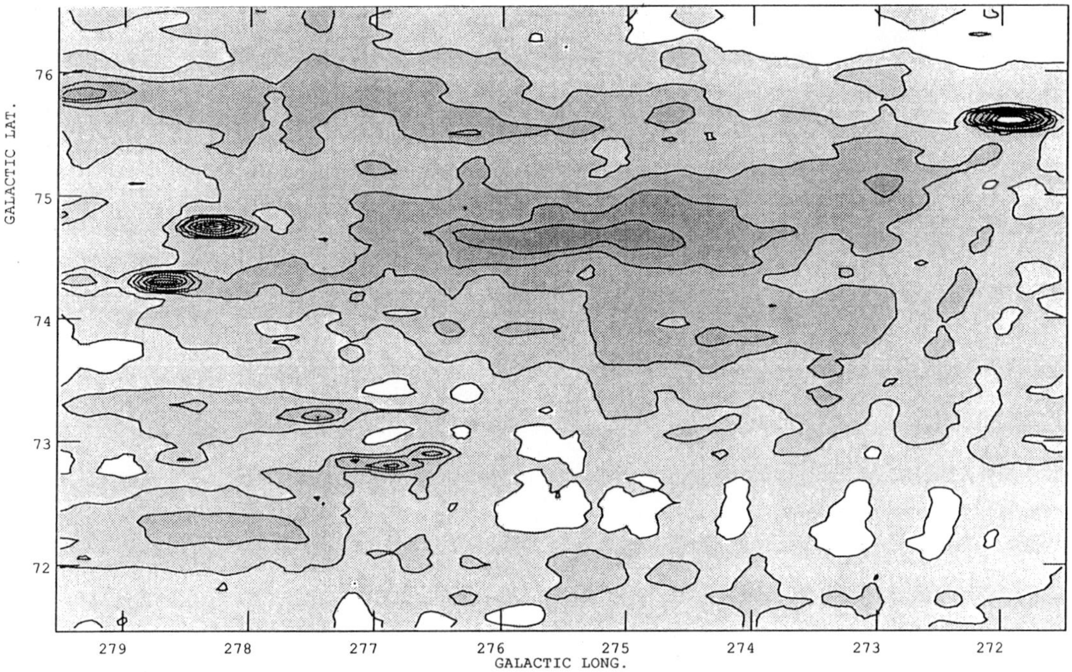

Figure 1. Dust emission at 100 micron from the vicinity of the cirrus feature "B", found in the cirrus detection paper by Low *et al.* (1984). The IRAS map has been smoothed to the 6-arcminute resolution of the Arecibo HI data.

The Dwingeloo data used here are from the Leiden-Dwingeloo survey by Hartmann and Burton, now nearing completion, of the sky north of δ=-30°. The half-power beamwidth of the 25-m telescope is 36 arcminutes at 21-cm wavelength. The spectra are being taken at grid spacings of 30 arcminutes in both b and l cos(b). The 21-cm receiver is used in conjunction with the new 1024-channel spectrometer sampling, at 1 km/s intervals, LSR velocities in the range $-500 < v < 500$ km/s. The system temperature is typically less than 40 K; the rms noise limit of the data is typically better than 0.07 K. The data as used here have not yet been corrected for stray-radiation contamination; this correction will be applied by P. Kalberla and colleagues at Bonn University, before the completed survey is presented in definitive form. A complementary survey of the southern sky, with essentially the same sampling and noise characteristics, is being undertaken by E. Bajaja and colleagues with the 100-foot radio telescope in Argentina.

3 KINEMATIC STRUCTURE OF SEVERAL REPRESENTATIVE CIRRUS FIELDS

A large variety of observational evidence must be gathered to specify the details of the nature of the gas-to-dust relationship in terms of all the relevant physical

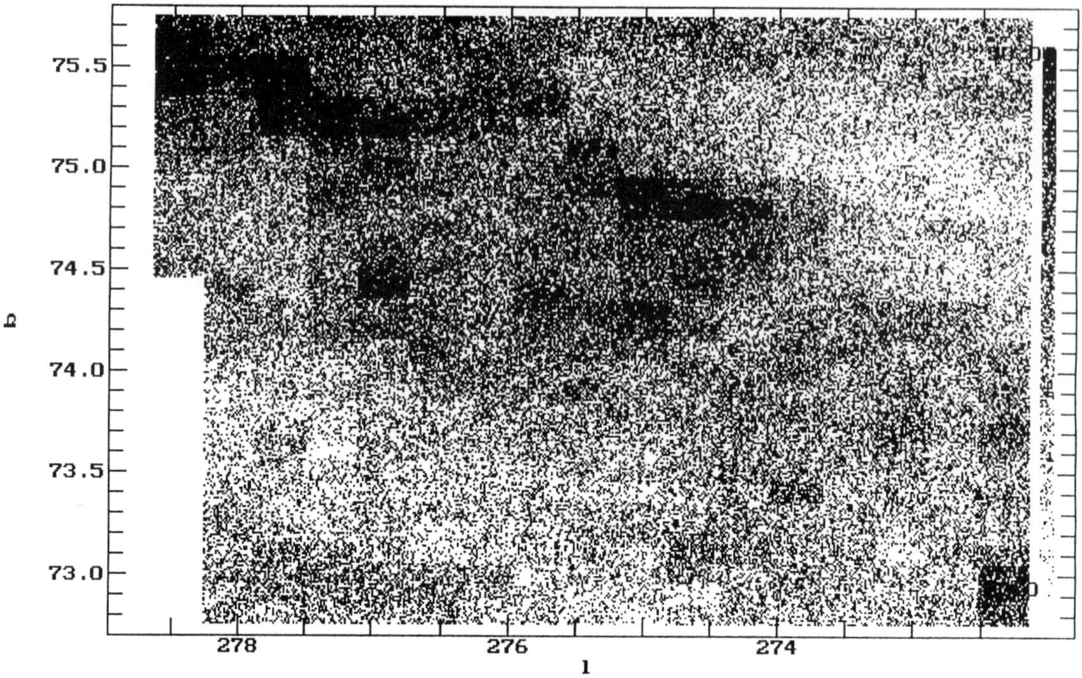

Figure 2a. HI column density corresponding to intensities in Arecibo observations integrated over the range $-5 < v < +10$ km/s. N(HI) at these "normal" velocities correlates with some of the dust structure, but not with the bulk of it.

parameters. Dust cirrus features are recognized as such essentially only at high $|b|$: dust emission from low $|b|$ is so heavily blended that separate features are rarely identified. It has become well established that dust features not uncommonly have gaseous counterparts. The gas associated with the more intense knots of dust emission may be in molecular form, and it appears (with only a few exceptions) that the associated molecular material is kinematically quiescent, located within a few km/s of zero, and with kinematic variations spanning only a few km/s. Most cirrus, however, have yet to be searched for molecular emission. The fields considered here, and, in general, cirrus fields whose infrared properties are dominated by emission at 100 μm, are not distributed as galactic molecular clouds or sites of star formation.

The kinematics of high-latitude atomic gas, on the other hand, is much more exuberant. A fiduciary HI profile can be formed by averaging observations taken at high $|b|$ over much of the sky: the shape of such a profile (see Figure 2 of Lockman and Gehman, 1991) shows a relatively narrow, relatively intense, component centered within a few km/s of zero, and a weaker, broader, shoulder extending to $|v| = 70$ km/s; this shape has been interpreted as due to several isothermal components, dominated

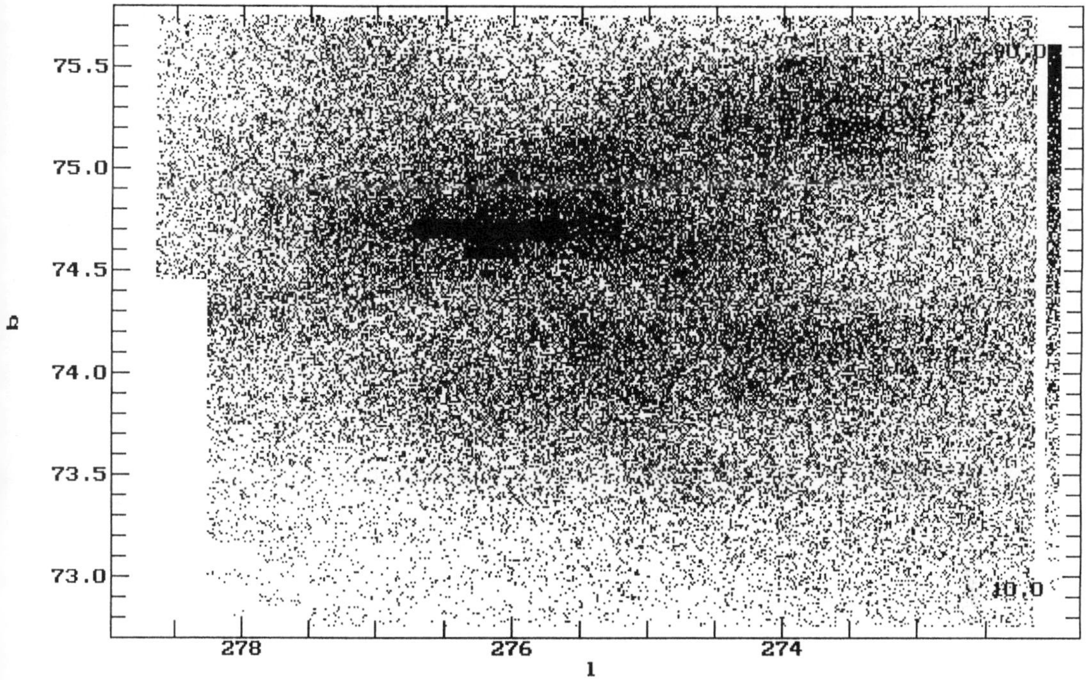

Figure 2b. HI column density corresponding to intensities in Arecibo observations integrated over the range $-40 < v < -25$ km/s. N(HI) at these anomalous velocities dominates the correlation with the cirrus dust feature.

by a cooler, slower gas, embedded in a warmer, faster envelope, both components being more or less volume filling (see Radhakrishnan and Srinivasan, 1980; Kulkarni and Fich, 1985). A profile shape composed of a narrow, cool-gas core, with broad, warm-gas wings, is not the signature of the gas associated with high-latitude cirrus: it remains to be seen how relevant this conclusion is to interpretation of the gas in the lower halo.

3.1 HI gas associated with the feature B cirrus cloud centered near 276°, 74.7

Figure 1 shows emission from dust observed by IRAS at 100 μm toward feature B, one of the dust clouds discussed in the cirrus detection paper by Low *et al.* (1984), smoothed to the 6-arcminute sampling interval of the Arecibo HI data. The dust cloud is quite sharply defined, showing peak emissivity near l, b = 276°.0,74°.7, with a two-pronged tail extending several degrees to lower longitudes. It is generally the case that IRAS cirrus dust features have gaseous counterparts; although the entire spatial extent of the HI filament which crosses this region is not spanned by the Arecibo data, it is obvious from the total HI column density, integrated over the entire band, that the dust is spatially concentrated where N(HI) also peaks.

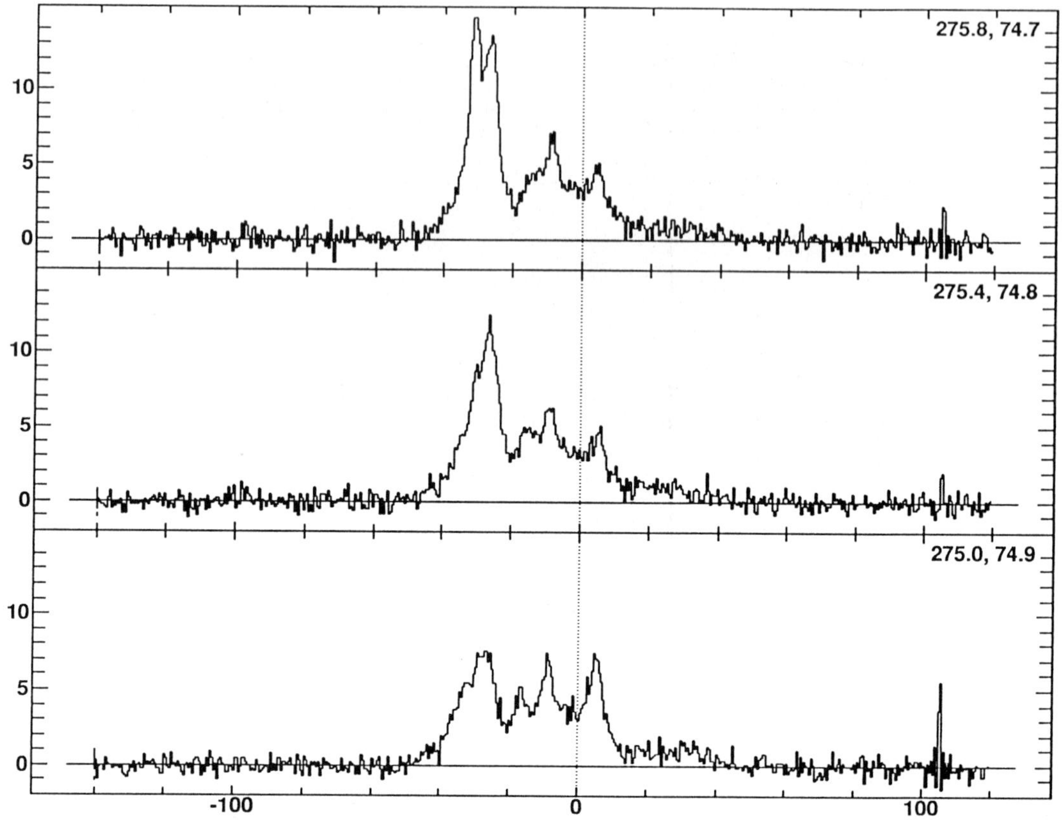

Figure 3. Arecibo 21-cm spectra at the indicated directions towards the cirrus feature B. The profile structure is complex, but dominated by anomalous velocities; the gas/dust correlations extend over some 40 km/s. (All spectra in this paper show Tb(K) plotted against LSR v(km/s).)

Deul and Burton (1990) showed that the kinematic structure of cirrus can be unravelled by considering the gas/dust relationship in different velocity intervals. Figure 2 shows the gas column density in the region of the dust cloud, for two such intervals. Figure 2a shows the HI column densities contributed from the range $-5 < v < +10$ km/s, corresponding to the velocities one would expect at these high latitudes for a well-behaved, quiescent gas layer. An HI filament does cross the region of the cirrus feature, and does bear some clear spatial relation to it, but the column densities shown in Figure 2b, corresponding to the range of velocities $-40 < v < -25$ km/s – certainly anomalous for these latitudes – bears a much tighter spatial similarity to the dust feature: the core structure of the dust feature near 276°, 74°.7 correlates with gas at these peculiar velocities, as do the two prongs of dust emission extending toward lower longitudes.

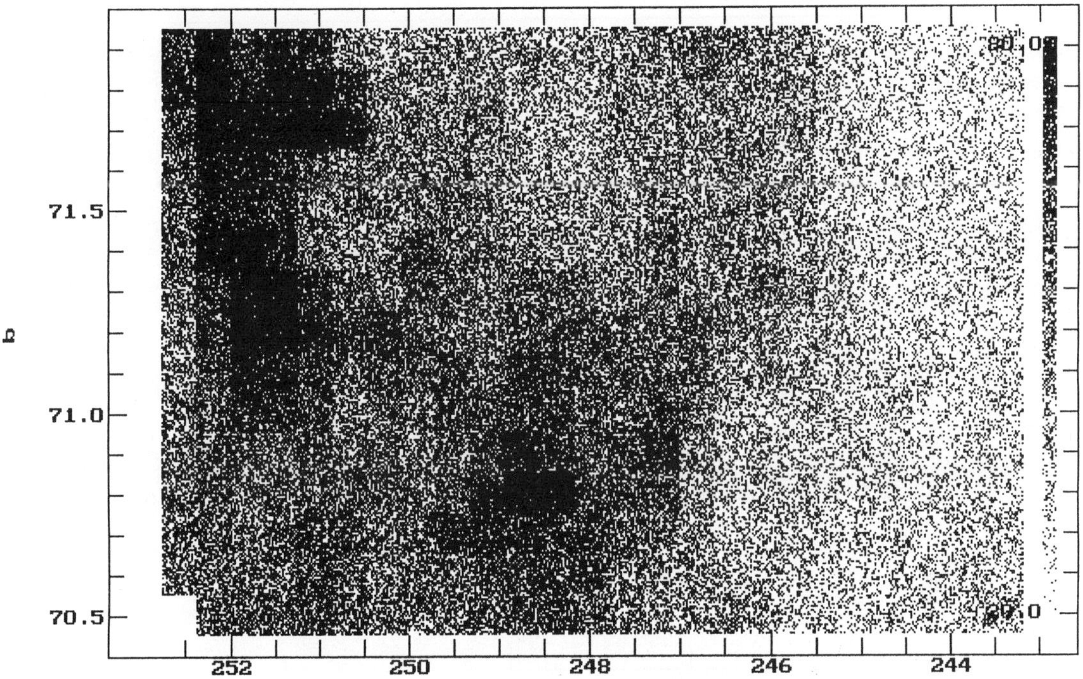

Figure 4a. HI column density corresponding to intensities integrated over the range $-10 < v < +5$ km/s in Arecibo observations toward the cirrus field, with a rather angular shape, centered near $l, b = 248°.7, 70°.8$. At these velocities, the higher-longitude portion of the angular cirrus feature has an HI counterpart; the gas/dust correlation is weak in normal velocities at the lower-longitude portion.

The N(HI) values are larger at the anomalous velocities than over the equivalent velocity extent at the quiescent velocities; the tighter gas/dust correlation at the anomalous velocities indicates that most of the dust likewise is moving rapidly. It is furthermore interesting to note that N(HI) at the quiescent velocities shows a local minimum in the direction of the core structure. Figure 3 shows representative Arecibo HI line profiles observed near the core of the cirrus feature B. The profile structure is complex; it changes over short angular scales; the emission at anomalous velocities is at least as intense as at normal velocities; the anomalous-velocity structures are distinct and rather narrow (certainly not what might be expected at these velocities from an extended envelope of warm, diffuse, fast gas); the total repertory of gas associated with the cirrus has a kinematic extent of more than 40 km/s.

These profiles fall within a single Dwingeloo beam of 36 arcminutes; note that the Bell Labs HI survey, currently the best available for study of IVC gas (and, by implication of the results presented here, of the cirrus/gas correlation), samples on a 2° x 3° grid, at $dv = 5.3$ km/s, missing some important spatial and kinematic information.

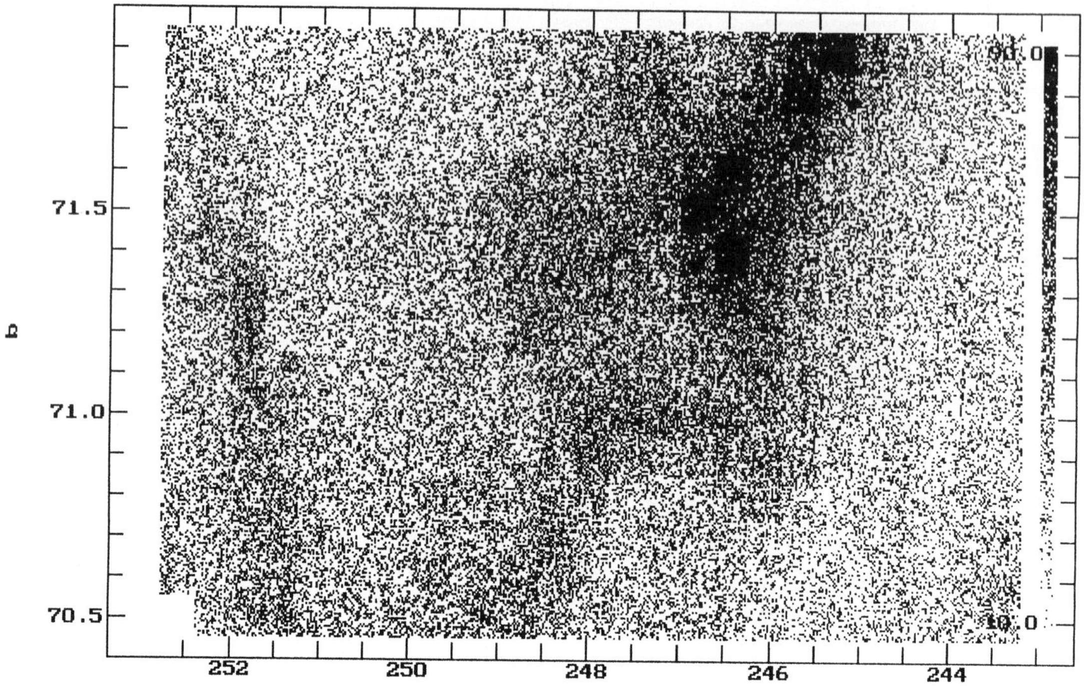

Figure 4b. HI column density corresponding to intensities integrated over the range $-35 < v < -20$ km/s in Arecibo observations toward the angular cirrus field. At these anomalous velocities, the lower-longitude branch of the angular dust feature has an HI counterpart, but not the higher-longitude branch.

The characteristics of the feature B dust/gas correlation are similar in many regards to those of the other high-latitude cirrus features discussed here. It is interesting to ask how generally these characteristics apply to cirrus clouds: do cirrus generally span such a range of velocity; do cirrus embrace the class of HI intermediate-velocity clouds; is the anomalous-velocity emission commonly associated with a disturbance to the normal-velocity gas?

3.2 HI gas associated with the angular cirrus cloud centered near 248°, 71°

Deul and Burton (1990) drew attention to the cirrus field centered near 248°, 71°, containing a dust and gas structure with an angular shape; in the portion which we have now mapped at Arecibo, the angular shape takes the form of a letter "V", with its vertex near 248°.7, 70°.8. Figure 4a shows the HI column depth contributed by the quiescent-velocity range $-10 < v < +5$ km/s. Here, the higher-longitude structure dominates; the lower-longitude branch is present, but weak. That situation is reversed in Figure 4b, which shows N(HI) contributed by gas moving at velocities in the range

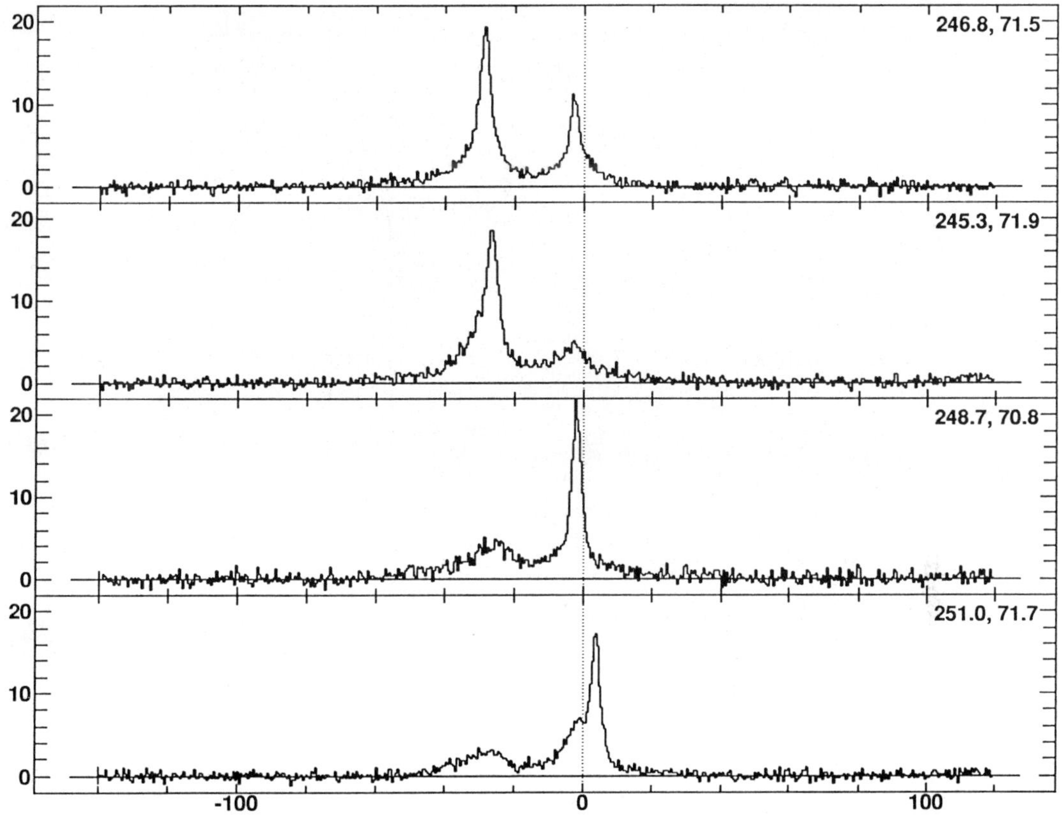

Figure 5. Arecibo 21-cm spectra at the indicated directions towards the cirrus feature showing an angular, two-pronged dust morphology. The total extent of the gas/dust correlation spans some 40 km/s; the two prongs of the feature occur at different velocities.

$-35 < v < -20$ km/s. The lower-longitude branch dominates, but connects smoothly in position on the sky, as well as in velocity, with the other branch at the vertex of the pattern.

Figure 5 shows representative Arecibo 21-cm spectra from the cirrus feature: the upper two spectra from the lower-longitude, higher $|v|$ branch, the lower two from the branch dominated by the normal velocities. It is important to establish whether or not the anomalous-velocity material is physically related to the structure observed at the more normal velocities; in this regard we note that the two branches meet in the same position on the sky, and show, in addition, a remarkable kinematic symmetry.

The HI features associated with the dust cloud are quite narrow in velocity; the feature centered near +4 km/s in the bottom spectrum can be fit by a Gaussian

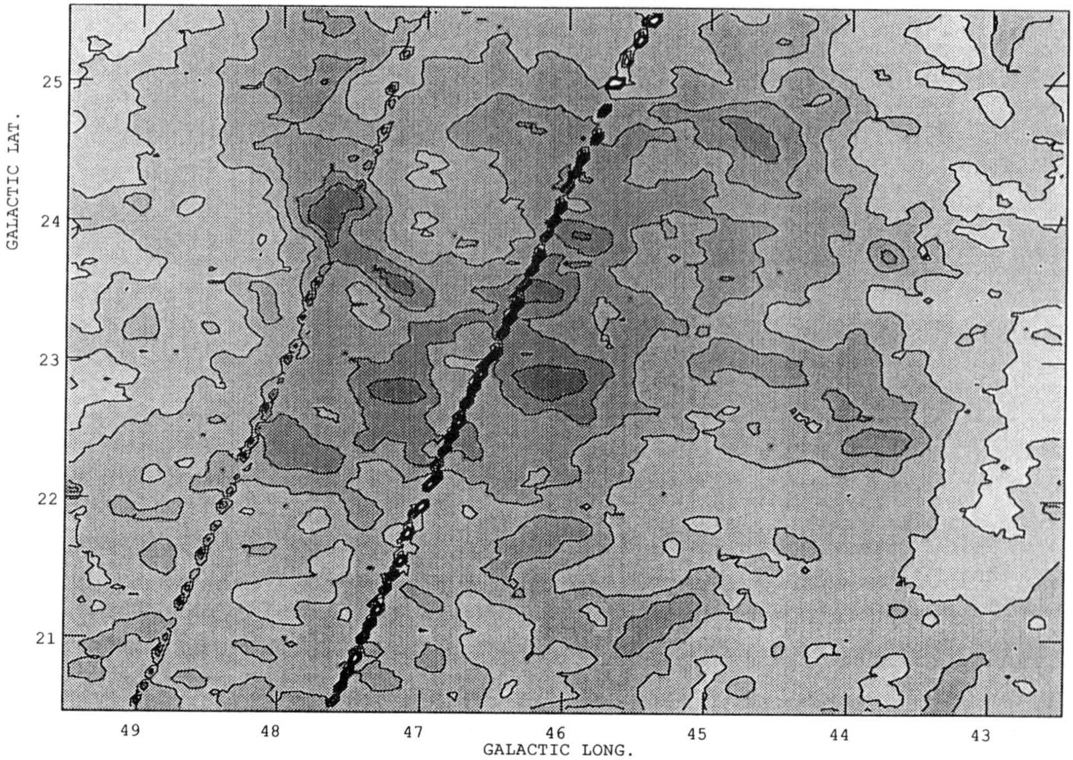

Figure 6. Dust emission at 100 micron from the vicinity of the cirrus feature considered by Deul and Burton (1990), centered near $l, b = 46°, 23°$. The IRAS map has been smoothed to the 6-arcminute resolution of the Arecibo HI data; the dust feature shows a filamentary morphology, centered on a knot of dust emission at 46°, 23°.

of dispersion 1.1 km/s which, if translated to a kinetic temperature, corresponds to $T_k < 150$ K. The commonly accepted picture of high-latitude HI (see e.g. Dickey and Lockman, 1990) attributes variations in N(HI) largely to gradients in a generally smooth, diffuse distribution. The profiles shown here, in particular the ones observed at Arecibo, seem to be more consistent with the concept of an ensemble of "clouds", characterized by a volume filling factor substantially less than 1, and an area filling factor also less than 1; particularly for what has been called the "high dispersion" gas, that at $|v| > 20$ km/s, it would be interesting to verify if the ensemble picture is a relevant one.

3.3 HI gas, including an IVC, associated with the cirrus cloud centered near 46°, 23°

Figure 6 shows IRAS emissivity at 100 μm from a cirrus dust cloud of coherent morphology, comprised of a number of filaments apparently focussed on a knot of infrared emission at 46°, 23°. These dust filaments have gaseous counterparts: the

6-arcminute sampling of the Arecibo HI data allows the associated gas to be traced in more detail than could be done by Deul and Burton (1990) using the 21-arcminute beam of the 140-foot Green Bank telescope. Figure 7a shows that the low-$|v|$ column depth traces some of the filaments, but that there is a deficiency of N(HI) at the position of the IR knot. Figure 7b shows the column depths contributed by gas in the region at velocities in the range $-70 < v < -50$ km/s. Located at the position of the knot of infrared emission, and, similarly, at the position of the deficiency of gas emission at about +5 km/s, there is a compact HI structure, centered near $v = -60$ km/s, with all of the characteristics of a prototypal IVC.

Figure 8 shows the Arecibo HI spectrum taken towards the dust-knot and HI-IVC feature at $46^\circ\!.0$, $23^\circ\!.0$, as well as two spectra taken at nearby positions. The IVC cloud is evidently associated with a deficiency of HI emission near about +5 km/s. (We note that the IVC lies outside the band of the Berkeley survey of Heiles and Habing (1974), and subtends a total area smaller than the effective beamwidth of the Bell Labs survey of Stark *et al.* (1991)).

A satisfactory description of the energetics of this cirrus field will have to account for a variety of information: a prototypal HI IVC has an IR counterpart; the IVC dust and gas structure is associated in position with a disturbance near zero velocity; at the location of the disturbance, there is a local deficiency in N(HI) which corresponds to the excess in column density at the IVC velocity; filamentary dust features apparently emanate from the location, and from the velocity, of the disturbance.

3.4 Gas and dust tracing the North Celestial Pole Loop

The feature called the North Celestial Pole Loop by Meyerdierks *et al.* (1991) is a prototypal example of a Heiles shell. It can be followed in a variety of tracers over a span of some 30° of longitude and some 20° of latitude. Figure 9, reproduced from the article by Meyerdierks *et al.*, shows IRAS 100-μm dust emission from the NCP Loop, as well as HI column densities integrated over the range $-25 < v < +41$ km/s in the Berkeley-survey data of Heiles and Habing (1974).

Figure 10 shows the IRAS 100-μm dust emission from the region of the Loop, smoothed spatially to a resolution of 30 arcminutes, approximately that of the Dwingeloo HI data. (The data of the Leiden-Dwingeloo survey are well suited to an investigation of the NCP Loop, by virtue of the large sky coverage and sensitivity of that survey, but also because of its extensive velocity coverage.) Figure 11a shows the HI column density derived from the Dwingeloo data over the velocity range $-4 < v < +21$ km/s. At these normal velocities, the HI shell structure follows the general shape of the dust

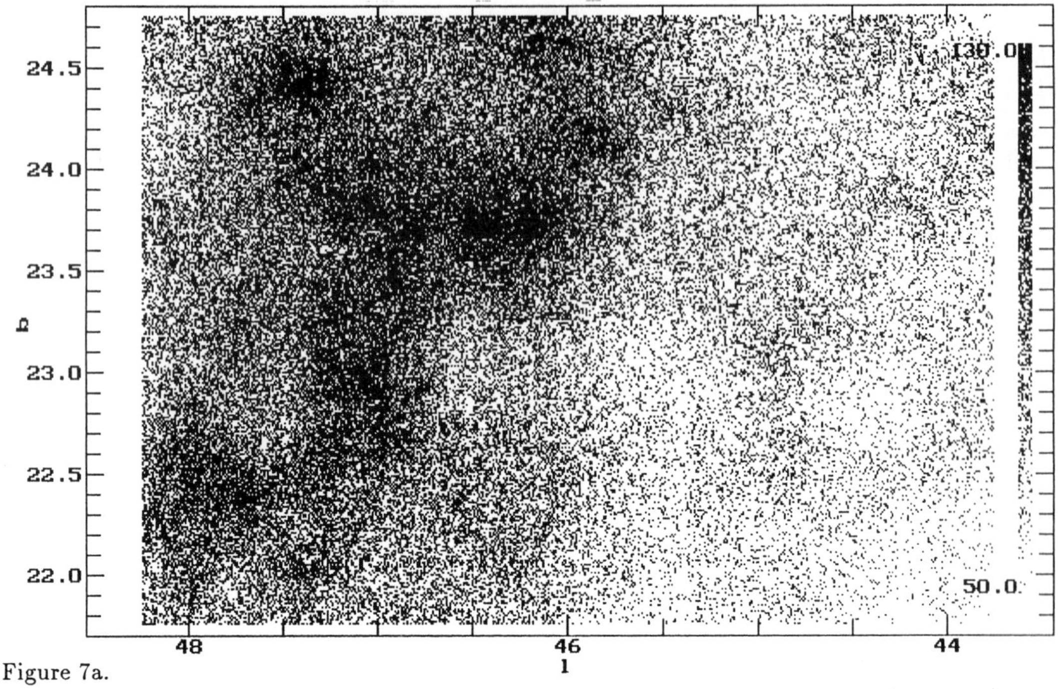

Figure 7a.

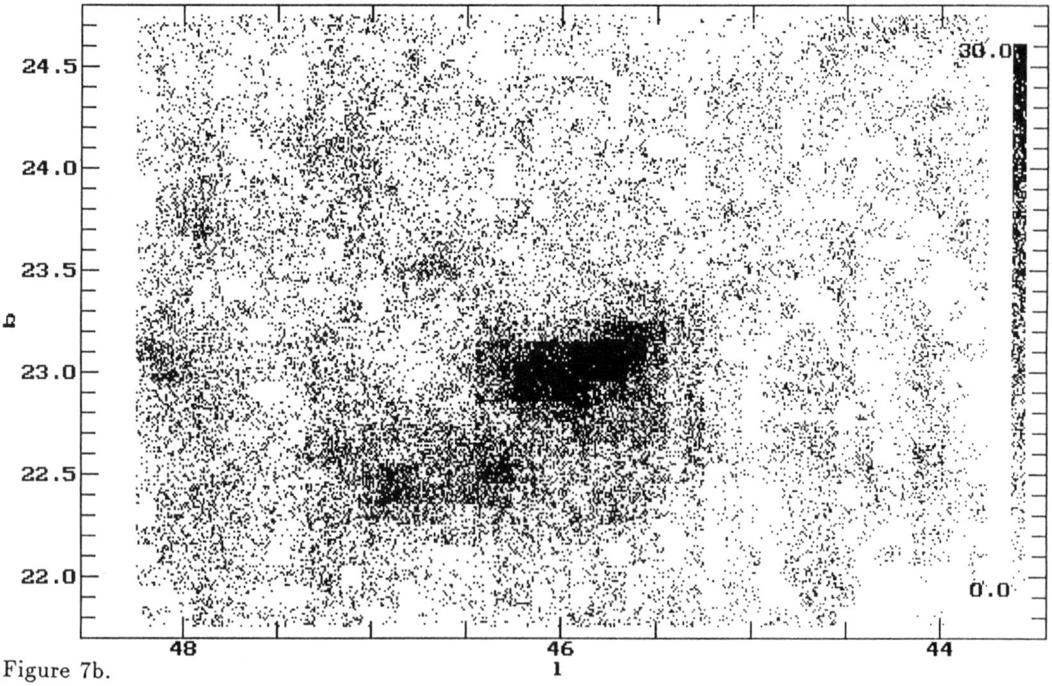

Figure 7b.

Figure 7a. HI column density corresponding to intensities integrated over the range $0 < v < +10$ km/s in Arecibo observations toward the cirrus field portrayed in dust emission in Figure 6. The low-$|v|$ N(HI) traces the filaments, but there is a deficiency of HI at the 46°, 23° IR knot.

Figure 7b. HI column density corresponding to intensities integrated over the anomalous-velocity range $-70 < v < -50$ km/s in Arecibo observations toward the Figure 6 cirrus field. The isolated HI structure at these velocities has the characteristics of a prototypal IVC; it has an IR counterpart in the knot of dust emitting at the same position. That the cirrus filaments, with velocities near zero, evidently are associated with the knot indicates that the cirrus kinematics embraces more than 65 km/s.

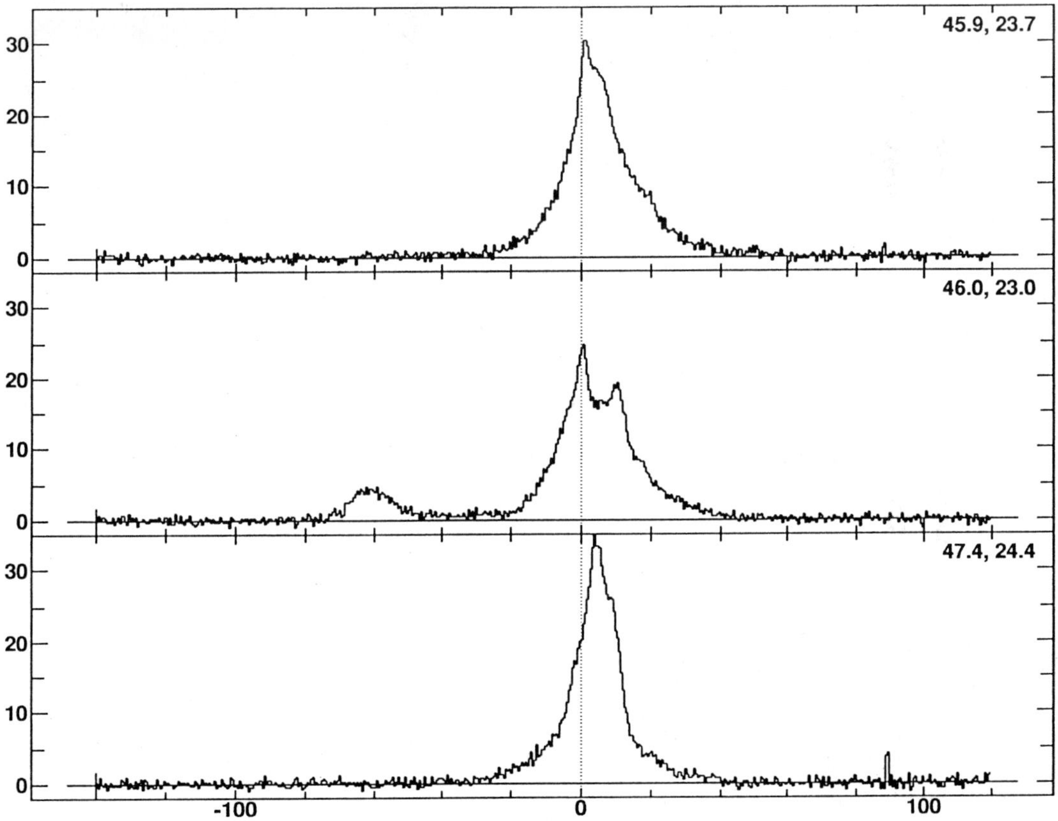

Figure 8. Arecibo 21-cm spectra at the indicated directions towards the cirrus feature showing the knot/filament dust morphology. It appears that the anomalous-velocity part of the cirrus is associated with a disturbance in the ambient gas disk.

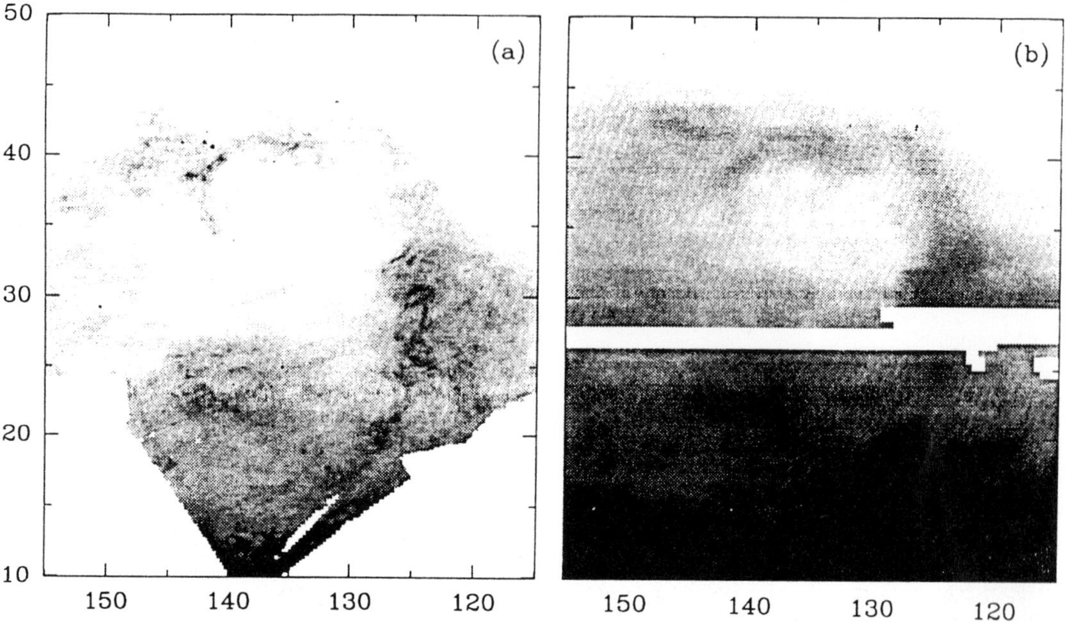

Figure 9. Distribution of the dust and gas in the NCP Loop as presented by Meyerdierks *et al.* (1991). The IRAS 100-μm data (left) has been smoothed to 6' resolution; the HI column densities correspond to integration over the velocity range $-25 < v < 40$ km/s in data from the Heiles and Habing (1974) survey.

loop. Figure 11b shows, on the other hand, the HI column depth integrated over the anomalous-velocity range $-72 < v < -56$ km/s. The shell feature can be followed to negative velocities of the magnitude spanned by IVC's; indeed, the negative-velocity branch of this loop has been known since the pioneering IVC work of Wesselius and Fejes (1973). Some aspects of this extensive IRAS dust loop can be associated with structures in the Figure 11a N(HI) map at velocities near zero; other aspects, with structures at the negative velocities. The total velocity extent of the shell seems to span more than 70 km/s, implying greater energetics than heretofore supposed for a gaseous shell (or for a dust cirrus) feature. (Regarding the matter of velocity coverage, we note in general that the parameters of the Heiles shell phenomenon have been largely based on the Berkeley survey, whose useful velocity extent is limited to about $|v| < 45$ km/s.)

Figure 12 shows representative Dwingeloo 21-cm spectra toward the NCP Loop. The intermediate-velocity-cloud emission near -60 km/s traces the same general morphology of the Loop as revealed by the IRAS 100-μm emissivity; the gas emitting at the moderate positive velocities similarly shows this same general morphology, although the details of the dust/gas correlation differ with location and velocity across the

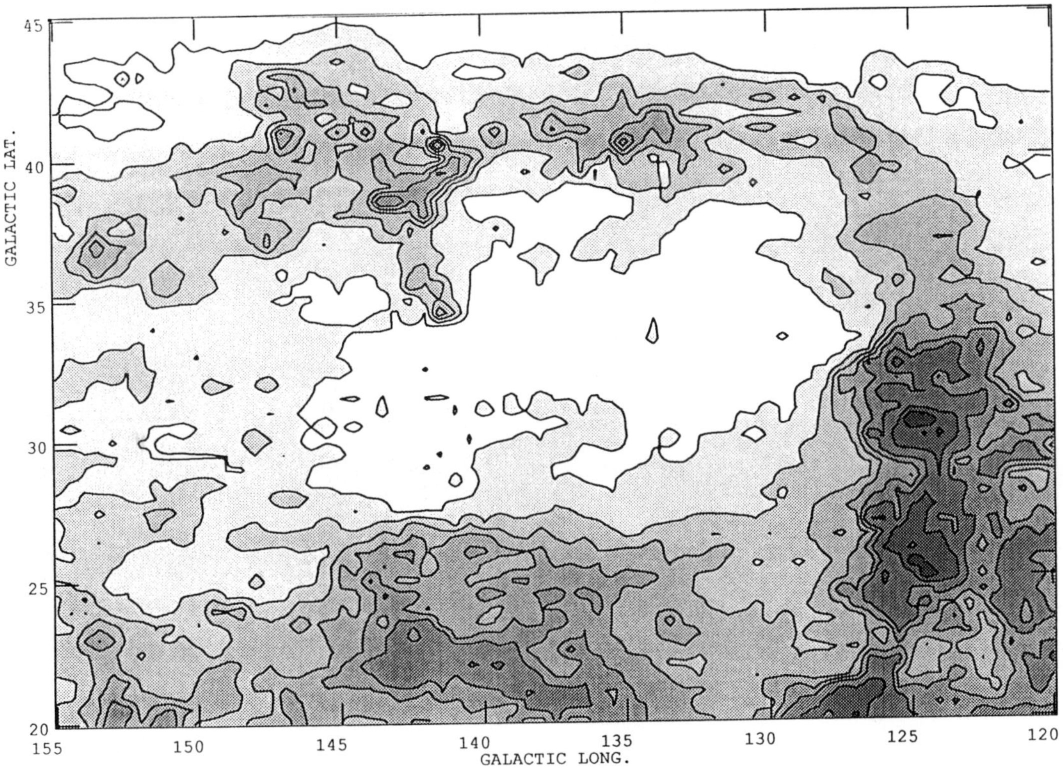

Figure 10. IRAS emission at 100 micron from dust associated with the NCP Loop, smoothed to the 36-arcminute resolution of the new Dwingeloo HI survey data.

large Loop. In addition to the IVC gas, which is correlated with the Loop, we note that the high-velocity-cloud Complex A crosses this part of the sky. There is no obvious relationship between the HVC complex and the IVC gas, and thus no obvious relationship of the HVC complex with the cirrus feature. Dust emission has not been detected from HVC's, although there is some evidence that HVC structures may be associated with disturbances in the galactic disk which do show dus peculiarities.

The intensities of the negative-velocity portion of the feature are generally less than those of the portion at positive velocities. But if we take zero velocity as the origin of expansion (see Meyerdierks *et al.*, 1991), we note that the momentum in the negative-velocity portion is equal to that of the positive-velocity portion. Figure 13 shows that the two portions together essentially account for all of the gas emission associated with the cirrus Loop. The figure shows the total N(HI) derived over the entire Dwingeloo band, after subtraction of the column depths in the ranges $-4 < v +21$ km/s and $-72 < v < -56$ km/s; the HI emission residual at $b < 30°$ shows the structural and

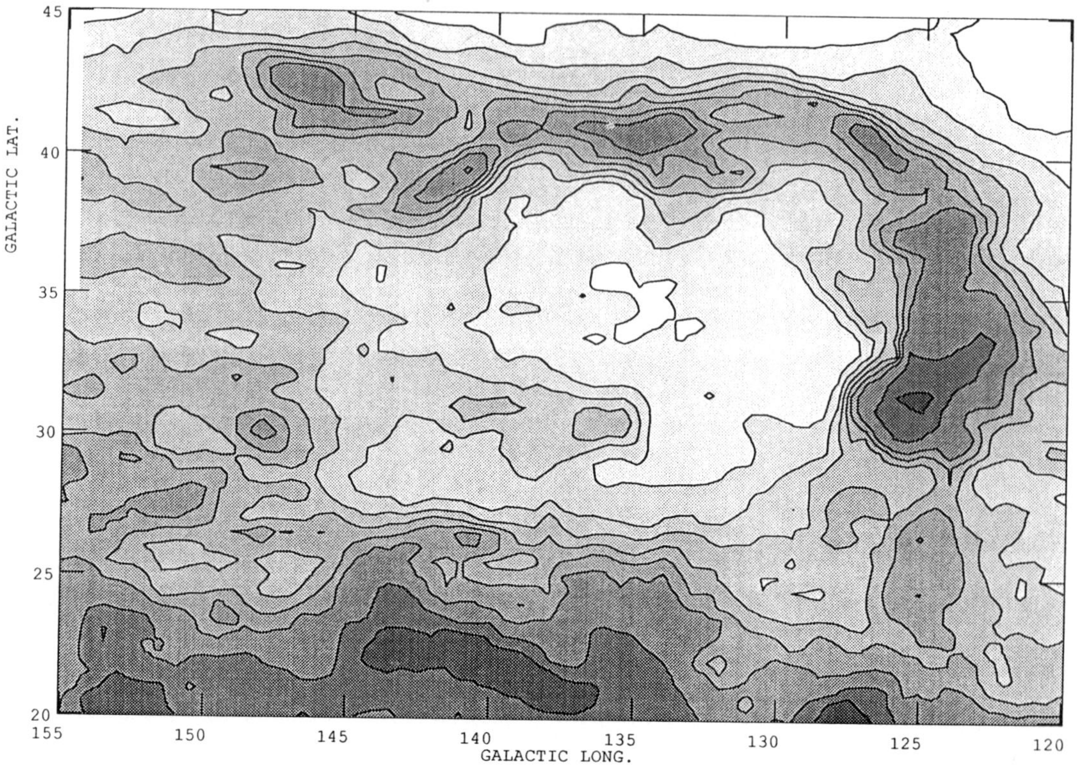

Figure 11a. HI column density corresponding to intensities integrated over the range $-4 < v < +21$ km/s in Dwingeloo observations toward the cirrus field portrayed in dust emission in Figures 9 and 10. The gas at these velocities traces only part of what has been identified as a prototypal Heiles shell.

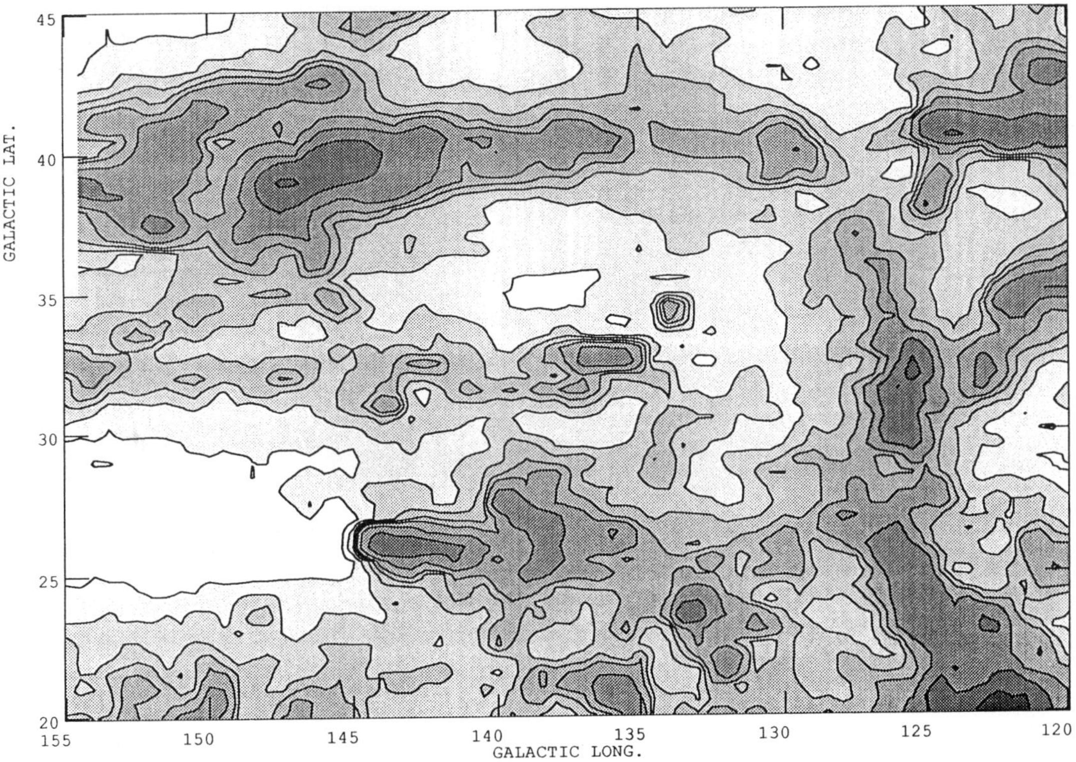

Figure 11b. HI column density corresponding to intensities integrated over the range $-72 < v < -56$ km/s in Dwingeloo observations toward the cirrus field portrayed in dust emission in Figures 9 and 10. That the gas at these anomalous velocities, outside the effective limits of the Berkeley survey, also traces aspects of the morphology NCP Loop feature, indicates higher energetics for this cirrus/shell feature than previously supposed.

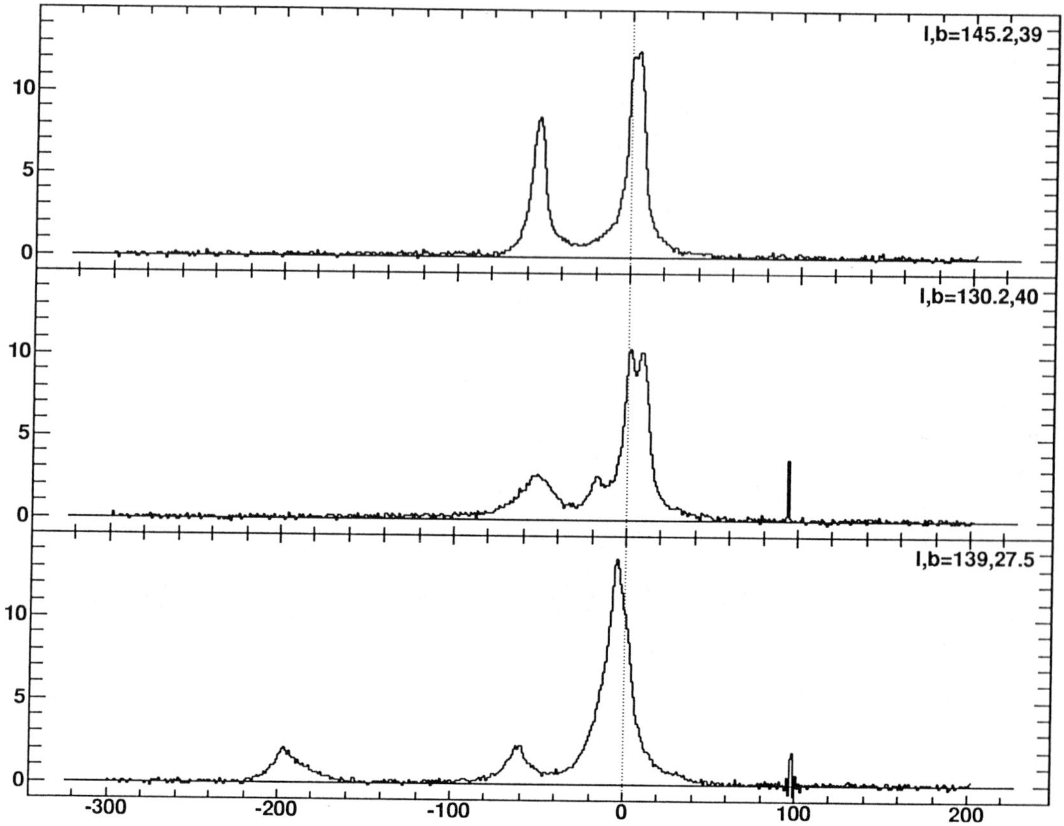

Figure 12. Dwingeloo 21-cm spectra at the indicated directions intersecting the NCP Loop feature. The gas emitting at the intermediate-negative-velocities, near -60 km/s, follows aspects of the morphology of the IRAS dust emission as well as of the gas column density at velocities near zero, and is thus evidently part of the same shell/supershell/bubble feature. The high-velocity-cloud Complex A traverses the region of the shell, but no physical relation between the HVC gas, on the one hand, and the intermediate- and normal-velocity gas and dust, on the other, has yet been established.

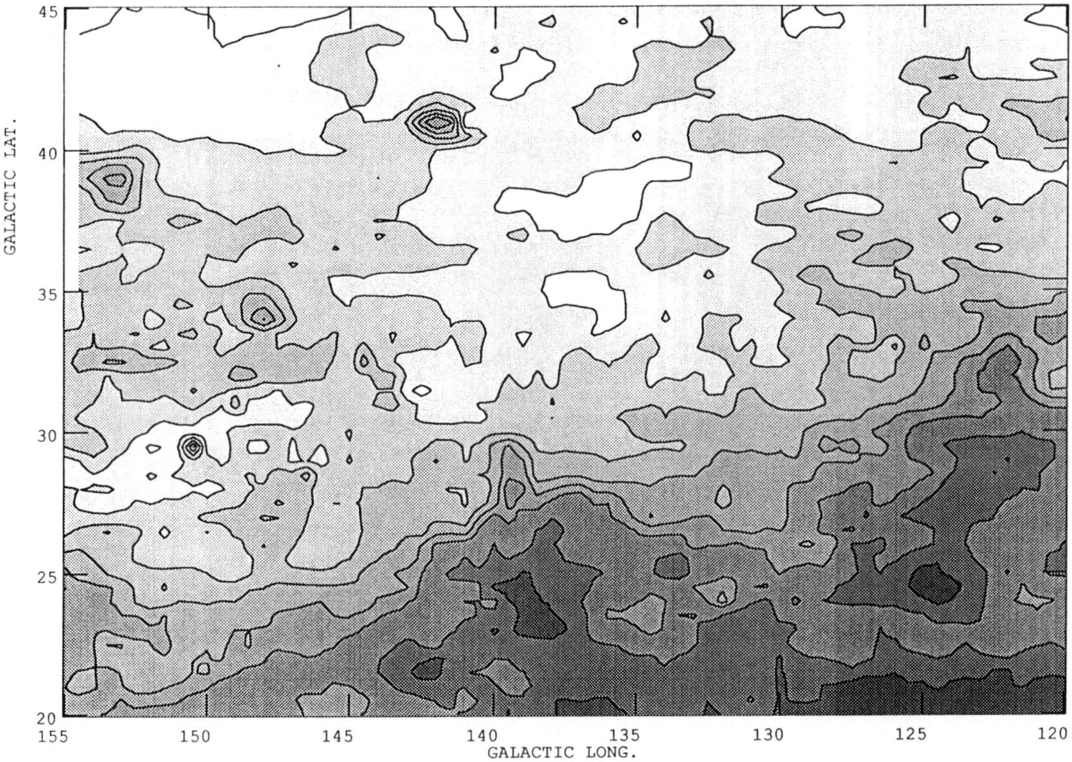

Figure 13. Residual N(HI) densities after subtraction of the two parts of the shell given by the integrated emission in the ranges $-4 < v < +21$ km/s and $-72 < v < -56$ km/s.

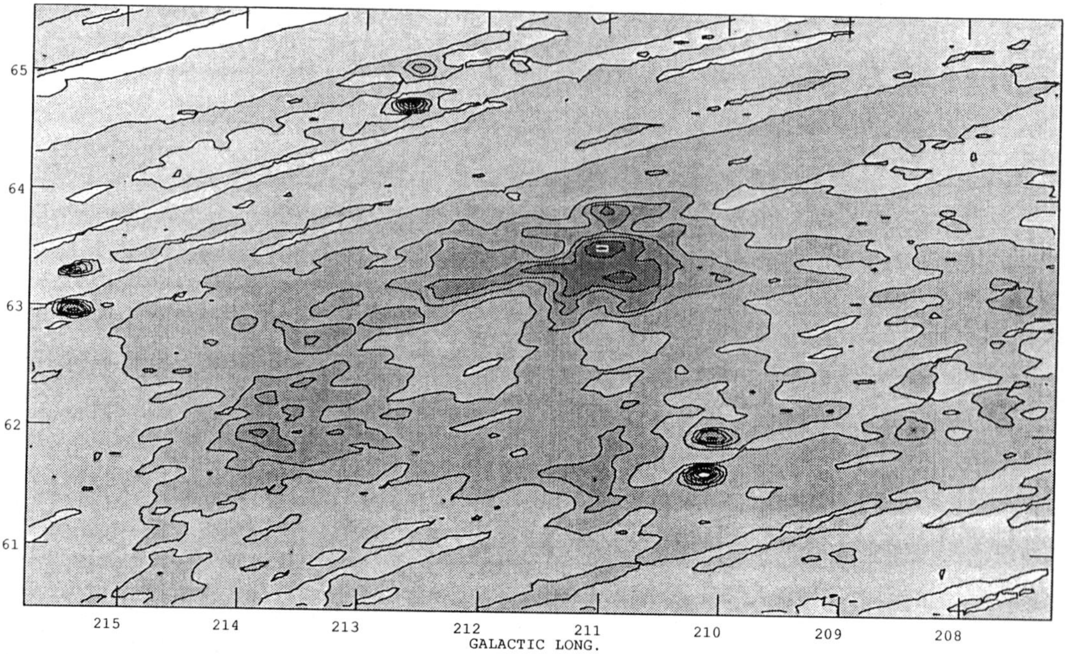

Figure 14. Dust emission at 100 micron from the vicinity of the cirrus feature in whose heart at $l, b = 211°, 63°$ Desert *et al.* (1990) found CO emission at −39 km/s. The IRAS map has been smoothed to the 6-arcminute resolution of the Arecibo HI data.

kinematic properties that one would expect for gas at large distances from the Sun lying in the galactic warp.

3.5 HI shell incorporating the compact cirrus featured centered near 211°, 63°

Figure 14 shows the IRAS 100-μm emission from a cirrus feature centered on 211°, 63°. At the heart of this cirrus feature, Desert *et al.* (1990) found CO emission from a molecular cloud at −39 km/s. Molecular clouds at high latitude and at velocities differing from zero by more than 25 km/s are rare. Only three such clouds are known: the Desert *et al.* cloud, G210.8+63.1−39; one discovered by Heiles *et al.* (1988) at G90.0+38.6−25; and one discovered by Mebold *et al.* (1989) at G135.3+54.5−45. Molecular gas associated with cirrus features has, of course, been observed frequently, especially in regions within cirrus showing particularly intense dust emission; but the kinematic picture which has been derived from such molecular observations is a quiescent one: molecular emission associated with cirrus knots is generally confined to within about 5 km/s of a centroid velocity, itself usually near zero.

We have observed HI in the vicinity of the 211°, 63° cirrus feature using the Arecibo telescope, and then – realizing how extensive the gas pattern associated with the dust one is – we have extracted the HI data from a larger region from the Leiden-Dwingeloo survey. Figure 15a shows N(HI) derived from Arecibo data over the range $-12.5 < v < +12.5$ km/s. The normal-velocity gas is exceptionally weak, and shows, furthermore, little correlation with the dust-emission cirrus pattern shown in Figure 14. The situation is different, however, at the anomalous velocities represented in Figure 15b, which shows N(HI) over the range $-50 < v < -25$ km/s. At these velocities, the gas traces a pattern on the sky well correlated with that of the dust emission. Just as the 100-μm emission does, the HI column density peaks near the location of the Desert *et al.* molecular feature.

Figure 16 shows representative Arecibo spectra from this region. Emission centered near −40 km/s is more intense throughout the region than emission near zero. Indeed, the emission at velocities expected from a quiescent ambient gas is exceptionally weak throughout the region. Adjacent to the IR knot at 211°, 63°, the total HI column depth approaches, furthermore, the exceptionally low values found in a different region by Lockman *et al.* (1986) and thought to be the lowest anywhere on the sky. The HI features associated with the IR knot are quite narrow: the spectral feature near −39 km/s in the 212°.3, 63°.0 spectrum has a Gaussian dispersion of 1.4 km/s; the feature near 0 km/s, 1.1 km/s. (We repeat that the spectra have not been corrected for stray-radiation contamination: such contamination will not invalidate our conclusions based on the isolated features typically more intense than 10 K, but does prevent quantitative use of the spectral information at levels below about 0.5 K. It is interesting to note thus that the weak intensities represented in the upper spectrum in Figure 16 are upper limits to the true situation.)

Intermediate-velocity HI features are seldom isolated; they more commonly trace extensive patterns over large angles on the sky (see e.g., Wesselius, 1972, and Wesselius and Fejes, 1973). Figure 17 shows the distribution of the total HI column depth (over the range $-100 < v < +100$ km/s) in a large area centered near the cirrus cloud at whose heart was found the G210.8+63.1−38.7 molecular line. The cirrus cloud shown in Figure 14 is part of a larger gaseous structure, which evidently traces a shell-like structure centered roughly around the direction 208°, 68°. Figure 18a shows the column densities contributed by the range $-12.5 < v < +12.5$ km/s; Figure 18b, by the range $-50 < v < -25$ km/s. The gaseous shell morphology dominates at the negative velocities, but cannot be followed in the velocity range centered at zero. In this regard, this cirrus region differs from the others whose kinematics we have attempted to unravel using HI data: the other features taken together lead to

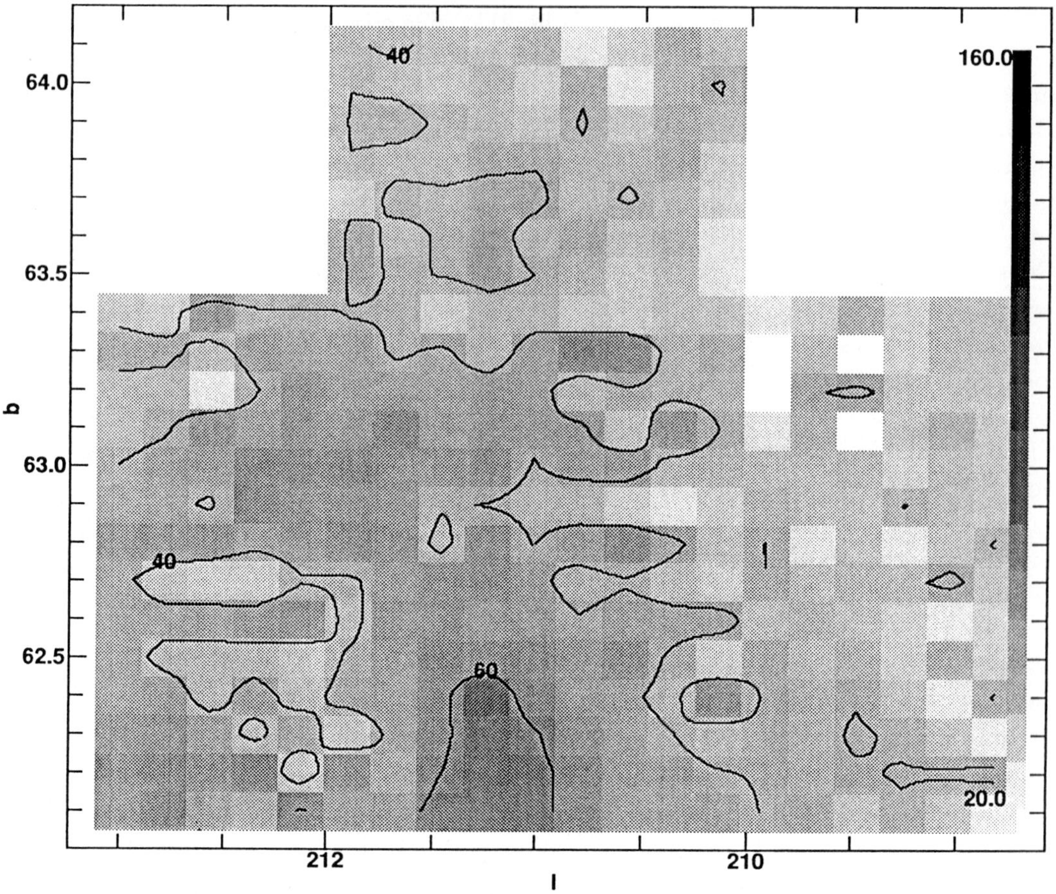

Figure 15a. HI column density corresponding to intensities integrated over the range $-12.5 < v < +12.5$ km/s in Arecibo observations toward the cirrus field portrayed in dust emission in Figure 14. There is little similarity between the dust morphology in this direction and the normal-velocity gas.

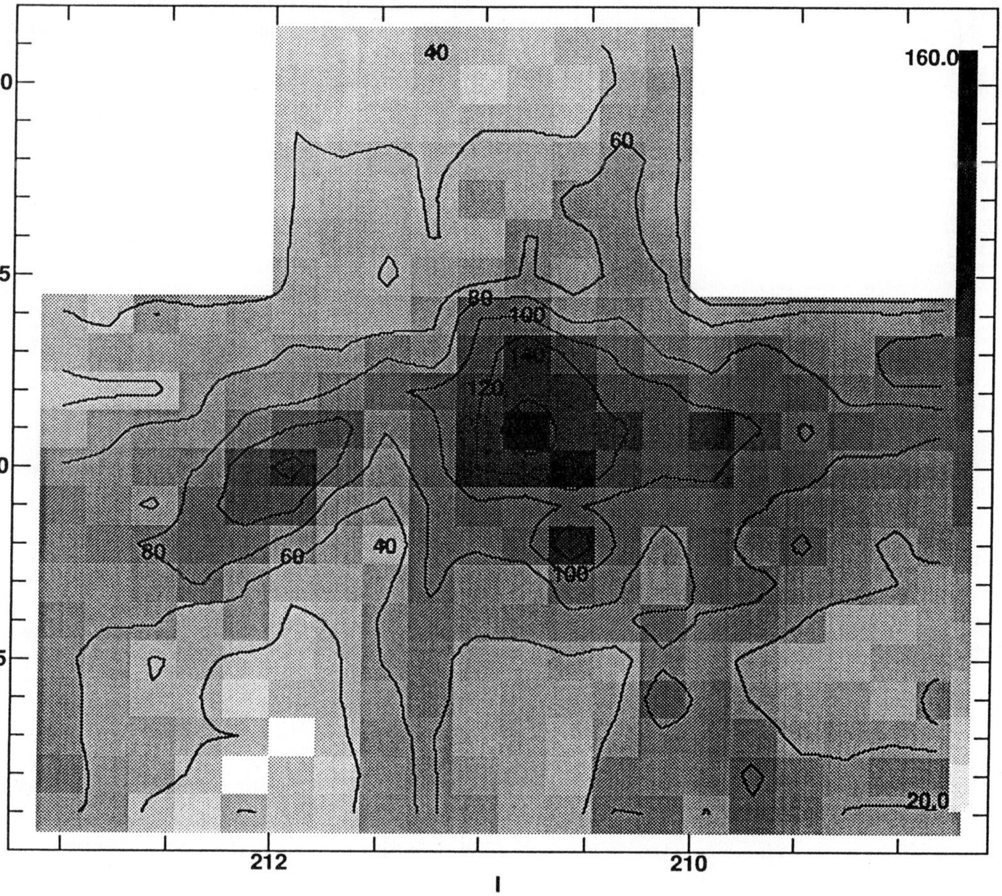

Figure 15b. HI column density corresponding to intensities integrated over the range $-50 < v < -25$ km/s in Arecibo observations toward the cirrus field portrayed in dust emission in Figure 14. Most of the HI column density in this region occurs at anomalous velocities, centered near -40 km/s; there is little gas at all near zero velocity.

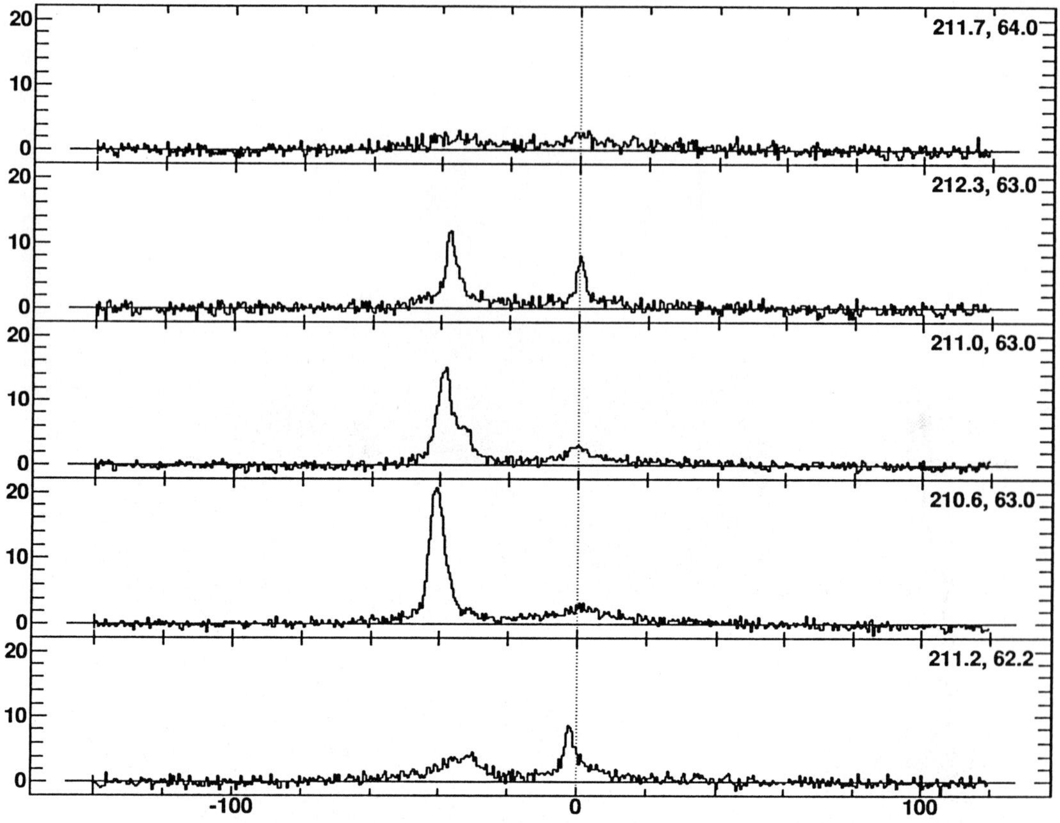

Figure 16. Arecibo 21-cm spectra at the indicated directions towards the cirrus feature containing the Desert *et al.* molecular cloud at G210.8+63.1−38.7. The gas emitting at the intermediate-negative-velocities, near −40 km/s, dominates the total gas density budget in the direction of the cirrus structure; adjacent to the cirrus feature is a region of exceptionally low total HI column density. The substructures in the HI lines are unusually narrow, and change substantially over short angular scales.

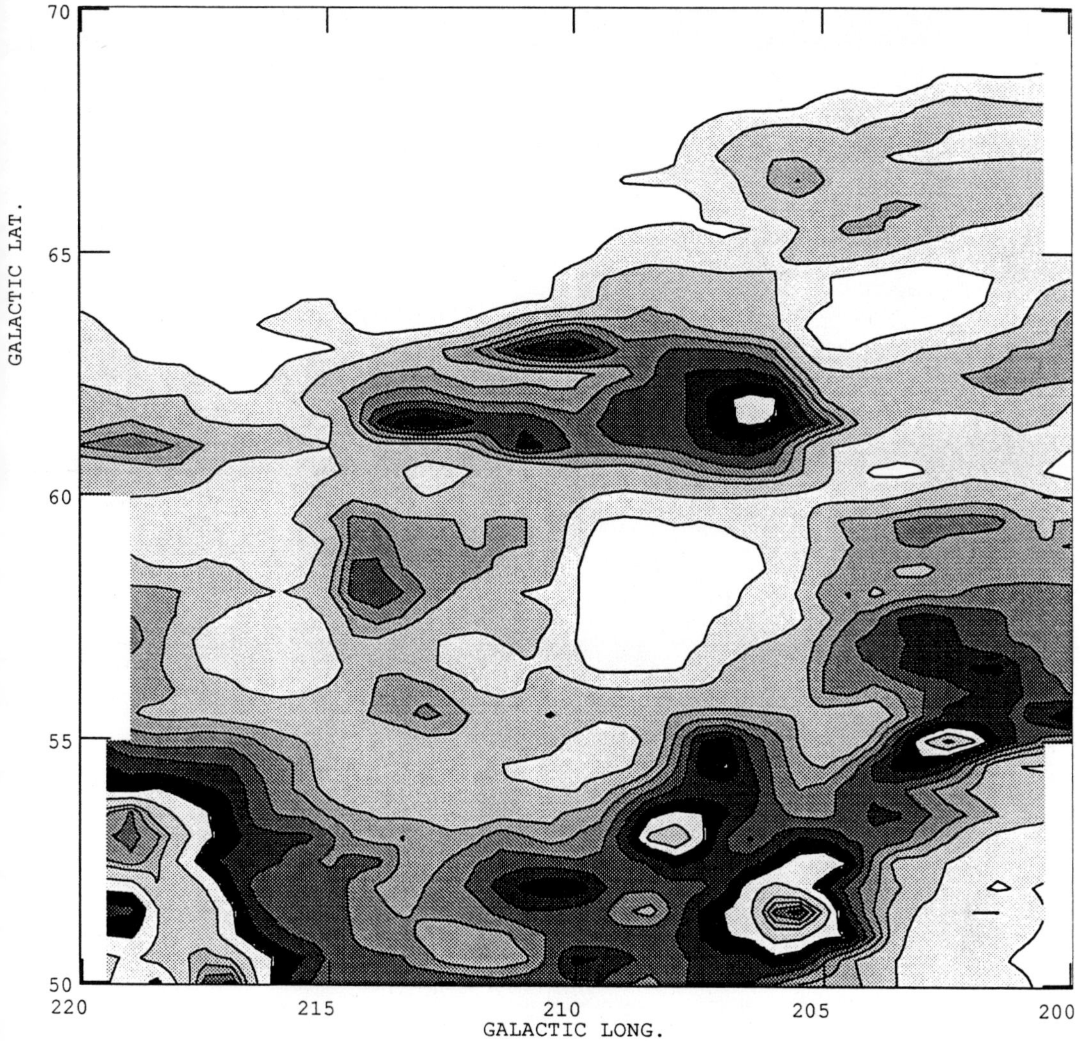

Figure 17. HI column density corresponding to intensities integrated over the broad range $-100 < v < +100$ km/s in Dwingeloo observations centered near the cirrus field represented in Figures 14 and 15, but covering a larger total area of sky. The cirrus feature is evidently part of a larger, shell-like structure centered near $l, b = 209°, 58°$ with a diameter of about 10°.

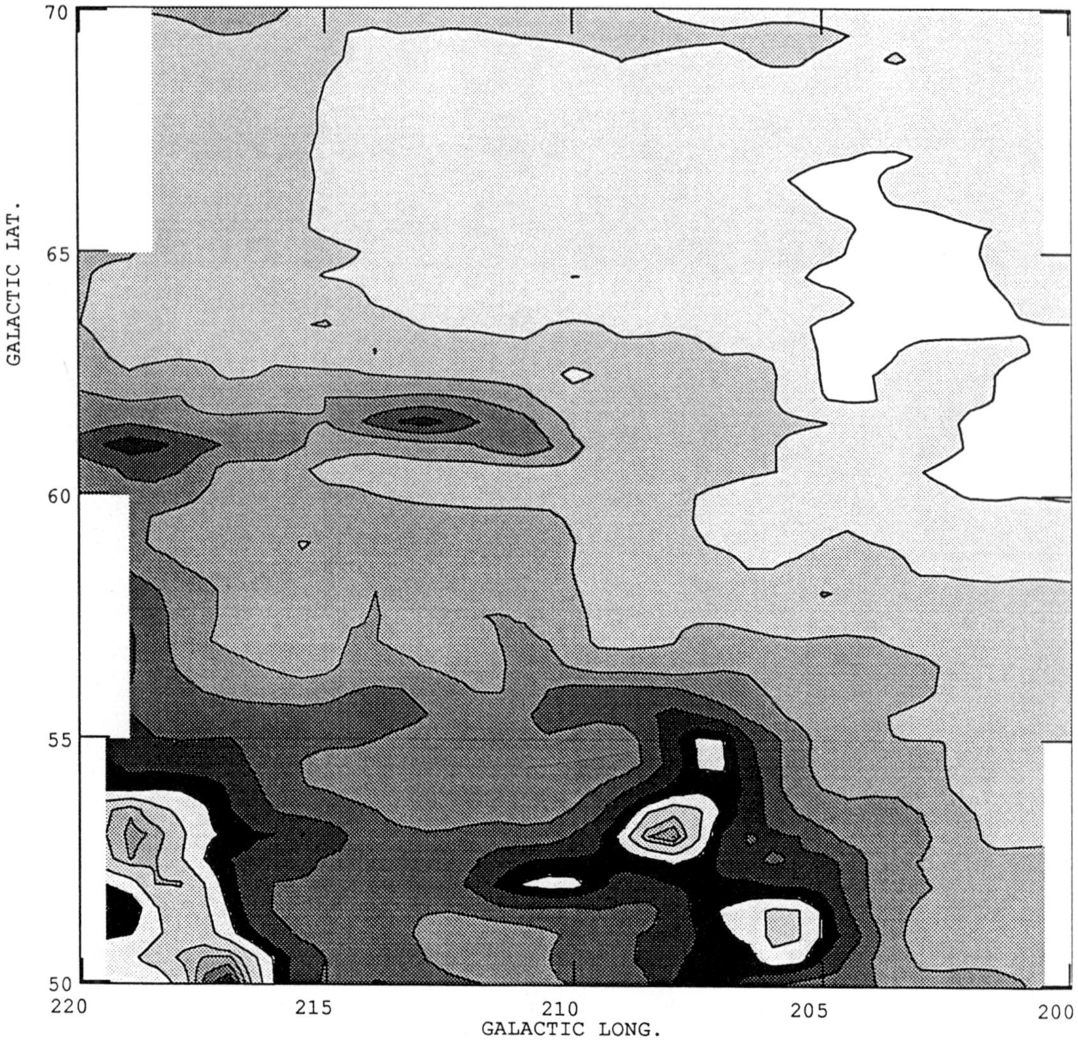

Figure 18a. HI column density corresponding to intensities integrated over the range $-12.5 < v < +12.5$ km/s in Dwingeloo observations as in Figure 17. The shell-like structure evident in the integration over the large velocity range represented in Figure 17 is not evident at the "normal" velocities.

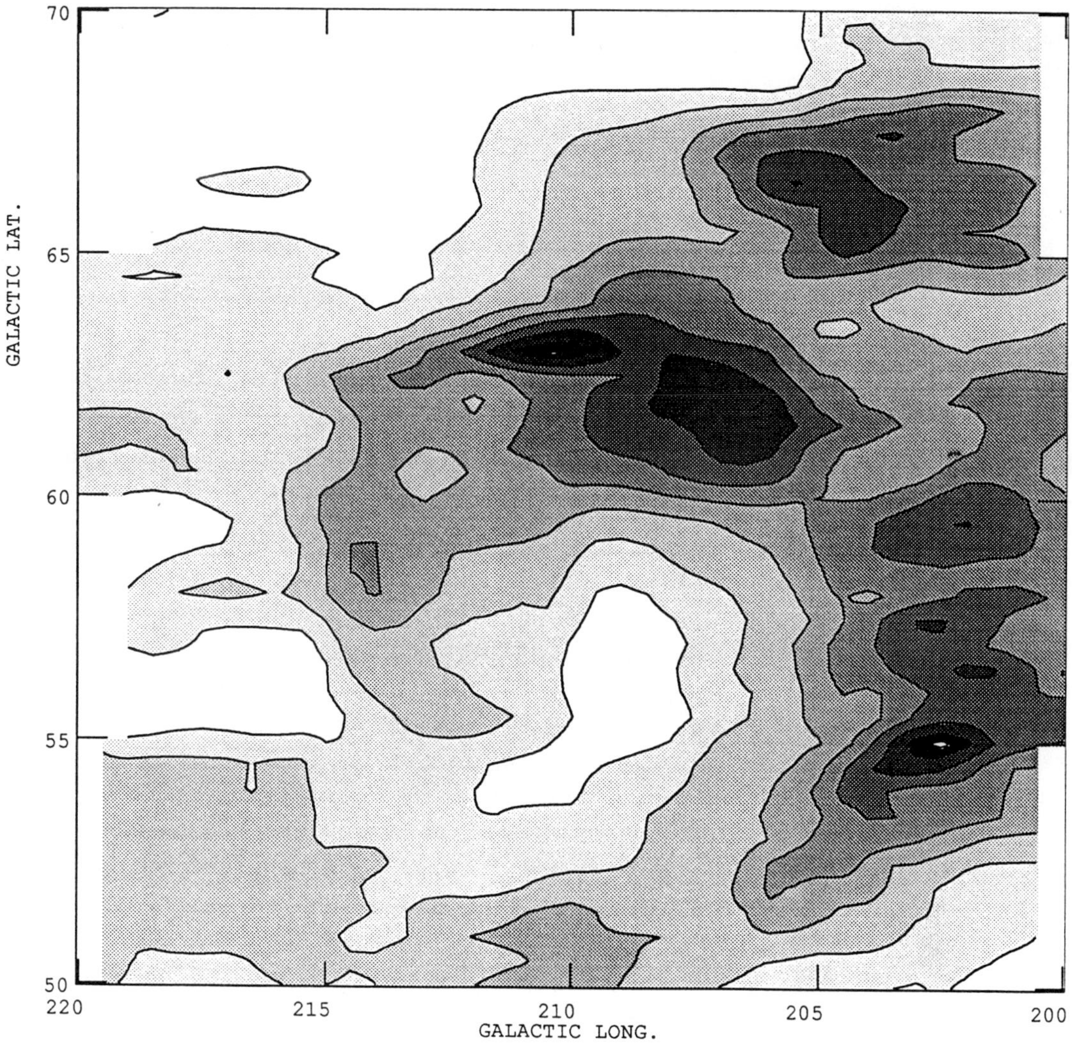

Figure 18b. HI column density corresponding to intensities integrated over the range $-50 < v < -25$ km/s in Dwingeloo observations as in Figure 17. The shell-like structure evident in the integration over the large velocity range represented in Figure 17 is largely accounted for by emission at these anomalous velocities.

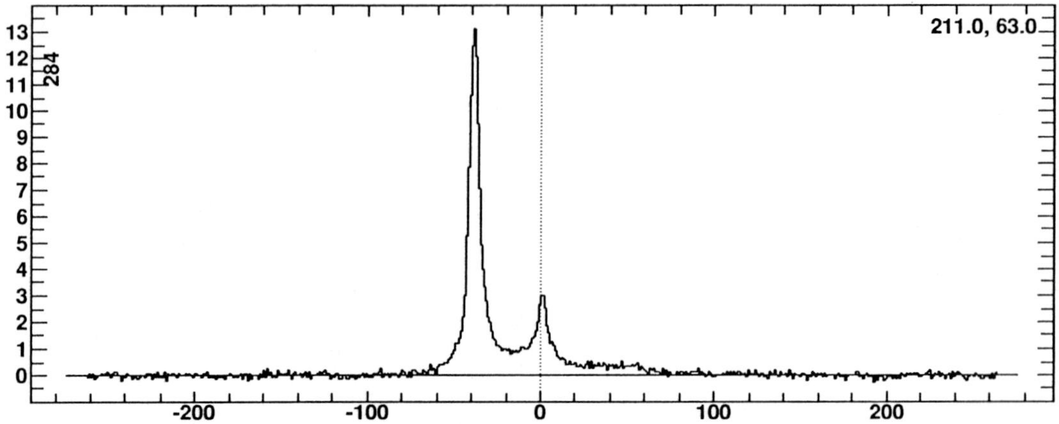

Figure 19a. Dwingeloo HI spectrum taken in the direction where Desert *et al.* found a CO line at −39 km/s.

the impression that individual cirrus general incorporate a range of velocities, not unusally extending over some 40 or 60 km/s, and therefore indicative of substantial energetics. The total velocity extent of the shell-like structure which contains the peculiar-velocity molecular cloud found by Desert *et al.* remains a puzzle.

Several Dwingeloo HI spectra representative of this region are shown in Figure 19. Figure 19a shows the 21-cm spectrum taken in the direction of the Desert *et al.* anomalous-velocity CO emission. Although this cirrus region displays the – evidently common – characteristic of having associated gas at intermediate negative velocities, it is the only one which we have investigated which shows such relatively weak HI emission at velocities near zero. Although only the negative-velocity portion of the shell structure has been followed in the HI data, the weak gas intensities observed near zero velocity suggest that a perturbation has disrupted the ambient material. In this sense, it could still be the case that the total kinematic extent requiring interpretation embraces the zero velocity emission, not only the narrow kinematic band around −40 km/s.

Figure 19b shows HI spectra taken adjacent to the compact cirrus feature. The total-velocity N(HI) levels in the adjacent region are exceptionally low. Because the intensities near zero velocity are weak, we plan to investigate the possible association of the zero-velocity gas with the cirrus feature, and the possible relation of the rather empty region adjacent to it, only after the stray-radiation correction has been applied by Kalberla and colleagues in Bonn.

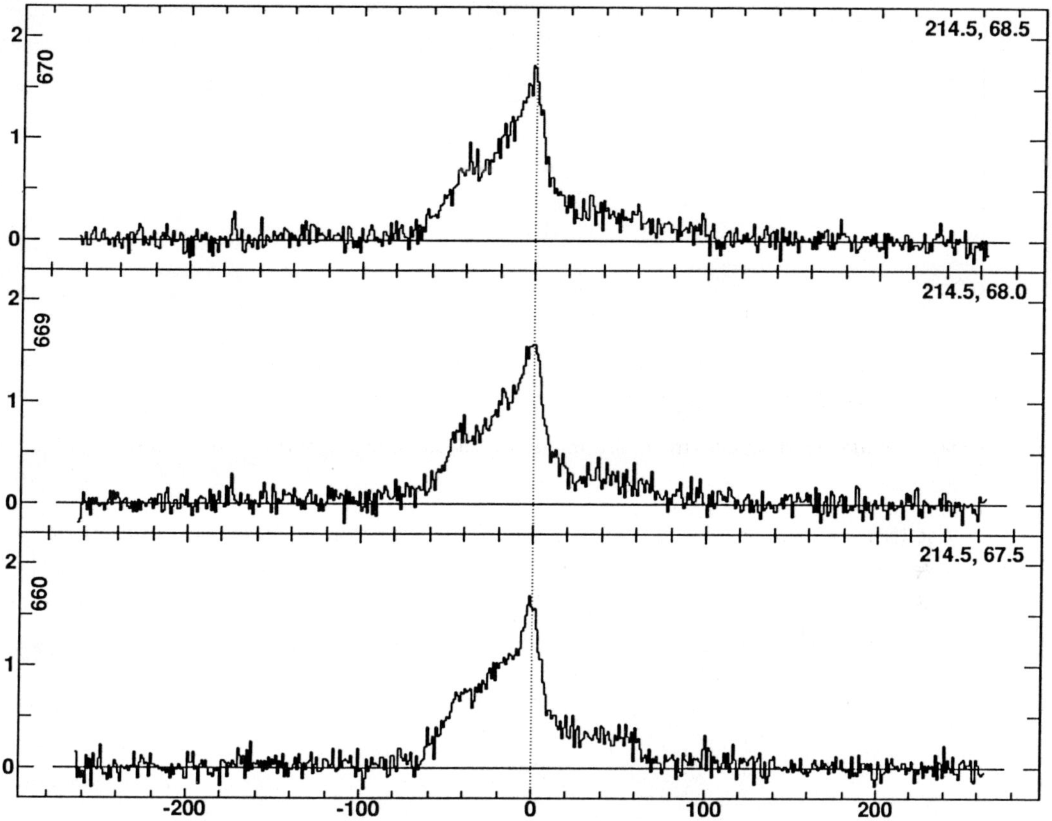

Figure 19b. Dwingeloo HI spectra taken in the region adjacent to the compact cirrus feature containing the Desert *et al.* molecular feature and to the larger HI shell in which the dust feature is evidently a local density concentration.

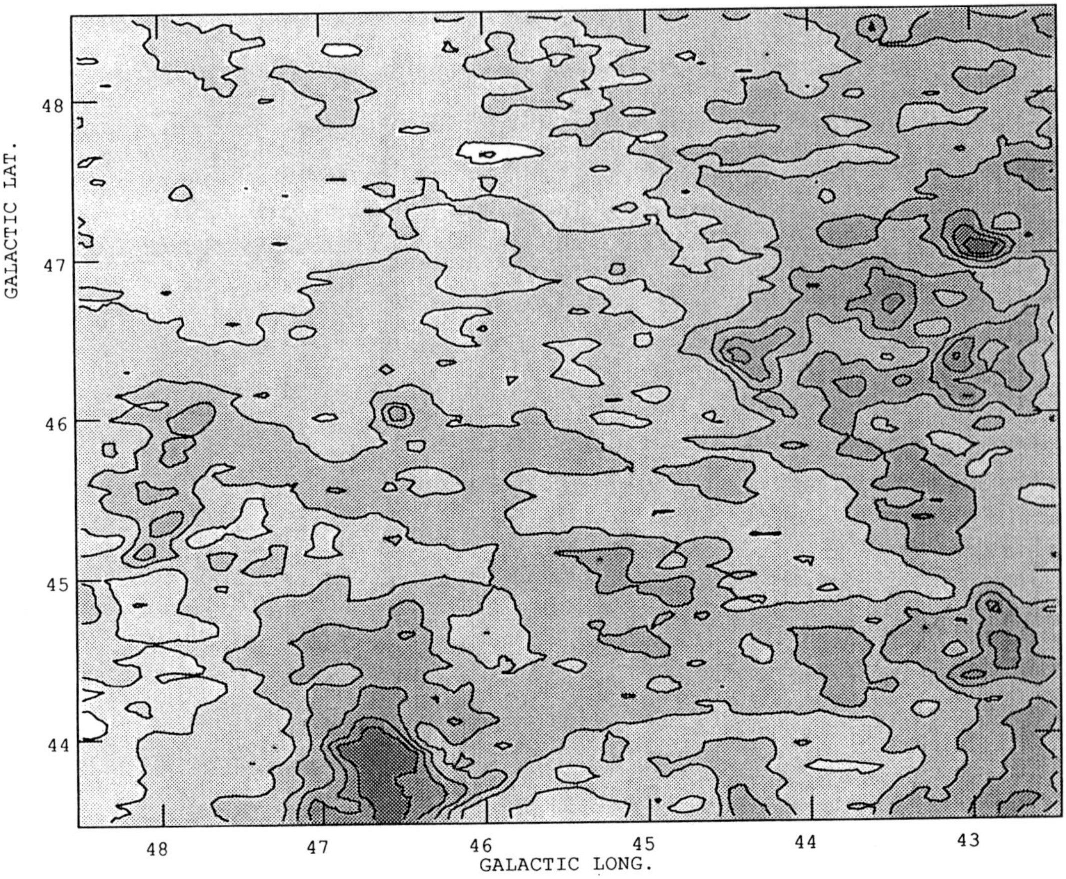

Figure 20. IRAS emission at 100 micron from the vicinity of l,b = 45°, 45°, smoothed to the resolution of the Arecibo HI data. This field contains weak, rather wispy, dust structures.

3.6 Small-scale variations in gas and dust near the field centered on 45°, 45°

All of the cirrus fields for which we have attempted kinematic unravelling using HI observations show relationships with the regime of intermediate-velocity gas. The associated gas also typically shows structural and kinematic variations on a small scale. Both of these characteristics are illustrated by observations of the rather inconspicuous cirrus field near l, b = 45°, 45°. The IRAS 100-μm data shown in Figure 20 pictures a clumpy, wispy, region of rather weak dust emission. Before observing it from Arecibo, we had been aware that the HI spectrum towards the "cardinal" direction 45°, 45° is simple, coming as close as any to showing what one would expect from a well-behaved, plane-stratified gas layer. Figure 21 shows the Arecibo spectrum in this direction, as well as several representative spectra from the near vicinity.

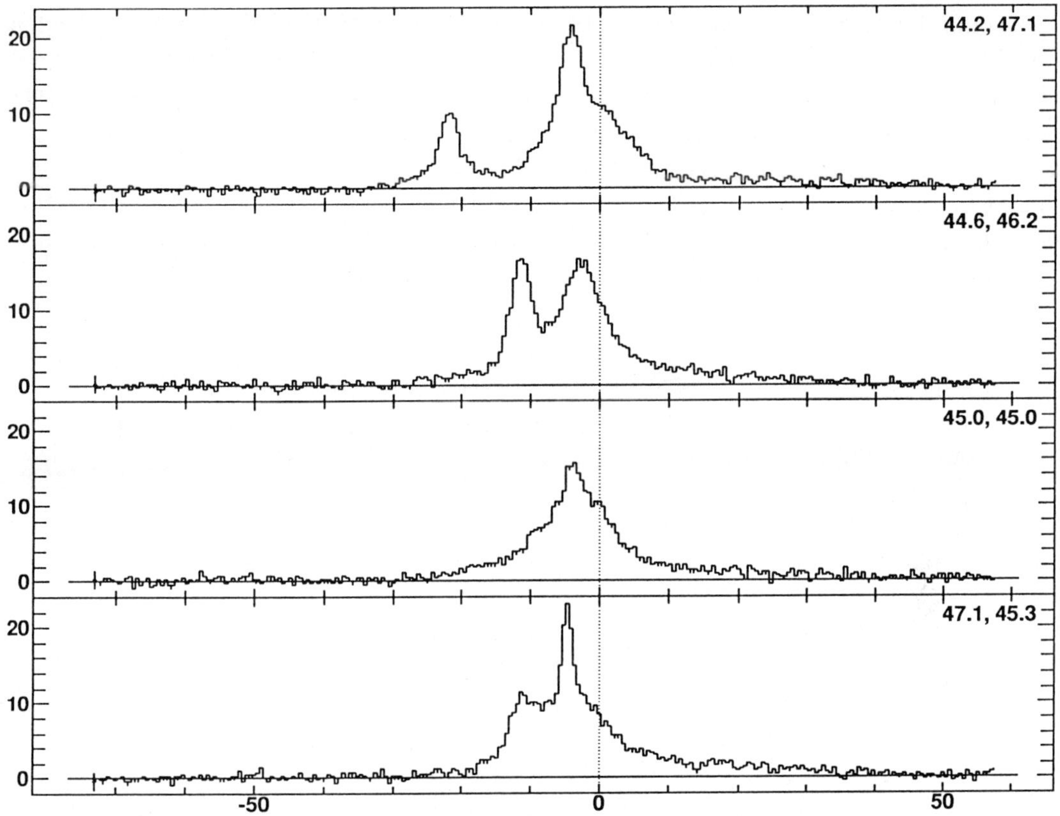

Figure 21. Arecibo 21-cm spectra at the indicated positions in the vicinity of the direction 45°, 45°. Arecibo HI mapping shows that in this region gas at anomalous velocities is prevalent and that it is correlated with the cirrus dust forms. The Arecibo beam reveals exceptionally narrow lines and abrupt positional changes, also in the "fast" HI. The Gaussian dispersion of the narrow feature in the bottom profile is 0.68 km/s, corresponding to an upper limit for the kinetic temperature of 57 K.

The HI situation towards this otherwise rather unremarkable cirrus field is quite interesting. The individual cirrus clumps typically have HI counterparts, found over a range of different velocities. There is substantial kinematic and spatial structure on all accessible scales.

4 SOME ADDITIONAL CONSIDERATIONS

In the directions of several IVC complexes, Danly (1989) found ultraviolet absorption lines at about −50 km/s in stars at distances between 400 and 700 pc, but not towards nearer stars. Acceptance of this large distance is why Lockman and Gehman (1991) reject velocities $|v| > 30$ km/s from their discussion of the role of turbulence in the lower-halo gas. Acceptance of the large Danly distance together with acceptance of the generalization of the association of cirrus clouds with IVC HI at, say −50 km/s, would place the cirrus phenomenon *also* at large distances. There seem a number of problems with attributing such large distances to the anomalous-velocity HI gas with associated dust emission: star-count distances (e.g., de Vries and Le Poole, 1985) and independent HI and cirrus scale-height measures are two separate indications of much more modest distances for the cirrus dust and its associated gas. The role of area filling factor should be investigated further in any case, but also because of its possible relevance to the resolution of the distance question.

Sampling with large beams, and, especially, smoothing over very large portions of the sky has resulted in a description of high-latitude HI, especially that at $|v| > 20$ km/s, as quite smoothly distributed. More detailed data suggest that the concept of a cloud ensemble might merit more attention, certainly for the HI gas associated with the cirrus fields, but perhaps generally at high latitude. The Arecibo mapping which we have been doing indicates that the volume filling factor of high-latitude HI is substantially less than unity. The kinematic variety characteristic of this gas suggests that the area filling factor must be considered as a function of velocity, and that, so defined, this factor is also less than unity.

The concept of motions of separate entities might well be more realistic than discussions of high-dispersion wings from a diffuse gas component (see also Verschuur and Schmelz, 1989). We do note, however, that the anomalous velocities associated with the cirrus, like those of the IVC's in general (including the shell/supershell/bubble structures), are dominated by velocities in the sense of infall, and thus not indicative of turbulence in the normal sense of that word. It should also be noted that these motions are *vertical* ones which have no *horizontal* equivalent: the dispersion of cloud-to-cloud motions have been measured in the galactic equator for both the HI population (Burton and Gordon, 1978) and the molecular-cloud ensemble (Liszt *et al.*, 1984), and in both cases the bulk random motions are characterized by dispersions less than about 5 km/s.

We stress finally that the "fast" clouds in the wing domain can in fact be quite narrow in velocity as well as in angular size. It is clear that even sampling fully with the Dwingeloo 36-arcminute beam – and certainly with the 2° x 3° effective beam of the Bell Labs survey – will hide much spatial information, but that velocity resolution approaching about 0.5 km/s is also an important consideration.

Acknowledgements: The Radio Observatory in Dwingeloo is operated by the Netherlands Foundation for Research in Astronomy, under contract with the Netherlands Organization for Scientific Research (NWO); the Arecibo Observatory is operated by the National Astronomy and Ionosphere Center under contract with the U.S. National Science Foundation (NSF). The Leiden-Dwingeloo HI survey represents the first use of the 1024-channel autocorrelator spectrometer built under the supervision of A. Bos. T.M. Bania acknowledges support from the NSF under grant 85-11844, Dap Hartmann from the NWO, and Tang Yuan from the Academia Sinica.

REFERENCES

Bania, T.M., and Lockman, F.J., 1984,*Astrophys. J. Suppl.* **54**, 5 13.

Burton, W.B., and Deul, E.R., 1987, in "The Galaxy", eds. G. Gilmore and R. Carswell, Reidel Pub. Co., Dordrecht, p. 141.

Burton, W.B., and Gordon, M.A., 1978, *Astron. Astrophys.* **63**, 7.

Danly, L., 1989, *Astrophys. J.* **342**, 785.

Deul, E.R., and Burton, W.B., 1990, *Astron. Astrophys.* **230**, 153.

De Vries, C.P., and Le Poole, R.S., 1985, *Astron. Astrophys.* **145** , L7.

Dickey, J.M., and Lockman, F.J., 1990, *Ann. Rev. Astron. Astrophys.* **28**, 215.

Heiles, C., and Habing, H.J., 1974, *Astron. Astrophys. Suppl.* **14** , 1.

Heiles, C., Reach, W.T., and Koo, B.-C., 1988, *Astrophys. J.* **332** , 313.

Kulkarni, S., and Fich, M., 1985, *Astrophys. J.* **289**, 792.

Liszt, H.S., Burton, W.B., and Xiang, D.-L., 1984, *Astron. Astrophys.* **140**, 303.

Lockman, F.J., Jahoda, K., and McCammon, D., 1986, *Astrophys. J.* **302**, 432.

Lockman, F.J., and Gehman, C.S., 1991, *Astrophys. J.*, in press.

Low, F.J., Beitema, D.A., Gautier, T.N., Gillet, F.C., Beichman, C.A., Neugebauer, G., Young, E., Aumann, H.H., Boggess, N., Emerson, J.P., Habing, H.J., Hauser, M.G., Houck, J.R., Rowan-Robinson, M., Soifer, B.T., Walker, R.G., and Wesselius, P.R., 1984, *Astrophys. J. Letters* **278**, L19.

Mebold, U., Herbstmeier, U., Kalberla, P.W.M., Souvatzis, I., 1989, in "Structure and Dynamics of the Interstellar Medium", Proc. IAU Coll. 120, eds. G. Tenorio-Tagle, M. Moles, and J. Melnick, Springer Lecture Notes in Physics, p. 424.

Meyerdierks, H., Heithausen, A., and Reif, K., 1991, *Astron. Astrophys.* **245**, 247.

Radhakrishnan, V., and Srinivasan, G., 1980, *J. Astron. Astrophys.* **1**, 47.

Stark, A.A., Gammie, C.F., Wilson, R.W., Bally, J., Linke, R.A., Heiles, C., and Hurwitz, M., 1991, *Astrophys. J. Suppl.*, in press.

Verschuur, G.L., and Schmelz, J.T., 1989, *Astron. J.* **98**, 267.

Wesselius, P.R., 1972, Ph.D. thesis, University of Groningen.

Wesselius, P.R., and Fejes, I., 1973, *Astron. Astrophys.* **24**, 15.

Neutral Hydrogen Observations of Extended Galactic Structures

N.V. BYSTROVA

Special Astrophysical Observatory, USSR Academy of Sciences, St Petersburg, Branch of the SAO

INTRODUCTION

For study of extended structures of different sorts in galactic interstellar neutral hydrogen, the most convenient presentation of HI observational data during the last decade has been the position-position contour maps of temperature generated at fixed velocities.

These maps can be directly compared with maps in other spectral regions, e.g. infrared, H_α, etc. Such comparisons give information about the velocity behaviour of common details and their environs.

2 TELESCOPES; METHODS OF OBSERVATIONS AND REDUCTIONS

Two transit radiotelescopes, the Large Pulkovo radiotelescope and later the RATAN 600 in Zelenchuskaja, were used for a systematic study of the galactic neutral hydrogen radiation in the 21-cm radio line. The parameters of the telescope in Pulkovo are: beamwidth at this wavelength, 6 arcminutes $\times$ 5 degrees; 10 spectral channels with of bandwidth 20 kHz. For the RATAN 600 the parameters are: 2.5 arcmin $\times$ 2.5 deg, 40 channels, 30 kHz. When additional panels were fixed on every element of the RATAN 600 in 1991, the vertical size of the beam became 1.6 deg.

The immobility of our telescopes could be overcome by changing the frequency adjustment of the receivers during the observations. This procedure was done often enough for keeping the frequency in every spectral channel constant relative to the local standard of rest.

The most rational method of spectral observations with our transit telescopes involved "drift curves" which lasted many hours at the same setting of the telescope at fixed declination. These drift curves formed the basis of maps of HI emission at fixed velocities, after correction for seasonal spill-over effects. The full extent of the beams of our two telescopes from zenith to horizon is still unknown, and the direct calculation of the stray radiation is rather impossible. But the sidelobes are of course present; their extent was partially revealed when the Sun was rising or setting behind the elements of the telescopes during observations of other objects.

An indirect method was used for excluding at least partially the seasonal spill-over effects. The HI emission consists of two components: the structural one is free from stray radiation in contrast to the "structureless" component (Bystrova,1980). Such a division into two components is quite easy when using the 24-hour drift curves and a wide set of declination, but involves problems when the observations last only a few hours. The spline interpolation is the first stage; the corrections of the positions of some nodes are made taking into account the Pulkovo sky survey where the scans lasted 24 hours.

3 HII FILAMENTS AND HI DISTRIBUTIONS

Several HII zones were observed with our telescopes. These zones were previously suggested to be associated with supernova remnants: the regions of th e Sco OB-2 association, of λ Orionis, and filaments in Cetus-Eridanus. In all of them, the HI behaviour was interpreted earlier in terms of shells expanding from a common center of expansion.

The marvelous observations made by J.P. Sivan revealed the faint filaments when the main HII zone was overexposed. And then it became obvious that these filaments are unambiguously connected in the HI distribution. These filaments play the dominant role in the HI distribution and rule its behaviour.

3.1 The region of Scorpius OB-2 association

Our results obtained with the Large Pulkovo radiotelescope were published in 1979. They showed the certain connectio n of HI distribution with outer filaments belonging to the S7 and S1 nebulae. No other observations of these filaments in the 21-cm line yet exists possibly because such details can be missed when the observations were made using the prearranged net of positions (Cappa de Nicolau et al., 1986). Our RATAN 600 observations confirmed the earlier results and gave some new details. The most prominent structures are those situated directly against the S7 nebula itself. These details could be registered only after diminishing the vertical size of the beam of RATAN to 1.6 deg. Preliminary positions and antenna temperature are given in the table.

Dec_{1950}	RA_{1950}	T_A
-22°	16^h 00.0^m	
-23	15 57.0	approxima-
-24	15 54.5	tely 3-5 K
-25	15 52.6	

note: the source at -22° is double

According to the photographs taken with shorter exposures by J.P. Sivan, this narrow HI ridge is connected with filaments in the nebula itself. The halfwidths along this unique ridge are less than one minute in RA. Its velocity is +2.4 km/s.

In this region the fact that the positions of the remarkable detail found at -12 km/s in Dwingeloo (Sancisi and van Woerden, 1970) and at approximately -6.1 km/s exclude each other remained unexplained. This was confirmed using Gauss analysis of the profiles observed in the Sco OB-2 region (Cappa de Nicolau, 1986).

3.2 The surroundings of λ Orionis

After the detection of HII filaments to the west from the overexposed HII region itself (Sivan,1974) our HI observations showed the gas in the next spectral channels to be connected with different HI loops and filaments (Bystrova,1980). Therefore, after the comparison of HI distribution with far-infrared dust emission the conclusion could be made that the HII filaments are possibly more distinctly connected with the neutral hydrogen distribution than with the dust details. The brightest and narrowest HI signal lies between the western border of the main HII zone and the bright extended dust detail more to the west. This HI signal occurs with $T_A \sim 12$ K, halfwidth in RA $\sim 1.2°$, at velocity -0.8 km/s.

3.3 Filaments in Cetus-Eridanus

Inside the suggested expanding huge HI shell (Heiles, 1976) exists neutral hydrogen at the same spectral channel as characterizes gas situated on the border of the shell. This border is shown by the long HI ridge in Bystrova and Rachimov 1976. Both ridges outside and inside the adopted shell are accompanied by long H_α filaments. These facts have not yet been explained.

4. OTHER EXTENDED STRUCTURES

Several structures of a different nature were also studied in the 21-cm line. There are Orion's great dust cloud, the surroundings of the nebula around ζ Ophiuchi, and the ring-shaped detail around the south galactic pole. Some preliminary conclusions are given below.

4.1 The Orion great dust cloud

The clear image in the far infrared of dust clouds, in particular of the dense ones, makes such pictures convenient material for studying together with the neutral hydrogen distribution at a range of velocities. Using the observations at Pulkovo we see that the HI contours cover the whole GDC at velocities from -6.1 to +9.7 km/s. This interval is broader for the western, brighter part of GDC; against this part of the cloud the HI contours project with velocities from +4.5 to +15 km/s.

Very remarkable is the behaviour of the HI at the borders of the GDC. At the velocity +25.6 km/s, between the declinations -2° and +12° , the main body of HI around the galactic plane begins exactly at the eastern border of the GDC and of the details connected with λ Orionis. To the south, from decl. -2° , this border joins the HI at +15 km/s. At the south-western direction to the GDC, HI cloud continues at -0.8 km/s until decl. -10° .

Exactly at the western border of the GDC the known bright cloud is situated which stretches itself for 12-13 degrees until declinat ion +3° , with T_A near 6 K and central velocity -11.3 km/s. Some concentrations of HI project themselves on this cloud at +9.7 and +15 km/s. Further to the north at this border of the GDC, gas is found at -6.1 km/s. The whole picture of the HI motions in the area studied will possibly be revealed using the RATAN observations after their full reductions.

4.2 The surroundings of the nebula around ζ Ophiuchi

A special set of observations on the RATAN 600 was made to study the HI distribution around S27 (ζ Ophiuchi). Almost 20 cross-sections with a one-degree interval in declination were made. The large size of this nebula needed the duration of the records for a period of not less than 3 hours. In the equatorial coordinate system the HI distribution is not symmetric relative to the center of the nebula. Many details are seen as projected on the HII zone.

4.3 The maps of HI distribution around the south galactic pole

Using the data obtained at Pulkovo, the polar map was generated in the galactic coordinate system up to latitudes $\pm$ 10° . The maps were made at 10 velocities from -21.8 to +25.6 km/s, in intervals of 4.2 km/s separately for the two components of the HI emission. One of the prominent details in the southern galactic hemisphere is "the ring at negative velocities" (Cleary et al.,1979). In Pulkovo approximately half of this ring is seen. The later is easily found on the maps at pages 7, 11, 15 and part of page 19 for the structural component of HI emission in Bystrova (1980). The values of radial velocities of the ring are from -16.6 to -6.1 and even to -0.8 km/s LSR. At -0.8 only part of the ring in the first quadrant remains; at -16.6 the parts in the second and third ones remain. The angular dimensions of this elliptical are approximately 100×50 degrees. The direction of the major axis is from the longitude 100 to 280° . These coordinates make the connection of the ring with Gould's Belt doubtful.

The -110 km/s details studied by Cohen in 1981 project themselves on the brightest part of the ring, its northern part. Therefore a special search with the RATAN was made to find some details having velocities around -110 km/s in the other parts of the ring extending to the southern limit of the ring's visibility. The data are being reduced now.

REFERENCES

Bystrova, N.V., Rakhimov, I.A., 1976, Pis'ma Astron. Zh.2, No9, 440.
Bystrova, N.V., 1980, Contour maps to Pulkovo Sky surv., "Nauka", Len-d.
Bystrova, N.V., 1979, Astrophys. Issled. (Izv.SAO), 11, 236 and 244.
Cappa de Nicolau, C.E., Pöppel, W.G.L., 1986, Astron. Astrophys. 164, 279
Cleary, M.N., Heiles, C., Haslam, C., 1979, Astron. Astrophys. Suppl., 36, 95.
Cohen, R.J., 1981, Mon. Not. RAS, 196, 835.
Heiles, C., 1976, Astrophys. J., 208, L137.
Sancisi, R., van Woerden, H., 1970, Astron. Astrophys., 5, 135.
Sivan, J.P., 1974, Astron. Astrophys. Suppl., 16, 163

Supergiant Shells in the Irregular Galaxy NGC 2366

DOMINIK J. BOMANS [1], ULRICH HOPP [2]

[1] Sternwarte der Universität Bonn, [2] MPI für Astronomie, Heidelberg

Huge loops and filaments of ionized gas with sizes of several hundred parsecs have recently been discovered in some Magellanic irregular galaxies (NGC 4449, 3738, 4214, 2366, Sex A) (Hunter & Gallagher 1990). Structures of this kind are known to exist in irregular galaxies since the deep Hα photographs of the Magellanic Clouds (Davies et al. 1976). Their production mechanism is under debate (Tenorio-Tagle & Bodenheimer 1988) as well as the ionization mechanism of the filaments (Hunter & Gallagher 1990).

Using the prime focus of the 3.5m telescope at the Calar Alto observatory, we observed the nearby (D = 3.5 Mpc) irregular galaxy NGC 2366 through a narrow band Hα filter and in the R and I band.

There are several diffuse H II structures visible in our continuum subtracted Hα frame of NGC 2366 (Fig. 1). Their sizes range from 200 to 900 pc. The most striking feature is a complex system of three interlocking supergiant shells in the north of the largest H II region. Each shell has an apparent diameter in the range of 400 pc. These features appear to be connected to the brighter structure in the northeast noted by Hunter & Gallagher (1990). If this connection is physical the supershell system has an extension of more than 1000 pc and is the second largest H II supergiant shell known so far. The interlocking of the three shells in the north is most probable a projection effect. We interpret these shells tentatively as the traces of consecutive energetic events (e.g. in time correlated SN explosions) inside the region hidden by the giant H II region. Roy et al. (1991) found an expanding bubble at the center of this giant H II region and signs of an outflow of gas to the north-northwest. At this position we find an incomplete low surface brightness loop, which is apparently the largest single supershell in NGC 2366.

Using resonable guesses about the density of the surrounding gas and the energy input (Roy et al. 1991) it is possible to estimate the age of the supergiant shell (Tenorio-Tagle & Bodenheimer 1988). It ranges from 1 to 3×10^7 yr. This is a much higher age than that of the bubble inside the giant HII region (Roy et al. 1991) and implies that the shells are an effect of an older starburst event, or that these objects reached their large diameters due to expansion into a low density halo of NGC 2366, a process which accelerates the expansion (Tenorio-Tagle & Bodenheimer 1988).

At three positions H II filaments seem to extend far from the main body of the galaxy. These may be chimneys according to the theory of Norman & Ikeuchi (1990) and therefore effects of the recent active starformation phase. Alternatively these may be parts of an gigantic very low surface brightness loop. This would be the relict of an older extreme burst of star formation, which included large parts of the bar of NGC 2366.

Comparing the size distribution of the supergiant shells in NGC 2366 with those in the LMC, we find a clear difference. While in the LMC no shell is known with sizes between 300 and 500 pc, in NGC 2366 this diameter range is well populated. The lack of shells with diameter larger than 700 pc in NGC 2366 may be an effect of incompleteness in our data, but can also be taken as additional hint for a physically different population of supergiant shells in the two galaxies. One prelimiary explanation for this finding is that the most recent burst of star formation started longer ago in the LMC than in NGC 2366. Another possible explanation is that NGC 2366 has a denser or less extended gaseous halo than the LMC.

References

Davies, R.D., Elliot, K.H., Meaburn, J. 1976, MmRAS, 81, 89
Hunter, D.A., Gallagher, J.S. 1990, ApJ, 362, 480
Norman, C.A., Ikeuchi, S. 1989, ApJ, 345, 372
Roy, J.-R., Boulesteix, J., Joncas, G., Grundseth, B. 1991, ApJ, 267, 141
Tenorio-Tagle, G., Bodenheimer, P. 1988, ARAA, 26, 145

Fig. 1. Continuum subtracted Hα frame of NGC 2366

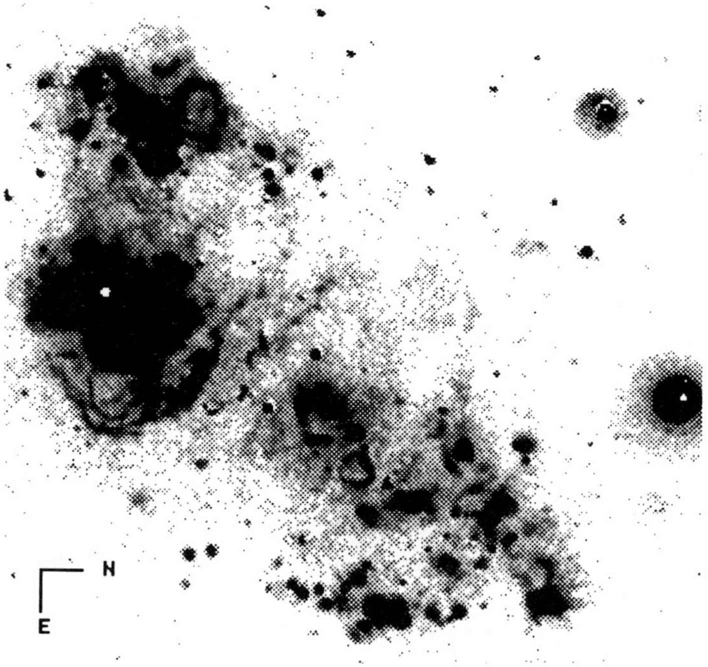

3D Simulations of Expanding Bubbles

JAN PALOUŠ

Center for Theoretical Study, Charles University, and Astronomical Institute, Czechoslovak Academy of Sciences

1 INTRODUCTION

Heiles(1979) discovered in HI of our Galaxy large expanding structures or superbubbles. The sizes range up to 1 kpc and their expansion velocities are $\sim 20\ km\ s^{-1}$ (see the table 2 in Heiles).

We decided to reobserve the region $75^o < l < 55^o$, $-5^o < b < 5^o$ containing at LSR radial velocities V_{LSR} : $-97\ km\ s^{-1} < V_{LSR} - 75\ km\ s^{-1}$ according to Heiles (1979) the expanding supershell GS064-01-97. This is a part of the new 21 cm survey of the Galaxy, being carried out with the radiotelescope in Dwingeloo by Hartmann and Burton.

The intensity distribution in this region in narrow, $2\ km\ s^{-1}$ thick, radial velocity intervals proved the existence of a feature centered at $l = 64^o$, $b = -1^o$, with the largest diameter of 6° at $V_{LSR} = -95\ km\ s^{-1}$. This is in nice correspondence with the indentification of Heiles. We reproduce one of our intensity maps in Fig. 1. The smooth shell structure, which is visible on the low galactic longitude side, is absent on the other side. Here, we detect another HI clump centered at $l = 71^o$, $b = -5^o$.

In this contribution we try to clarify if the 3D simulations of expanding bubbles using the infinitely thin shell approximation can give similar HI density distribution in relevant radial velocity intervals.

2 THIN SHELL IN 3D

Stars more massive than $7\ M_\odot$ of an OB association stronly influence the ambient interstellar medium: the ionizing photons, stellar winds and clumped supernovae insert several times 10^{53} erg over 10^7 years. This energy is partly converted into the thermal energy, and partly into the kinetic energy of the expanding blastwave.

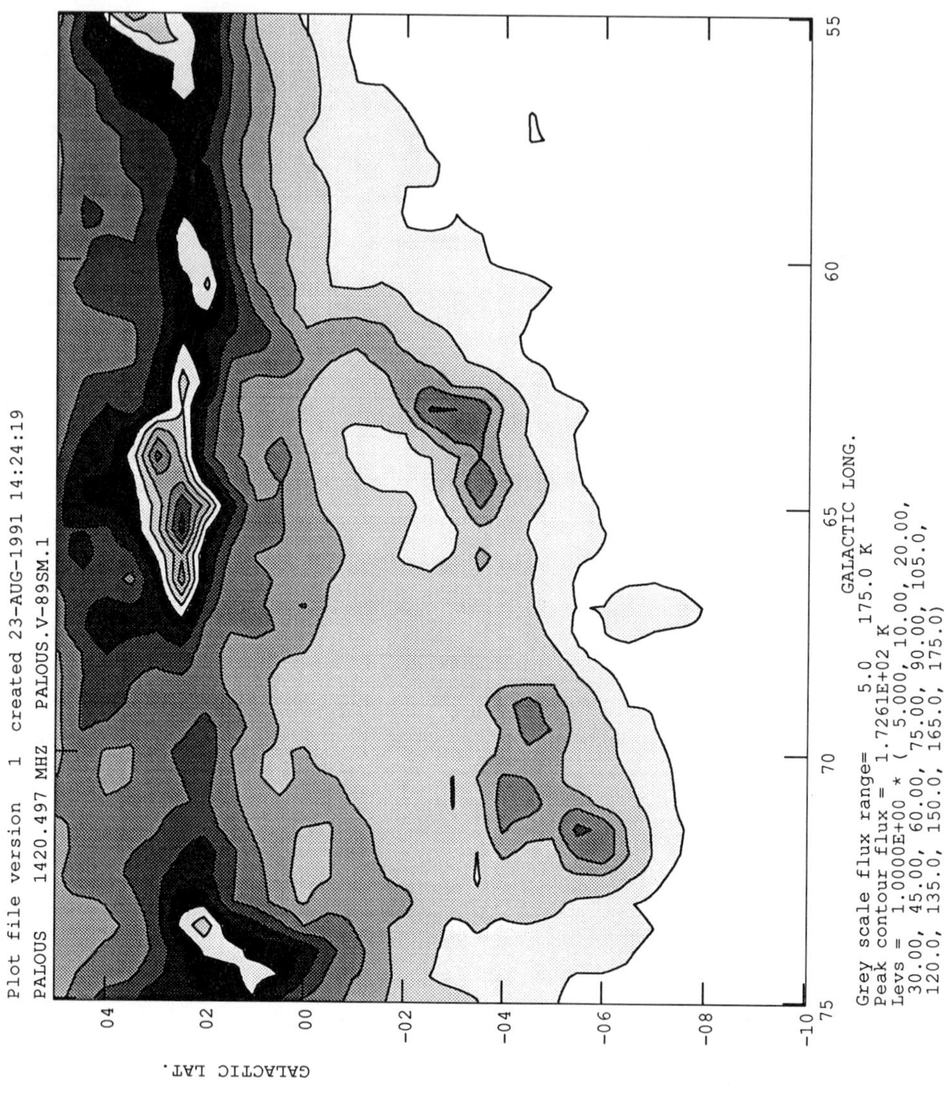

Fig. 1. Intensity of the HI radiation in the LSR radial velocity interval $-89.0\ km\ s^{-1}$ to $-87.0\ km\ s^{-1}$ from the new 21 cm survey of the Galaxy by Hartmann and Burton.

Several regimes can be distinquished (Ostriker and McKee, 1988): 1. the free expansion, when the mass of the stellar remnant is larger than the swept up mass; 2. the adiabatic expansion, when mass of the remnant is smaller than the swept up mass, but the radiative looses are insignificant; 3. the momentum conservation phase, when the ambient medium is accumulated into the thin, cold and expanding shell.

In our simulations only the third phase is considered. The simulation starts when the cavity has certain finite size, which is assumed as completely spherical. The thin shell expands supersonically into the ambient medium, which is accumulated to it. The equation of motion of an element of the bubble reads

$$\frac{d}{dt}(Mv_S) = dS[(P - P_o) + \rho_o v_o(v_S - v_o)] - M(t)g(R, z) \qquad (1)$$

where P, P_o are the external and internal pressures, $dS, M(t), v_S$ are the surface density, mass, and expansion velocity of the element, v_o, ρ_o are the velocity and density of the ambient medium and $g(R, z)$ is the gravitational force of the Galaxy. The mass of an element grows with time

$$\frac{dM}{dt} = v_\perp \rho_o dS \quad if \quad v_\perp > v_{sound} \qquad (2)$$

where $v_\perp$ is the component of v_S perpendicular to the surface and v_{sound} is the sound velocity in the ambient medium.

The simulations are quite sensitive to the vertical stratification of the disk. We consider that it is composed of three components (Dickey and Lockman, 1990):

$$\rho = \rho_C + \rho_W + \rho_E. \qquad (3)$$

ρ_C and ρ_W are the cold and warm components with a gaussian vertical distribution, ρ_E is the thick exponential component. In the solar vicinity, the following values are addopted for maximum densities and the disk thicknesses: $\rho_{C,o} = 0.395\ cm^{-3}$, $\sigma_C = 127\ pc$, $\rho_{W,o} = 0.107\ cm^{-3}$, $\sigma_W = 318\ pc$, $\rho_{E,o} = 0.064\ cm^{-3}$, $\sigma_E = 403\ pc$. We also consider that the HI galactic plane can be displaced from $b = 0^o$. The amplitude of the Galactic warp, which depends on the galactic longitude and on the galactocentric distance, is taken from Burton (1988).

The blowout or the expansion of bubbles to several kpc from the galactic plane is possible only if the thick exponential HI disk component is absent. The exponential component is able to decelerate the expansion within 1kpc from the plane for all plausible expansion energies (Palouš, 1990).

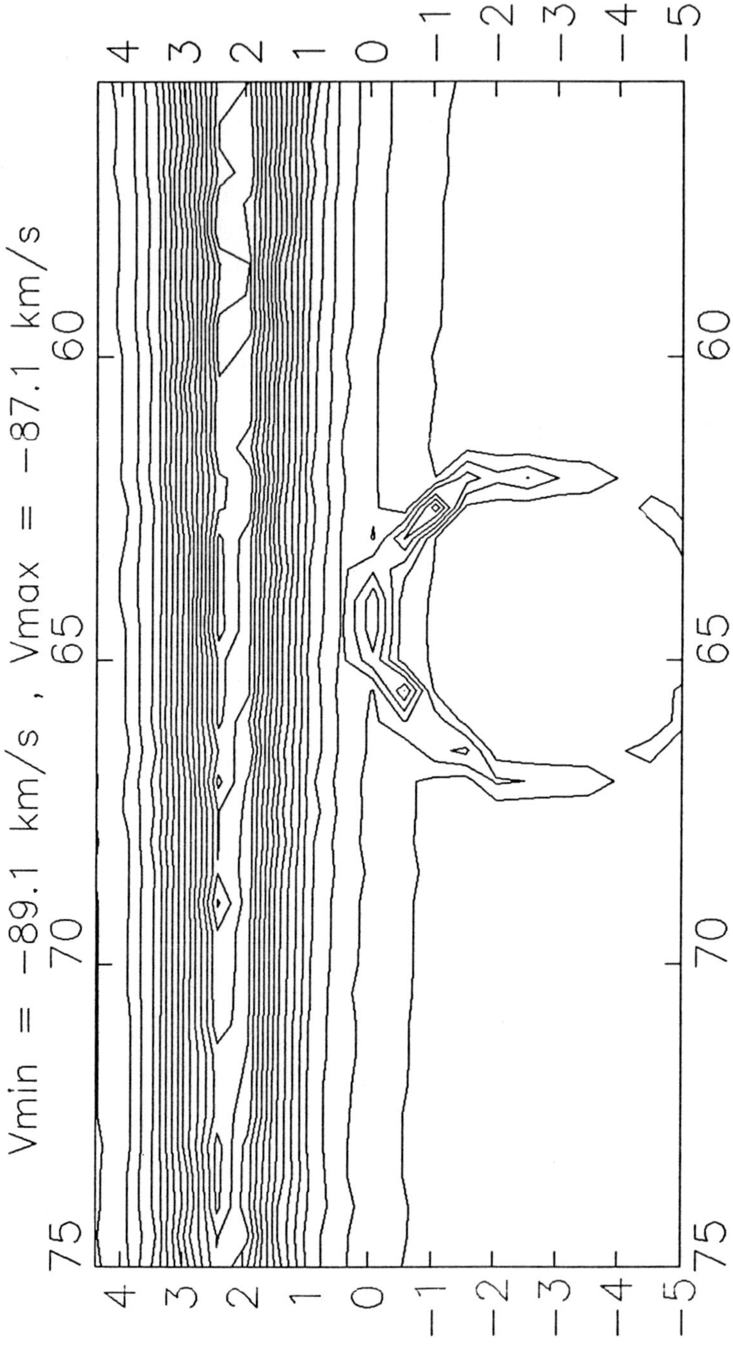

Fig. 2. The HI column density distribution according to the best model in the similar LSR radial velocity interval represented in Fig. 1. The horizontal axis is the galactic longitude, the vertical axis is the galactic latitude.

The two components of the galactic gravitational force are introduced: the radial, according to the rotation curve (Wouterloot et al., 1990)

$$\Theta(R) = \Theta_o(R/R_o)^{0.0382}\ km/s, \tag{4}$$

$\Theta_o = 220\ km/s$, $R_o = 8.5\ kpc$, and the z component of the galactic force, K_z, which is computed from the formula given by Kuijken and Gilmore(1989)

$$-K_z = 2\pi G\sigma_D \frac{z}{\sqrt{z^2 + z_D^2}} + 4\pi G\rho_H z. \tag{5}$$

σ_D and z_D are the stellar disk surface density and scale height and ρ_H is the effective density of the galactic halo. For $R_o = 8.5\ kpc$ we adopt $\sigma_D = 46\ M_\odot pc^{-2}, z_D = 300\ pc$.

3 THE EXPANDING FEATURE GS064-01-97

The maximum radius of the feature is according to our observations at LSR radial velocity $-95\ km\ s^{-1}$. Using the rotation curve (4), this corresponds to the galactocentric distance 17 *kpc*. At this distance the vertical HI scale height is about 2.6 times larger than near the Sun(Wouterloot et al., 1990). Thus, the scale heights of the three HI disk components are: $\sigma_C = 327\ pc, \sigma_W = 818\ pc, \sigma_E = 1036\ pc$.

The total HI surface density at $R = 17\ kpc$ is according to Wouterloot et al.(1990) $3.31\ M_\odot pc^{-2}$. The division of HI between the gaussian and exponential components is left as the free parameter in our simulations. Another free parameter is the effective halo density ρ_H, which is used in formula (5) describing the galactic K_z force.

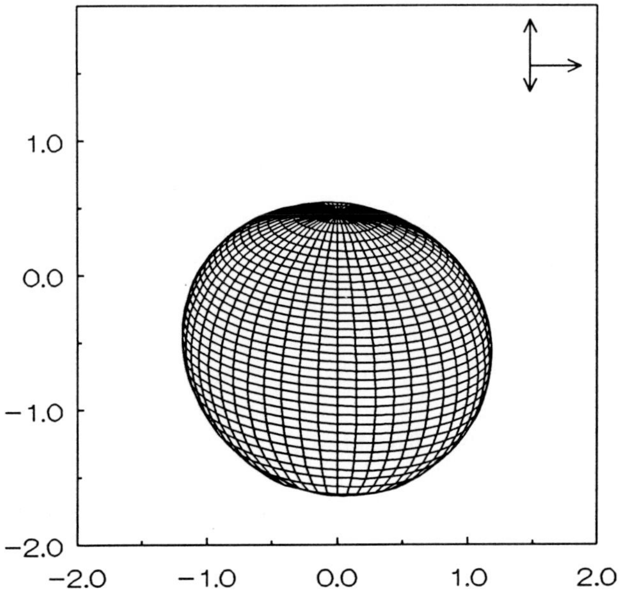

Fig.3. The best model: 3D shape of the superbubble.

The best model is represented in Figs 2 and 3. Fig. 2 gives the HI density distribution in the same LSR radial velocity interval as Fig. 1, Fig. 3 shows the spatial 3D shape of this expanding bubble.

In the best model, HI is divided so that the maximum value of HI density in the thick exponential component is 0.05 cm^{-3}. This is about 5 times more than that would be proportional to the HI distribution in the solar vicinity. On the other hand, the effective halo mass ρ_H in the formula (5) is quite low: it is in this case 0.015 $M_\odot\ pc^{-3}$.

4 CONCLUSIONS

We did not covered all possibilities of the division between the three HI components. We also did not try all possible values of ρ_H. But, this is the usual problem with the numerical simulations. However, we can discriminate some implausible variants:

1. At $R = 17kpc$ the fraction of HI in the thick exponential component can not be as low as in the solar vicinity. The total preexisting mass within the observed cavity would be too low to explain the HI mass seen in the supershell. We conclude from our simulations that at $R = 17\ kpc$ the fraction HI in the thick disk must be 3 to 5 times higher than near the Sun.

2. The effective mass of the halo should not be very high. High ρ_H would increase the K_z force, which would be able to decelerate the shell without accumulation of HI. Our second conclusion is that the effective halo density at $R = 17\ kpc$ can not be higher than $0.03 M_\odot pc^{-3}$.

REFERENCES

Burton, W.B.: 1988, in Galactic and Extragalactic Radio Astronomy, eds. G. Verschuur and K.I.Kellermann, New York:Springer- Verlag, p.295

Dickey, J., Lockman,F.:1990, Ann.Rev.Astron.Astrophys.28,215

Heiles,C.:1979, Astrophys.J. 229, 533

Kuijken,K.,and Gilmore,G.:1989, Mon.Not.Roy.Astron.Soc.239,571

Ostriker,J.P.,McKee,Ch.F.:1988, Rev.Mod.Phys. 60, 1

Palouš,J.: 1990, in The Interstellar Disk-Halo Connection in Galaxies, ed.H.Bloemen,

Sterrewacht Leiden, p.101

Wouterloot, J.G.A., Brand,J., Burton,W.B.,and Kwee,K.K.:1990, Astron.Astrophys. 230, 21

ACKNOWLEDGEMENTS

I like to thank Butler Burton and Dap Hartmann of Sterrewacht Leiden for providing previously unpublished data and for preparing Fig. 1.

3D SUPERSHELLS

SERGEY A. SILICH

Main Astronomical Observatory, Ukrainian Academy of Sciences 252127, Kiev - 127, Goloseevo, Ukraina

1 INTRODUCTION

Correlated supernovae explosions in OB-associations blow out interstellar gas from large space regions and create huge expanding supershells (see reviews Tenorio-Tagle and Bodenheimer, 1988; Silich, 1990). The evolution of supershells mainly depends on two factors: the gas density distribution and the effects of differential galactic rotation. Dynamics of supershells in a plane-stratified gas distribution have been investigated in 2D by Tomisaka and Ikeuchi (1986), Mac Low and McCray (1988), Bisnovatyi-Kogan et al. (1989), Igumentshchev et al. (1990), and others. Expansion of 2D remnants in the differentially rotating galactic disk has been considered by Tenorio-Tagle and Palous (1987), and Palous et al. (1990). But 3D calculations are necessary for adequate simulations of the evolution of supershells. First results of 3D calculations have been presented by Palous (1990 and 1991). The evolution of 3D adiabatic supershells in a magnetized ISM has been studied by Tomisaka (1990). Bisnovatyi-Kogan and Silich (1990) have proposed a 3D scheme based on a thin layer approximation for the interpretation of the recent observations of the adiabatic SNRs. Here we extend this scheme to take into account gravity, continuous energy input into cavity from the explosions and energy losses from the shock front. We concentrate on the evolution of superbubbles in the galactic disk with the same parameters as in the solar vicinity.

2 THE MODEL

For the calculations we use Cartesian galactocentric coordinates (X, Y, Z). The vertical structure of the disk has been described by the composite Lockman et al. (1986) and Reynolds (1989) atmosphere

$$n(Z) = n_0 \left[\frac{0.7}{\exp(Z/H_1)^2} + \frac{0.24}{\exp(Z/H_2)} + \frac{0.06}{\exp(Z/H_3)} \right], \tag{1}$$

where $n_0 = 0.45\ cm^{-3}$, $H_1 = 190\ pc$, $H_2 = 500\ pc$, $H_3 = 1500\ pc$. Bearing in mind that the temperature distribution is pooly known, we have used the extreme case $T(Z) = n_0^2 T_0/n^2(Z)$, where $T_0 = 10^4 K$. The rotation velocity was set equal to $V_R = 250\ km\ s^{-1}$ in all calculations. We assume that the undisturbed gas rotates around the galactic center under the influence of the gravitational accelaration given by $g_x = -xV_R^2/R^2$ and $g_y = -yV_R^2/R^2$. To examine the role of the Z-component of the gravitational field we have considered two cases with the same initial parameters but with different g_z. The value of g_z for every space point has been calculated by linear interpolation of the Allen (1976) data. We have considered the three cases presented in Table 1.

Table 1

N	$R(kpc)$	L_{38}	g_z	$t_m(Myr)$	$Z_m(pc)$	$M_s(M_\odot)$	M_d/M_s
1	10	2	g_1	35.9	1700	7.7×10^7	0.45
2	10	0.5	g_1	35.2	969	2.7×10^7	0.45
3	10	0.5	g_2	98.0	2380	2.0×10^7	—

Here R is the galactocentric radius, L_{38} is the energy input rate, g_z is the Z-component of the gravitational field, t_m is the time at the end of calculations. Parameters of the shells are shown in the three last columns of the Table. Z_m is the distance of the remnant's pole from the galactic plane, M_s is the mass of the shell, and M_d/M_s is the ratio of the mass that concentrated in the galactic plane to that of fully swept up.

We have used Weaver et al.(1977) analytic results for a spherical bubble of radius R= 50 pc at 10 kpc from the galactic center as initial conditions for our calculations. At the onset of calculations the initial hemisphere has been split into N = 40x40 = 1600 Lagrangian elements by the planes parallel and perpendicular to the galactic plane. We defined the Lagrangian coordinates at the beginning of calculations as follows:

$$x = R_0 + R_e sin\theta cos\varphi,\ y = R_e sin\theta sin\varphi,\ z = R_e cos\theta \tag{2}$$

Every Lagrangian element has been labelled by the two numbers, I and J. I corresponds to the Lagrangian angle θ, J corresponds to the Lagrangian angle φ. We assume that mass accumulation stops for the elements with expansion velocities less than the sound velocity and that their motion continues in a collisionless way. It follows from Table 1 that the mass of the shell in case 3 less than the mass of the shell in case 2, despite the fact that the size of shell 3 is larger. The reason for this is that in case 3 the expansion velocity near the equatorial plane of the shell becomes less than the sound velocity and mass accumulation stops earlier than in case 2.

3 RESULTS

We have considered expansion only for the upper parts of the remnants (with respect to the galactic plane). The shape of the supershell for case 2 after 35.2 Myr is shown in

Figure 1 for different positions of the observer. Because of the density stratification, the superbubble is pinched at its equator but spread out to large radii above the galactic plane. After the shell takes such form, both the gravitational field and the pressure of the hot gas inside the cavity act on the lower parts of the shell in the same direction. The Z-component of the expansion velocity slows down and becomes negative. The shell begins to fall down to the galactic plane. In the equatorial plane of the remnant an expanding cusp is formed. The onset of the cusp formation strongly depends on the value of the gravitational field.

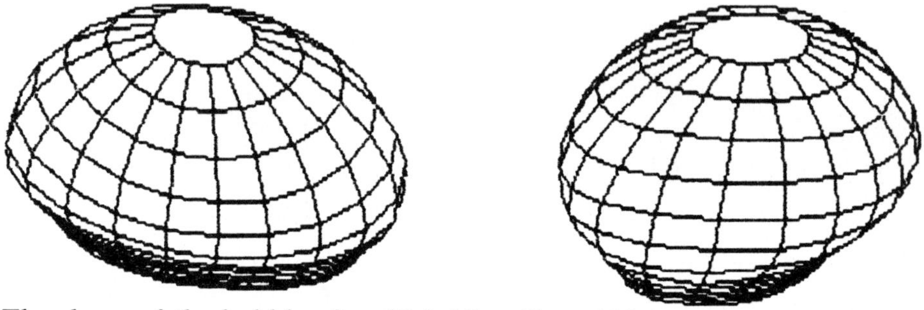

Fig.1 The shape of the bubble after 35.2 Myr, $Z_m = 969pc$.

In cases 1 and 2, approximately 45 % of the mass of the shell falls on the galactic plane after 35 Myr. In case 3, this process begins only after 98 Myr. The variation of the Z-component of the expansion velocity with the distance from the galactic plane is illustrated in Figure 2b. In cases 1 and 2, after 35 Myr almost all of the shell moves towards the galactic plane. In case 3, even after 72.8 Myr only the lower parts of the shell have negative velocities. The time evolution of Z-components of top velocities are shown in Figure 2a. An acceleration phase of expansion exists for all cases. It lasts approximately 10 Myr.

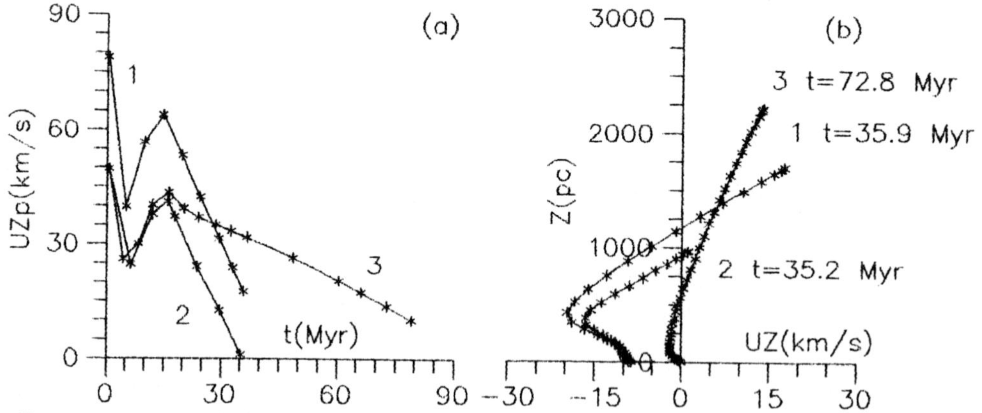

Fig.2 Z–component of the expansion velocity: a. as a function of time; b. as a fuction of the distance from the galactic plane.

The distributions of the column density along the equators of supershells and between theirs top and bottom in a time before cusp formation are shown in Fig.3a,b. Two peaks in the column density distributions are formed at the tips of the remnants in agreement with the 2D calculations of Tenorio-Tagle and Palous (1987), and Palous et al. (1990). But for cases 1 and 2, the inhomogeneity in the column density distribution remains small up to the onset of accumulation of the shell in the galactic plane. At the same time column densities at the remnant bottom exceed column densities at the remnant top by a several orders of magnitudes.

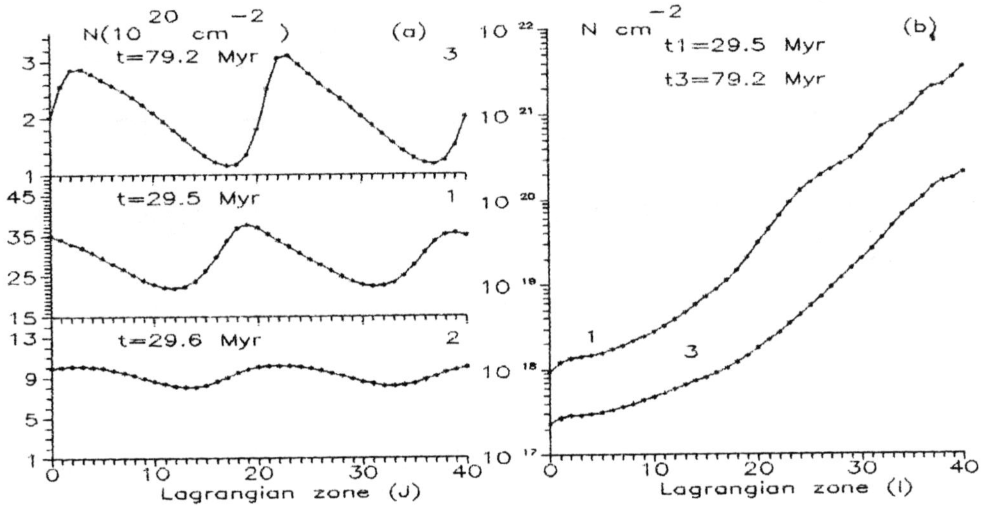

Fig.3 The column density distribution: a. along the equator; b. between the top and bottom of the remnant.

4 CONCLUSIONS

1. The 2.5-dimensional numerical hydrodynamic scheme based on a thin layer approximation is proposed to describe the propagation of blast waves in plane-stratified, differentially-rotating galactic disks.

2. The results of calculations of the evolution of supershells in the vicinity of the solar circle were presented. It was shown that in the late stages of evolution supershells begin to fall down to the galactic plane under the influence of the gravitational field. Expanding cusps occur in the equatorial planes of supershells after approximately 30 Myr. This process restricts the lifetime of superbubbles and does not allow the remnants to become highly elongated inside the solar circle.

ACNOWLEDGEMENTS

I wish to thank P.P. Berczik for his 3D graphic programme and for his help with the preparation of the 3D pictures.

REFERENCES

Allen C.W., 1976, Astrophysical Quantities, London

Bisnovatyi-Kogan G.S., Blinnikov S.I., Silich S.A., 1989, Astrophys. Space Sci. 154, 229

Bisnovatyi-Kogan G.S., Silich S.A., 1991, Astron. Zh. (in press)

Igumentshchev I.V., Shustov B.M., Tutukov A.V., 1990, Astron. Astrophys. 234, 396

Lockman F.J., Hobbs L.M., Shull J.M., 1986, Ap.J. 301, 380

Mac Low M.-M.,McCray R.,1988, Ap.J. 324, 776

Palous J., 1990, IAU Symp. 144, poster book, H. Bloemen (ed.), Leiden Observatory, p. 101

Palous J., 1991, CTS Workshop No 1 (this volume)

Palous J., Franco J., Tenorio-Tagle G., 1990, Astron. Astrophys. 227, 175

Reynolds R.J., 1989, Ap.J. 339, L 29

Silich S.A., 1990, in "Structure and evolution the regions of star formation", Kiev, p. 161-202

Tenorio-Tagle G., Palous J.,1987, Astron. Astrophys. 186, 287

Tenorio-Tagle G.,Bodenheimer P., 1988, Ann. Rev. Astron. Astrophys. 26, 145

Tomisaka K., Ikeuchi S., 1986, PASJ, 38, 697

Tomisaka K., 1991, Ap.J. (in press)

Weaver R., McCray R., Castor J., Shapiro P., Moore R., 1977, Ap.J. 218, 377

Numerical simulations of multiple supernova remnants in a galactic disk with a clumpy interstellar medium

R. KUNZE[1], H.W. YORKE[2], R. SPURZEM[1]

[1] Institut für Theoret. Physik und Sternwarte, Univ. Kiel

[2] Institut für Astron. und Astrophysik, Univ. Würzburg

ABSTRACT

Massive stars in galactic disks are clumped into associations rather than having a smooth distribution. This leads to a clumpy energy input and heating by supernova explosions. In an OB association exploding stars form a large bubble of tenuous hot gas surrounded by a shell of dense, cool matter on a time-scale of $\approx 10^7$ yr. This multiple supernova remnant (MSR) may eventually break out of the galactic disk (see Różyczka 1989 for a review).

We numerically simulate the evolution of a MSR in a clumpy medium whereby the interstellar matter in a galactic disk is modelled by two dynamically different components: cold clouds with mean mass $\langle m \rangle$ and radius $\langle a \rangle$ and a warm or hot intercloud medium. Interaction between clouds and intercloud medium (cloud formation and growth, evaporation, acceleration, crushing and collisions) are taken into account.

The results of our numerical simulations are discussed in terms of cloud formation in the MSR shell, the kinematics and properties of this clouds and differences between earlier models and our calculation.

1 INTRODUCTION

The interstellar medium in the galactic disc is inhomogeneous. Regions of enhanced density are embedded in a more tenuous medium which itself can vary in temperature and degree of ionization. This regions are usually denoted as clouds or clumps depending on their mass. Their masses, density contrasts to the surrounding medium, and spatial extents range over several orders of magnitude. The number distribution of clouds per unit mass follows a mass spectrum with a power law index between -1.5 and -2.0 (*e.g.* Dickey *et al.* 1989). Clouds are not stationary - they can be destroyed by interactions with the surrounding medium (*e.g.* shock fronts, ionizing radiation) or by internal star formation. Their internal parameters (density, mass, size, temperature) vary due to cloud collisions, evaporation in a hot ambient medium or mass agglomeration due to cooling ambient gas. Last not least clouds can form due to instabilities [gravitational, magnetic, thermal or combinations of the above (Elmegreen 1990)] in the tenuous medium.

Here we investigate the formation of clouds in the cooling shell of supernova remnants. Since massive stars being type II supernova progenitors spend their lifes in OB associations or open clusters their explosions form a common remnant enveloping the original cluster. The hydrodynamical evolution of such multiple supernova remnants was followed numerically by a number of authors (Różyczka 1989 and references therein, Tenorio-Tagle *et al.* 1990). In the publications mentioned above the interstellar medium was assumed to be homogeneous. The various components of the interstellar medium (molecular and neutral clouds, warm and hot intercloud gas) were only taken into account through their different density profiles and scale-heights perpendicular to the galactic plane (see *e.g.* Bloemen 1987).

Incorporating the interstellar clouds into the numerical calculations as one or more *dynamically* distinct components has the advantage not only to follow the evolution of the cloud system during the expansion of a MSR due to formation and destruction of clouds but also to make some quantitative statements *e.g.* about the number density of clouds at high galactic latitude.

2 THE MODEL

We divide the interstellar medium into two main components: clouds and an intercloud medium (ICM). There is no further division between atomic and molecular clouds in our picture since we deal mostly with small and medium size clouds of atomic hydrogen during the expansion of a MSR (see McKee and Ostriker (1977), Figure 2). The interaction of a supernova explosion with a molecular cloud is described in Yorke *et al.* (1989).

While the evolution of the ICM follows the Eulerian equations of hydrodynamics we use momentum equations derived from the collisionless Boltzmann equation to describe the evolution of the cloud system. Collisions and interactions are included subsequently in the moment equations (as sources or sinks of mass, momentum and energy). The "zero heat flux" approximation is used as closure condition. A justification of this method and a more detailed derivation of the momentum equations is given elsewhere [Yorke *et al.* this proceeding, see also Scalo and Struck-Marcell (1984)].

In our model both components (ICM and the cloud system) are initially at rest. The individual clouds, however, possess an internal velocity dispersion, which has the function of a kinetic temperature and keeps the cloud system in hydrostatic equilibrium . We expect clouds formed in the cooling shell of the MSR to move with its expansion velocity (some 10 km s^{-1}). Thus, there are two dynamically quite different types of clouds. Mixing them in one cloud component would result in too low expansion velocities of the newly formed clouds if their total mass is smaller than the mass of the preexisting clouds in the same volume. To avoid the above "mixing error" we separate the clouds condensing out of the MSR's shell in an own component. Therefore we have to deal with one ICM and two cloud components.

For each cloud component we get a set of variables (density ρ, stream velocities u, v and velocity dispersions σ_z^2, σ_r^2 and σ_ϕ^2) which describe the evolution of the whole *system* of clouds and variables as cloud mass $\langle m \rangle$, radius $\langle a \rangle$ (spherical cloud symmetry) and temperature T which are mean values of *individual* clouds (independent variables are the two cylindrical spatial coordinates z and r and the time t). All interactions between either the clouds and the ICM or the two cloud components show up as source terms in the equations and are parameterized as functions of both global variables and mean values of indivual clouds.

2.1 Interactions between the components

A more detailed description of the interactions between the components listed below is given in Spurzem *et al.* (1991). Therefore we here only give some brief information on the source terms; only the cloud formation is described in more detail. We have taken into account

- evaporation (and growth) of clouds due to electron thermal conduction including effects of saturation taken from Cowie *et al.* (1977).
- cloud acceleration by drag of the onstreaming ICM and by crushing due the shockfront of the MSR. Erosion or destruction of clouds due to the latter process is not yet included.
- cloud-cloud collisions which lead to changes in $\langle m \rangle$, $\langle a \rangle$ and lower the velocity dispersion of the cloud components. We assume negligible mass transfer from clouds to ICM due to this process.
- formation of clouds in the MSR shell triggered by the isobaric thermal instability. This instability occurs when a region of enhanced density can cool faster than its surrounding. The instability criterion $d\Lambda_{\rm eff}/dT < 0$ (Field 1965) can be translated into the requirement that ICM density and pressure have to exceed certain critical values $\rho_{\rm crit}$ and $P_{\rm crit}$, provided $\Lambda_{\rm eff}(T) = \Gamma - \Lambda$ and the equation of state are known. Since the fastest growing mode of the thermal instability has a length scale of $\approx c_s\, t_{\rm cool}$ where c_s is the isothermal sound speed and $t_{\rm cool}$ the cooling time-scale of the unstable blob we can expect small clouds to form in the MSR shell. Having parameterized the thermal instability with the above argument we derive the exchange terms for the

equations of continuity, momentum and energy balance in the following way:

$$\left(\frac{\partial \rho_2}{\partial t}\right)_{\rm cond} = \chi \frac{(\rho_1/f_1) - \rho_{\rm crit}}{t_{\rm cool}} \, , \left(\frac{\partial P_1}{\partial t}\right)_{\rm cond} = -\frac{P_1}{\rho_1} \cdot \left(\frac{\partial \rho_2}{\partial t}\right)_{\rm cond}$$

$$\left(\frac{\partial \rho_2 u_2}{\partial t}\right)_{\rm cond} = u_1 \left(\frac{\partial \rho_2}{\partial t}\right)_{\rm cond} \, , \qquad \left(\frac{\partial \rho_2 v_2}{\partial t}\right)_{\rm cond} = v_1 \left(\frac{\partial \rho_2}{\partial t}\right)_{\rm cond}$$

$$\langle a_2 \rangle = c_s \, t_{\rm cool}, \qquad \langle m_2 \rangle = \frac{4\pi}{3} (1.1\rho_1) \langle a_2 \rangle^3 .$$

Here ρ_1 denotes the ICM density averaged over the entire volume, f_1 its filling factor; hence ρ_1' is the physical ICM density in the intercloud space. u_1 is the ICM velocity component in z-direction and v_1 in r-direction. P_1 is the ICM pressure. Subscript 2 points to the condensing cloud component, subscript 3 to the preexisting cloud component. The switch χ is set to one if P_1 exceeds the critical pressure $P_{\rm crit}$ and is zero otherwise. The condensing clouds are assumed to gain no velocity dispersion. After being formed the clouds are able to contract when they cool. Both cloud components and the ICM are allowed to cool by (optically thin) radiation. We implicitly assume that equilibrium between heating processes (Γ) and cooling (Λ) is attained at $T_{icm} = 6000\,$K and $T_{cl} = 100\,$K for ICM and clouds, respectively.

3 NUMERICAL CALCULATIONS

The numerical calculation have been performed on a cylindrical grid with 384 grid-points in z-direction (perpendicular to the galctic plane) and 192 in r-direction (in the plane) resulting in a spatial resolution of 2.6 pc.

The initial density profiles of both ICM and preexisting clouds in z-direction were Gaussians with scale heights of 120 and 100 pc, respectively. Densities in the plane were chosen to $\rho_{1_0} = 6 \cdot 10^{-25}\,{\rm g\,cm^{-3}}$ and $\rho_{3_0} = 1.67 \cdot 10^{-24}\,{\rm g\,cm^{-3}}$. Both components are at their equilibrium temperatures of 6000 and 100 K, respectively. The mean mass of the preexisting clouds is $\langle m \rangle = 10\,{\rm M_\odot}$. Such a value is reasonable to cover the most abundant species of clouds in the ISM.

The expansion of the MSR is triggered by 50 supernova-explosions over a time scale of 10^7 yr. We model this explosions as continuous energy input into a small volume around the origin of the grid (see *e.g.* Różyczka 1985).

4 RESULTS

The cooling time-scale for the gas swept up in the shell of the MSR can be estimated to be $t_{\rm cool} \approx 3 \cdot 10^5$ yr. Cloud formation in the shell therefore occurs early in the evolution of the MSR. In Figures 1a and b we show the densities of ICM and shell cloud component after $4 \cdot 10^5$ yr. The mass density of the cloud component can be translated into a (mean) number density since the mean mass of individual clouds is known. We get number densities of $n_{\rm cl} \geq 100\,{\rm pc}^{-3}$. The clouds itself have masses of $\langle m_2 \rangle \leq 10^{-3}\,{\rm M}_\odot$ and radii of $\langle a_2 \rangle \approx 10^{-3}\,{\rm pc}$. Thus one can speak of very small clumps wich are embedded in the more teneous (ICM) gas of the shell with a density of about $10^{-24}\,{\rm g\,cm}^{-3}$. The density contrast between the clumps and the surrounding gas is ≈ 100. Due to their small column density of $N_{\rm cl} \approx 10^{18}\,{\rm cm}^{-2}$ the clumps should remain atomic (insufficient self-shielding).

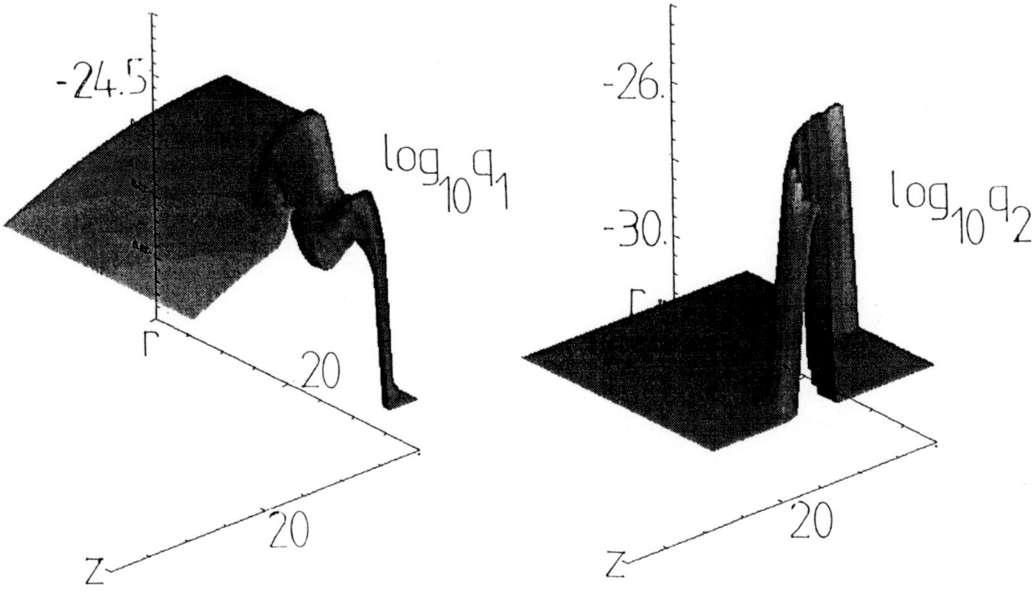

Fig. 1 – The logarithm of the ICM density ρ_1 is shown in Fig. 1a as surface plot over the spatial coordinates z and r. In Fig. 1b the logarithm of the density ρ_2 of the cloud component in the MSR shell is plotted. The evolutionary time is $4 \cdot 10^5$ yr, shortly after the formation time of the clouds.

Since there is sufficient uv-radiation from the cluster stars the MSR shell should be at least partially ionized. Different from earlier calculation we expect about $10^5\,{\rm M}_\odot$ of neutral material in form of the above mentioned clumps in the shell at early evolutionary states of the MSR. A more detailed numerical calculation including radiative transfer of the ionizing radiation will be dedicated to future work. While about 50% of the shell mass of the MSR condensed into clouds, in its interior evaporated 10^4 $M_\odot$ of the preexisting cloud component. The latter value corresponds to 1% of the initial total mass in this component.

After their formation the kinematics of the clumps in the MSR shell agreed quite well with the ICM gas being there. But during the ongoing expansion of the MSR we observed that the clouds began to lag behind the ICM shell. This is presumably due to the collisions of the shell clumps with the clouds of the interstellar medium surrounding the MSR which were initially at rest. A forthcoming investigation on the collisions of small clumps with high internal density with more diffuse clouds of different masses and sizes at rest will be necessary to clear the details. A more comprehensive version of the present investigation will be published elsewhere.

ACKNOWLEDGEMENT
This work was partially supported by DFG grant Yo 5/5–2.
The numerical calculations have been performed on a CRAY Y-MP at the HLRZ Jülich.

REFERENCES

Bloemen, J.B.G.M.: 1987, Astrophys. J. **322**, 694.
Cowie, L.L. and McKee, C.F.: 1977, Astrophys. J. **211**, 135.
Dickey, J.M. and Garwood, R.W.: 1989, in "Structure and Dynamics of the Interstellar Medium", *IAU-Coll. No. 120, G. Tenorio-Tagle, M. Moles and J. Melnick (eds.), Dordrecht, Reidel, p. 511.*
Elmegreen, B.G.: 1990, in "The evolution of the Interstellar Medium", *Astron. Soc. Pac. Conf. Ser.* **12**, *L. Blitz (ed.), p. 247.*
Field, G.B.: 1965, Astrophys. J. **142**, 531.
McKee, C.F. and Ostriker, J.P.: 1977, Astrophys. J. **218**, 148.
Różyczka, M.: 1985, Astron. Astrophys. **143**, 59.
Różyczka, M.: 1989, in "Structure and Dynamics of the Interstellar Medium", *IAU-Coll. No. 120, G. Tenorio-Tagle, M. Moles and J. Melnick (eds.), Dordrecht, Reidel, p. 463. Scalo, J.M. and Struck-Marcell, C.: 1984, Astrophys. J.* **276**, 60.
Spurzem, R., Kunze, R.; Yorke, H.W.; : 1991, Astron. Astrophys., in prep.
Tenorio-Tagle, G., Różyczka, M. and Bodenheimer, P.: 1990, Astron. Astrophys. **237**, 207.
Yorke, H.W., Tenorio-Tagle, G., Bodenheimer, P. and Różyczka, M.: 1989, Astron. Astrophys. **216**, 207.
Yorke, H.W., Spurzem, R., Kunze, R.: 1991, this proceeding, in press.

SUPERNOVAE IN A POROUS INTERSTELLAR MEDIUM: SINGLE AND MULTI-SUPERNOVA REMNANTS

J. FRANCO[1], P. BODENHEIMER[2], G. TENORIO-TAGLE[3], AND M. RÓŻYCZKA[4]

[1]Instituto de Astronomía–UNAM, Apdo. Postal 70–264, 04510 México D. F., México.
[2]University of California Observatories/Lick Observatory, Board of Studies in Astronomy and Astrophysics, University of California, Santa Cruz, CA 95064, USA.
[3]Instituto de Astrofísica de Canarias, 38200 La Laguna, Tenerife, Spain.
[4]University of Warsaw Observatory, Al. Ujazdowskie 4, PL-00478 Warszawa, Poland.

ABSTRACT

Supernova remnants do not evolve into a constant-density interstellar medium but rather into one which has been modified by bubbles caused by precursor stellar winds on small scales or by previous supernova explosions on large scales. This paper discusses the effect of these bubbles on the structure and appearance of individual supernova remnants, and considers the general effects of multisupernova explosions from OB associations on the overall structure and dynamics of galaxies.

1 INTRODUCTION

Stellar winds and supernovae ejecta from massive stars inject large amounts of mechanical energy into the interstellar medium. Thus, the collection of OB associations in any gaseous galaxy represents a rich energy source which may be controlling the main properties of the general interstellar medium (*e. g.* Cox and Smith 1974; Salpeter 1976; McKee and Ostriker 1977; Weaver *et al.* 1977; Abbott 1982; Cioffi and Shull 1991). The strong mechanical energy input generates hot and overpressured regions which drive shock waves into the ambient medium (see review by Ostriker and McKee 1988). The resulting wind-driven bubbles and supernova remnants, either from a single progenitor or from an entire association, are believed to generate most of the structuring observed in a gaseous galactic disk (*e. g.* Reynolds and Ogden 1978, 1979; Cowie *et al.* 1979; Heiles 1979; McCray and Snow 1979). Actually, many observed structures in the Milky Way and in external galaxies have been ascribed to this stellar energy injection (*e. g.* Heiles 1979, 1984; Brinks and Bajaja 1986; Deul and Hartog 1990; see review by Tenorio-Tagle and Bodenheimer 1988).

The expanding shocks, aside from sweeping the ambient gas, have been suspected of inducing star formation (*e. g.* Herbst and Assousa 1977; Dopita *et al.* 1985), providing a feedback control to the star forming activity at large scales (*e. g.* Mueller and Arnett 1976; Gerola and Seiden 1978; Franco and Shore 1984; Dopita 1985, 1988; see review by Franco 1991), and of accelerating cosmic rays (*e. g.* Blandford and Ostriker 1978). Also, the energy injection can drive the mass exchange among the different gas phases (*e. g.* Tenorio-Tagle and Palouš 1987; Palouš *et al.* 1990) and, depending on the star formation rate, may even exchange mass with the halo and create "fountains" or winds at galactic scales (*e. g.* Shapiro and Field 1976; Chevalier and Oegerle 1979; Bregman 1980; Cox 1981; Ikeuchi *et al.* 1984; Corbelli and Salpeter 1988; Norman and Ikeuchi 1989; Heiles 1990; Houck and Bregman 1990).

If stellar activity is truly responsible for the bulk properties of the general interstellar gas, then, the resulting medium could be viewed as a "porous" one, containing a large collection of cavities with different sizes. Hence, aside from its importance in the study of individual interstellar features, the evolution of either single or composite wind-driven bubbles and supernova remnants represents a fundamental tool in our understanding of the structure and activity of star-forming galaxies. Here we describe the main features of SN evolving in the cavities of this porous interstellar medium.

2 SN IN WIND-DRIVEN CAVITIES

Beginning with the classical work of Sedov (1959), the evolution of adiabatic and radiative wind-driven bubbles (*e. g.* Weaver *et al.* 1977) and single supernova remnants has been studied over several decades (see review of blast waves by Ostriker and McKee 1988). Single-remnant evolution has been explored under a variety of ambient conditions including constant density gas with typical disk and molecular cloud conditions (*e. g.* Chevalier 1977; Shull 1980; Wheeler *et al.* 1980; Cioffi *et al.* 1988), breaking out of molecular clouds (*e. g.* Tenorio-Tagle *et al.* 1985; Yorke *et al.* 1989), a cloudy medium (*e. g.* McKee and Cowie 1975; Cowie and McKee 1977; McKee and Ostriker 1977; McKee *et al.* 1978; Cowie *et al.* 1981; Tenorio-Tagle and Różyczka 1986; Różyczka and Tenorio-Tagle 1987), a dusty interstellar medium (*e. g.* Draine 1981; Dwek 1981, 1987), power-law ambient density distributions (*e. g.* Cavaliere and Messina 1976; Cox and Franco 1981; Cox and Edgar 1983; Edgar and Cox 1984), power-law distributions for both the atmosphere of the progenitor and the ambient medium (*e. g.* Chevalier 1982; Nadyozhin 1985; Chevalier and Liang 1989), clumpy ejecta with power-law stratifications (Hamilton 1985), and a magnetized medium with a constant field (*e. g.* Bernstein and Kulsrud 1965; Kulsrud *et al.* 1965; Chevalier 1974; Slavin and Cox 1991; Insertis and Rees 1991).

The evolution of SNR inside cavities, on the other hand, began only in the early 80's but much progress has certainly been achieved (see review by Franco *et al.* 1991a). The initial interstellar cavity can be generated by a single star, via winds and photoionization (*e. g.* Weaver *et al.* 1977; McKee *et al.* 1984; McKee 1988), or by the combined effect of stellar winds and previous SN (see next section). Analytic treatments of SNR inside cavities with a power-law stratification have been presented by Cox and Franco (1981), Chevalier (1982), Cox and Edgar (1983), Nadyozhin (1985), and Shull *et al.* (1985). Chevalier and Liang (1989) considered analytically the case of the SN ejecta in the cavity created by a blue supergiant wind. Several numerical codes, under the assumption of spherical symmetry, have been used to study the problem, including work by Fabian *et al.* (1983), Itoh and Fabian (1984), Dickel and Jones (1985), Band and Liang (1988), Ciotti and D'Ercole (1988), and Tenorio-Tagle *et al.* (1990a). These studies show that the ejecta essentially evolve in free expansion as they progress down the density gradient in the inner part of the cavity, and that they are observationally undetectable in this phase. At the later stages, when the remnant encounters the dense shell at the edge of the cavity, the interaction results in intense emission in the soft X-ray and optical regions of the spectrum. Ejecta thermalization lead to multiple shocks which propagate through the shell, and to a complex velocity field within the remnant. The mass of the shell turns out to be an important parameter (Tenorio-Tagle *et al.* 1990a). When it is at least 50 times that of the ejecta, radiative losses become important after shell interaction and there is no adiabatic (Sedov) evolutionary phase of the remnant. The intense optical emission occurs earlier, but the remnant ages faster than it would in a medium of constant density. If the shell mass is low, there also is strong emission generated at the moment of interaction, but the supernova shock overruns the shell before it has undergone significant radiative loss, and evolves through the Sedov phase essentially as it would in a constant-density medium. A two-dimensional hydrodynamic numerical calculation has been carried out by Tenorio-Tagle *et al.* (1991) with the ejecta assumed initially to be spherically symmetric. The results look similar in many respects to those of the 1-D calculations, except that chaotic, turbulent motions occur inside the bubble, leading to mixing between the ejecta and the gas in the wind. The generation of a Rayleigh-Taylor instability at the interface between the wind and the swept-up interstellar gas leads to deformation of the original shell and some further mixing. In the case of high-density environments, above 10^5 cm^{-3}, such as those expected in dense molecular clouds or in the central parts of galaxies, the strong radiative phase occurs at very early times, with large shock speeds, and the rapidly cooling shock generates a significant photoionized region (Terlevich *et al.* 1991). The emitted spectrum covers a wide range in photon energies, which mimics a power-law, and the resulting photoionized region displays features which are similar to those observed in AGN's.

An additional important issue is the possibility of having dense clumps (or fragments) in the supernova ejecta. Observational evidence for clumpy ejecta is growing, and has been found in a number of different remnants including Cas A (*e. g.* Braun *et al.* 1987, and references therein), Tycho (*e. g.* Seward *et al.* 1983; Hamilton 1985), Puppis A (*e. g.* Winkler *et al.* 1988), Kepler (*e. g.* Bandiera and van den Bergh 1991), SN 1987A (see Arnett *et al.* 1989 and Lucy *et al.* 1989), and some other extragalactic objects (see Hamilton 1985). This partial list of examples already indicates that ejecta fragmentation could be a quite common process, and Rayleigh-Taylor instabilities seem to be its likely origin (*e. g.* Falk and Arnett 1973; Chevalier and Klein 1978). Detailed numerical simulations in 2 and 3-D show that R-T instabilities can develop shortly after the explosion and can appear in any type of supernova (Arnett *et al.* 1989; Fryxell *et al.* 1991; Müller *et al.* 1991; Hachisu *et al.* 1991).

The effects produced by the resulting fragments depend on their size and ambient conditions. Hamilton (1985) derived similarity solutions for power-law stratifications in both clump distribution and ambient density, and Tenorio-Tagle *et al.* (1991) have obtained 2-D simulations for the interaction of large fragments with a pre-existing wind-driven bubble (see review by Franco *et al.* 1991a). The high-density fragments move almost unimpeded inside the bubble cavity, but they are rapidly thermalized at impact with the external shell. The fragment-shell interaction drives strong shocks at the places of impact which, aside from generating a variety of time-dependent features, can puncture and distort the original shell. Similar effects are produced when a clump of ejecta interacts with ambient clouds (McKee 1983).

3 SN IN MULTI-SN REMNANTS

The combined action of winds and supernova explosions from OB associations may be responsible for many of the observed large scale "holes" and "superbubbles" (see review by Tenorio-Tagle and Bodenheimer 1988). These interstellar structures have sizes ranging from 10^2 to 10^3 pc, a gas mass in excess of 10^4 $M_\odot$, and energies reaching up to several times 10^{53} erg. Except for the largest and most energetic structures, which can be better ascribed to collisions of infalling high-velocity clouds with the gaseous disk (*e. g.* Tenorio-Tagle *et al.* 1987b; Franco *et al.* 1988; van der Hulst and Sancisi 1988; Alfaro *et al.* 1991), numerical models of expanding "multi-supernova" remnants (MSR) driven by typical young stellar groups can indeed produce cavities and shells with properties corresponding to the observed ones.

The evolution of MSR has been explored with both analytical formulations and numerical simulations, covering a wide range of ambient conditions (*e. g.* Bruhweiler *et al.* 1980; Tomisaka *et al.* 1981; Tomisaka and Ikeuchi 1986; McCray and Kafatos 1987; Tenorio-Tagle *et al.* 1987a; Mac Low and McCray 1988; Mac Low *et al.* 1989; Igumentshchev *et al.* 1990; Palouš *et al.* 1990; Tenorio-Tagle *et al.* 1990a; Ferriére *et al.* 1991). These models include the disk density stratification along the z-axis and, in a few cases, an external B-field or the galactic differential rotation. Under these circumstances, MSR evolution can lead to the formation of a rich variety of gaseous structures (see reviews by McCray 1988 and Różyczka 1989) associated with flow instabilities in cooling shocks, accelerations due to decreasing density gradients, or to the shear generated by galactic rotation. In addition to these environmental effects, the details of shell evolution also depend on the way the matter and energy are injected into the expanding remnant. For instance, a pulsed mass and energy input has been already included in numerical models. This allows one to simulate the injection from repeated supernova explosions. The properties of a wind-driven bubble, however, are not very different from those of MSR, and a logical extension of these studies is to explore the effects of clumpy ejecta in the evolution of MSR.

Franco *et al.* (1991b) have addressed this important question with an analytical approximation to the fragment evolution inside a hot cavity. Aside from providing a deeper physical insight about the relevant processes governing fragment-environment interactions, an analytical approach represents a much needed and useful tool to test the results of simulations and, perhaps more important, it allows one to bypass the numerical problems arising when one has to follow the evolution over several orders of magnitude in scale (*i. e.,* from the initial fragment size to the dimensions of MSR). The results indicate that MSR can also be punctured and distorted by fragment bombardment during, at least, the first 10^6 yr of evolution but the effects of the early impacts remain for longer time scales. Thus, as in the case of single SN remnants, the early evolutionary phases and the final shape of MSR can be severely affected by fragment interactions.

4 FINAL REMARKS

As sketched in the Introduction, wind-driven shells and MSR are continuously perforating the interstellar medium and can be pictured as a collection of holes with a wide range of sizes. Thus, a large fraction of Type II supernovae should occur inside this porous medium. This view is certainly not new, but the effects associated to this porosity have been explored only recently. Many details in the evolution of supernova remnants depend on the initial ambient and ejecta conditions. In particular, the very existence of interstellar bubbles, whether they were carved by a single star or by an entire stellar association, already indicates that the resulting SN remnant should be a composite one. The growth of a composite remnant begins with the dimensions of the original bubble, and the external shell is excited and re-accelerated by the supernova ejecta. The ejected plasma, on the other hand, can be clumpy. This destroys the symmetry of the ejecta-ambient medium interaction, and generates a series of position-dependent and time-dependent effects. Clump bombardment excites, and can puncture, the shell at the particular places of impact, and this generates distortions in the growth rate at these same places. Thus, depending on the initial density stratifications of both bubble and ejecta, different parts of the ejecta are thermalized in different regions of the bubble, and shell excitation is in general anisotropic.

The bubble size, or shell mass, is also important in the evolutionary scheme. Small shells are easily overtaken by the supernova shock and the energy lost while traversing the shell is usually negligible (unless the density is higher than about 10^5 cm^{-3}). After shell interaction, which results in a rapid thermalization of the ejecta, the shell can be destroyed by instabilities whereas the outward shock simply follows the corresponding adiabatic track imposed by the ambient interstellar medium. In the case of "medium size" shells (*i. e.,* large wind-driven shells or small MSR) the supernova shock is not able to overrun the original shell but fragment bombardment can certainly produce distortions. For very large MSR, on the other hand, ejecta thermalization (including also dense fragments) occurs inside the shell and simply add mass and pressure to the bubble interior. Thus, the initial stages of MSR could be dominated by fragment collisions, causing a different growth rate at the places of impact. These distortions are long lasting, and the details of the structure of large MSR are basically determined, aside from cooling instabilities, by their early history and large scale environmental effects (*i. e.,* blow-out due to density gradients or distortions due to galactic shear).

ACKNOWLEDGMENTS

JF and MR thank the hospitality of the Instituto Astrofísico de Canarias, Spain. JF acknowledges the support given to this project by DGAPA-UNAM through the grant IN103991. MR was supported by the grant G-MEN-156/90 from the Polish Ministry of Education.

REFERENCES

Abbott, D. C. 1982, *ApJ,* 263, 723.
Alfaro, E. J., Cabrera-Caño, J. & Delgado, A. J. 1991, *ApJ,* 378, 106.
Arnett, D., Fryxell, B. & Müller 1989, *ApJ,* 341, L63.

Band, D. L. & Liang, E. P. 1988, in *Supernova Remnants and the ISM, IAU Coll. 101*, ed. R.S. Roger and T.L. Landecker (Cambridge Univ. Press: Cambridge), p. 69.

Bandiera, R. & van den Bergh, S. 1991, *ApJ,* 374, 186.

Bernstein, I. B. & Kulsrud, R. M. 1965, *ApJ,* 142, 479.

Blandford, R. D. & Ostriker, J.P. 1978, *ApJ (Letters),* 221, L29.

Braun, R., Gull, S. F. & Perley, R. 1987, *Nature*, **327**, 395.

Bregman, J. N. 1980, *ApJ,* 236, 577.

Brinks, E. & Bajaja, E. 1986, *A& A,* 169, 14.

Bruhweiler, F. C., Gull, T., Kafatos, M. & Sofia, S. 1980, *ApJ (Letters),* 238, L27.

Cavaliere, A. & Messina, A. 1976, *ApJ,* 209, 424.

Chevalier, R. A. 1974, *ApJ,* 188, 501.

Chevalier, R. A. 1977, *Ann. Rev. Astron. Ap.*, **15**, 175.

Chevalier, R. A. 1982, *ApJ,* **258**, 790.

Chevalier, R. A. & Klein, R. I. 1978, *ApJ,* 219, 994.

Chevalier, R. A. & Liang, E. P. 1989, *ApJ,* 344, 332.

Chevalier, R. A. & Oegerle, W. 1979, *ApJ,* 227, 39

Cioffi, D. F., McKee, C. F. & Bertschinger, E. 1988, *ApJ,* 334, 252.

Cioffi, D. E. & Shull, J. M. 1991, *ApJ,* 367, 96.

Ciotti, L. & D'Ercole, A. 1988, *A& A,* 215, 347.

Corbelli, L. & Salpeter, E. E. 1988, *ApJ,* 326, 551.

Cowie, L. L. & McKee, C. F. 1977, *ApJ,* 211, 135.

Cowie, L. L., McKee, C. F. & Ostriker, J. P., 1981, *ApJ,* 247, 908.

Cowie, L. L., Songaila, A. & York, D. G. 1979, *ApJ,* 230, 469.

Cox, D. P. 1981, *ApJ,* 245, 534.

Cox, D. P. & Edgar, R. 1983, *ApJ,* 265, 443.

Cox, D. P. & Franco, J. 1981, *ApJ,* 251, 687.

Cox, D. P. & Smith, B. W. 1974, *ApJ (Letters),* 189, L105.

Deul, E. & Hartog, R. H. 1990, *A& A,* 229, 362.

Dickel, J. R. & Jones, E. M. 1985, *ApJ,* 288, 707.

Dopita, M. 1985, *ApJ (Letters),* 295, L5.

Dopita, M. 1988, *Supernova Remnants and the ISM, IAU Coll. 101*, eds. R. Roger and T. Landecker, (Cambridge Univ. Press: Cambridge), p. 493.

Dopita, M. A., Mathewson, D. S. & Ford, V. L. 1985, *ApJ,* 297, 599.

Draine, B. T. 1981, *ApJ,* 245, 880.

Dwek, E. 1981, *ApJ,* 246, 430.

Dwek, E. 1987, *ApJ,* 322, 812.

Edgar, R. & Cox, D. P. 1984, *ApJ,* 283, 833.

Fabian, A. C., Brinkman, W. & Stewart, G. C. 1983, in *Supernova Remnants and their X-ray Emission (IAU Symp. 101)*, ed. J. Danziger and P. Gorenstein (Dordrecht:

Reidel), p. 119.
Falk, S. W. & Arnett, D. 1973, *ApJ (Letters),* 180, L65.
Ferriére, K. M., Mac Low, M. M. & Zweibel, E. 1991, *ApJ,* 375, 239.
Franco, J. 1991, in *Chemical and Dynamical Evolution of Galaxies* ed. F. Ferrini, J. Franco and F. Matteucci,(ETS Editrice; Pisa), p. 506.
Franco, J. & Shore, S. N. 1984, *ApJ,* 285, 813.
Franco, J., Tenorio-Tagle , G., Bodenheimer, P., Różyczka , M. N. & Mirabel, I. F. 1988, *ApJ,* 333, 826.
Franco, J., Tenorio-Tagle , G., Bodenheimer, P. & Różyczka , M. N. 1991a, *PASP,* 103, 803.
Franco, J., Ferrara, A., Tenorio-Tagle , G. & Różyczka , M. N. 1991b, in preparation.
Fryxell, B., Arnett, D. & Müller, E. 1991, *ApJ,* 367, 619.
Gerola, H. & Seiden, P. 1978, *ApJ,* 223, 129.
Hachisu, I., Matsuda, T., Nomoto, K. & Shigeyama, T. 1991, *ApJ (Letters),* 368, L27.
Hamilton, A. J. S. 1985, *ApJ,* 291, 523.
Heiles, C. 1979, *ApJ,* 229, 533.
Heiles, C. 1984, *ApJS,* 55, 585.
Heiles, C. 1990, *ApJ,* 354, 483.
Herbst, W. & Assousa, G. E. 1977, *ApJ,* 217, 473
Houck, J. C. & Bregman, J. N. 1990, *ApJ,* 352, 506.
Igumentshchev, I.V., Shustov, B.M. & Tutukov, A.V. 1990, *A&A* **234**, 396.
Ikeuchi, S., Habe, A. & Tanaka, 1984, *MNRAS,* 207, 909.
Insertis, F. M. & Rees, M. J. 1991, *MNRAS*, in press.
Itoh, H. & Fabian, A. C. 1984, *MNRAS,* 208, 645.
Kulsrud, R. M., Bernstein, I. B., Kruskal, M., Fanucci, J. & Ness, N. 1965, *ApJ,* 142, 491.
Lucy, L.B., Danziger, I.J., Gouiffes, C., Bouchet, P. 1989, in *Structure and Dynamics of the Interstellar Medium (IAU Coll. 120),* ed. G. Tenorio-Tagle, M. Moles and J. Melnick (Springer: Berlin), p. 164.
Mac Low, M. M. & McCray, R. 1988, *ApJ,* 324, 776.
Mac Low, M. M., McCray, R. & Norman, M. L. 1989, *ApJ,* 337, 141.
McCray, R. 1988, in *Supernova Remnants and the ISM, IAU Coll. 101),* ed. R.S. Roger & T.L. Landecker (Cambridge Univ. Press: Cambridge), p. 447.
McCray, R. & Kafatos, M. 1987, *ApJ,* 317, 190.
McCray, R. & Snow, T. P. Jr. 1979, *Ann. Rev. A. Ap.,* 17, 213.
McKee, C. F. 1983, in *Supernova Remnants and their X-ray Emission (IAU Symp. 101),* ed. J. Danziger and P. Gorenstein (Dordrecht: Reidel), p. 87.
McKee, C. F. 1988, in *Supernova Remnants and the ISM, IAU Coll. 101),* ed. R.S. Roger and T.L. Landecker (Cambridge Univ. Press: Cambridge), p. 205.

McKee, C. F. & Cowie, L. L. 1975, *ApJ,* 195, 715.
McKee, C. F., Cowie, L. L. & Ostriker, J. P. 1978, *ApJ (Letters),* 219, L23.
McKee, C. F. & Ostriker, J. P. 1977, *ApJ,* 218, 148.
McKee, C. F., van Buren, D. & Lazareff, B. 1984, *ApJ (Letters),* 278, L115.
Müller, E., Fryxell, B. & Arnett, D. 1991, in *Chemical and Dynamical Evolution of Galaxies*, ed. F. Ferrini, J. Franco and F. Matteucci, (ETS Editrice: Pisa), 394.
Mueller, M. & Arnett, W. D. 1976, *ApJ,* 210, 670.
Nadyozhin, D. K. 1985, *ASS*, **112**, 225.
Norman, C. & Ikeuchi, S. 1989, *ApJ,* 345, 372.
Ostriker, J. P. & McKee, C.F. 1988, *R. M. Phys.* **60**, 1.
Palouš , J., Franco, J. & Tenorio-Tagle , G. 1990, *A& A,* 227, 175.
Reynolds, R. J. & Ogden, P. M. 1978, *ApJ,* 224, 94.
Reynolds, R. J. & Ogden, P. M. 1979, *ApJ,* 229, 942.
Różyczka , M. 1989, in *Structure and Dynamics of the ISM*, ed. G. Tenorio-Tagle , M. Moles and J. Melnick, (Springer: Berlin), 463.
Różyczka , M. & Tenorio-Tagle, G. 1987, *A& A,* 176, 329.
Salpeter, E. E. 1976, *ApJ,* 229, 533.
Sedov, L. I. 1959, *Similarity Solutions & Dimensional Methods in Mechanics*, (Academic; New York).
Seward, F., Gorenstein, P. & Tucker, W. 1983, *ApJ,* 266, 287.
Shapiro, P. R. & Field, G. B. 1976, *ApJ,* 205, 762.
Shull, J. M. 1980, *ApJ,* 237, 769.
Shull, P. Jr, Dyson, J. E., Kahn, F. D. & West, K. A. 1985, *MNRAS,* 212, 799.
Slavin, J. D. & Cox, D. P. 1991, preprint.
Tenorio-Tagle, G. & Bodenheimer, P. 1988, *Ann. Rev. A. Ap.,* 26, 145.
Tenorio-Tagle, G., Bodenheimer, P., Franco, J. & Różyczka , M. N. 1990a, *MNRAS,* 244, 563.
Tenorio-Tagle, G., Bodenheimer, P. & Różyczka , M. N. 1987a, *A& A,* 182, 120.
Tenorio-Tagle, G., Bodenheimer, P. & Yorke, H. W. 1985, *A& A,* 145, 70.
Tenorio-Tagle, G., Franco, J., Bodenheimer, P. & Różyczka , M. N. 1987b, *A& A,* 179, 219.
Tenorio-Tagle, G. & Palouš , J. 1987, *A& A,* 186, 287.
Tenorio-Tagle, G. & Różyczka , M. 1986, *A& A,* 155, 120.
Tenorio-Tagle, G., Różyczka , M. & Bodenheimer, P. 1990b, *A& A,* 237, 207.
Tenorio-Tagle, G., Różyczka, M., Franco, J. & Bodenheimer, P. 1991, *MNRAS,* 251, 318.
Terlevich, R., Tenorio-Tagle, G., Franco, J. & Melnick, J. 1991, *MNRAS*, in press.
Tomisaka, K., Habe, H. & Ikeuchi, S. 1981, *ASSc,* 78, 273.
Tomisaka, K. & Ikeuchi, S. 1986, *PASJ,* 38, 697.
van der Hulst, T. & Sancisi, R. 1988, *AJ,* 95, 1354.

Weaver, R., McCray, R., Castor, J., Shapiro, P. & Moore, R. 1977, *ApJ,* 218, 377.
Winkler, P. F., Tuttle, J. H., Kirshner, R. P. & Irwin, M. J. 1988, in *Supernova Remnants and the ISM, IAU Coll. 101),* ed. R.S. Roger and T.L. Landecker (Cambridge Univ. Press: Cambridge), p. 65.
Wheeler, J. C., Mazurek, T. J. & Sivaramakrishnan, L. 1980, *ApJ,* 237, 781.
Yorke, H. W., Tenorio-Tagle, G., Bodenheimer, P. & Różyczka , M. 1989, *A& A,* 216, 207.

MHD Shock Interactions with Interstellar Clouds

M.-M. MAC LOW[1], R. I. KLEIN[1], C. F. MCKEE[1,2],
M. L. NORMAN[3], AND J. M. STONE[3]

[1]Astronomy Department, University of California at Berkeley
[2]Physics Department, University of California at Berkeley
[3]Astronomy Department, University of Illinois at Urbana-Champaign

1 INTRODUCTION AND COMPUTATIONS

Shock waves driven by supernova explosions are the most important means of transferring energy into the interstellar gas aside from direct radiation from the stars. How these shock waves interact with clouds will shape the appearance of SNRs, may trigger star formation by sweeping up clouds, and mediates the transfer of energy to the gas. We here compute the effect of magnetic fields on these interactions.

Klein, McKee, & Colella (1990, herafter KMC) calculated field-free models of shocked clouds using an adaptive mesh refinement code called AMR. Since this code used an exact Godunov shock resolver, however, adding in magnetohydrodynamic (MHD) shocks remains a difficult technical challenge. Instead, we use ZEUS-2D (Stone & Norman 1991*a*, 1991*b*), a well-tested astrophysical MHD code using a Von Neumann artificial viscosity to resolve shocks.

We first compared the results from the two codes in the field-free case to understand the effects of resolution. We then computed ZEUS-2D models having initial magnetic fields parallel to the shock velocity, with a ratio of thermal to magnetic pressure ahead of the shock, $\beta_0 = 1$ and 5. All of our runs have a Mach 10 shock hitting a spherical cloud with a density ratio $\chi = 10$ to the background. A useful time scale is the time for the shock to cross the cloud, called the cloud-crushing time, $t_{cc} = \chi^{1/2} R/v_s$, where R is the initial cloud radius and v_s is the shock velocity.

2 RESULTS

Without a magnetic field, calculations with the AMR code by KMC showed that a strong shock passing over a cloud caused extreme fragmentation of the cloud. Without a cloud, a plane shock running along straight magnetic field lines will not affect the field at all. Both of these results change dramatically in our calculations.

The fragmentation in the absence of a field occurs when the slip surface between the cloud and its surroundings becomes Kelvin-Helmholtz unstable and winds up into a series of vortexes. In the presence of a field, this process begins in the same fashion, but the field gets concentrated by the vortexes, and by the surrounding flow, limiting the amount of fragmentation that can occur. After $8t_{cc}$, the amount of fragmentation is reduced by $\sim 25\%$ for $\beta_0 = 5$, and by $\sim 50\%$ for $\beta_0 = 1$.

When the shock sweeps past the cloud it curves and forces field lines passing through the cloud towards the axis. The flow concentrates the field on the axis until the magnetic pressure of the field becomes dynamically important, that is $\beta = 1$. Thus, the presence of the cloud allows the field to be compressed by a factor of the Mach number rather than just $(\gamma + 1)/(\gamma - 1)$.

The strong fields on the axis and in the vortices contribute to enhanced synchrotron emission. We find that the peak enhancement will last for the order of 10 t_{cc}. The distribution is quite similar to that of the field as shown in the Figure.

REFERENCES

Klein, R. I., McKee, C. F., and Colella, P. 1990. in *The Evolution of the Interstellar Medium*, ed. L. Blitz (Astronomy Society of the Pacific: San Francisco), p. 117.
Stone, J. M., and Norman, M. L. 1991*a, b*, *Ap. J. Suppl.*, in press.

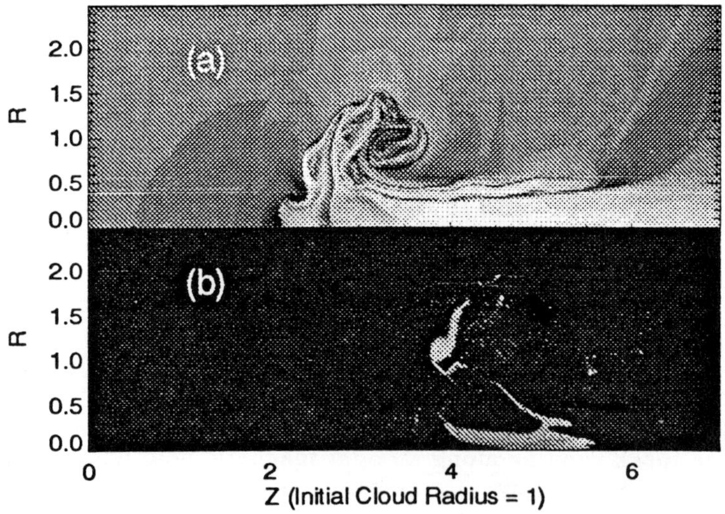

Figure 1. Cloud with density ratio $\chi = 10$ hit by a Mach 10 shock. A uniform field giving $\beta_0 = 1$ ahead of the shock lies parallel to v_s. *(a)* Log of the magnetic pressure at time $3t_{cc}$, and *(b)* Log of the density at time $7.5t_{cc}$.

Supernova Remnants as Sites of Star Formation

N. JUNKES[1,2], E. FÜRST[2]

[1]Institut für Theoretische Physik und Sternwarte, Olshausenstr. 40, W-2300 Kiel 1, FRG

[2]MPI für Radioastronomie, Auf dem Hügel 69, W-5300 Bonn 1, FRG

ABSTRACT

The far-infrared (FIR) radiation of compact sources with IR spectra of cool dust ($T_d < 50\ K$) is a sensible tracer for protostars or young stellar objects (YSOs). We investigated the distribution of such sources from the IRAS point source catalogue in the direction of all known supernova remnants (SNRs) in the northern hemisphere. We found a statistical excess of IRAS sources for several SNRs in relation to reference areas. For some of these SNRs (like W44 and IC443) their interaction with the interstellar medium (ISM) had previously been established by molecular line observations.

In case of the extended SNR G54.4-0.3 we found a CO emission shell at about 40 km/s radial velocity in striking alignment with the radio continuum shell. The correspondence in position and size gives strong evidence for a physical connection and allows to derive the kinematic distance as well as mass and density of the CO shell.

High-resolution CO maps of small areas around some of the IRAS point sources in the direction of G54.4-0.3 show that an association of the IR sources to compact small-diameter CO clouds is probable, with their radial velocities coincident to those of the CO shell around the SNR. One of these clouds (labelled IRAS10) shows a distinct bipolar structure with a linear extent of about 4 pc.

For G54.4-0.3 the number of IR sources and their positional coincidence with compact CO clouds give a strong hint that star formation takes place in condensed parts of the CO shell around this SNR.

Additional SNRs with a significant excess of cool IRAS point sources, candidates for similar star formation scenarios, are targets for forthcoming investigations.

1. THE SNR - IRAS CONNECTION

We have correlated the positions of all SNRs north of $-26°$ declination with the IRAS Point Source Catalogue (IRAS, 1985) to establish candidates for the interaction of SNRs with the interstellar medium. The FIR emission in the direction of SNRs gives conclusions for the distribution of dust which has been heated by the expanding shock wave of the SNR (Braun, 1987). In case of IC443 a connection between low-temperature IRAS point sources and CO gas, emitting velocity-broadened lines, has been suggested (Huang et al., 1985). In general, two explanations are possible for a connection between compact FIR sources and SNRs: They could be local density enhancements or clumps in the dust which have been heated by the SNR shock wave. The luminous FIR sources could also be explained with protostars or YSOs which were formed by the impact of the shock wave on the interstellar medium (triggered star formation, originally suggested by Öpik, 1953). IRAS point sources in the direction of SNRs have been selected by the following criteria:

1) $S_{100} > 20$ Jy
2) $S_{100}/S_{60} = 1.2 \ldots 6.0$ *or* $T_{\rm d} = 24 \ldots 45$ K
3) $S_{60}/S_{25} > 1$
4) $Q_{60} + Q_{100} \geq 4$
5) $D_{\rm IRAS} = 2D_{\rm Radio}$

These criteria provide a selection of strong sources (1), a limitation to cool dust temperatures according to the IR spectrum (2,3) and to confirmed detections ($Q_\lambda = 2, 3$) from the IRAS catalogue (4). The search area is chosen larger than the radio diameter to include the complete surface of asymmetric SNRs (5). The "physical" selection criteria 2 and 3 allow a classification of different types of IR sources. According to statistical results from Emerson (1987) they include values for YSOs (e.g. T Tau stars) as well as for dense cores of molecular clouds, and exclude cool stars and planetary nebulae. However, extragalactic objects cannot be distinguished from molecular cloud cores. The IRAS flux ratios are almost identical for both classes of objects. To correct for the amount of extragalactic sources, we checked reference areas in the Galactic plane and considered only SNRs with a statistical surplus of IRAS sources.

Table 1: *SNRs with excess number of IRAS point sources*

SNR	IRAS Number	Density [deg^{-2}]	Other name
G18.95-1.1	8	8.4	
G22.7-0.2	11	18.7	
G23.3+0.3	9	14.2	
G34.7-0.4	17	12.2	W44
G40.5-0.5	7	16.6	
G49.2-0.7	15	19.1	W51
G54.4-0.3	16	9.1	
G78.2+2.1	28	8.9	γ Cyg
G109.1-1.0	4	5.9	CTB109
G117.4+5.0	16	5.1	
G189.1+3.0	5	6.4	IC443
G211.7-1.1	13	4.1	

A total of 94 SNRs was investigated. In the search area of 56 SNRs we found IRAS sources, 12 of them showing a statistical excess of these sources (Table 1). As the number of IRAS sources in reference areas is significantly lower in the direction to the

Galactic anticenter (2nd and 3rd Galactic quadrant), the number density of IRAS sources for the selected SNRs is lower in those regions.

Fig. 1 shows the distribution of IRAS sources with respect to one source from table 1, W44. The SNR was previously investigated for interaction with the ISM (Wootten, 1977). One of the IR sources in the area of W44 was interpreted as YSO embedded in a molecular cloud which is associated with the SNR (Wootten, 1978).

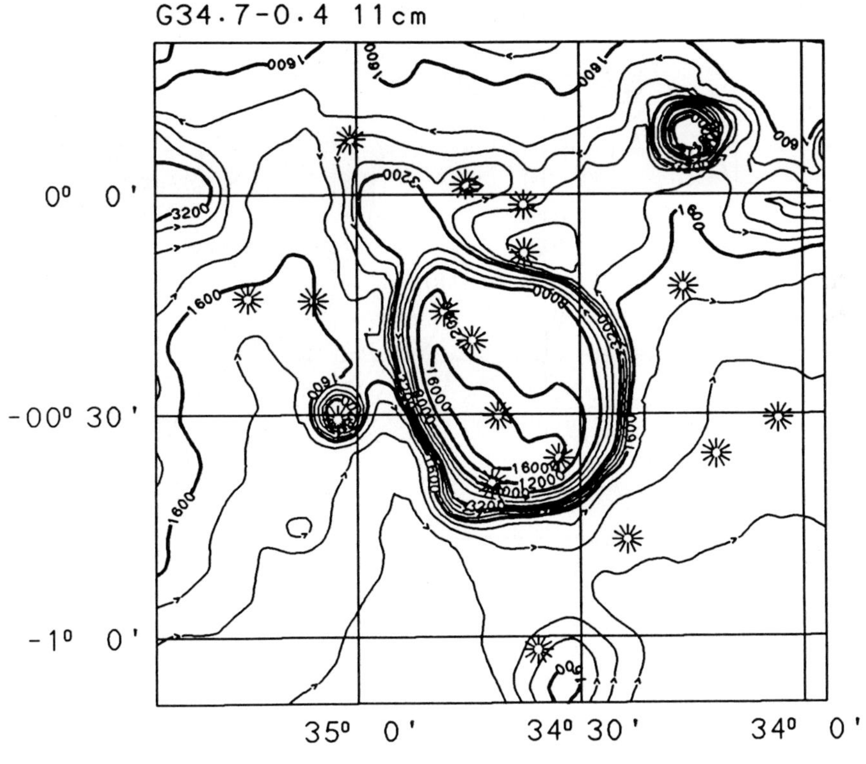

Fig. 1: *Positions of cool IRAS point sources relative to an SNR. Contour lines of 11 cm radio continuum are shown for SNR G34.7-0.4 (W44), IRAS sources are marked with asterisks.*

We considered G54.4-0.3 as object for further investigations because it is clearly identified as shell-type SNR with non-thermal spectral index (-0.5) in an interarm region of the Galaxy. Fig. 2 shows the correspondence of IRAS sources taken from the point source catalogue with the IRAS 60μ map. A more detailed investigation of IRAS sources in the area of G54.4-0.3 is given in Junkes et al. (1991).

2. G54.4-0.3: A CO-SHELL AROUND AN SNR

Medium-resolution CO observations of SNR G54.4-0.3 with the KOSMA-3m-telescope reveal a shell structure in CO at a central velocity of about 40 km/s (v_{lsr}) which shows a striking coincidence with the radio continuum feature.

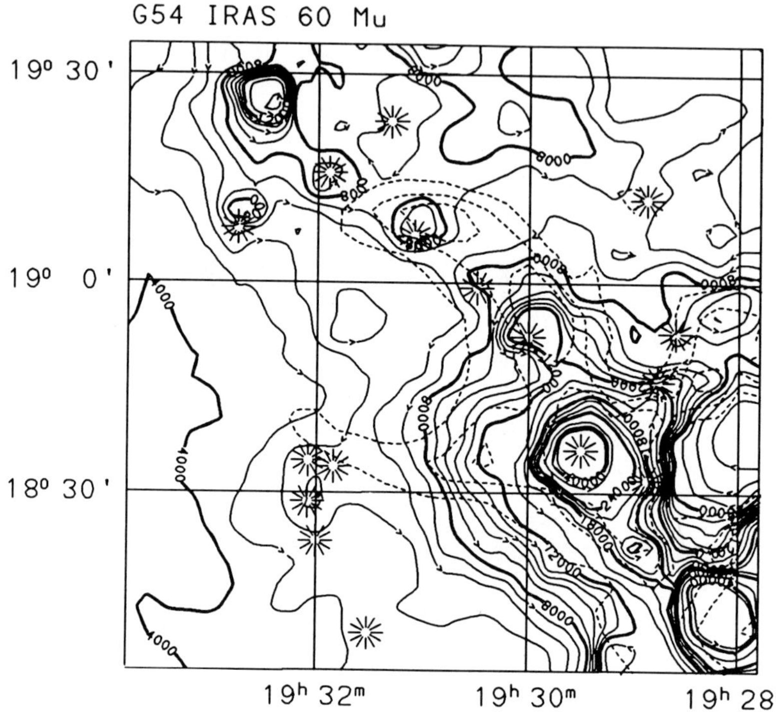

Fig. 2: *IRAS 60μ map of the area of SNR G54.4-0.3 (contour lines scaled in $10^4 Jy/ster$). Selected IRAS point sources are marked with asterisks, the 11cm radio feature of the SNR is indicated in dashed contours.*

Fig. 3 shows the central part of the CO shell (39-41 km/s), parts of this shell are visible in CO channel maps from 36 to 44 km/s. The radial velocity of the CO shell allows to determine a kinematic distance (3 or 7 kpc). The closer distance value of 3 kpc is supported by additional observational results (Junkes et al., 1991).

The line integral of the CO shell allows to estimate the column density of molecular matter in the shell, the total mass ($5 \cdot 10^4$ $M_\odot$) and the volume density (30 cm^{-3}) of the molecular shell.

3. IRAS10: A BIPOLAR OUTFLOW IN THE ENVELOPE OF AN SNR

Small areas around five IRAS sources in the region of SNR G54.4-0.3 have been mapped in CO with the IRAM-30m-telescope. We have detected CO condensations or small CO clouds at these positions, with radial velocities similar to the CO shell around the SNR. Masses and densities of these clouds for an assumed distance of 3 kpc can be estimated to $200 - 400\ M_\odot$ and $100 - 300\ cm^{-3}$, respectively.

The most interesting result is obtained for the position IRAS10, with a distinct bipolar structure (Fig. 4) and self-absorption in ^{12}CO at the central position. From position and radial velocity we can associate this source with the CO shell around the SNR.

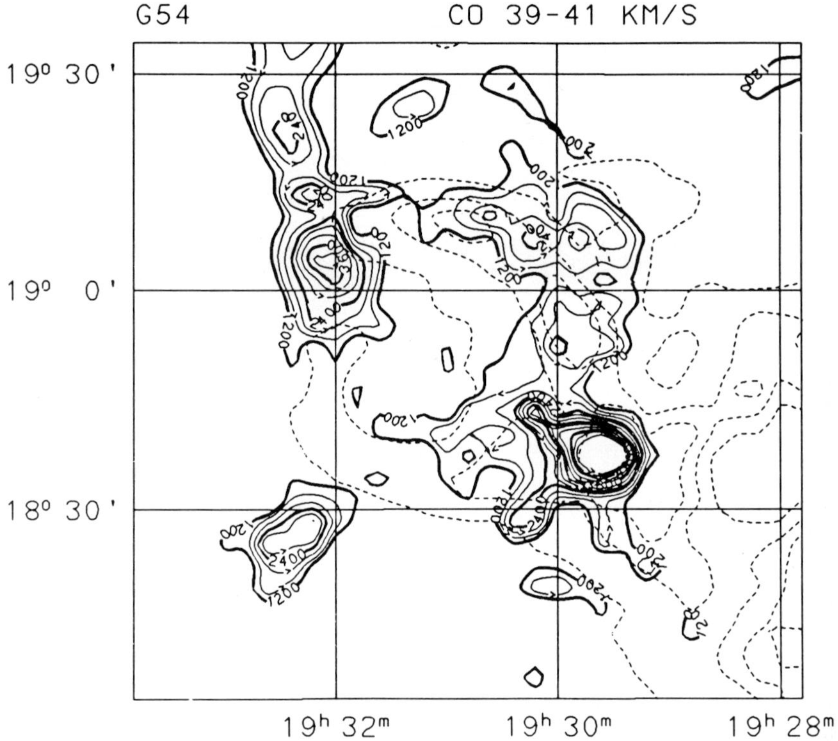

Fig. 3: *CO shell around SNR G54.4-0.3. The line integral for 39-41 km/s radial velocity shows the CO distribution (solid lines) with respect to the radio continuum structure of the SNR (dashed lines).*

4. SOME CONCLUSIONS

High-velocity wings as found at the IRAS10 position indicate protostars or YSOs still hidden in dust and molecules where molecular gas is blown out by the star (Bally and Lada, 1983). The region around SNR G54.4-0.3 shows signposts for recent (or on-going) star formation. It contains an OB-association and several HII-regions in a common distance of 3 kpc. At the positions of cool IRAS sources we find molecular cloud cores, probably YSOs which may have developed from condensations in the CO shell. Calculations on the basis of SNR- or stellar wind expansion models suggest that the formation of the CO shell results from a stellar wind and that the SNR has expanded into the wind bubble located inside the shell

(Junkes et al., 1991). This stellar wind could provide a triggering mechanism for the formation of YSOs in the dense CO shell ("Stellar wind-induced star formation").

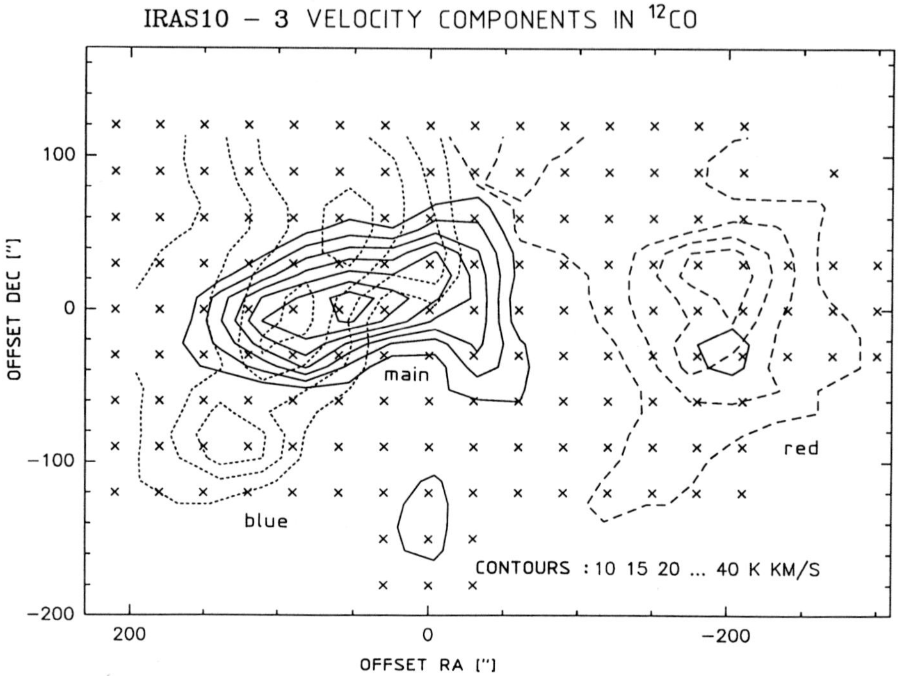

Fig. 4: *Bipolar outflow at the position of IRAS10. Line integrals of the main velocity component (36-40 km/s) and red- and blue-shifted wings are shown separately.*

The correlation of cool IRAS point sources with SNRs results in a number of candidate SNRs for similar scenarios (Table 1). A closer view on the IRAS sources could help to confirm their identification as protostars or YSOs and to establish the rôle of SNRs as sites of star formation.

REFERENCES:

Bally, J., Lada, C.J.: 1983, Astrophys. J. **265**, 824

Braun, R.: 1987, Astron. Astrophys. **171**, 233

Emerson, D.P.: 1987, in: *Star Forming Regions*, IAU Symp. 115, M. Peimbert, J. Jugaku (eds.), Reidel, Dordrecht, p. 19

Huang, Y.-L., Dickman, R.L., Snell, R.L.: 1985, Astrophys. J. **302**, L63

IRAS, 1985: IRAS Catalogues and Atlasses (Explanatory Supplement), US Government Printing Office

Junkes, N., Fürst, E., Reich, W.: 1991, Astron. Astrophys. (submitted)

Öpik, E.J.: 1953, Irish Astron. J. **2**, 219

Wootten, H.A.: 1977, Astrophys. J. **216**, 440

Wootten, H.A.: 1978, The Moon and the Planets **19**, 163

Galactic Fountains and Instabilities

A. FERRARA AND G. EINAUDI

Dipartimento di Astronomia e Scienza dello Spazio, Università di Firenze, Largo E. Fermi, 5 50125 Firenze, Italy

1 INTRODUCTION

In the last years the presence of a hot component of the interstellar medium at high Galactic latitudes has been suggested by several types of observations. The most diffuse techniques to monitor this hot coronal gas are represented mainly by soft X-ray observations (Nousek *et al.* 1982; Marshall *et al.* 1984) and by measurements of the UV absorption lines produced by highly ionized metals (Savage and Massa 1987). Unfortunately, different theoretical interpretations of these observations are possible and a clear picture of the structure of the coronal gas has not yet been drawn. For this reason, it is important to refine the theoretical models in order to deduce further implications of the presence of the hot gas, which can eventually be tested observationally.

The sequential occurrence of many supernovae explosions in the Galactic disk creates large-scale structures (superbubbles) filled with very hot gas. The surrounding cool shell will become Rayleigh-Taylor unstable at the top during the expansion phase under typical disk-halo interface conditions, thus allowing the internal, hot and rarefied, gas to flow freely into the halo because of its larger gravitational scale-height. This hot gas may undergo further thermal instability (Field 1965) leading to the formation of cool, non-gravitational condensations which will fall back onto the disk. The above process was included in the Galactic Fountain model (Shapiro and Field 1976; Bregman 1980) where the falling gas was identified with the observed high velocity HI clouds (HVCs) discovered in the halo (van Woerden *et al.* 1985).

In this paper we study the dynamical and thermal properties of the hot gas breaking out of the disk into the halo in order to find the physical conditions eventually leading to the formation of condensations through a thermal instability. Furthermore different instabilities may grow in the hot flow and their effects are also explored.

2 HYDRODYNAMICS OF THE HOT GAS

We are interested, for the aim stated above, in making a linear stability analysis of the MHD equation for the flow; thus, first we have to find the solutions describing the equilibrium configurations to which small amplitude perturbations are subsequently applied. In addition, as will be shown in the next Section, the steady-state approximation provides a great deal of information about the processes governing the physics of the flow.

2.1 Steady-state solutions

The relevant hydrodynamical equations for the flow can be written (Einaudi and Ferrara 1991) in a one dimensional form as follows:

$$\frac{d}{dz}(\rho A v) = 0, \tag{2.1}$$

$$\frac{1}{2}\frac{dv^2}{dz} = \frac{M^2}{M_0^2(M^2-1)}\left[\frac{T}{A}\frac{dA}{dz} - \frac{dT}{dz} - \frac{r_0}{L_g}g\right], \tag{2.2}$$

$$\rho^\gamma v \frac{d}{dz}\left(\frac{p}{\rho^\gamma}\right) = \tau_{cv}\left[\frac{1}{\tau_{cd}}\Lambda_{cd} - \frac{1}{\tau_{rd}}\Lambda_{rd}\right], \tag{2.3}$$

$$p = \rho T. \tag{2.4}$$

The z coordinate is measured in the direction perpendicular to the disk, which is supposed to be axisymmetric. The flow has a variable cross section $A(z)$ and M_0 is its Mach number at the disk level; the functions Λ_{rd} and Λ_{cd} represent the radiative cooling and thermal conduction terms respectively, τ's being a measure of the time scales of the corresponding processes at the disk level; r_0 is a fiducial scale length, L_g is the gravitational scale height, and the other symbols have the usual meaning. A detailed discussion on the validity of the 1-D approximation along with a full description of the physical assumptions made can be found in Einaudi and Ferrara (1991). Here we limit ourselves to the consideration that, if there is a component of the Galactic magnetic field perpendicular to the disk, (i) the streamlines tend to be aligned with the magnetic lines, and (ii) the thermal conduction perpendicular to the field is strongly inhibited, thus reducing to a good degree of approximation the dimensionality of the problem to one. The set of equations (2.1)-(2.4) can be solved when four initial conditions, i.e., density, velocity, temperature and heat flux at the base of the flow ($z = 0$) are given.

Depending on the base parameters of the flow three different classes of solutions are found (Einaudi and Ferrara 1991) which can be well classified in terms of the Mach number: (1) Critical solutions, i.e., solutions in which a transition takes place between two subsonic ($M < 1$) and supersonic ($M > 1$) dynamical regimes; (2) Solutions which are subsonic everywhere; and (3) Solutions which are supersonic everywhere. A typical example of a critical solution is shown in Fig. 1 for a set of initial conditions which are thought to be suitable for the hot gas flowing out of the superbubble cavity. We point out that, given the functional form of Eq. (2.2) which determines the position of the critical point z_c, every z can be a critical point for an infinite (but not necessarily physical) number of solutions as can be seen by making a Taylor expansion of the system (2.1)-(2.4). The importance of these solutions resides in the fact that they can produce supersonic motions and shocks in the halo which can be observed. Moreover, from an inspection of the energy terms giving the importance of the various physical processes (radiative losses, convection, thermal conduction, and gravity), it comes out that the energetics of the flow are always dominated by conductive and convective terms.

2.2 Linear stability analysis

A criterion for the growth of a thermal instability in a flow was derived by Field (1965). He considered a homogeneous and static background flow taking into account the effects of radiadive losses, mechanical heating and thermal conduction in the energy equation; also he supposed that the small perturbations applied to the system are periodic in space along with their derivatives. These assumptions allowed him to write the generic perturbation in the form $e^{i(\mathbf{kr}+nt)}$ and to derive the following dispersion relation:

$$n^3 + \left(K_T + \frac{k^2}{K_K}\right) c_s n^2 + c_s^2 k^2 n + \frac{c_s^3 k^2}{\gamma}\left(K_T - K_\rho + \frac{k^2}{K_K}\right) = 0, \qquad (2.5)$$

where K_T and K_ρ are wavenumbers of soundwaves whose angular frequencies are equal to the growth rates of isothermal and isochoric perturbations, respectively. Eq. (2.5) has three roots, two of which correspond to sound waves and the third to a condensation mode which is unstable, under the assumptions made, if the relation

$$\left(\frac{\partial \Lambda_{rd}}{\partial T}\right)_\rho - \frac{\rho_0}{T_0}\left(\frac{\partial \Lambda_{rd}}{\partial \rho}\right)_T < 0, \qquad (2.6)$$

holds. The inequality expressed by Eq. (2.6) is often referred to as the isobaric criterion for the thermal instability because this mode is characterized by the fact that pressure variations are negligible.

There are several limitations in the applicability of Field's results (Ferrara and Einaudi, 1991) due to the strict assumptions made:

— the criterion has a local applicability in a non-homogeneous medium whereas in a real galactic environment the flow is homogeneous only under very special circumstances;

— when the periodicity hypothesis is relaxed Field's results are not directly applicable when a velocity field is present. Also, the choice of the boundary conditions (BCs)

is crucial in a stability analysis. If, as it is likely, an external medium is present, its interaction with the flow influences strongly the properties of the flow itself and this effect becomes relevant at the point where the flow pressure becomes comparable to the external one. In order to take into account this effect we consider BCs simulating the presence of the external medium;
— The role of thermal conduction is underestimated by the assumption of a homogeneous equilibrium, because in this case several terms disappear from the linearized energy equation.
Furthermore we would like to study whether different types of instabilities may develop in the flow when velocities comparable to the sound speed are present.

Assuming perturbations of the form

$$\delta\zeta(z,t) = \zeta(z)e^{i\omega t}, \tag{2.7}$$

we have obtained the system of linearized equations describing the behaviour of the perturbations:

$$v_0\rho' + \rho_0 v' + (i\omega + \Gamma v_0 + v_0')\rho + (\Gamma\rho_0 + \rho_0')v = 0, \tag{2.8}$$

$$T_0\rho' + \rho_0 v_0 v' + \rho_0 T' + \left(v_0 v_0' + \frac{1}{L_g} g_z + T_0'\right)\rho + (i\omega\rho_0 + \rho_0 v_0')v + \rho_0' T = 0, \tag{2.9}$$

$$\begin{aligned}
-\frac{1}{\tau_{\rm cd}} T_0^{\frac{5}{2}} T'' + (v_0 T_0 - \gamma v_0 T_0)\,\rho' + (v_0\rho_0 - 5\frac{1}{\tau_{\rm cd}} T_0^{\frac{3}{2}} T_0')T' + \\
+ \left(i\omega T_0 - i\gamma\omega T_0 + v_0 T_0' + 2\rho_0 T_0^f\right)\rho + \\
+(\rho_0 T_0' + T_0\rho_0' - \gamma T_0 \rho_0')v + \\
+ \left\{i\omega\rho_0 + v_0\rho_0' - \gamma v_0 \rho_0' + f\rho_0^2 T_0^{f-1} - \frac{1}{\tau_{\rm cd}}\left[\frac{15}{4} T_0^{\frac{1}{2}} T_0'^2 + \frac{5}{2} T_0^{\frac{3}{2}} T_0''\right]\right\} T = 0,
\end{aligned} \tag{2.10}$$

where the subscript 0 refers to the unperturbed quantities, the prime indicates the z-derivative and $\Gamma = 1/A(dA/dz)$. This is an eigenvalue problem where the eigenvalue is represented by the growth rate ω (normalized with the inverse of the radiative time). According to the previous discussion, in order to resemble the effects of an external medium, of the four boundary conditions required to solve the system (2.8)-(2.10), two are imposed at the base and two at the top of the flux tube, at a point z_ℓ where the two pressures become comparable. In most of the runs the two boundaries are supposed to behave as mass reservoires and thermostats for the perturbations, but other possibilities have been explored (Ferrara and Einaudi 1991).

Adding an appropriate constant heating term to our energy equation, a homogeneous solution for the problem is admitted. This allows us to compare our results with

Field's ones and to check the finite difference method code used to solve the problem numerically. In the homogeneous case we find that, differently from Field's results because of the different BCs adopted, the spectrum of the eigenvalues is infinite but a condensation mode is clearly individuated as the fastest isobaric mode with a pure imaginary growth rate ω. Also, the growth time of the thermal mode is, as obtained by Field (1965), very close to the radiative one. The main difference with respect to Field (1965) is that the critical lengths we find are systematically larger, i.e., some configurations of the flow are stabilized by the BCs we have adopted.

We now move to a realistic case using for the background flow the solutions obtained from the steady state study performed in Sec. 2.1. The results we discuss here are the most typical ones and are obtained for $n^* = 0.6 \times 10^{-3}$ cm^{-3}, $T^* = 4.1 \times 10^6$ K, but several other cases have been computed (Ferrara and Einaudi 1991). We follow the eigenvalue corresponding to the condensation mode as the heating term is smoothly decreased and we maintain a very low Mach number at the base ($M^* \simeq 10^{-3}$). This produces a progressive enhancement of the gradients in the steady state which amplify the stabilizing role of the thermal conduction until the mode is completely stabilized. If we now increase M^*, the system becomes unstable again and the modulus of the growth rate is an increasing function of M^* in the subsonic regime, that is, the system becomes more and more unstable, whereas it decreases in the supersonic one. However, a relevant feature is that the critical solution is *fully stable*; this behaviour is shown in Fig. 1.

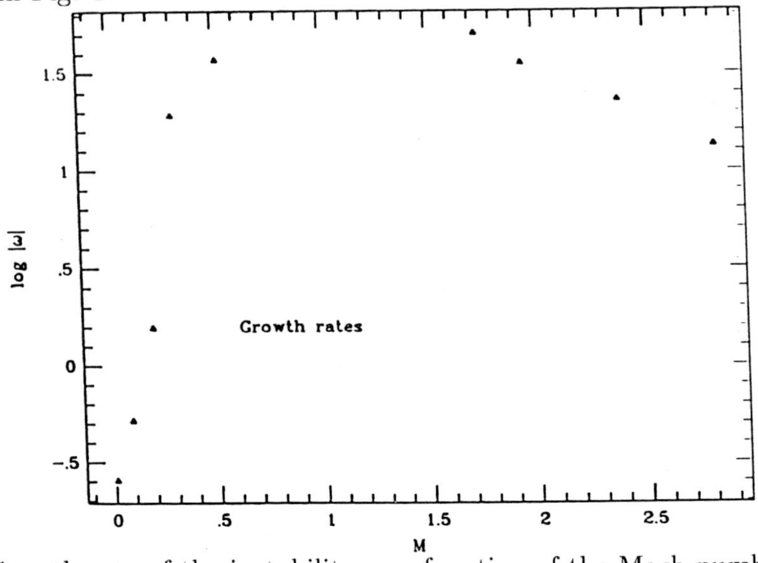

Fig. 1. Growth rate of the instability as a function of the Mach number of the flow.
Can this instability be classified as a thermal instability similar to that we discussed in the homogeneous and low-M^* cases ? There are several suggestions leading to the

conclusion that this instability has a different nature.

- Typical growth rates ω of the instability are 20-40 times larger than the radiative growth rate, depending on the parameters of the unperturbed flow.
- The perturbations are no more isobaric as in the case of the pure thermal instability;
- The energetics of the flow are dominated by the convective and conductive terms (see Fig. 2); thus radiative losses are not the driving agents of the instability anymore.

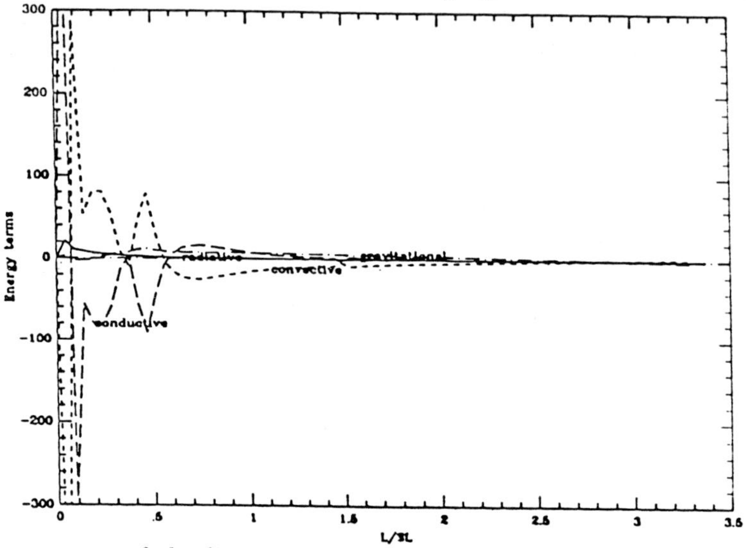

Fig. 2. Energy terms of the linearized energy equation (arbitrary units) for the case $M^* = 0.3$, $|\omega| = 19.2$, and $n^* = .6 \times 10^{-3}$ cm^{-3},$T^* = 4.1 \times 10^6$ K.

Therefore, it seems more suitable to refer to it as a convective instability rather than a thermal one. However, in order to derive the subsequent evolution of the instability a full non-linear treatment is required (Ferrara and Einaudi, in preparation), so the analysis performed in this paper represents the first step towards the comprehension of the stability of the system.

3 CONCLUSIONS

The results we have obtained show that, differently from the Galactic Fountain model predictions, HVCs are hardly formed as a result of a thermal instability occurring in the hot flows produced by the blowout of the disk superbubbles into the halo. Suitable conditions for the formation of cool condensations are actually found only if the halo can be considered static and some mechanical heating is provided to the flow. However, on the one hand Danly (1990) has clearly demonstrated that systematic motions are present in the Galactic halo, and on the other, the heating provided by high latitude Type I supernova events may have a disruptive effect on the flow.

A different class of (convective) instabilities has been recognized to be at work in the system. Their role has still to be fully understood, but we suggest temptatively that these instabilities reflect the tendency of the system to relax on a critical (transonic) solution by rearranging the base parameters of the flow. Following the above conclusion, it seems more likely that rather than cloud formation, transonic Galactic winds are generated through this mechanism. The presence of a wind in the central regions of the Milky Way has been argued by Lockman (1991) who has detected a lack of the HI halo in the inner 2.5 kpc. Furthermore, these convective modes may lead to turbulent motions in the flow.

Given the above discussion, the origin of the HVCs remains quite uncertain, so that different processes need to be examined. Some of the observed high velocity features may have an extragalactic origin (Bajaja *et al.* 1989); a "photolevitation" mechanism by which diffuse clouds are raised at high Galactic latitudes has been suggested by Franco, Ferrini, Ferrara and Barsella (1991); finally, ballistically moving dense blobs of cool material may be created as a result of the supershells fragmentation. The last two processes may provide a suitable explanation of the low latitude HVCs.

REFERENCES

Bajaja, E., Cappa de Nicolau, C.E., Martin, M.C., Morras, R., Olano, C.A. and Pöppel W.G.L. 1989, Astron. Astrophys. Suppl. Ser., 78, 345

Bregman, J.N. 1980, Ap.J., 236, 577

Danly, L., 1991, in *The Interstellar Disk-Halo Connection in Galaxies*, IAU Symp. 144, ed. H. Bloemen, (Reidel: Dordrecht), p.53

Einaudi, G. and Ferrara, A. 1991, Ap.J., 371, 571

Ferrara, A. and Einaudi, G. 1991, Ap.J., submitted

Field, G.B. 1965, Ap.J., 142, 531

Franco, J., Ferrini, F., Ferrara, A., and Barsella, B. 1991, Ap. J., 336, 443

Lockman, F.J., 1991, in *The Interstellar Disk-Halo Connection in Galaxies*, IAU Symp. 144, ed. H. Bloemen, (Reidel: Dordrecht), p.37

Marshall, F.J. and Clark, G.W. 1984, Ap.J., 287, 633

Nousek, J.A., Fried, P.M., Sanders, W.T., and Krushaar, W.L. 1982, Ap.J., 258, 83

Savage, B.D. and Massa, D. 1987, Ap.J., 314, 380

Shapiro, P.R. and Field, G.B. 1976, Ap.J., 205, 762

van Woerden, H., Schwarz, U.J. and Hulbosch, A.N.M. 1985, in IAU Symposium 106,The Milky Way Galaxy, eds. H. van Woerden, R.J.Allen and W.B.Burton, (Dordrecht:Reidel), p.387

Magnetic Field Geometry Towards the High Velocity Clouds (HVCs)

A.I. GÓMEZ DE CASTRO[1,2]

R.E. PUDRITZ[2], P. BASTIEN[3]

[1]IUE Observatory - VILSPA, [2]McMaster University, [3]Université de Montreal

1 INTRODUCTION

The origin of the High Velocity Clouds is still controversial. The most widely accepted theory proposes that the HVCs are cool clouds that rain back onto the galactic plane (Bregman, 1980) having been formed from the hot material ejected into the Galactic Halo by explosive events or hot winds. However, as suggested by Ferrara in these proceedings, there may be some difficulties in cooling the ejected material in the tenuous halo. In this contribution we present an alternative view, namely, that the HVCs represent gas that was initially driven out of the galactic plane by the magnetic buoyancy associated with a Parker Instability, and is currently falling back to the plane along these magnetic field lines. A thorough analysis of the data available in the literature suggests that there is a connection between the local magnetic field, the HVCs and Gould's Belt (a local dynamical entity that contains a large percent of the young material, OB stars, gas, molecular clouds, in a radius of 500 pc around the Sun). This connection is suggestive of a Parker Instability operating locally. A detailed description of all these features is deferred to a further paper (Gómez de Castro and Pudritz, in preparation).

In this contribution, we report the preliminary results of an ongoing observational campaign to investigate the magnetic field geometry towards the high velocity clouds. Our purpose is to investigate a possible connection between the local magnetic field and the HVCs stream falling (in projection) over Orion.

2. A CONNECTION BETWEEN THE HVC'S AND GOULD'S BELT

The OB stars, molecular clouds and young stellar associations in a radius of $\simeq 500$ pc around the Sun are distributed in two planes, the galactic plane and Gould's Belt. Gould's Belt is tilted respect to the galactic plane, $\simeq 20^{o}$, and the objects that constitute it, also share a peculiar kinematics whose origin is not well understood. Neither streaming (Eggen, 1989) nor expansion from a central source (Lesh, 1968; Frogel and Stothers, 1978; Olano, 1982) are able to account for the complex kinematics of the system (see Westin, 1985).

The intersection of Gould's Belt, as defined from the OB stars, and the Galactic Plane is given by the line of nodes. This line coincides approximately with the galactic longitude axis $l_{II} = 115° - 295°$ (Stothers and Frogel, 1974). Therefore the vertex line is approximately defined by the galatic longitude axis: $l_{II} = 25° - 205°$. This axis agrees reasonably well with that defined by the distribution of molecular clouds (Taylor et al., 1987).

The geometry of the local magnetic field as described by Mathewson (1968) is displayed in Figure 1. The data are based on optical polarization observations of 1800 stars. Note that the field shows a very symmetric structure with respect to the axis $l_{II} = 25° - 205°$, defined by the vertex line. It is interesting to note that the measurements of the line-of-sight field by means of the Faraday rotation towards nearby pulsars, also suggests this symmetry (Rand and Kulkarni, 1989).

Let us now compare these data with the chain of H I clouds falling onto the anticenter region. As shown in Figure 2 (taken from Franco et al., 1988), the stream is falling (in projection) onto the Orion-Monoceros complex. In fact, there is direct evidence of a shock occurring at the location where the chain impinges upon the complex (Mirabel, 1982). Note that the stream runs almost parallel to the local magnetic field direction as indicated in Mathewson's (1968) map.

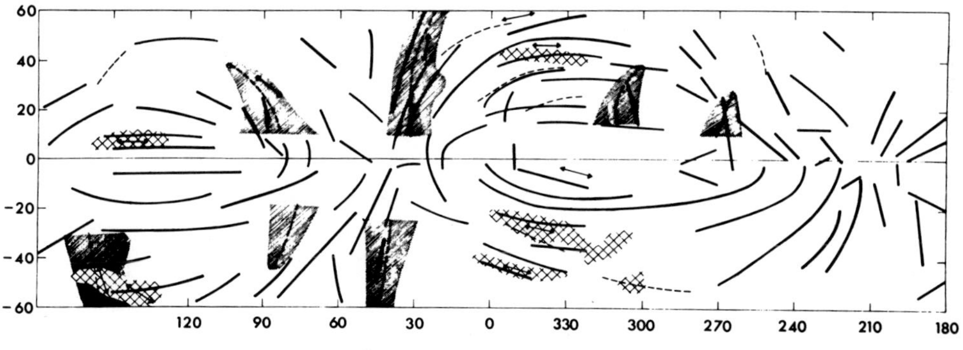

Figure 1. Geometry of the local magnetic field as inferred from optical polarization vectors (from Mathewson, 1968). The plot is in galactic coordinates. The heavy black lines are a schematic representation of the magnetic field geometry.

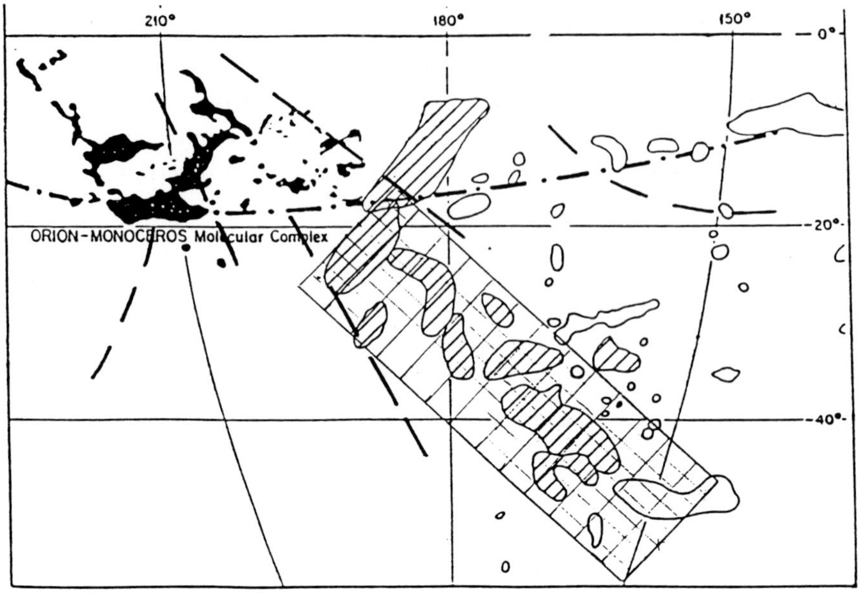

Figure 2. Map of the Orion region (from Franco et al., 1988). The Orion-Monoceros cloud complexes are shown as black filled structures. H I high velocity clouds with $V_{LSR} \leq -150$ km/s are filled with diagonal lines, and the HVC's with -80 km/s $\geq V_{LSR} \geq -150$ km/s are delineated by thin solid lines. The thick dot-dashed line shows the approximate location of Gould's Belt (from Stothers and FRogel, 1974). The approximate direction of the local magnetic field is indicated by thick dashed lines (from Mathewson, 1968). The grid we are covering with the polarization measurements is also sketched.

3. THE PARKER INSTABILITY

The connection between the magnetic field and the HVCs strongly suggests that a Jeans-Parker Instability may be acting locally. Measurements indicate that magnetic and gravitational energies are comparable in the local interstellar medium. The instability pushes some of the cool gas from the Galactic Plane up into the halo the driving agent being magnetic buoyancy. The gas eventually, falls back along the field lines due to the action of gravity. Theoretical models (Blitz and Shu, 1980; Elmegreen, 1982) show that molecular cloud complexes can form as a consequence of Parker instability in the magnetized HI disk of the galaxy. Gas in the magnetic "valleys" becomes self gravitating initiating the formation of molecular clouds.

We employ the usual linear instability analysis for Parker instability in a magnetic gas layer containing cosmic rays. The growth time of the instability is $\simeq 4.3 \times 10^7$ years, and the scale, $\simeq 1.2$ Kpc (Elmegreen, 1982). These values agree with the basic parameters of the Gould's Belt. The age estimated for the constituents is $\leq 6 \times 10^7$ years (Westin, 1985) and the size of the region $\simeq 1$ Kpc. Also, the dynamical time scales inferred from expansion/streaming models of the kinematic agree with this estimate, $T_{exp} \simeq 5-6 \times 10^7$ years. Moreover, the magnetic field is very buoyant, with loops rising up to 2 kpc. The velocity of material in free-fall from this scale heights is $\simeq 200$ km/s, which is approximately what is observed in most of the HVCs.

Recent numerical computations of the non-linear growth of the odd mode of Parker instability, show that slight inclinations (up to 20°) of the gas layer are also expected to occur, as it is the case of the Gould's Belt (see Figure 3 from Matsumoto et al., 1989).

Figure 3. Numerical simulation of the non-linear growth of the odd mode of the Parker instability (from Matsumoto et al., 1989). The thin lines represent the field configuration, the arrows the gas motions, and the thick dotted area the "corrugated" gas layer.

4. THE EXPERIMENT

If the HVCs represent gas that is magnetically dominated and is raining back to the galactic plane following the field lines, then there should be a relation between the geometry of the magnetic field and the "streaming" direction of the HVCs. One might expect that such a correlation should be present if the instability is operating. In order to test this idea we have selected one of the most outstanding chains of HVCs, the complex that falls over Orion (see Figure 2). The chain is approximately aligned in the West-East direction and subtends 45° in the sky.

The geometry of the magnetic field is defined by the direction of the optical polarization vector. The polarization is assumed to be due to dichroic extinction of the radiation from unpolarized background sources, produced by dust grains aligned with the magnetic field. The observations were carried out with a photopolarimeter at the 1.5 m Mount Megantic Telescope (Canada) and at the 1.23 m Calar Alto Telescope (Almería, Spain). For the time being we have only covered a 10% of the complex but the preliminary results are extremely interesting.

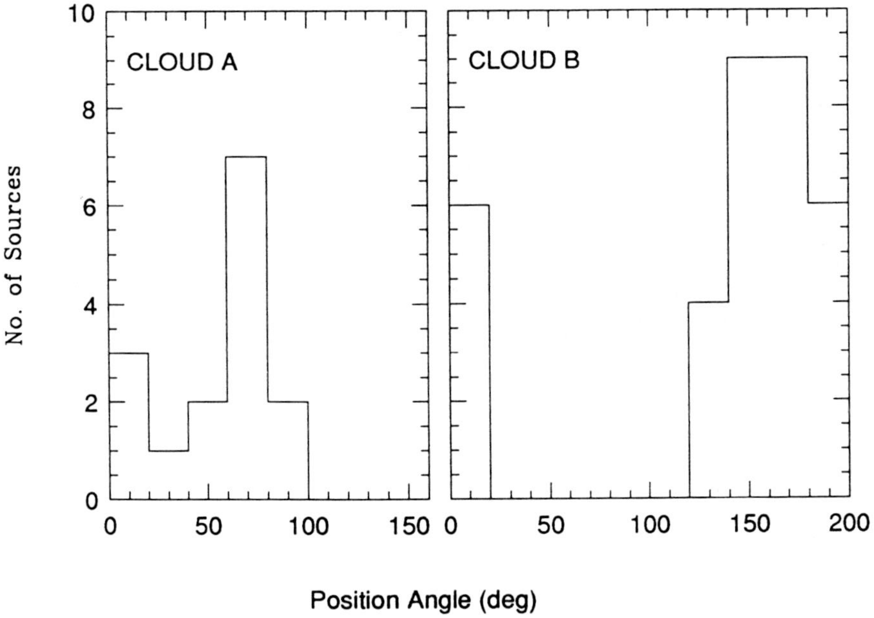

Figure 4. Histograms showing the predominant orientation of the magnetic field in the direction to cloud A (close to the galactic plane) and cloud B (high galactic latitudes).

In Figure 4 we present a histogram of the distribution of the polarization vectors in cloud A (part of the stream close to galactic plane) and cloud B (high galactic latitude stream). The distribution is bimodal. The peaks correspond to a magnetic field geometry that is essentially parallel to the stream close to the plane and perpendicular to it up in the galactic halo. The observed alignment is especially remarkable when one realizes that the typical polarizations we are measuring are $\simeq .6\%$. This suggests that while the material close to the plane is falling along the magnetic field, this may not be the case for high galactic latitudes. It is interesting to note that the largest velocity gas, with $v \simeq 200 - 300$ km/s is concentrated at these high latitudes.

Acknowledgements: We thank K. Patel for her help during the observing campaign at Mont Mégantic. A. I. Gómez de Castro wants to thank the Ministery of Education of Spain for a fellowship of "reincorporación". This work was partially supported by NSERC of Canada. We thank the staffs at Mont Megantic and Calar Alto for their assistence with the observations.

5. REFERENCES

Blitz, L., Shu, F.H., 1980; *Ap.J.*, **238**, 148

Bregman, J.N., 1980; *Ap.J.*, **236**, 577

Eggen, O.J., 1989; *Fund. Cosmic Phys.*, **13**, 1

Elmegreen, B.G., 1982; *Ap.J.*, **253**, 655

Franco, J., Tenorio-Tagle, G., Bodenheimer, P., Różyczka, M., Mirabel, I.F., 1988; *Ap.J.*, **333**, 826

Frogel, J.A., Stothers, R., 1977; *A.J.*, **82**, 890

Giovanelli, R., 1980; *A.J.*, **85**, 1155

Lesh, J.R., 1968; *Ap. J. Supp. Ser.*, **17**, 371

Mathewson, D.S., 1968; *Ap. J. Lett.*, **153**, L47

Matsumoto, R., Horiuchi, T., Hanawa, T., and Shibata, K., 1989; *Pub. Astr. Soc. Japan*, **40**, 171

Mirabel, I.F., 1982; *Ap. J.*, **256**, 120

Olano, C.A., 1982; *A. Ap.*, **112**, 195

Rand, R.J., Kulkarni, S.R., 1989; *Ap.J.*, **343**, 760

Stothers, R., Frogel, J.A., 1974; *A.J.*, **79**, 456

Taylor, D.K., Dickman, R.L., Scoville, N.Z., 1987; *Ap. J.*, **315**, 104

Westin, R.N.G., 1985; *A. Ap. Supp. Ser*, **60**, 99

The Collision of Large High Velocity Clouds with the Galactic Disk

F. COMERÓN

Departament d'Astronomia i Meteorologia, Universitat de Barcelona - Spain

Observations of HI complexes in the halo of our Galaxy falling towards the galactic plane imply that the disk- halo interaction sometimes takes place in a violent way. Large structures observed in ours and other galaxies have been interpreted as consequences of HVC impacts on the galactic disk (Mirabel, 1982; Heiles, 1984; Franco et al., 1988; Tenorio-Tagle and Bodenheimer, 1988).

Sizes and masses of HVCs are very poorly known as a consequence of the lack of precise distance determinations. Nevertheless, rather massive clouds would be required to form the largest observed HI supershells, and the shock structure in the galactic anticenter reported by Mirabel (1982) may actually involve cloud masses in the range 10^5 - 10^6 $M_\odot$. A gigantic HI complex has been observed by van der Hulst and Sancisi (1988) projected on the disk of M 101 with a mass of up to 10^8 $M_\odot$.

We have studied the consequences of a large HVC impact on a galactic disk mostly filled by the warm, neutral ISM. Tenorio-Tagle (1981), Tenorio-Tagle et al. (1987) and Franco et al. (1988) have discussed the case of a front collision. To study a more general case, we have included in our numerical simulations a velocity component parallel to the galactic plane, thus obtaining an interesting, qualitatively different final distribution of gas in the affected zone. As in the case of the front collision, a strongly shocked layer develops in the interface between the cloud and the unperturbed galactic disk. The shocked layer rapidly evolves to the cool, dense phase and a region appears where further star formation is likely. In the case of an oblique collision, the velocity component parallel to the galactic disk causes a significant stream of shocked material flowing out of the region encomprised between the shock fronts propagating into the cloud and the galactic disk. This material finally remains distributed roughly along the path followed by the infalling cloud, thus giving rise to a dense layer of material tilted with respect to the galactic plane. In our case of a cloud with $M = 10^6$ $M_\odot$ and a characteristic length of 500 pc, this dense layer can be several hundred parsecs in diameter.

We find a remarkable agreement between our model results and the distribution of early type stars and molecular clouds in the solar neighbourhood, which tend to be concentrated along the Gould Belt (see e.g. Lindblad, 1974; Westin, 1985; Dame et al., 1987). This supports the idea that the HVC- galactic disk collision was the mechanism responsible for the Gould Belt formation (Gomez de Castro, elsewhere in this volume). Further work on the characteristics, kinematics and evolution of the resulting structures in the ISM is currently in progress.

Acknowledgements: I am pleased to acknowledge long and fruitful discussions with Drs. J. Torra and A.E. Gomez. This work was begun with the support of the CIRIT and is currently supported by the CICYT under contract ESP88-0731 and Acciones Integradas Hispano- Francesas.

REFERENCES

Dame, T.M., Ungerechts, H., Cohen, R.S., de Geus, E.J., Grenier, I.A., May, J., Murphy, D.C., Nyman, L.Aa., and Thaddeus, P.: 1987, *Ap. J.* **322**, 706.

Franco, J., Tenorio-Tagle, G., Bodenheimer, P., Rozyczka, M., and Mirabel, I.F.: 1988, *Ap. J.* **333**, 826.

Heiles, C.: 1984, *Ap. J. Supp. Ser.* **55**, 585.

van der Hulst, T., Sancisi, R.: 1988, *Astron. J.* **95**, 1354.

Lindblad, P.O.: 1974, in *Stars and the Milky Way System*, ed. L.N. Mavridis, Springer- Verlag.

Mirabel, I.F.: 1982, *Ap. J.* **256**, 112.

Tenorio-Tagle, G.: 1981, *Astron. Astrophys.* **94**, 338.

Tenorio-Tagle, G., Bodenheimer, P.: 1988, *Ann. Rev. Astron. Astrophys.* **26**, 145.

Tenorio-Tagle, G., Franco, J., Bodenheimer, P., and Rozyczka, M.: 1987, *Astron. Astrophys.* **179**, 219.

Westin, T.N.G.: 1985, *Astron. Astrophys. Supp. Ser.* **60**, 99.

The Effect of Cosmic Rays on Superbubble Dynamics

A.M. BYKOV

A.F. Ioffe Institute of Physics and Technology, SU 194021, Leningrad, USSR

ABSTRACT

We study the generation of nonthermal particles in an OB-association. A model of CR acceleration is proposed which includes the interaction of charged particles with an ensemble of shocks and large scale subsonic plasma motions inside the bubble. The CR acceleration appears to be very efficient, especially after explosions of a few supernovae in the association. The CR acceleration time is of the order of 3×10^5 yr. Some tens of percent of the bubble kinetic energy can be converted to the CR acceleration.

INTRODUCTION

Stellar winds from OB stars and supernova explosions provide the dominant sources of energy, momentum, and ionizing radiation in the interstellar medium (ISM). Young massive stars which produce strong winds and explode as type II supernovae are known to be spatially and temporally correlated in the OB associations. These associations are of special interest as they could play an important role in the formation of HI shells and supershells. They can regulate the mass, energy, and magnetic field input to the halo of the Galaxy (see, e.g., Heiles 1990). Studies of superbubbles supported by stellar winds and multiple supernova explosions in an OB association have received considerable attention (see Tenorio-Tagle and Bodenheimer 1988, for review).

As shown by the two-dimensional hydrodynamical simulations of Mac Low *et al.* (1989), superbubbles can break through the disk to the halo on a time scale $\sim 10^7$ yr. On the other hand, three-dimensional adiabatic magnetohydrodynamical simulations by Tomisaka (1990), in which the effect of the disk regular magnetic field has been included, reveal that the magnetic field can suppress the flow in the halo. The disk break can occur only if the 'mechanical luminosity' exceeds some critical value which depends on the disk magnetic field.

The studies mentioned above ignore the effect of nonthermal particles, which should be generated inside the bubbles by electromagnetic field fluctuations associated with large-scale plasma motions. Here we discuss acceleration of nonthermal, charged particles by an ensemble of large-scale plasma motions, taking into account the shock waves generated by OB stellar winds and supernovae.

Various models of CR acceleration by terminal shocks of young stars have been discussed (see, e.g., the reviews by Axford (1987), Cesarsky and Montmerle (1983), and Völk (1983)) in connection with the problem of gamma-ray sources. A distinctive feature of our model is that it includes the effect of large-scale motions which generate hard-spectra particles up to GeV-energies.

PARTICLE ACCELERATION MODEL

The most massive O-stars begin to explode as type II supernovae in $(1 - 3) \times 10^6$ yr after an OB-association is formed. The shocks created by the first supernova explosions interact with stellar winds of OB stars and with one another. As a result, strong supersonic turbulence is created which consists of smooth large-scale motions of hot (10^6 K), tenuous ($n \sim 10^{-2} - 10^{-3}$ cm^{-3}) plasma with an ensemble of discontinuities. The basic energy-containing scale of the turbulence is approximately the mean distance between strong shocks (of the order of 3 - 10 pc). The mean square amplitude of the velocity fluctuations u in the scale l is about $(1 - 3) \times 10^8$ cm s^{-1}.

The evolution of the magnetic field in a superbubble is quite complicated. It is affected by two basic processes. The first one is a relatively rapid generation of fluctuating fields, with scales smaller than l, by the turbulent motions. The energy density of these fields can reach a substantial fraction of the mechanical energy density (we base our estimates on $< B > \sim 3 \times 10^{-5}$ G). The second effect is that it sweeps up the regular ISM magnetic field from the superbubble on longer time scales, of the order of 10^7 yr.

The system is surrounded by a relatively dense and massive HI shell with a possibly enhanced magnetic field. The shell slowly expands into the ambient ISM, at velocities of tens km s^{-1}.

Now let us analyse the evolution of nonthermal charged particles - low energy CR's - in the superbubble interior where violent large-scale motions (as described above) are available. The CR distribution has an intermittent structure because of the presence of strong shocks. The kinetic equation for the CR distribution function, with account taken for the intermittency, has been obtained by Bykov and Toptygin (1990) (hereafter BT). The regimes of particle acceleration by an ensemble of shocks are determined by the dimensionless parameter

$$\psi = \frac{ul}{\kappa(p)}, \tag{1}$$

where $\kappa(p)$ is the longitudinal diffusion coefficient (with respect to the regular magnetic field) produced by small-scale resonant magnetic inhomogeneities.

If $\psi > 1$, the particle distribution function $N(\boldsymbol{r}, p, t)$ averaged over the ensemble of turbulent motions satisfies the transport equation (see BT):

$$\frac{\partial N}{\partial t} + \frac{N}{\tau_e} = \left(\frac{1}{\tau_{sh}} + B\right)\hat{L}N + \frac{1}{p^2}\frac{\partial}{\partial p}p^4 D\frac{\partial N}{\partial p} + A\hat{L}^2 N + 2B\hat{L}\hat{P}N + Q(p), \tag{2}$$

where $\hat{L}$ and $\hat{P}$ are the operators given by

$$\hat{L} = \frac{1}{3p^2}\frac{\partial}{\partial p}p^{3-\alpha}\int_0^p \mathrm{d}p'\, p'^{\alpha}\frac{\partial}{\partial p'}, \quad \hat{P} = \frac{p}{3}\frac{\partial}{\partial p}. \tag{3}$$

In this case $\alpha = 3\Delta/(\Delta - 1)$ is the power-law exponent of the 'universal' CR spectrum on an individual strong shock front, and Δ is the shock compression ratio. The kinetic coefficients A, B, D, τ_{sh}, and τ_e are expressed through the spectral functions which describe the correlations between the parameters of large-scale turbulent motions and of the shock ensemble (see BT for details). Particle adiabatic deceleration due to shell expansion and synchrotron losses for electrons (energies less than 10 GeV) can be neglected.

Let us suggest that the injection of suprathermal particles takes place at the shock fronts. Energetic particles, as shown by Ellison and Eichler (1984), are naturally extracted from the thermal pool with high efficiency by the same scattering process which is responsible for the formation of collisionless shock structure. This suggestion is in good quantitative agreement with observations (see the review by Blandford and Eichler, 1987).

In our case, the source function in Eq.(2) can be expressed through the quasi-stationary production rate of suprathermal particles q. The latter can be calculated by averaging the 'microscopic' shock injection rate over the shock ensemble:

$$q = < \eta n u_{sh} \Sigma_{sh} V_{sb}^{-1} >, \tag{4}$$

where η is a fraction of particles incident on the shock surface (Σ_{sh}) to be injected into the acceleration, and V_{sb} is the system volume.

$$Q(p) = (\alpha - 3) q \frac{p_i^{\alpha-3}}{p^{\alpha}} \theta(p - p_i). \tag{5}$$

Another important parameter is p_i, the momentum of injected particles whose typical value is likely to be 'several' times larger than the momentum of the downstream thermal particles ($p_i > mu$) (see, e.g., Drury *et al.* (1989)).

A solution of Eqs.(2), (4) for the particle distribution function ($p_i < p < p_{max}$) inside the superbubble with scale $R_{sb} \gg l$ can be written as

$$N(p,t) \approx \frac{(\alpha-3)n\eta}{9p_i^3} \begin{cases} \left(\frac{p_i}{p}\right)^3 \left[1 - \frac{\exp(-\sigma(t+t_p))}{1-[3\alpha/(\alpha-3)(2\alpha-3)]}\right], & p < p_*, \\ \left(\frac{p_i}{p}\right)^{\alpha} \exp\left[\frac{t}{\tau_{sh}}\right], & p_* < p < p_{max}, \end{cases} \tag{6}$$

where $p_* = p_i \exp[(t/\tau_{sh})/(\alpha - 3)]$, $t_p = \alpha\tau_{sh} \log(p/p_i).\sigma = 3/(\tau_{sh}(2\alpha - 3))$.

Note that the maximum momentum p_{max} of the spectrum is determined by the condition $\psi(p_{max}) \sim 1$. The equation for the distribution function at $p > p_{max}$ has been obtained by Bykov and Toptygin (1979).

If the shocks are strong, we have $\alpha \approx 4$, $\tau_{sh} \approx l/(1 - 3) \times 10^4$ yr. In this case typical energies of the injected protons are of the order of 20 keV; corresponding energies of the electrons are probably lower. For $t \sim 10^6$ yr p_* becomes as high as p_{max}, and the hard particle spectrum extends up to p_{max}. Then cp_{max} becomes $\sim$ (1 - 10) GeV, depending on the spectrum of the small-scale magnetic field.

The principal parameter of the CR distribution given by Eq.(5) is the injection efficiency η. From direct observations of protons accelerated at shocks in the heliosphere one can estimate $\eta_p \sim 10^{-4}$ (see, e.g., Blandford and Eichler 1987). This value seems to be rather high for the superbubbles since in this case the CR energy density $\eta n E(p_{max})$ can greatly exceed the kinetic energy density. It is important that η depends on the energy density of CR and can be reduced due to shock front mediation by CR. A reasonable estimate for the proton injection rate seems to be $\eta_p \sim 10^{-6}$. In this case the CR energy density is of the order of the system kinetic energy density. This can be tested by the observation of the shells γ -ray emissivity.

Using our model, we can calculate the injection of electrons. As follows from the catalogue of radio sources in the LMC observed by McGee *et al.* (1978) at 3.4 cm, many extended HII regions (which can be connected with OB associations) have spectral indices close to zero and radiofluxes of the order of Jy. It is easy to show that the thermal radio emission of hot tenuous gas inside the superbubble cannot explain the observed radio fluxes (with account for the X-ray emission, see Chu and Mac Low 1990). However, some contribution to the radio emission may come from the shell and clouds. Let us assume that the fluxes are produced by the synchrotron emission of nonthermal electrons. The electrons whose spectrum ($\propto p^{-3}$) is given by Eq.(6) would generate radio emission with zero spectral index. The radio flux index corresponding to the electron spectrum $\propto p^{-\alpha}$ is close to -0.5. From the standard theory of synchrotron emission we obtain $\eta_e < 10^{-7}$ for the electrons ($B = 3 \times 10^{-5}$ G). In this case, the energy density of the nonthermal electrons appears to be a few percent of the kinetic energy density of the system and a few percent of the CR energy density (determined by protons). For a supernova exploded in the medium without pre-generated CR, the injection rate can be substantially high: $\eta_e > 10^{-6}$.

Ionization of HI shells by CR ,γ - emissivity of superbubbles, and the effect of CR on the bubble evolution will discussed elswhere.

CONCLUSIONS

(i) We have shown that a hard spectrum of low energy CR should be generated if the standard scenario of the evolution of OB associations is adopted.

(ii) CRs gets a significant fraction of the total kinetic energy; this should affect the dynamics of superbubbles.

(iii) The suppression of the injection rates near the shocks under the effect of pre-generated low energy CR can be the reason why the OB associations are not highly luminous sources of nonthermal electromagnetic emission.

Aknowledgements. The author is indebted to G.D. Fleishman and I.N.Toptygin for helpful discussions and to J.Palous for the invitation to the Prague CTC Workshop.

REFERENCES

Axford, W.I. 1987, *Proc. 20th ICRC (Moscow)* **8**, 120.
Blandford, R., Eichler, D. 1987, *Phys. Rep.* **154**, 2.
Bloemen, H. 1989, *Ann. Rev. Astron. Astrophys.* **27**, 469.
Bykov, A.M., Toptygin, I.N. 1979, *Proc. 16th ICRC (Kyoto)* **2**, 66.
Bykov, A.M., Toptygin, I.N. 1990, (BT) *Sov. Phys. - JETP* **71**, 702.
Cesarsky,C.J., Montmerle, T. 1983, *Space Sci. Rev.* **36**, 173.
Chu, Y.-H., Mac Low, M.M. 1990, *Astrophys. J.* **365**, 510.

Drury, L.O'C., Markiewicz, W.J., and Völk, H.J. 1989, *Astron. Astrophys.* **225**, 179.
Ellison,D.C., Eichler,D. 1984, *Astrophys. J.* **286**, 691.
Heiles, C. 1990. *Astrophys. J.* **354**, 483.
Mac Low, M.M., McCray, R., and Norman, M.L. 1989, *Astrophys.J.* **337**, 141.
McGee, R.X., Newton, L.M., and Butler, P.W. 1978, *Mon. Not. Roy. Astr. Soc.* **183**, 799.
Tenorio-Tagle, G., Bodenheimer, P. 1988, *Ann. Rev. Astron. Astrophys.* **26**, 145.
Tomisaka, K. 1990, *Astrophys. J.* **361**, L5.
Völk, H.J. 1983, *Space Sci. Rev.* **36**, 3.

Evolution of SNR's with Cosmic Rays

E.A. DORFI

Institut für Astronomie, Türkenschanzstr. 17, A–1180 Wien, Austria

INTRODUCTION

It is known that cosmic rays have a considerable astrophysical significance for various interstellar processes (e.g. the development of a Parker-type instability, heating and ionization of interstellar clouds, driving of galactic winds, production of γ-ray photons above 100 MeV by decay of neutral pions, nonlinear modifications of shock waves, etc.). Moreover, the interstellar space is filled with high energy particles having a typical energy density of about $1\,\mathrm{eV\,cm^{-3}}$. Most of the energy is carried by particles in the range from 10^{10}eV/nuc up to 10^{15}eV/nuc where the particle distribution function $f(T)$ exhibits a simple power law in kinetic energy T with a differential spectral index of $\alpha \sim 2.65$. At the upper end of this energy spectrum we observe a break towards a slightly steeper spectral index indicating a possible extragalactic component of cosmic rays at energies higher that about 10^{15}eV/nuc. The biannual proceeding of the *International Cosmic Ray Conference* (and references therein) can serve as a detailled up-to-date source of information on all these aspects.

It is commonly accepted that the galactic component of the cosmic rays is accelerated in shock waves originating from supernova explosions (e.g. Blandford 1988, Drury et al. 1989). These charged particles can gain energy in a first-order Fermi process (Axford et al. 1977, Krymsky 1977, Bell 1978a,b, Blandford & Ostriker 1978) which can be modelled by an additional energy equation for the cosmic rays if averaged over the particle distribution function (e.g. Drury 1983). In order to use the appropriate physics the time variation of the effective adiabatic index γ_c of the cosmic rays as well as the mean diffusion coefficient $\bar{\kappa}$ have to be estimated during the evolution of a SNR (e.g. Drury et al. 1989). The presented models of SNR's include the process of particle acceleration in shock waves, the injection of suprathermal particles, heating by dissipation of Alfvén waves in regions of large cosmic ray gradients and radiative cooling of the thermal plasma. The resulting system of nonlinear PDE's is solved fully implicit on an adaptive grid to ensure the necessary numerical accuracy during the computions (Dorfi 1990, 1991).

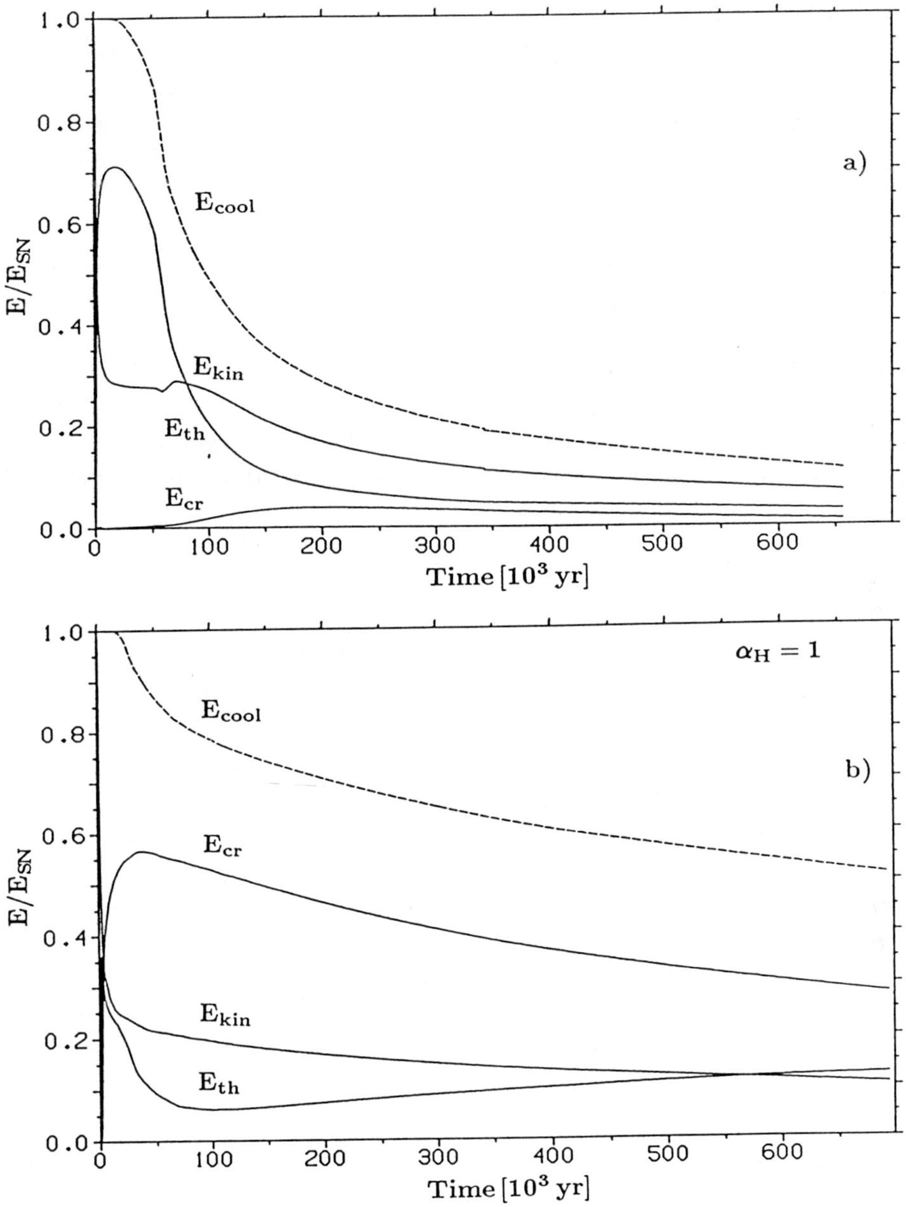

Figure 1: The temporal evolution of the total energies as a function of time in units of the initial SN explosion energy E_{SN} in the case a) of low cosmic ray production and b) for a higher conversion of E_{SN} into E_{cr}.

GLOBAL ENERGETICS

In this short contribution we describe basically two SNR models with different efficiencies for the conversion of E_{SN} into cosmic rays depending on physical parameters like the injection rate and the initial magnetic field. The supernova explosion energy is set to $E_{SN} = 10^{51}$erg, the external density of $n_0 = 0.3\,\mathrm{cm}^{-3}$ and the temperature of $T_{ext} = 8000\,\mathrm{K}$ correspond to the warm interstellar medium.

The time evolution of the different energy contributions is shown on a linear scale up to $7\,10^5$yr in Fig. 1. The case depicted in Fig. 1a is characterized by a low conversion efficiency from the explosion energy E_{SN} to the total cosmic ray energy E_{cr} and E_{cr} stays below $0.04\,E_{SN}$ during the entire evolution. We emphasize the importance of radiative cooling where the most of the SNR thermal energy content is radiated away after about $5\,10^4$yr and E_{cool} denotes the remaining SNR energy. Fig. 1b exhibits the temporal dependence of the energies in the case of a more efficient cosmic ray production. This model includes also heating of Alfvén waves in the precursor region of the forward shock (written like $\alpha_H v_A|\nabla P_c|$, $\alpha_H = 1$) which yields an increase of the thermal energy E_{th} although radiative cooling decreases the total energy of the remnant.

RADIAL STRUCTURES AND GAMMA-RAY LUMINOSITIES

Radiative cooling alters also the radial structure in the cooling region behind the forward shock and can lead to the development of so-called secondary shocks (Falle 1975, 1981). In the case of energetic particles we have an additional pressure which limits catastrophic cooling and allows a smooth expansion of the forward shock (cf. Fig. 3b). In Fig. 2 we present the density a), the velocity b), the cosmic ray energy c) and the gas temperature d) in the shock frame. We want to emphasize the density enhancement due to cooling and the occurrence of complicated structures in the downstream region of the shock wave which yield also fluctuations in the collisional production of neutral pions and the associated γ-ray fluxes (Fig. 3). However, in all computations carried out the compression ratio r in the radiative shoch wave remains below $r < 70$ limited by the downstream cosmic ray pressure.

Collisions of cosmic rays with the thermal particles lead to the production of neutral pions and these π^0's decay into observable γ-ray photons above 100 MeV. Adopting the simple formula of Higdon & Lingenfelter (1975) we can convert the energy density of the cosmic rays into a γ-ray flux integrated over the entire remnant. The results are presented in Fig. 3 where the three curves label the background γ-ray flux $F_{\gamma,b}$, the γ-ray flux $F_{\gamma,s}$ integrated up to the shock radius R_s and the flux $F_{\gamma,p}$ contained within the precursor radius R_p. We can clearly distinguish three reverse shocks reducing the γ-ray luminosity and at the time of $t = 4.05\,10^{11}$s a secondary shock triggered

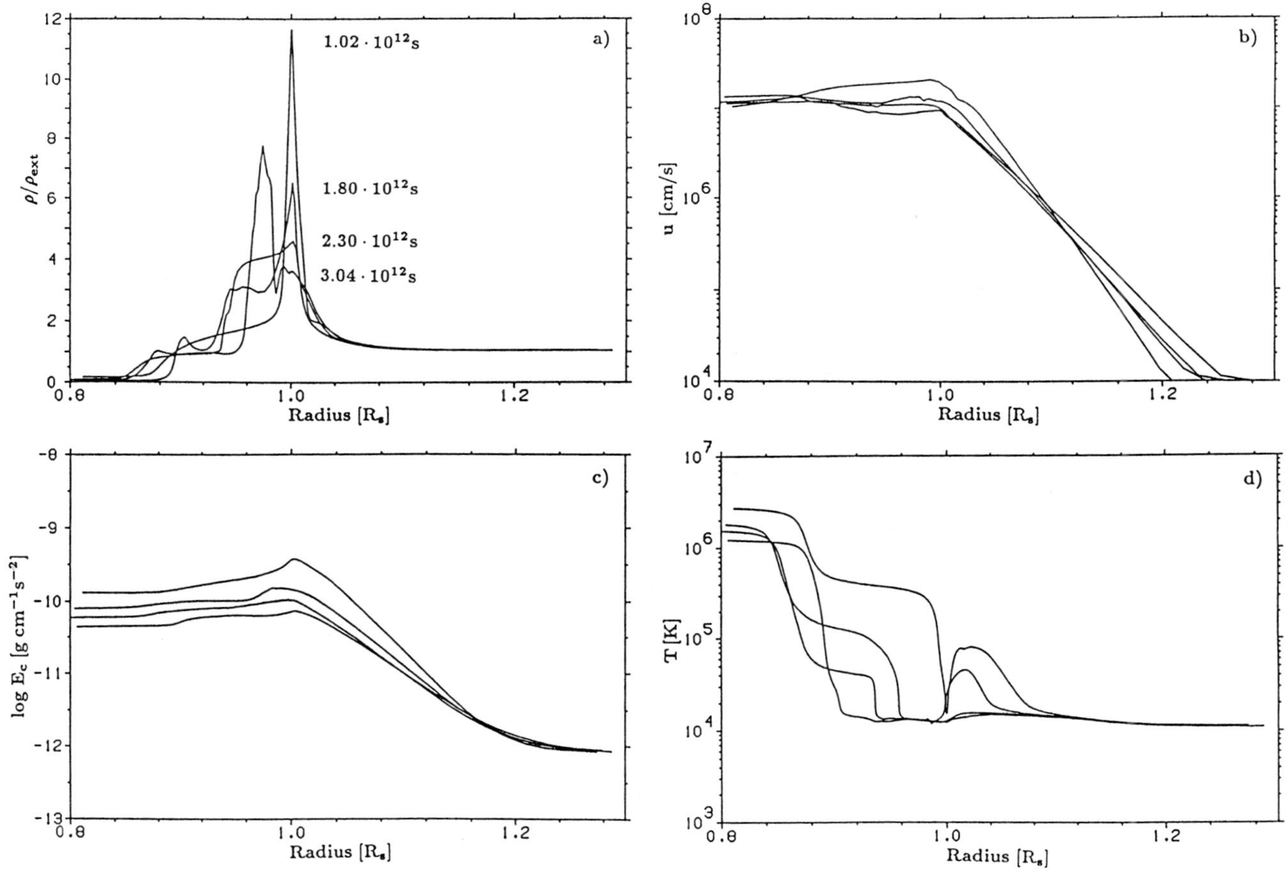

Figure 2: The radial structure of the physical variables in the shock frame between 10^{12}s and $3\,10^{12}$s showing the development of downstream features due to radiative cooling.

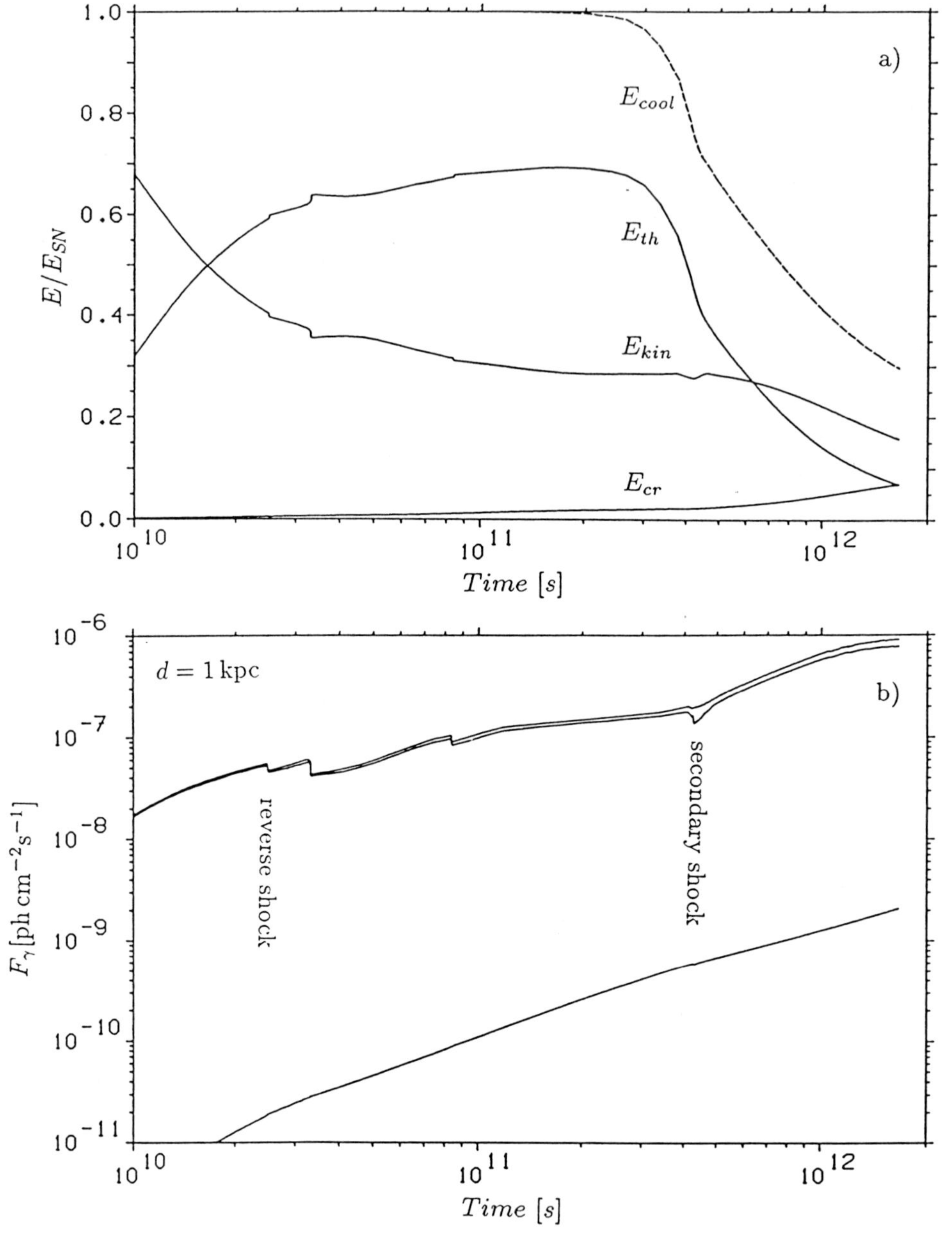

Figure 3: The time evolution of the total energies in units of $E_{SN} = 10^{51}$erg and the γ-ray fluxes for an external density of $n_0 = 5\,\mathrm{cm}^{-3}$ normalized to a source distance of $d = 1\,\mathrm{kpc}$.

by radiative cooling decreases the cosmic ray energy density in the vicinity of the forward shock yielding a lower γ-ray flux $F_{\gamma,s}$. Note that in this case of an ambient density of $n_0 = 5\,\mathrm{cm}^{-3}$ the γ-ray flux reaches about $10^{-6}\mathrm{ph\,cm^{-2}s^{-1}}$ although the total cosmic ray energy remains at about $E_{cr} \leq 0.07\,E_{SN}$.

The shock radius $R_s(t)$ is a smooth function of time as can be inferred from the lowest curve in Fig. 3b, and we find that radative cooling does not affect the motion of the shock front so drastic compared to simulations without cosmic rays (Falle 1975, 1981). The γ-ray fluxes from π^0-decay yield $10^{-10}\,\mathrm{ph\,cm^{-2}s^{-1}}$ for a typical remnant ($n_0 = 0.3\,\mathrm{cm}^{-3}$) with a small amount of cosmic ray energy at a time of 300 years and increase almost like R_s^3 due to geometric effects. Remnants with higher fractions of cosmic rays or remnants evolving into a medium of higher ambient density produce γ-ray fluxes of up to about $10^{-6}\,\mathrm{ph\,cm^{-2}s^{-1}}$ (normalized to a SNR at a distance of 1 kpc). In the latter case the γ-ray fluxes are almost constant between 10^4 and 10^6 years.

CONCLUSIONS

In summary we want to emphasize that radiative cooling as well as Alfvénic heating are important to determine the amout of E_{SN} which can be converted into cosmic rays. Typical values are in the range between $0.1\,E_{SN}$ and $0.3\,E_{SN}$ and can explain the power needed to maintain the observed cosmic ray energy density. Radiative cooling and acceleration of cosmic rays in shock waves need accurate numerical methods to resolve the physical length scales in order to obtain correct results.

REFERENCES

Axford, W.I., Leer, E., Skadron, G.: 1977, Proc. 15th Int. Cosmic Ray Conf., Plovdiv, **11**, 132

Bell, A.R.: 1978a, Mon. Not. Roy. Astron. Soc. **182**, 147

Bell, A.R.: 1978b, Mon. Not. Roy. Astron. Soc. **182**, 443

Blandford, R.D.: 1988, in: 'Supernova Remnants and the Interstellar Medium', eds. R.S. Roger and T.L. Landecker, Cambridge Univ. Press, Cambridge, p. 309

Blandford, R.D., Ostriker, J.P.: 1978, Astrophys. J. **221**, L29

Dorfi, E.A.: 1990, Astron. Astrophys. **234,** 419

Dorfi, E.A.: 1991, Astron. Astrophys., in press

Drury, L.O'C.: 1983, Rep. Prog. Phys. **46**, 973

Drury, L.O'C., Markiewicz, W., Völk, H.J.: 1989, Astron. Astrophys. **225**, 179

Falle, S.A.E.G.: 1975, Mon. Not. Roy. Astron. Soc. **172**, 55

Falle, S.A.E.G.: 1981, Mon. Not. Roy. Astron. Soc. **195**, 1011

Higdon, J.C., Lingenfelter, R.E.: 1975, Astrophys. J. **198,** L17

Krymsky, G.F.: 1977, Dokl. Nauk. SSR **234,** 1306, (also Sov. Phys. Dokl. **23,** 327)

Infrared Properties of Supernova Parent Galaxies

P. POLECHOVÁ[1], P. MAYER[2]

[1]Štefánik Observatory, Petřín 205, 118 46 Praha 1, Czechoslovakia
[2]Department of Astronomy and Astrophysics, Švédská 8,
150 00 Praha 5, Czechoslovakia

1 INTRODUCTION

If a higher infrared relative brightness of a galaxy means a higher stellar production rate, then this higher brightness should be accompanied also by a higher supernova rate. In order to compare the infrared properties of supernova parent galaxies (SPGs) with galaxies without SNe, we used galaxies from the Uppsala General Catalogue (Nilson 1973; UGC) up to 14.5 m_{pg}. The recent supernova catalogue by Barbon et al. (1989) was supplemented by 44 SNe found by the end of March 1991, as they were reported in the IAU Circulars.

The infrared data has been taken from Cataloged Galaxies and Quasars Observed in the IRAS Survey (hereafter CGQ; Lonsdale et al. 1985). We did not consider the galaxies the identification of which has been ambiguous, i.e. when an IR source was identified with two galaxies, as well as when UGC refers the galaxy as a double (or multiple) system. The final list consists of 3280 non-SPGs and of 198 SPGs. Among them, there are 57 galaxies where SNe of type II occurred (hereafter SPG II).

We have formed the infrared index IR as $IR = MAG + 2.5\,[\,log\,(FIR) + 8\,]$, where MAG and $log\,(FIR)$ are taken from the respective columns of CGQ, and the number 8 is chosen for a convenience of having the indices close to zero. The $log\,(f100/f60)$ results again from the respective columns of CGQ.

2 THE EVALUATION OF DATA

The selection effects which could affect the results are as follows:
a) The relative frequency of SPGs decreases with the increasing magnitude.
b) Both characteristics and SPG frequency depend on the Hubble type; therefore, the samples were split so that always only those Hubble types which show similar infrared properties were joined for the tests. Since the Hubble types in which SPGs are more frequent are also the more infrared ones and those with higher $f100/f60$ ratio,

without the splitting one would find SPGs more infrared and cooler than non-SPGs.

All the statistical tests were used as the bilateral ones. As the distribution of characteristics is not normal due to sensitivity limit in the infrared, for testing the null hypothesis we have used mainly the nonparametric Mann-Whitney-Wilcoxon's test. (The null hypothesis claims that the distributions being compared are drawn from the same parent distribution.) For the test of independency of the presence of SNe on the presence of bars we have used χ^2 contingency table test, and, for very weakly populated fields of the tables, we have used the exact Fisher factorial test.

3 THE RESULTS

Neither a statistically significant difference in *IR* indices among SPGs and non-SPGs, nor a difference between $log\,(f100/f60)$ of SPGs and non-SPGsP was found.

The only difference we have found concerns the frequency of SNe II in normal and barred spirals of the Sa to Sm type. When SPGs are divided according to their infrared index, than there is a strong deficiency of SNe II in the SPGs with $IR > 0.9$: among 275 normal galaxies, there are 15 SPGs with SNe II, while among 141 barred galaxies, there is only 1 such SPG (8 would be expected). According to the Fisher's test, the probability of false rejecting the hypothesis of independency is 1.1 % . This effect does not manifest itself in the sample of 533 SA and 252 SB galaxies of Sa to Sm type with smaller or undefined *IR* (where there are 19 and 5 SPGs, respectively).

4 THE DISCUSSION

We have found that there are no differencies among SPG and non-SPG in terms both of their *IR* indices and $log\,(f100/f60)$ flux. This may be explained as follows: for the galaxies M82 (Kronberg et al. 1985), NGC 253 (Antonucci and Ulvestad 1988) and NGC 3448 (Noreau and Kronberg 1987) it was found that SNe can be discovered as radio sources - but not optically. The dust responsible for the infrared brightness is also responsible for a high optical extinction. This may be true in other galaxies, too.

References

Antonucci R. R. J., Ulvestad J. S. 1988, Astrophys. J. **330,** L97

Barbon R., Capellaro E., Turatto M. 1989, Astron. Astrophys. Suppl. **81**, 421

Bothun G., Lonsdale, C.J., Rice, W. 1989, Astrophys. J. **341**, 129

Kronberg P. P., Biermann P., Schwab F. R. 1985, Astrophys. J. **291**, 693

Lonsdale C. J., Helou G., Good J. C., Rice W. 1985, Cataloged Galaxies and Quasars Observed in the IRAS Survey, JPL, Pasadena

Nilson P. 1973, Uppsala General Catalog of Galaxies, Uppsala

Noreau L., Kronberg P. P. 1987, Astron. J. **93**, 1045

DYNAMICAL EVOLUTION OF DUST GRAINS: DUSTY FLOWS IN GALAXIES

J. FRANCO[1] AND A. FERRARA[2]

[1]Instituto de Astronomía – UNAM, Apdo. Postal 70–264, 04510 México D. F., México.
[2]Dipartimento di Astronomia, Università di Firenze, Largo E. Fermi 5, 50125 Firenze, Italy.

ABSTRACT

Radiation pressure on dust grains is very efficient in driving a variety of dusty flows. Starting inside the gaseous disk, starlight can "photolevitate" small interstellar clouds to high galactic latitudes, driving a soft galactic fountain composed of neutral gas and dust, and can maintain a neutral gas column density of the order of 10^{20} cm^{-2} above the main gaseous disk. Bare grains located at the base of a galactic halo receive a strong acceleration but their final fate (*i.e.,* destruction in the halo or expulsion from host galaxy) depend mainly on their radiation pressure coefficients: graphites move faster than silicates and are less affected by sputtering, whereas silicates behave as a "fragile" grain population. Both silicates and graphites, however, can be expelled under a wide variety of conditions. The efficiency for grain destruction in the halo increases with increasing halo gas temperature, decreasing galaxy mass, and decreasing radiation pressure coefficients. The presence of halo magnetic fields restrict the grain paths and, hence, observations of dust lanes in galactic haloes can provide information about the morphology of B-fields at high-latitudes.

1 INTRODUCTION

There is strong observational evidence for the presence of dust, mixed with gas, at high galactic latitudes in the Milky Way. The extinction of high–z stars (*e.g.* Teerikorpi 1990) and the distribution of refractory elements with distance above midplane (Edgar and Savage 1989), indicates the presence of dust with a scale height in excess of 400 pc. The diffuse infrared cirrus is smoothly distributed on the whole galactic disk and extends to high galactic latitudes, revealing the existence of dust grains well beyond the main gaseous scale height (*e.g.* Hauser 1988; Désert *et al.* 1988; Burton 1990). Evidence for solid particles outside the disk of other spiral galaxies, as well as in the intergalactic medium, also exists. Sofue (1987, 1990) has found a collection of dusty filaments emerging from the disk of NGC 253 and NGC 7331, and additional examples of dusty filaments are found in M 82 (Mathis 1973; Telesco *et al.* 1989), NGC 1808 (Veron-Cetty and Veron 1985; Dahlem *et al.* 1990), and M 81 (M. Peimbert, personal communication). Evidence for intergalactic dust, both in galaxy clusters and also in regions apparently disconnected to any galaxy aggregation, has been recently discussed by Rudnicki *et al.* (1989).

All these evidences show that the dusty interstellar phases are efficiently, and continuously, removed from the gaseous disk. This, in turn, opens a series of important questions regarding the driving agents of dust expulsion, the fate of grains outside the gaseous disk, and the possible effects of a population of solid particles in the intergalactic medium. Here we describe some recent results on these issues.

2 DUST EVOLUTION

2.1 Expulsion of dusty clouds from the disk

Franco *et al.* (1991) have proposed a mechanism for a quiet expulsion of the cool gas phases to high galactic latitudes. Radiation pressure from the stellar disk exerts an upward force on dust grains and this force can push, via dust–gas coupling, small diffuse clouds to high galactic latitudes. This coupling operates over a wide range of conditions and dust grains are usually forced to evolve with the gas under typical disk conditions. In the absence of a magnetic field, and assuming cosmic abundances, an adequate viscous coupling requires a maximum gas column density of

$$N_{max} \simeq 10^{19} \left(\frac{a}{10^{-5}\,\mathrm{cm}}\right) \left(\frac{\rho_d}{2\,\mathrm{g\,cm^{-3}}}\right) \mathrm{cm}^{-2}, \tag{1}$$

where a is the grain radius, and ρ_d mass density inside a dust grain. Thus, the gas in clouds with column densities in excess of this value shares the effects of the radiation pressure and is accelerated by starlight.

This acceleration, however, occurs only at moderate opacities and operates within a restricted range of column densities. The optical dust cross–section per gas particle depends on the dust–to–gas ratio, the particle size distribution, and assumed grain composition. For clouds with cosmic abundances and a "normal" dust–to–gas ratio, a simple and useful approximation to the minimum total gas column density required to an optical depth of $\tau = 1$ in the visible can be expressed as (Franco and Cox 1986)

$$N_{\tau} \simeq 5 \times 10^{20} \left(\frac{a}{10^{-5}\,\mathrm{cm}}\right) \left(\frac{\rho_d}{2\,\mathrm{g\,cm^{-3}}}\right) \mathrm{cm}^{-2}. \tag{2}$$

Such a value indicates that the process operates better inside the "standard" diffuse cloud population, which can receive a net acceleration. Clouds located above mid-plane can be pushed by the radiation field from the disk to even larger heights, and can even be ejected from the disk. The maximum height reached by these photolevitated clouds depends on the radiation field and dust–to–gas ratios. Clouds located above luminous stellar clusters or near spiral arms with intense star formation can reach several hundred parsecs in height.

2.2 Grain evolution in a hot gaseous halo

Thus, starlight can drive a "soft" galactic fountain composed of neutral gas and dust. Once the dust particles are located above the disk, radiation pressure can also expel bare grains out of the Galaxy (*e.g.* Chiao and Wickramasinghe 1972; Greenberg *et al.* 1987; Barsella *et al.* 1989; Ferrara *et al.* 1989). Magnetic forces and particle destruction in the hot halo also play an important role and, under certain circumstances, can dominate the bare grain evolution and may prevent such an expulsion (Ferrara *et al.* 1991). Grain interactions, drag and sputtering, with the hot halo material play a crucial role in the evolution, but they depend on the electrostatic grain potential (*i.e.*, Draine and Salpeter 1979).

An adequate knowledge of a series of different galactic parameters are required to properly compute grain evolution. These include the mass and luminosity distribution of the underneath galaxy, the structure of the gaseous halo, and the strength and morphology of the halo B-fields. Except for the luminosity distribution, there is very little information on these issues and one has to explore the allowed parameter space. This problem has been recently addressed by Ferrara *et al.* (1991), and they have followed grain evolution assuming a co-rotating and isothermal gaseous halo in hydrostatic equilibrium and a poloidal B-field. Two detailed galaxy models (mass and luminosity in the spectral range 1400 and 8000 Å) were included as test cases, corresponding to the Milky Way and the spiral NGC 3198, and a wide range of different halo gas temperatures ($3 - 10 \times 10^5$ K) and base densities ($10^{-1} - 10^{-3}$ cm^{-3}) were considered. In addition, models without magnetic field (corresponding to the case of a quiet radio halo) or with a constant 1 μG B-field in the z–direction were explored. Assuming the corresponding steady–state values of the grain potentials ($U \sim 5$ Volt for silicates with $a = 0.1$ μm, and $U \sim 3$ Volt for graphite grains with $a = 0.01$ μm), the continuous reduction of grain dimensions due to sputtering was computed. This, which can eventually disintegrate the particle, changes the grain response to radiative forces and pollutes with heavy elements the regions crossed by the grains.

The two test galaxies belong to different Hubble types (and have different mass and luminosity distributions: the Milky Way is of an earlier Hubble type and is more massive and luminous), but the general features of dust circulation are roughly similar in both cases. The details of particular evolutionary tracks, however, are sensitive to the initial grain conditions, to the morphology of the radiative and gravitational fields, and to sputtering from the hot gaseous halo. Grain velocities achieve larger values, and hence reach outer positions in shorter time scales, in the case of the Milky Way. This is due to its stronger radiation field, and hence larger radiative force. Another difference is associated to the dimensions of the gaseous halo. The scale height of the gaseous halo increases with decreasing galaxy mass and smaller galaxies have more extended gas envelopes. Thus, they are more efficient in destroying dust grains. This effect is more pronounced in grain compounds with smaller radiation pressure coefficients because they move slower, and silicates are more easily destroyed than graphites. In any case, however, both silicates and graphites can reach velocities in excess of 10^2 km s^{-1}, and grain expulsion seems to be a common phenomenon in spiral galaxies.

The main dynamical effect of the inclusion of the magnetic field, on the other hand, is to constrain the motion of the dust particle and the field lines simply act as tracks for the grain motion (*i.e.,* the Larmor radius is very small compared to the path length along the field lines). Thus, charged grains can be good tracers of the magnetic field configuration in the halo. If the halo field is open or connected with an intergalactic field, the grains escape following their configuration. Otherwise they are forced to remain inside the hot halo and are finally destroyed. Thus, the final outcome of grain evolution through the gaseous halo is either ejection from the host galaxy or particle destruction by sputtering, which in turn modifies the local chemical abundances.

3 FINAL REMARKS

The overall effect of radiation pressure on the existing interstellar phases depends on the gas–grain coupling, and are obviously more pronounced in the vicinity of strong radiation fields (see Franco *et al.* 1991). Radiation pressure from starlight represents a very efficient engine which drives a variety of flows. Here we have described its role in the generation a soft galactic fountain and in the circulation of dust particles through the halo of spiral galaxies, but it is also the driving agent of massive winds from cool giant stars (*e.g.* Salpeter 1974; Kwok 1975), internal holes in HII regions (*e.g.* Krishna Swamy and O'Dell 1967; Mathews 1967; Cochran and Ostriker 1977), ionized loops around stars (*e.g.* O'Dell, York and Henize 1967), and even large interstellar shells (*e.g.* Elmegreen and Chiang 1982).

In the case of bare grain flows, the grain optical properties are the most important parameters in defining the final state of the circulating particles. Graphites have a stronger response to radiative forces and reach larger velocities than silicates. These "fast" grains lose less mass per unit time and reach the outskirts of the halo in shorter time scales. Thus, graphite grains retain a good fraction of their mass and can be easily expelled into the intergalactic medium, whereas silicate grains behave as a "fragile" dust population which can be detroyed in the halo. This, in turn, indicates that the chemical enrichment of the halo is different from that of the intergalactic medium. The halo is mainly polluted by gas phase refractory elements mainly eroded from the surface of silicate-type grains, whereas the intergalactic medium is preferentially polluted by the solid leftovers of the erosion process. Thus, *intergalactic dust should be mainly composed by small-sized grains.*

Given the decrease in surface luminosity, the inner regions of spiral galaxies generate more vigorous flows than the external parts. This, in turn, indicates that the radiatively driven mass flow from the disk into the halo decreases with increasing galactocentric distance and could be extremely important in the context of active galactic nuclei. This latter issue has not been well explored yet, but is relevant in dust obscuration of QSOs (see Ostriker *et al.* 1990) and may be a dominant process in the epoch of galaxy formation: dust expulsion at these early times contaminate primordial gas and can trigger the formation of further protogalaxies.

REFERENCES

Barsella, B., Ferrini, F., Greenberg, J. M. and Aiello, S. 1989, Astr. Ap., 209, 349.

Burton, W. B. 1990, in Chemical and Dynamical Evolution of Galaxies, ed. F. Ferrini, J. Franco and F. Matteucci, (ETS Editrice: Pisa), 665.

Chiao, R.Y. and Wickramasinghe, N. C. 1972, M. N. R. A. S., 159, 361.

Cochran, W. D. and Ostriker, J. P. 1977, Ap. J., 211, 392.

Dahlem, M., Koribalski, B., Mebold, U. and Wielebinski, R. 1990, in The Interstellar Disk-Halo Connection in Galaxies, IAU Symp. 144, Poster Proceedings, ed. H. Bloemen, in press.

Désert, F. X., Bazell, D. and Boulanger, F. 1988, Ap. J., 334, 815.

Draine, B. T. and Salpeter, E. E. 1979, Ap. J., 231, 77.

Elmegreen, B. G. and Chiang, W. H. 1982, Ap. J., 253, 666.

Edgar, R. J. and Savage, B. D. 1989, Ap. J., 340, 762.

Ferrara, A., Franco, J., Ferrini, F. and Barsella, B. 1989, in Structure and Dynamics of the Interstellar Medium, IAU Colloquium 120, ed. G. Tenorio–Tagle, M. Moles and J. Melnick (Springer Verlag: Berlin), p. 454.

Ferrara, A., Ferrini, F., Franco, J. and Barsella, B. 1991, Ap. J., in press.

Franco, J. and Cox, D. P. 1986, P. A. S. P., 98, 1076.

Franco, J., Ferrini, F., Ferrara, A. and Barsella, B. 1991, Ap. J., 336, 443.

Greenberg, J. M., Ferrini, F., Barsella, B. and Aiello, S. 1987, Nature, 327, 214.
Hauser, M. G. 1988, Astron. Lett. Comm., 26, 249.
Krishna Swamy, K. S. and O'Dell, C. R. 1967, Ap. J., 147, 529.
Kwok, S. 1975, Ap. J., 198, 583.
Mathews, W. G. 1967, Ap. J., 147, 965.
Mathis, J. S. 1973, Ap. J., 183, 41.
O'Dell, C. R., York, D. G. and Henize, K. G. 1967, Ap. J., 150, 835.
Ostriker, J., Vogeley, M. and York, D. 1990, Ap. J., 364, 405.
Rudnicki, K., Wszolek, B., Masi, S., De Bernardis, P., and Salvi, A. 1989, Comm. in Ap., 13, 171
Salpeter, E. E. 1974, Ap. J., 193, 585.
Sofue, Y. 1987, Pubbl. Astr. Soc. Jap. 39, 547.
Sofue, Y. 1990, in The Interstellar Disk-Halo Connection in Galaxies, IAU Symp. 144, ed. H. Bloemen, (Reidel: Dordrecht), in press.
Teerikorpi, P. 1990, Astron.Ap. 235, 362
Telesco, C. M., Decher, R. and Joy, M. 1989, Ap. J. Lett., 343, L13.
Veron–Cetty, M. P. and Veron, P. 1985, Astr. Ap., 145, 425.

POSSIBLE DETECTION OF THE D I 327 MHz LINE

CARL HEILES

University of California, Berkeley, U.S.A.

The 327-MHz D I line is so weak that it has never been detected despite repeated attempts. Weinreb (1962) pioneered the efforts with a very sensitive search towards the continuum radio source Cas A. We have repeated Weinreb's observations of Cas A with the 85-foot Hat Creek radio telescope for $\sim 2 \times 10^6$ seconds and have made a tentative detection of D I at a velocity of –47 km/s (Figure 1). We purposely used a high frequency resolution in anticipation of the possibility that the D I line would coexist with a narrow *molecular* line, not the much broader *atomic* 21-cm line.

Atomic D might exist in H_2 clouds because H_2 has sufficient column density to shield itself from UV photodissociation (Jura 1975; Brusten *et al.* 1982), while most other molecules are shielded only by dust. Current astrochemical theory finds that, for $100 \lesssim n_H \lesssim 10^4$ cm^{-3}, the fraction of H tied up in molecules increases with increasing n_H while the fraction of D tied up in molecules remains essentially zero (see the succinct summary in Figure 13 of Tielens and Allamandola 1987). D I that resides in a molecular cloud is much easier to detect than that which resides in an atomic cloud. The 327 MHz line opacity $\tau_{DI} \propto N_{DI}/(T\Delta v)$. For the *atomic* gas towards Cas A the temperature $T \sim 40$ K and the line width $\Delta v \sim 10$ km s^{-1}; for the *molecular* gas N_{DI} is probably at least twice as high, $T \sim 10$ K, and $\Delta v \sim 1$ km s^{-1}. Thus τ_{DI} for the molecular gas should be ~ 80 times larger than that for the atomic gas.

Clearly, the result in Figure 1 needs to be confirmed. During the summer of 1991 we are again performing a long, 5-month integration to increase the signal-to-noise ratio on Figure 1 by a factor of two. The first month's observation nicely confirms the result in Figure 1, but the second month's carries absolutely no hint of the result. This capricious behavior is psychologically damaging for the observer, and if manages to remain sane during the five month observation period he hopes to determine definitively whether the possible result is, in fact, real.

REFERENCES

Brusten, P., Audouze, J., Vidal-Majar, A., and Laurent, C. 1982, *Ap.J.*, **243**, 161.

Jura, M. 1975, *Ap. J.*, **197**, 581.

Tielens, A.G.G.M. and Allamandola, L.J. 1987, in *Interstellar Processes*, ed. D.J. Hollenbach and H.A. Thronson, Jr., p. 397.

Weinreb, S. 1962, *Nature*, **195**, 367.

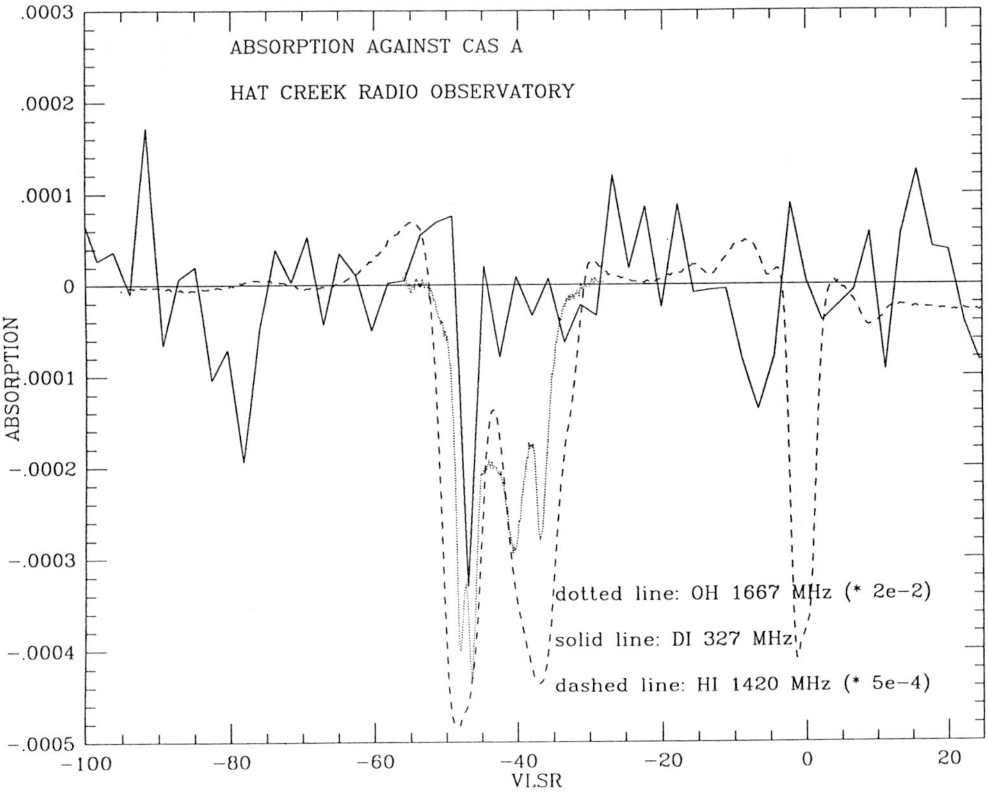

Figure 1. Possible detection of the 327 MHz line of D I in absorption against Cas A. The D I spectrum is shown together with the absorption spectra of the 21-cm H I line and the 1667 MHz OH line. The OH line also has a zero-velocity component, as does the H I line, but it is not shown here.

Models of the deuterium enrichment in ISMC dust grains

V. VANÝSEK

Institute of Astronomy, Charles University Švédská 8 150 00 Praha, Czechoslovakia

ABSTRACT

The deuterium-to-hydrogen ratio in many molecular species in dense interstellar clouds is enhanced by a factor of 100 or more relati ve to the D/H ratio in diffuse ISM. Much larger deuterium enhance ment could be expected in mantle of the core-mantle interstellar grain. This effect could be explained by the gas-phase or by gas-grain chemistry in dense ISM clouds. Preliminary results are here mentioned.

DEUTERATION PROCESSES

The main channels for deuteration in interstellar molecules are ion-neutral reactions with large reaction rates. This and related subsequent reactions are main channels for enhancement of deute rium (relative to hydrogen) in interstellar molecules (see for references: Watson 1976; D. Smith, 1987; Herbst, 1988; Geiss 1988; Millar, 1988). Thielens (1983) first considered the reactions involving deuterium atoms on the grain surface in the steady-state cases. The most important studies concerning the deuterium frac tionation in dense ISM were published by Millar, Bennet and Herbst (1989), R. D. Brown. and Rice (1986), for gas-phase and by Paul D. Brown and Millar (1989a, 1989b). Millar's et al. (1989) pseudo-time-dependent gas-phase models for temperature range 10 to 70 K indicate that significant fractionation can be expected even at 70 K. The results are in agreement (although some differences exist) with finding published by Brown and Rice (1986). A set of quasi-time dependent models for deuteration of HD, HDO, DCN and other species were computed by Brown and Rice (1986). For models compa rable with dark clouds as TMC-1 was found that the time required for saturation of all considered species by deuterium is about 10 years with an D/H enrichment by a factor of 100 or more relative to the average D/H ratio in diffuse ISM. These results are consis tent with observed high ratio of deuterated molecules in dense cool clouds.The time-dependent models of the gas-grain-deuterium chemistry in a dense quiescent interstellar cloud, including accretion on the dust grains, were recently computed by Brown and Millar (1989a,b) Detailed calculation show that two types of reac tions contribute to the formation of deuterated molecular analogs on the grains surface: hydrogenation of an accreted deuterium-bearing molecule and reaction of the accreted radicals with atomic D. The grain surfaces reactions may produce large deuterium enhan cement in the solid phase abundances of deuterated water, ammonia and methane.

EARLY DEUTERATION PHASE

Relatively fast ion-neutral molecule reactions and gas-dust inter action play important role in the deuterium redistribution among the various species. In dense and cool ISMC the species with hig her D/H ratio might condense on the dust particles. Since the deu teration process is virtually terminated by the burial of the molecules in grains, the enrichment of condensates or organic refractory by the deuterium depends on time scales of the satura tion as well as of the condensation process.

The initial deuteration phase of the grain mantles depends in some extend on the abundance of already accreted deuterated molecules. These species, however are produced in the gas-phase, Thus, the detailed study of very early stage of relevant processes in the cool dense interstellar clouds are of some importance. Our study is based on models assuming that the in the mantle are accreted only gas-phase products. The gas-phase chemistry incorporates 2606 reaction between 120 species containing H,C,N,O. and a repre sentative (fictive) metal (Mg). The relative abundances are (part ly depleted): $D = 10^{-5}$, $He = 0.085$, $C = 6.6 \times 10^{-5}$, $N = 9.1 \times 10^{-5}$, $O = 3.9 \times 10^{-4}$, $Met = 7 \times 10^{-5}$, $CO = 2.6 \times 10^{-4}$, $H_2 = 0.3$, and $HD = 7,5 \times 10^{-6}$. Density $n_H = 3 \times 10^4$, ratio grain/H $= 3 \times 10^{-13}$, $T = 10$ K, sticking coeff. $S = 1$, (for H $S = 0.3$). Grain radius = O.15 μm. CR ionization rate 1.3×10^{-17} sec .The of the reaction network and relevant computing pro gram was compiled by Moravec (1991). The most of reaction rates coefficients were adopted from Prasad and Huntress (1980) paper. The dust grains were chemically inert and without charge.The quasi-time dependent models for deuteration indicate that the con densable species as H_2O, HCN, H_2CO, CH are highly deuterated. The refore, the abundance of D could be significantly enhanced in man tles of dust grains in dense cool interstellar clouds. These models suggest that the D/H ratio in primitive molecular condensa tes (as ice) should be about 10^{-3} or more - which corresponds with deuterium enrichment by a factor of 100 - will be reached in a surprisingly short time-scale of about 10^3 to 10^4 years.

REFERENCES

Brown, R.D. and Rice, E.H.N., 1986, Month. Not. RAS 223, 429

Brown, P.D., Charnley, S.B., and Millar, T.J. 1988, Month. Not. RAS 231, 409

Brown, P.D., and Millar, T.J. 1989a, Month. Not. RAS 237, 661

Brown, P.D., and Millar, T.J. 1989b, Month. Not. RAS 240, 25p

Geiss, J. 1988, in Review of Modern Astronomy 1 (Cosmic Chemistry) ed. G. Klare, Springer, Heidelberg, p. 1

Millar, T.J. 1989, Proc. 22nd ESLAB Symposium ESA SP-290, 109

Millar, T.J., Bennet, A., and Herbst, E. 1989, Asrophys. J. 340, 906

Moravec, Z. 1991, Thesis, Charles University Math-Phys.Fac., Prague

Prasad S. S., and Huntress W.T. 1980, Astrophys. J. Suppl 43, 1

Smith, D. 1987, Phil. Trans. Roy. Soc. London A 323, 269

Thielens, A.G.G., 1983, Astron. and Astrophys. 119, 177

Watson, W.D., 1976, Rev. Mod. Phys. 48, 513

Abundances from Planetary Nebulae: The Past ISM

J. KÖPPEN[1], A. ACKER[2], M. SAMLAND[1], & B.STENHOLM[3]

1) Institut für Theoretische Astrophysik, Im Neuenheimer Feld 561, D-6900 Heidelberg, (FRG)

2) Équipe "Populations stellaires", URA 1280, Observatoire de Strasbourg, 11 rue de l'Université, F-67000 Strasbourg, (F)

3) Lund Observatory, Box 43, S-221 00 Lund, (S)

Abstract: The chemical compositions of planetary nebulae tell us that the abundance gradient of interstellar gas in the solar neighbourhood about 5 Gyrs ago was almost as steep as it is now observed in HII regions.

1. INTRODUCTION Planetary nebulae (PN) are important objects for studies of the chemistry of galaxies, as they can be observed at large distances. Also, they represent the late stage of evolution of stars with masses from about 1 $M_{\odot}$ to 8 $M_{\odot}$. Thus their chemical properties reflect – apart from He and N, which could be affected by products of stellar nucleosynthesis – the state of chemical evolution of the interstellar gas at the time of birth of the progenitor stars, i.e. a few Gyrs in the past.

In the present interstellar medium steep abundance gradients are found by HII region studies. Using abundance data compiled from various sources, Faúndez-Abans and Maciel (1987) detected in disk PN abundance gradients of the same steepness.

Since 1984, two of us (A.A., B.S.) have conducted a spectrophotometric study of all PN known in our Galaxy (for details see Acker et al. 1989)). The spectra cover the wavelength range from λ3990 to 7400 Å with a resolution of 7 Å. From this homogeneous material, 90 objects with good quality spectra were selected for an analysis of their chemical compositions.

2. ABUNDANCE ANALYSIS The spectra are analyzed automatically by two different methods: A simple but fast analysis is based on the measurement of density and temperature of the electrons by forbidden line ratios, and employs empirical correction factors (ICF) to account for unseen ionic stages (cf. Köppen et al., 1991). Unfortunately, some crucial lines are inherently weak, and ICFs are reliable only for some elements. This can be improved by constructing complete photoionization models to match the observed spectrum. Though this is more labourious, it also is physically more consistent (Samland, 1990). For our statistical study, both methods give the same results.

3. RESULTS Table 1 lists the solar abundances, the constants a and b in the expression $\lg(A(R)) = b + a(R - R_\odot)$ (with the galactocentric distance R in kpc, $R_\odot = 7.8$kpc), the correlation coefficient r, the errors s_b, s_a, and the number of nebulae used. To select old disk objects, nebulae were chosen with a deviation from the local circular motion of less than 60 km s^{-1}, a helium abundance of less than 11.10, and a lg(N/O) ratio of less than -0.3. For comparison, the results of Faúndez-Abans and Maciel (1987) (FAM) are also shown.

Table 1. Abundances from Galactic Planetary Nebulae

Elem.	Sun	b(FAM)	a(FAM)	b	a	r	s_b	s_a	No.
N	7.99	7.98	-0.072	8.05	-0.047	0.42	0.12	0.025	19
O	8.92	8.82	-0.072	8.81	-0.014	0.20	0.08	0.016	21
S	7.23	6.91	-0.098	6.89	-0.071	0.53	0.12	0.032	15
Ar	6.57	6.42	-0.056	6.39	-0.014	0.46	0.05	0.010	10
N/O	-0.93			-0.70	-0.022	0.33	0.07	0.015	19
S/O	-1.69			-1.97	-0.053	0.61	0.08	0.020	15
Ar/O	-2.35			-2.49	-0.028	0.73	0.05	0.009	10

Our results from a homogeneous set of observational data analyzed in the same way, confirm the findings of Faúndez-Abans and Maciel.

The average chemical composition of disk PN in the solar neighbourhood is solar. Thus for these PN the progenitor stars were slightly more massive than the Sun, and were born about 5 Gyrs ago. At that time, abundance gradients in the interstellar gas were already existing, though maybe not as steep as found in the present ISM.

Acknowledgements: Financial support from the Deutsche Forschungsgemeinschaft (SFB 328) is gratefully acknowledged.

References

Acker, A., Köppen, J., Stenholm, B., Jasniewicz, G.: 1989, *Astron.Astrophys.Suppl.* **80**, 201

Faúndez-Abans, M., Maciel, W.: 1987, *Astrophys. Space Sci.* **129**, 353

Köppen, J., Acker, A., Stenholm, B.: 1991, *Astron.Astrophys.*, in press

Samland, M.: 1990, Diploma thesis, Universität Heidelberg

Evolutionary Effects in Bipolar Planetary Nebulae

J.A. LOPEZ

Instituto de Astronomia, UNAM. Apartado Postal 877, Ensenada, B.C., 22830, México

Bipolar planetary nebulae (BPN) are characterized by their "hourglass" shape. Bipolar structures appear as early as the proto-planetary nebula phase indicating that the shaping process occurs since the very early stages of formation. There is an apparent "range of bipolarity" among BPN and a number of classification schemes have been proposed in the past in an attempt to link the apparent morphological sequence with evolutionary stages. In general terms, loose bipolar structures are identified with planetary nebulae in advanced evolutionary stages, whereas compact BPN are related to young PN. It is clear, however, that morphology alone is insufficient to determine a reliable evolutionary sequence, for elements like projection effects, sizes, etc. are difficult to compensate for in many instances; besides, any envisaged chronological development of shapes must conform with a physical theory of evolution, which in turn should be able to respond to different observational tests. Thus, a morphological classification needs to be complemented and reinforced by as many adittional observable parameters as possible to yield reliable physical grounds to the definition of an evolutionary sequence.

A number of models have been advanced in the past to explain the origin of bipolar structures in planetary nebulae. The conditions under which bipolarity is produced and retained throughout most of the planetary phase remain still to be understood. Nevertheless, the availability in recent years of new and better data for these objects in various wavelength domains permits now a better understanding of the correlation between the differences in shape and other observable parameters that are expected to trace the effects of the evolutionary path in BPN.

In the first stages of development of a BPN its morphology is indeed quite compact with its lobes well delineated. The central object is enshrouded by a thick dust shell and the infrared luminosity is in consequence high, presenting also a rich molecular infrared spectrum, as demonstrated by IRAS. Molecular emission in the envelope is rich from infrared to radio wavelengths. The transit time between the proto-planetary nebula to planetary nebula stage is very short, of the order or 10^3 years; during this period the object suffers drastic changes in its observable parameters, for the dust shell starts to dissipate, molecules are destroyed by the stellar radiation field and the infrared spectrum varies substantially. The nebular spectrum is here of low excitation and the kinematical field is usually composed of modest expansion velocities with a simple spatial structure, though in some cases high velocity collimated flows have been observed. As the planetary grows older, the bipolar morphology is relaxed by expansion, mainly due to the pressure exerted by the stellar wind which by now has already switched from a low velocity high density wind into a hot and fast, low density one. The infrared luminosity and molecular emission decrease both substantially; the infrared spectrum becomes dominated by atomic lines with practically only H_2 remaining observable in the periphery; ocassionally CO is (weakly) observed in some intermediate cases. As the central star moves quickly towards higher temperatures the nebular spectrum turns of high excitation, and the internal kinematical field becomes spatially complex with the presence of localized high velocity filaments.

Two contrasting cases are presented here that exemplify the observable parameters one expects in BPN in different evolutionary stages according to the description given above, namely: the evolved bipolar planetary nebula NGC 2899 (López et al, 1991) and, the fairly young bipolar Mz-3 (López and Meaburn, 1983).

NGC 2899 has a loose bipolar morphology, complex kinematical field, high excitation nebular spectrum, very low infrared emission and peripheral shocked H_2 emission. No CO or other molecular emission has been observed in this object. On the other hand, Mz-3 presents a compact bipolar structure. Its nebular spectrum is of low excitation with high internal extinction. H_2 has been observed in its core (Webster *et al*, 1988.). CO has been detected in it (Bujarrabal and Bachiller, 1991) and it is also a very bright infrared source at IRAS wavelenghts.

These two objects are part of a larger program on BPN whose details will be published elsewhere.

References

Bujarrabal,V.,Bachiller,R.: 1991, *Astron. Astrophys.* **242**, 247.

López,J.A.,Falcón,L.,Ruiz,M.T.,Roth, M.: 1991, *Astron. Astrophys.* **241**, 526.

López,J.A.,Meaburn,J.: 1983, *Mon. Not. R. astr. Soc.* **204**, 203.

Webster,L.,Payne,P.,Storey,J.,Dopita,M.: 1988, *Mon. Not. R. astr. Soc.* **235**, 533.

THE STUDY OF PLANETARY NEBULA VARIABILITY

E.B.KOSTYAKOVA

Sternberg State Astronomical Institute, Moscow, Russia

Photometric and spectral studies of planetary nebula variability have been carried out for 7 selected objects during the period 1968-88.

The photoelectric UBV-observations were begun at the Crimean Station of Sternberg Institute and at the Skalnate Pleso Observatory (Czechoslovakia). The results were summarized in two catalogues: Sov. Astron. J. Lett., vol. 16, No. 12, 1990, p. 1085, and Publ. of Sternberg Astron. Inst., vol. 62, 1991, p.143.

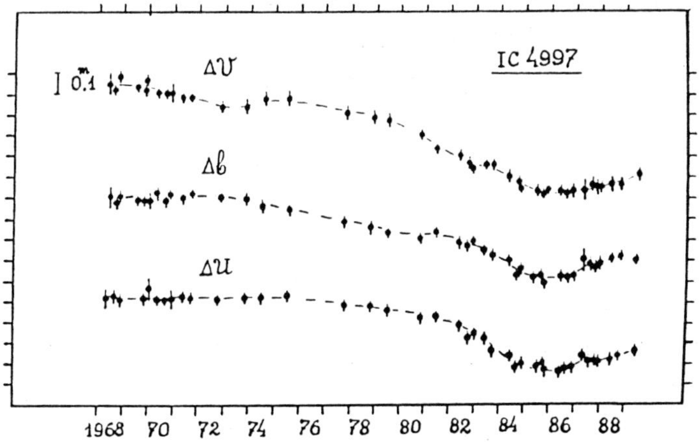

Fig.1. Run of the integral ubv-brightness of the planetary IC 4997.

During 20 years of photoelectric observations, 5 planetaries namely NGC 6572, IC 4997, Hu2-1, NGC 6891, and NGC 6543, have undergone noticeable variations of the integral ubv- brightness in the range 0.1^m-0.6^m. The nebulae NGC 6720 and IC 3568, suspected earlier to be variable, showed no changes of integral brightness exceeding 0.1^m.

The most remarkable photometric behaviour was found for the well-known young nebula IC 4997, which showed a gradual decrease of the integral ubv-brightness in 1968-85, which amounted (in total) to 0.4^m-0.6^m, then a sudden stop occurred in

1985-86, followed by an obvious brightening since 1987 (Fig.1).

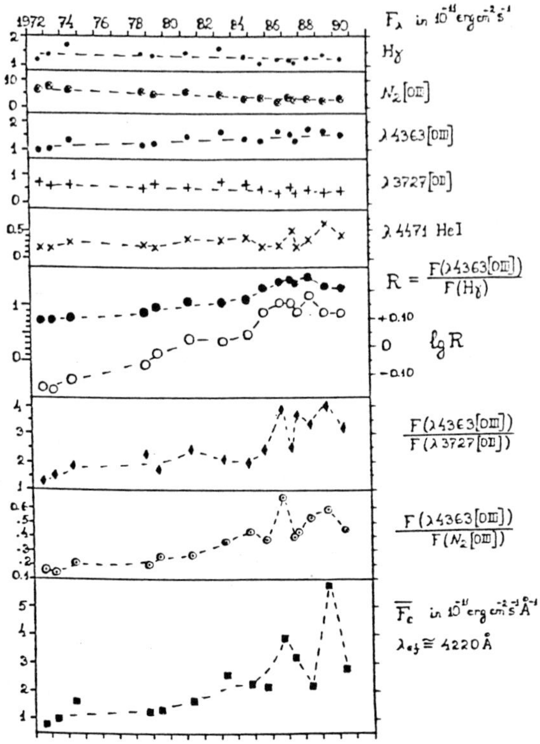

Fig.2. Spectral variations of IC 4997.

The systematic spectral study (since 1972) of this nebula also showed noticeable variations in its spectrum, both in certain emission lines and in the continuum of the central star (Fig.2). Although the energy flux in hydrogen lines during the observation period remained almost unchanged, the nebular lines and the line λ 3727 [OII] became systematically weaker, whereas the auroral line λ 4363 [OIII] and the helium line λ4471Å became rather stronger.

The often-used ratio R = F(λ 4363 [OIII])/F(H_γ), the ratios F(λ 4363)/F(λ 4959) [OIII], F(λ 4363)[OIII])/ F(λ 3727 [OII]), and the central star radiation flux in λ_{ef} = 4220Å, showed marked rises until 1986, and then sudden strange "jumps" near the minimum of the nebular integral ubv-brightness.

Preliminary analysis of the nebula IC 4997 suggests that the revealed variations reflect certain changes in the central star radiation field (perhaps fluctuations of stellar effective temperature).

The N-O Relation in the Interstellar Medium

M. WILMES & J. KÖPPEN

Institut für Theoretische Astrophysik der Universität Heidelberg, Im Neuenheimer Feld 561, D-6900 Heidelberg 1, F.R.G.

1. HOMOGENEOUS CHEMICAL EVOLUTION

Conventional models which assume that the interstellar gas is always perfectly well-mixed are known to fail in reproducing the observed relation between nitrogen and oxygen: Whereas oxygen is synthesized primarily, i.e. from hydrogen within one stellar generation, nitrogen is expected to be made secondarily from carbon and oxygen in the CNO cycle. This predicts $N \propto O^2$ which is steeper than observed with HII regions (e.g. Edmunds, 1989). One possible explanation is the primary production of nitrogen by burning freshly produced carbon, brought up from the stellar core. How efficient these dredge-up processes are, is still subject to debate.

2. AN INHOMOGENEOUS MODEL

An alternative explanation is provided by the chemical evolution with incomplete mixing, e.g. because spatial chemical inhomogeneities are allowed to form and evolve. We consider the test volume to be subdivided into a number (100) of chemically homogeneous subvolumes in which we follow the chemical evolution separately. Star formation is known to be localized in distinct regions. We simulate this by concentrating the star formation onto only a small number of subvolumes which are chosen randomly. The threshold T for the probability that a subvolume becomes "active", is taken as a model parameter. Mixing of synthesized elements to the surrounding interstellar gas occurs instantly and immediately after stars are formed, but over only a finite pollution radius l. The star formation rate in each subvolume is taken as proportional to the density of the available gas. Infall of gas from outside the test volume or outflows are not taken into account.

3. RESULTS

The figure shows our results for the [N/O] vs. [O] relation with an infinite mixing length and a threshold $T = 0.25$. We note that in the model at every time one gets a distribution of abundances while homogeneous models give only one value for every abundance.

The threshold T is only responsible for the width of the distribution: a small value for T leads to great scatter. The mixing length l determines the [N/O] vs. [O] relation: For small l all metals are kept in the subvolume and every subvolume acts as a homogeneous closed box model. In this case we get the same relation as predicted by homogenous models. For large l the subvolumes are strongly coupled because most of the metals in a subvolume entered the subvolume from outside. When l reaches infinity (as shown in the figure) one gets

$$N \propto O^{1.3}$$

because even in this case localized star formation leads to inhomogenities.

As the relation of primary to secondary abundances is independent from the value of T, only minor deviations from homogenity are necessary to change the homogeneous behaviour: The relation of primary to secondary abundances is very sensitive to inhomogenities.

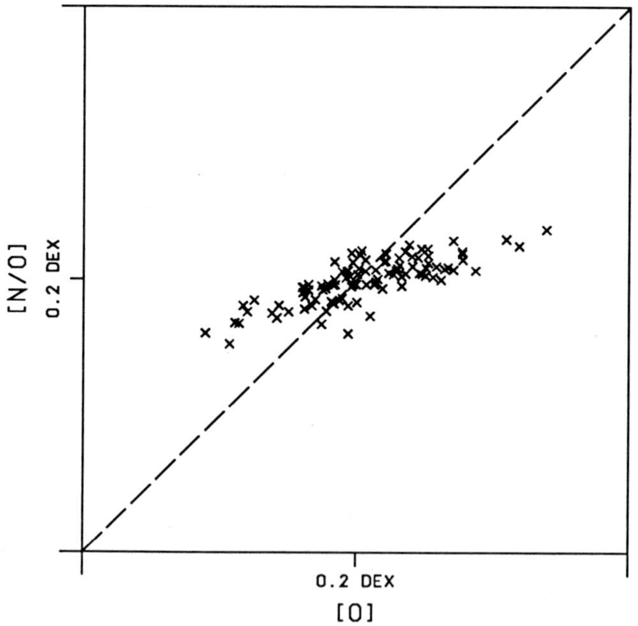

Figure 1. The [N/O] vs. [O] relation for $T = 0.25$ and $l = \infty$. The dashed line indicates the behaviour predicted by homogenous models while the crosses show the model abundances for the different subvolumes.

Visual Interstellar Extinction in the Galactic Disk

J. GUARINOS

Observatoire Astronomique de Strasbourg, France

1 AIM OF THE PRESENT STUDY

When one computes the visual absorption (A_V) from the color excess E_{B-V}, this means that one deals with the uniformely distributed material in the Galactic disk. If the highest values of the ratio of total to selective extinction (R) detect the darkest clouds, the laws generally used in statistical analyses show the effect of a more uniformely distributed material, hence dust and HI clouds. This is the subject of this paper.

The main purpose of this work is to study the general variation of the visual absorption with the distance, when all the directions are taken into account, provided we stay within the Galactic disk. Another purpose of this work is to show up the contribution of the circumstellar extinction.

2 BUILDING THE SAMPLE OF STARS

As a basis, we used a catalogue carefully constructed especially for this purpose. This catalogue contains UBV photometric measurements and MK types, computed from previous compiled catalogues and meticulously checked. Special attention has been paid to the consistency between photometric data and spectroscopic data. A cross-identification has been performed between the UBV Photoelectric Photometric Catalogue (Mermilliod, 1987), and several MK types catalogues, the two major ones being: The Michigan Spectral Catalogue (Houk and Cowley, 1975; Houk, 1978; Houk, 1982; Houk and Smith-Moore, 1988) and the Catalogue of Selected Spectral Types (Jaschek M., 1978). The final sample of stars has about 16 500 stars with consistent UBV photometry, chosen MK type, galactic coordinates, equatorial coordinates, derived intrinsic colors, derived absolute magnitudes, computed visual absorption and computed distance. For each star, the consistency between the MK type and the observed colors (taking into account the extinction) has been checked.

A dedicated software computes the visual extinction, assigning an absolute magnitude chosen in a (MK type/Absolute Magnitude) table which can be either Schmidt-Kaler's (1982), either Corbally and Garrison's (1983), or the Grenier et al. one (1985). The intrinsic color $(B-V)_o$ is also coming from such a table (Schmidt-Kaler, 1982) or computed using the Q method (Johnson and Morgan, 1953) which needs the observed colors. We used Olson's formula (1975) for computing the absorption: $R = \frac{A_V}{E_{B-V}} = 3.25 + 0.25(B-V)_o + 0.05(B-V)$

3 RESULTS

Mean visual absorption in 25 pc thick rings have been computed up to about 1200 pc in the Galactic disk, whatever is the direction, for $B4$ to K stars, and then for the rest of the stars. The curve A_V Versus *Distance* is shown for each of the 2 groups of stars.

According to many authors, the circumstellar enveloppes of massive O to B3 stars and massive M stars are responsible for a large part of the reddening towards them. This is why we separated these stars from the others.

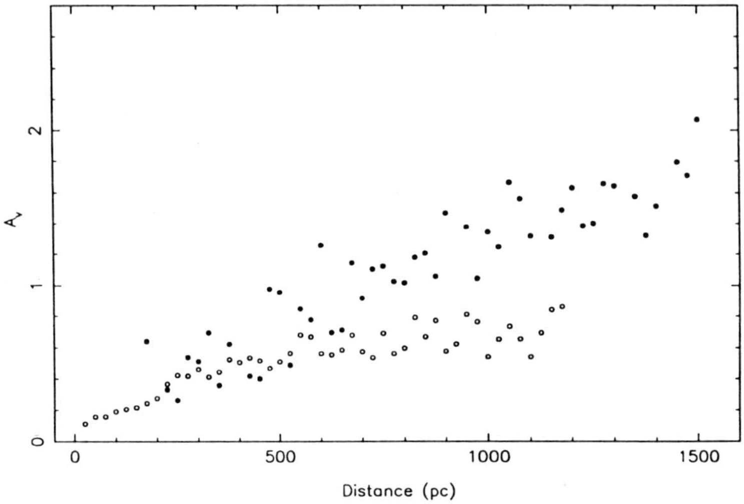

Comparing the two curves, it appears that the difference of mean extinction in the 25 pc thick rings is systematically larger for O-B3 and M stars (filled circles) than for the other stars (open circles), with a difference ranging from 0.1 to almost 1.0 magnitude. This is the mean contribution of the circumstellar enveloppes to the total extinction in the visual range.

Since we limited the sample of stars up to distances for which there is no observational bias, we can say that these curves are good estimates of the mean extinction as a function of the distance in the Galactic disk.

REFERENCES

- Corbally C. J. and Garrison R. F. 1983, The MK Process and Stellar Classification, Garrison ed., David Dunlap Observatory, Univ. of Toronto, p. 277.
- Grenier S., Gómez A. E., Jaschek C., Jaschek M., and Heck A. 1985, Astron. Astrophys. 145, 331.
- Houk N. and Cowley A. P. 1975, Univ. of Michigan, Ann Arbor, Michigan.
- Houk N. 1978, Univ. of Michigan, Ann Arbor, Michigan.
- Houk N. 1982, Univ. of Michigan, Ann Arbor, Michigan.
- Houk N. and Smith-Moore M., 1988, Univ. of Michigan, Ann Arbor, Michigan.
- Jaschek M. 1978, Bull. Inform. CDS 15, 121.
- Johnson, H. L. and Morgan, W. W. 1953, Astrophys. J. 117, 313.
- Mermilliod J. C. 1987, Astron. Astrophys. Suppl. Ser. 71, 413.
- Olson, B. I. 1975, Pub. Astr. Soc. Pac. 87, 349.
- Schmidt-Kaler T. 1982, Landolt-Börnstein, group VI, Vol. 2b.

Peak Position and Width of the 2175 Å Bump in the MRN Model

D. PROGA[1], R. SZCZERBA[2]

[1] N. Copernicus University, Institute of Astronomy,Toruń, Poland,
[2] N. Copernicus Astronomical Center, Laboratory of Astrophysics, Toruń, Poland

1 INTRODUCTION

The MRN model of interstellar dust assumes that dust is composed of bare silicate and graphite grains with the power-law size distribution:

$$n_i(a) = A_i \, n_H \, a^{-p_i} \; ; \qquad for \;\; a_i^- < a < a_i^+$$

where: i stands for silicate (Si) or graphite (C), n_H is the number density of H nuclei and a_i^- and a_i^+ are the lower and upper cutoffs. For each dust component normalization of distribution is given by the integral $\int_{a_i^-}^{a_i^+} n_i(a)da$, which is equal to the total number density of particles of type i with radii in the interval $[a_i^-, \; a_i^+]$.

One of the objection to graphite as a carrier of the 2175 Å bump, comes from the fact that the observed peak position of the bump-λ_0 remains almost constant, while its width-γ shows significant variations (Fitzpatrick & Massa, 1986-hereafter FM86, Cardelli & Savage, 1988). It is argued, that different size distributions of graphite grains could, in principle, produce the 2175 Å feature of various width, but at the same time λ_0 bump will change significantly its position. This conclusion is based on the well known fact that the bump position shifts with graphite grain size. However, if we are dealing with a power law distribution characterized by power law index p_C around -3.5, then λ_0 depends mainly on strongly preffered small particles.

2 RESULTS

We have checked numerically that the theoretical position of λ_0 seems to be stable over a quite wide range of graphite grain parameters. Figure 1 shows λ_0 versus upper cutoff to the graphite grains distribution. The calculations were performed, assuming $a_C^- = 0.005\mu m$, for various values of p_C. Each line in this Figure is described by its power law index. We can see that the central wavelengths concentrate around

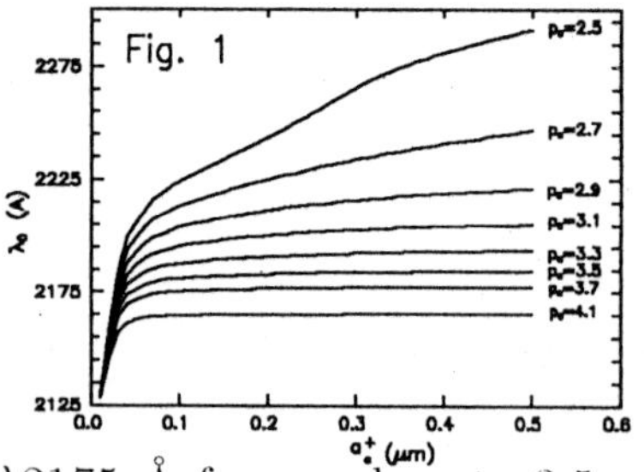

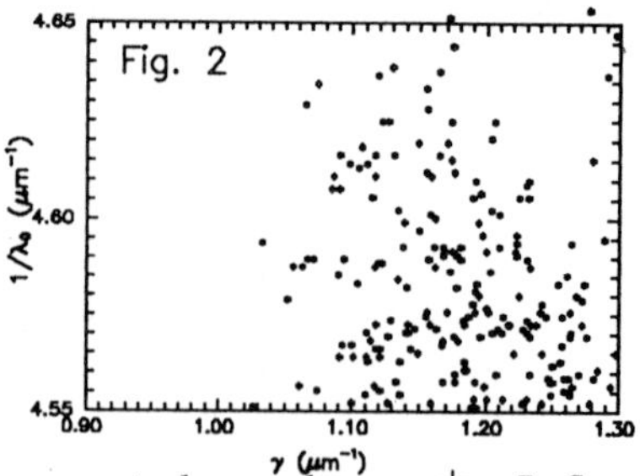

λ2175 Å for p_C close to 3.5, and almost does not depend on a_C^+. Influence of the upper cutoff is lager only for smaller power law indices ($p_C \leq 2.7$). This result can be explained by a preferential treatment of small particles in comparison with the larger ones. Because bump profiles of small graphite grains have λ_0 close one to another, and because the number of small particles is much higher than the number of larger ones, then small particles dominate contribution to the final extenction curve.

Next, we have calcutated theoretical extinction curves, taking into account different size distributions of both MRN components (i.e. graphite and silicate grains). Each resulting curve was parametrized around the bump ($1/\lambda$ changes from 3.3 to $5.9\mu m^{-1}$) according to the proposition of FM86, i.e. by the FWHM of the bump-γ, the position of the bump peak-λ_0, the strength of the bump-a_3 and coefficents of the straight line a_1 and a_2. From the obtained extinction curves we have chosen only these, parameters of which belong to the same ranges like the parameters of the observed curves (see FM86). In Figure 2 we have plotted the bump width γ versus its central wavelength $1/\lambda_0$, for chosen sample of the synthetic extinction curves. We can see that there is possible to explain the lack of correlation between γ and $1/\lambda_0$. Points represented in Figure 2 have power law indices in range from 3 to 4 and a_C^+, a_{Si}^+ in range from 0.1 to $0.3\mu m$. It means that the bump position in the MRN model is close to 2175 Å (value of $1/\lambda_0 \simeq 4.6\mu m^{-1}$) for quite wide range of dust parameters. It seems to be a strong argument in favour of the considered model. However, there are no theoretical curves with $\gamma \leq 1.0\mu m^{-1}$ (whereas in the FM86 sample are many bumps with γ less than $1.0\mu m^{-1}$, among them the narrowest one has γ=$0.768\mu m^{-1}$). This problem has been discussed by Draine (1989) and a suggestion how to solve it is also given there.

3 REFERENCES

Cardelli, J.A., Savage, B.D.: 1988, *ApJ*, **325**, 864
Draine, B.T.: 1989, *in IAU Symp. 135, Interstellar Dust*, eds. L.J. Allamandola, A.G.G.M. Tielens, Kluwer Academic Publishers, p.313
Fitzpatrick, E.L., Massa, D.: 1986, *ApJ*, **307**, 286
Mathis, J.S., Rumpl, W., Nordsieck, K.H.: 1977, *ApJ*, **217**, 425

SiC and Iron Rich Grains in Cometary Dust

M. ŠOLC, H. DANIELCZYK

Charles University Prague, Institute of Astronomy Švédská 8, CS-150 00
Prague 5

Elemental and isotopic abundances in cometary dust, as measured by the mass spectrometers PUMA/Vega and PIA/Giotto in comet Halley, show similar bulk composition as CI chondrites (Jessberger et al., 1989). Most of the cometary dust grains are complex conglomerates of more mineral species, usually mixed with some organic stuff. However, chemically homogenous material can be expected in very small grains. In the search for such particles in PUMA 1 archive several mass spectra with only few features were found. The total number of ions in those spectra was low and thus the masses of the corresponding grains were only sligthly above the lower limit of the instrumental mass range (femtogramms).

The main aim of the present paper was to select the most simple spectra from the set of about 1200 mass spectra obtained by the experiment PUMA 1 in the highly compressed mode, and to look for C, Si and iron rich grains without magnesium. Among well intepretable spectra of mixed grains consisting from silicate core and organic (CHON) mantle we found about 30 spectra with dominant Fe lines. For these particles, mass and bulk density was determined using the procedure given by Kissel and Krueger (1987a,b).

The time-of-flight spectrometers PUMA 1, PUMA 2 and PIA were designed to analyse the quantitative chemical and isotopic composition of cometary dust particles. The high velocities of spacecrafts with respect to the cometary material (69 km/s by GIOTTO and 78 km/s by VEGA) led to evaporation of all dust grains that entered the instrument and hit the metal target. (The platinum target was doped with silver natural isotopes 107 and 109). Positively charged ions from the impact point were accelerated by electrostatic field and received the same kinetic energy. Flight times of ions along 1 m path in the instrument depended thus only on the mass of ions, the initial energy distribution of ions was corrected by an electrostatic reflector placed between two drift tubes. Light ions like hydrogen completed this path in about 4 μs, heavier ions (silver) in about 40 μs. Hence almost every mass spectrum contains both lines originating from the ionized atoms (or molecular fragments) from the impacting dust grain and silver lines from the target. First step in evaluation of a spectrum was to identify lines of some frequent elements - mainly C, Fe and Ag - and to convert the time-of-flight scale to the mass scale. Following step was the identification of other elements and isotopes. The elemental abundances were determined by dividing the number of ions in the corresponding spectral lines by ionization efficiencies. These ion yields are in principle similar to those found by SIMS - Secondary Ion Mass Spectrometry (Kissel and Krueger, 1987a,b). Finally the bulk density of the particle was determined from the ratio of particle to target ions. The impact of hypervelocity particles is a complex process, but the more dense and compact the particle is, the more target ions are thrown out from the impact crater on the target (with respect to the number of particle ions). A compact and dense particle hollows larger crater than an equally massive but fluffy particle. Mass of the particle follows then from number of particle ions released during the impact. The semi-empirical formulae given by Krueger were confirmed by the laboratory tests made in 1990 with iron and carbon particles on the Dust Particle Accelerator in Max-Planck- Institute for Nuclear Physics in Heidelberg. Some numerical constants in these formulae were modified to fit these recalibration results of the PIA/Giotto instrument.

The accuracy of determining the density is better than 50Higher values of density are more precise because in the laboratory tests carbonyl iron spheres of density 7.8 g/ccm were used. Figure 1 (upper and middle part) shows the composition of the most dense grains that consist mainly of Fe (56 and 54 amu), S (32 amu) in form of troilit FeS, Si (28 amu) and C (12 amu) resp. organic stuff CHON. The highest densities found for these grains suggest the presence of highly reduced metal iron in particles, even if they can be very porous. According to the formula by Krueger, the spectrum 53924/PUMA1 gives the highest density found 4.70 g/ccm and the mass 4.35 fg. Other iron rich grains are 53435/PUMA1 with 3.01 g/ccm and 29.0 fg; 53220/PUMA1 with 2.18 g/ccm and 40.5 fg; 54221/PUMA1 with 1.29 g/ccm and 76.2 fg. The origin of iron metallic grains in comets deserves further study.

Carbon grains without magnesium were reported in all previous studies of Halley's dust (e.g. Jessberger 1989). However, several spectra contain also line 40 amu together with dominant lines of masses 12 and 28 amu. It cannot be excluded that the species responsible for the mass 40 is silicon carbide.

The growth of solid grains by condensation from cool plasma was studied both theoretically and in laboratories for more than thirty years. On the other side, laboratory and in-situ mass spectrometry is now available of some interplanetary and cometary grains that are believed to be very pristine, almost unaltered by thermal or geological processes active in the time of formation of the solar system. Zinner found several

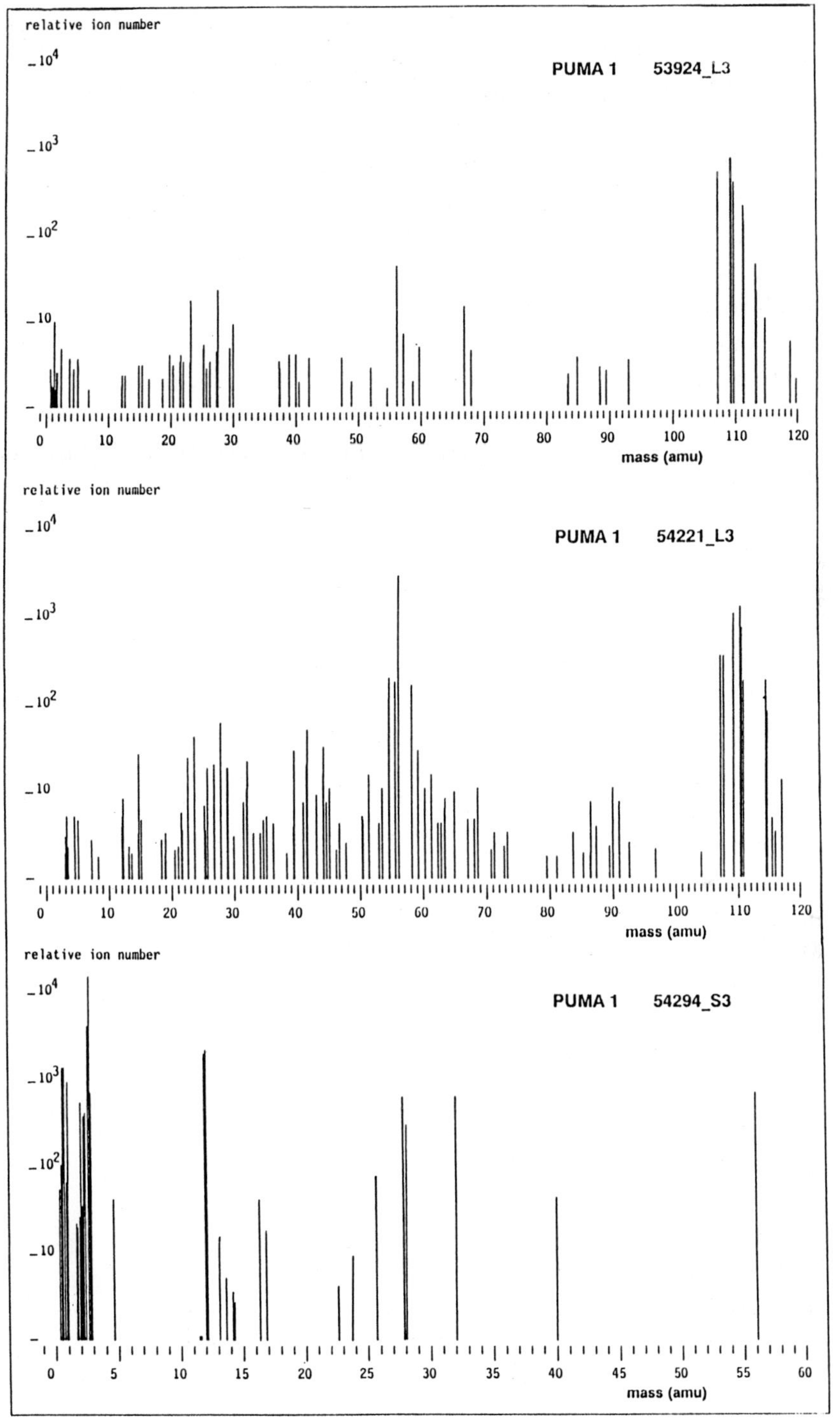
relative ion number
10^4
10^3
10^2
10
PUMA 1 53924_L3
mass (amu)
relative ion number
PUMA 1 54221_L3
mass (amu)
relative ion number
PUMA 1 54294_S3
mass (amu)

Figure 1

tiny graphite inclusions in carbonaceous chondrites CI. They show onion-like shells as a consequence of condensation from cool plasma. These carbon grains have not been processed since their origin in carbon rich late-type or AGB stars (Fegley 1991). Similarly, several silicon carbide inclusions have been found in CI chondrites which might be also of pre-solar origin, some of them possibly originated in supernova according to their isotopic composition (Zinner 1991 a,b).

The existence of mass line 40 in most simple spectra cannot be explained by mineralogy, if calcium 40 is assumed as the element responsible for it, because chemical bond between Ca and Si or Ca and C is not probable. The line at 40 amu can be interpreted rather as silicon carbide SiC instead of calcium, since there are several spectra where the presence of line 40 does correlate with the amount of carbon and silicon atoms. Figure 1 (lower part) shows one example of such grain where silicon cannot be assigned to silicates because of lack of magnesium; in typical silicate grains the abundances of magnesium and silicon ions are approximately of the same value. Pure grains of this composition are unique in the set of more than 1000 analysed spectra and they are rather small. However, it cannot be excluded that such grains are present in larger composite fluffy particles, because there is no way how to separate mass lines of various constituents (with exception of well recognizable organics CHON). Thus the analogy between some interplanetary dust particles, meteoritic inclusions, cometary grains and stellar condensates becomes more and more profound.

REFERENCES

Fegley B.: Condensation of graphite and refractory carbides in stellar atmospheres, 54th annual meeting of the Meteoritical Society, Monterey 1991 (Abstract 60)

Jessberger E.K., Christoforidis A., Kissel J.: Aspects of the major element composition of Halley's dust. Nature 332, 691, 1989

Kissel J., Krueger F.R.: Die Physik der Massenspektrometrischen Staubanalyse beim Kometen Halley. Phys. Bl. 43, 5, 1987a

Kissel J., Krueger F.R.: Ion formation by impact of fast dust particles and comparison with related techniques. Appl. phys. A 42, 69, 1987b

Zinner E., Amari S., Lewin R.: Silicon carbide from a supernova, 54th annual meeting of the Meteoritical Society, Monterey 1991 (Abstract 261)

Part II

INTERSTELLAR CLOUDS AND STAR FORMATION

Large Scale Properties of Giant Molecular Clouds

LEO BLITZ

University of Maryland

1 INTRODUCTION

It has been more than 15 years since it was established that molecular clouds are the sites of essentially all of the present day star formation in the Milky Way. The largest molecular clouds are the Giant Molecular Clouds (GMCs); they contain most of the mass of molecules in the Galaxy. GMCs are generally understood to be the birthsites for the vast majority of stars that are currently forming; small molecular clouds are responsible for only a small fraction of new stars. Thus, the global, galaxy-wide questions about the nature of star formation reduce to questions of how, where and why GMCs form. Furthermore, the study of the star formation in other galaxies requires a clear understanding of how the process works in our own Galaxy. To understand how GMCs form, it is necessary in turn to understand their large scale properties so that we may deduce how they have arisen out of the diffuse gas. It is therefore with the aim of illuminating how GMCs formed that this paper describes the large-scale properties of Giant Molecular Clouds.

2 OVERALL PROPERTIES OF GMCS

2.1 General Considerations

It is rather amazing that 15 years since the identification of giant molecular clouds, there is no generally accepted definition of what a GMC is. There seems to be little disagreement about the classification of the largest clouds as GMCs, but an all inclusive definition of what a GMC is has proven elusive. A large part of the problem is that the various studies of the mass spectrum of molecular clouds indicate that the spectrum is well fit by a power law (see below) and there is consequently no natural size or mass scale for molecular clouds. What we call a GMC is therefore largely a question of taste. For the purposes of this paper, I will arbitrarily define any molecular cloud that has a mass $\gtrsim 10^5$ $M_\odot$ definitely to be a GMC, a cloud with a mass 10^5 $M_\odot$ $\gtrsim$ M(cloud) $\gtrsim$ 10^4 $M_\odot$ probably to be a GMC, and a cloud with a mass $\lesssim 10^4$ $M_\odot$ probably not to be a GMC. In most cases, this fuzziness does not cause serious problems.

Giant molecular clouds have been studied as a whole largely through their CO emission in the radio portion of the spectrum. The studies have been mainly of two kinds. *1)* Observations of individual objects, where a cloud is identified, often by its association with a visible HII region, and then mapped to its outer boundaries. *2)* Surveys of the Galactic plane in CO, where GMCs are identified through some objective criterion. In the second case, GMCs are generally defined down to some contour level because of the possibility of confusion with other gas along the line of sight. Surprisingly, the general properties of GMCs defined in this way are not significantly different from those identified by the first method. Table 1 gives the properties derived for clouds in the solar vicinity (*e.g.* Blitz 1987b). Inner Galaxy molecular clouds may be somewhat denser and more opaque (see McKee 1989; Solomon, *et al.* 1987), but there is no evidence that they form a separate population from the local clouds. This important conclusion suggests that detailed studies of the molecular clouds near the Sun can tell us about the ensemble properties of GMCs everywhere in the disk. An exception is likely to be in the innermost regions of the Galaxy such as the molecular disk within 400 pc.

TABLE 1. Global Properties of Solar Neighborhood GMCs

Mass	$1 - 2 \times 10^5$ $M_\odot$
Mean diameter	45 pc
Projected surface area	2.1×10^3 pc^2
Volume	9.6×10^4 pc^3
Volume averaged $n(H_2)$	~ 50 cm^{-3}
Mean $N(H_2)$	$3 - 6 \times 10^{21}$ cm^{-2}
Local surface density	~ 4 kpc^{-2}
Mean separation	~ 500 pc

An example of a local GMC is shown in Figure 1 (Blitz 1978). The contours are the CO emission from the GMC associated with Mon OB1, the OB association that contains the young cluster NGC 2264. The GMC appears to consist of two clouds; the one on the left is associated with the cluster NGC 2264, and the one on the right with an association of reflection nebulosity, Mon R1. The molecular clouds are elongated parallel to the Galactic plane, a feature common to many GMCs (Blitz 1978; Stark and Blitz 1978). Also, the cloud on the right has a large scale velocity gradient. This can be seen from Figure 2 which shows channel maps of the emission from different velocites. The subject of gradients, and their interpretation as rotation are discussed in Section 4. Below, inferences made from studies of local GMCs and inner Galaxy GMCs are considered separately.

2.2 Local GMCs

From the study of Local GMCs, the following general conclusions may be drawn:

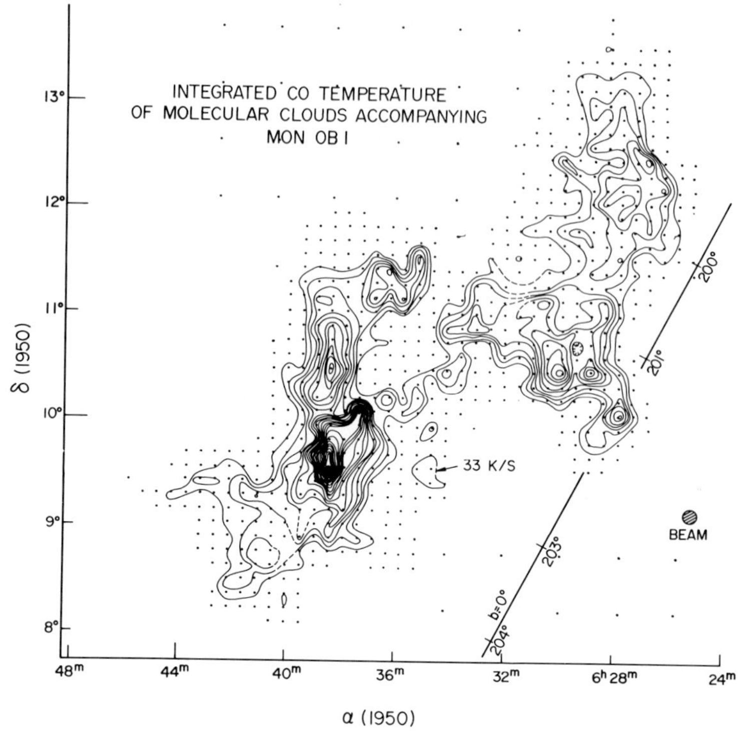

Figure 1. Map of the CO emission from the GMC associated with Mon OB1. The cloud on the left is associated with the young cluster NGC 2264 which is centered near the peak of the CO emission. The cloud on the right is associated with Mon R1, a reflection nebula association. Mon R1 is located near the peaks of the CO emission at the southernmost part of the clouds. The dots represent all of the sampled points.

1) GMCs are discrete objects with well defined boundaries (see *e.g.* Blitz and Thaddeus 1980 for a quantitative discussion of this point - see especially Appendix B). The well defined boundaries suggest that there is a phase transition at the edges of a molecular cloud.

2) GMCs are not uniform entities, but are always composed of numerous dense clumps and have small volume filling fractions. These clumps appear to have a range of geometries from spherical to highly filamentary (see the maps in the references to Table 4 as well as Bally *et al.* 1987). Nevertheless, maps of the CO emission from

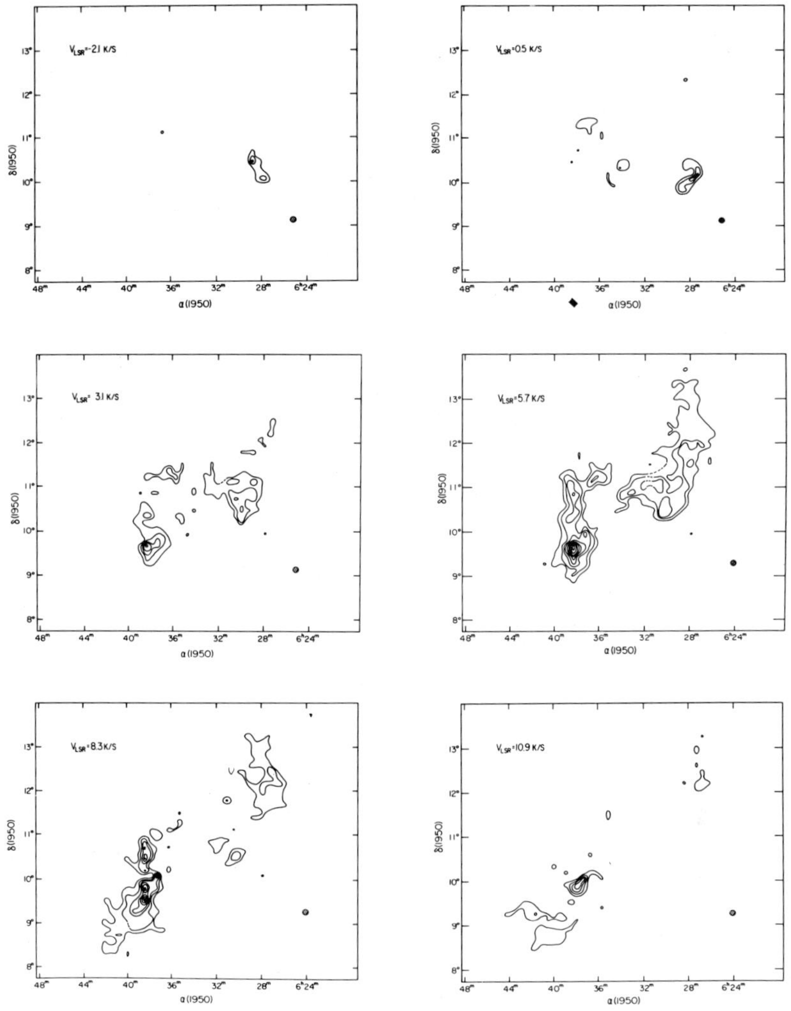

Figure 2. Channel maps of the emission shown in Figure 1. Velocities increase in steps of 2.6 $km\,s^{-1}$ from left to right and from top to bottom. Note that the centroid of the emission progresses steadily northward for the cloud on the left, but is essentially unchanged for the cloud on the right, indicating a velocity gradient for only the cloud associated with Mon R1.

the complexes suggest that the surface filling fraction of the clumps is almost always near unity, where unity is defined as one or more clumps along the line of sight (Blitz 1980 and references therein).

3) The CO and ^{13}CO line widths of the clumps are *always* wider than the thermal widths implied by the excitation temperature. The linewidths are usually interpreted as the result of bulk motions associated with turbulence and/or magnetic fields.

4) GMCs are gravitationally bound (*e.g.* Kutner *et al.* 1977; Elmegreen *et al.* 1979; Blitz 1980). GMC masses are orders of magnitude larger than a Jeans mass (see *e.g.*Bertoldi and McKee 1992).

5) All OB associations form from GMCs; thus GMCs are the nucleation sites for nearly all star formation in the Milky Way (Zuckerman and Palmer 1974; Blitz 1978, 1980 and references therein).

6) Within the uncertainties, the cloud-to-cloud velocity dispersion of local GMCs (with typical masses of $\sim 2 \times 10^5$ $M_\odot$) is the same as that of the small molecular clouds (with typical masses of ~ 50 $M_\odot$) found at high galactic latitude (Blitz 1978, Stark 1984, Magnani, Blitz and Mundy 1985). The velocity dispersion of local molecular clouds therefore appears to be independent of mass over three or four orders of magnitude. If clouds are formed by collisional agglomeration of smaller independently moving clouds, then cloud velocity dispersions would be proportional to $M^{-0.5}$, unless there were some way to selectively re-energize the more massive clouds, an unlikely prospect. Furthermore, the velocity dispersion of molecular clouds seems to vary only weakly with galactocentric distance (Liszt and Burton 1983, Stark 1984, Clemens 1985) implying that the constancy of velocity dispersion with mass is likely to hold everywhere in the disk of the Galaxy. For the most massive clouds, Stark (1983) has shown that the scale height, and by implication the vertical velocity dispersion of GMCs, exhibits a decrease in the inner Galaxy: cloud-cloud collisions may therefore be important for the largest, most massive clouds. All of these results taken together suggest that collisional processes are not important in the formation of GMCs except possibly in the inner Galaxy at the highest masses ($\sim 10^6$ $M_\odot$).

7) The survey of local molecular material within 1 kpc of the Sun (Dame *et al.* 1986) finds no GMCs without star formation. In fact, within 3 kpc of the Sun, only one GMC is found without evidence of star formation (Maddalena and Thaddeus 1986). This cloud is discussed in more detail below. In the solar vicinity at least, molecular clouds without star formation are quite rare. It therefore appears that star formation is quite rapid after the formation of a GMC.

8) The above result, when combined with calculations of the destructive processes associated with the formation of massive stars, implies that the GMCs are quite young, $\sim 3 \times 10^7 y$ (Blitz and Shu 1980). That is, since all GMCs appear to be sites of star formation, especially massive star formation, the clouds do not appear to be able to survive the birth of more than a few generations of massive stars, an argument that is consistent with various other lines of reasoning (Blitz and Shu 1980). Direct observational evidence for lifetimes of this order was first presented by Bash, Green and Peters (1977), and subsequently by Leisewitz, Bash, and Thaddeus (1986) from the association (or lack thereof) of molecular material with young clusters of vaying ages. Cohen (*et al.* 1980), have come to the same conclusion from the confinement of the molecular arms in the outer Galaxy to narrow velocity ranges.

9) A large fraction of the stars formed in GMCs do not return their material to the interstellar medium in a Hubble time. Therefore, either the GMCs were more numerous in the past, or the interstellar medium is replenished by infall or inflow from the outer reaches of the Galaxy (*e.g.* Lacey and Fall 1985).

2.3 Inner Galaxy GMCs

Studies of GMCs based on inner Galaxy CO surveys run into the problem of finding an objective way to identify the molecular clouds. The degree of blending of the spectral lines, the large amount of foreground and background emission, and the necessity of using kinematic criteria for identifying the clouds add a significant amount of uncertainty to the identification of GMCs (see *e.g.*Adler 1988, Blitz 1987a). Nevertheless, a number of properties of GMCs are inferred from the Galactic CO surveys, and the searches for molecular clouds in the outer Galaxy. Some of the more important of these are the following:

1) There appears to be a linewidth-size relation for GMCs in the inner Galaxy. From various studies, the results may be given as follows:

TABLE 2. Linewidth - Size Relation for GMCs

Dame *et al.* (1986)	$\Delta_V = 1.20R^{0.50}$
Solomon *et al.* (1987)	$\sigma_V = 1.0S^{0.50}$
Scoville *et al.* (1987)	$\sigma_V = 0.31D^{0.55}$
Sanders *et al.* (1985)	$\Delta_V = 0.88D^{0.62}$

where the quantity on the left is either the full width at half maximum or the one dimensional velocity dispersion in km s^{-1}, and the quantity on the right is some measure of the mean linear size of a GMC in pc. The scatter in all of the studies is not terribly large, but note that the second and the third determinations are obtained from the same data set. The close agreement between the various observers in the value of the power law exponent in the linewidth-size relation is so striking, that unless there is a selection effect common to all of the studies (see *e.g.* Blitz 1987a), it suggests that the relation underlies some fundamental property of GMCs.

Nevertheless, both the reality and the implications of the linewidth- size relation have been challenged by several observers. Adler and Roberts (1992) have argued on the basis of N-body simulations that many of the clouds identified in Galactic surveys are chance superpositions of unrelated clouds. Nevertheless, these spurious clouds produce a linewidth-size relation with the proper slope. The reality of the relation has been questioned by Issa *et al.* (1990) from an analysis of published survey data. These authors showed that if the Solomon *et al.* (1987) method of analysis is used on their own survey data, but at *a*) positions not centered on their clouds, and *b*) at random locations in the survey, one obtains a linewidth-size relation similar to that of the original analysis. This result is puzzling unless the linewidth-size relation is an artifact of the data set itself, rather than of the dynamical state of the clouds. In addition, Issa *et al.* also lowered the threshold of their analysis to a point where one expects clouds to be blended together into apparent clouds that are unrelated to one another. Again, Issa *et al.* obtained a linewidth-size relation indistinguishable from that of Solomon *et al.* suggesting that there is no physical significance to the linewidth-size relation. Its application to clouds to obtain, for example, their distances, may be flawed. Analysis of the Bell Labs ^{13}CO Galactic plane survey, should be able to resolve this issue because of the higher contrast in the ^{13}CO data compared to that of CO.

2) Solomon *et al.* (1987) have shown that for GMCs in the inner Galaxy there is a correlation between CO line luminosities and virial masses, and thus that *a*) the CO line is a good tracer of the mass of a GMC, and *b*) GMCs are in approximate virial equilibrium. Maloney (1988, 1990) has challenged this interpretation on two counts. First, he argues that the CO luminosity-virial mass correlation can be an artifact of the linewidth-size relation alone. Second, he argues that one obtains a linewidth-size relation indistinguishable from the data if one models an ensemble of pressure confined clouds. Since the CO luminosity-virial mass correlation may be the result of pressure confined clouds and not gravity, it need not imply that GMCs are virialized. Furthermore, because gravity may not be what determines the linewidth-size relation, one cannot calibrate the CO/H_2 conversion ratio using the CO luminosity-virial mass correlation. The subject of the CO/H_2 ratio and its variation has been the subject of considerable controversy, and a review of the arguments related to it is beyond the scope of this paper.

3) Independent estimates of GMC masses from gamma-ray observations (Bloemen

et al. 1984, 1986) and from IRAS observations (*e.g.* Boulanger and Perault 1988), suggest values for the CO/H_2 conversion ratio which imply that GMCs are gravitationally bound. If the clouds are in virial equilibrium, then

$$(\Delta V)^2 = \alpha GM/R.$$

where α is a constant near unity. If $(\Delta V) \sim R^{0.5}$, then $M \sim R^2$ which in turn implies that the mean H_2 column density of GMCs is constant. This simple conclusion is presumably telling us something important about either how GMCs form or how they regulate themselves. McKee (1989) has assumed the latter in his recent theory of photo-regulated star formation in GMCs.

4) GMCs in the outer Galaxy appear to have lower excitation temperatures than GMCs in the inner Galaxy (Mead and Kutner 1988). It is unclear at present whether this is due to a lower external heating rate, decreased star formation rate or both. In any event, the outer Galaxy GMCs have lower CO luminosities than those in the solar vicinity (Digel *et al.* 1990), which suggests that the CO/H_2 conversion ratio is a function of galactocentric distance.

5) The distribution of masses for GMCs in the inner Galaxy is found to be (for masses in excess of 10^5 $M_\odot$), $dN(M)/dM \propto M^{-1.5}$ (Solomon *et al.* 1987). The power law exponent is the same as that found for the distribution of clump masses in a GMC If one tries to determine at what cloud mass half of the H_2 mass in the Galaxy resides, the various analyses of the inner Galaxy CO give results that fall within a factor of two of the range $1 - 2 \times 10^5 M_\odot$ when account is made of the different assumptions used and assuming thatt the power law can be extrapolated to very low masses. This is very close to the mean value in the solar vicinity (Stark and Blitz 1978).

2.4 Future Work

One of the glaring deficiencies in Galactic studies of GMCs is a quantitative study of how the properties of GMCs vary with galactic radius. For example, we might expect that to be stable against the larger tidal forces and the larger energy density of dissociating radiation, that inner Galaxy clouds would be denser than the clouds at larger galactocentric distance. Such a conclusion was reached from an indirect analysis of Liszt, Burton, and Xiang (1984), however, no direct confirmation has been made. Such a study would be particularly useful in trying to understand how different Galactic environments affect the formation and evolution of GMCs. It has been known since the Altenhoff *et al.* (1970, 1978) 5 GHz surveys of the Galactic plane, for example, that giant HII regions are much more common in what later became known as the molecular ring than they are at larger Galactic radii. Surely the greater efficiency with which the inner Galaxy GMCs convert molecular gas into O stars must be a reflection of differing molecular cloud properties, but which properties? There have been a few studies of molecular clouds in the outer Galaxy that suggest that there are differences between them and the inner Galaxy clouds (*e.g.* Mead and Kutner 1988; Digel *et al.* 1990), but other than these, systematic studies of the gradients of the large scale properties of GMCs as a function of Galactic radius are almost totally lacking. It would seem imperative that even a rudimentary understanding of extragalactic CO emission would require an understanding of how GMC properties vary within the Milky Way.

3 DO ALL GMCS FORM STARS?

As mentioned above, various surveys of the molecular gas in the vicinity of the Sun have turned up only one GMC within 3 kpc that is devoid of star formation. On the other hand, Mooney and Solomon (1988) argue that at least 25% of the GMCs in the inner Galaxy are devoid of OB star formation. Therefore, on the basis of local observations, the answer to the question posed in the heading is essentially yes, and on the basis of inner Galaxy observations, the answer would be no. Strictly speaking, the solar vicinity and inner Galaxy observations are not incommensurate; care must therefore be taken in defining the question so that answers are not biased by studies with different sensitivity limits.

Consider, for example, the Taurus and Ophiuchus clouds. Each has a mass of about $10^4 M_\odot$ (Ungerechts and Thaddeus 1988; de Geus *et al.* 1990), and from the definitions given Section 2.1, these are probably not GMCs. Nevertheless, both clouds have substantial associated star formation, but no associated HII regions. The star formation efficiency in the core of Ophiuchus, for example, may be as high as 25% (Lada and Wilking 1984). Although these cloud complexes both have stars as early as B associated with them, it would be very difficult to determine whether clouds in the inner part of the Milky Way that otherwise appear to be devoid of signs of star formation are like these clouds. Ophiuchus, for example, has a 100 μm infrared luminosity of about 7000 $L_\odot$, and the infrared luminosity is consistent with it being reradiated starlight from embedded sources (Greene *et al.* 1989). The overall star formation efficiency of the Ophiuchus complex (zero age main sequence and pre-main sequence stellar mass divided by the mass of H_2) is about 2-3 % (Wilking, Lada and Young 1989; Lada personal communication), equal to the mean in the plane of the Galaxy (Myers *et al.* 1986). This value must be typical of GMCs in the Milky Way if most star formation takes place in GMCs. The level of activity found in Ophiuchus, even if it were scaled up for clouds with masses which would here be considered definitely to be GMCs, that is, masses greater than 10^5 $M_\odot$, would not be recognizable from studies such as those of Mooney and Solomon (1988).

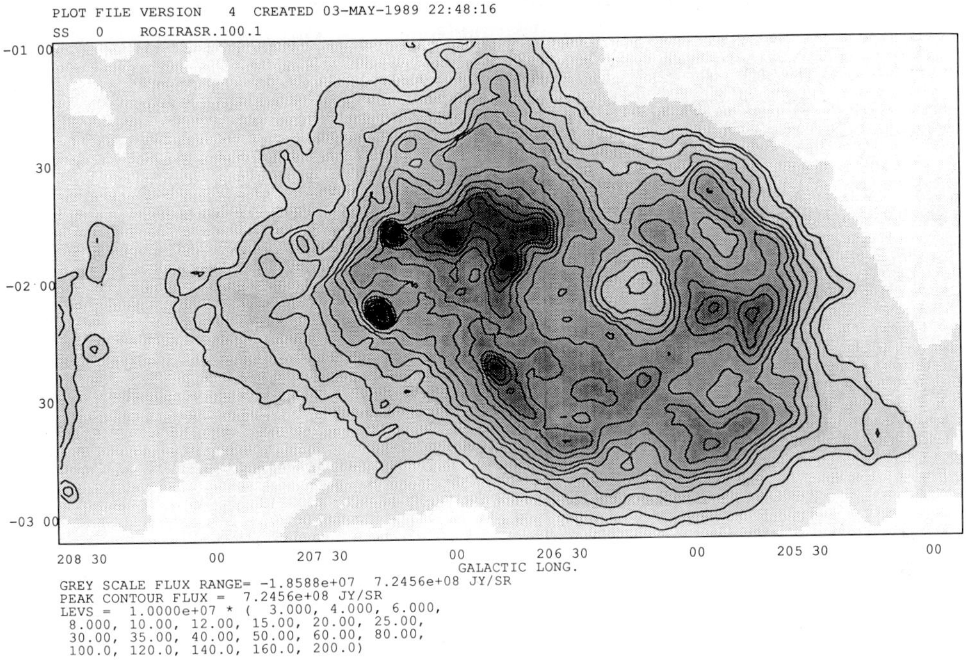

Figure 3. IRAS emission associated with the Rosette Molecular Cloud Complex. The map has had a background subtracted. Note that the diffuse peaks are related to peaks of the CO emission in the molecular cloud and that the hole in the emission is very nearly centered on the HII region.

However, we run into difficulties even if we try to determine whether all OB star formation would be detectable in Galactic surveys, where we define OB star formation as any star formation that gives rise to an HII region. To see this, see Figure 3 which shows the 100 μm surface brightness of the cloud complex associated with the Rosette Nebula as observed by IRAS. The peak brightness of the map, which has had a background removed, is 720 MJy sr^{-1}, and the infrared luminosity of the cloud is $\sim 5 \times 10^5$ L$_\odot$.

Mooney and Solomon find that there is a class of IR-quiet molecular clouds, many with masses $> 10^5$ M$_\odot$, which have peak 100 μm surface brightnesses only 1/20 that of the IR-strong clouds. The IR-strong clouds are defined by strong localized IR emission which, in most cases, seems to result from the action of an HII region. If it were located in the inner Galaxy, the Rosette Molecular Cloud (RMC) would probably be classified as infrared quiet based on the definition for IR-strong clouds given by Mooney and Solomon. That is, an IR-strong cloud must have a peak $100\mu m$ surface brightness more than a factor of two above the local backgrond. In the longitude range $40° < l < 0°$, the mean surface brightness of the galactic plane between $0.75° < b < -0.75°$ is never less than 1000 MJy sr^{-1} (Sodrowski *et al.* 1987), thus in a typical location in the inner Galaxy, the Rosette Molecular Cloud would not meet the brightness criterion. Incidentally, the Rosette HII region itself is actually a minimum on the infrared maps. Furthermore, its peak 100 μm surface brightness is only 3.5 times that of the mean of the IR-quiet clouds, $\sim$ 200 MJy sr^{-1}. Nevertheless, the RMC is particularly rich in O stars even though its size and mass are typical of GMCs in the solar vicinity. The nebula itself is illuminated by 5 O stars and the stellar association has 13 stars of spectral type O9.5 or earlier. The association is about as rich as the Orion OB1 association, the most spectacular star-forming region within 1 kpc of the Sun. Therefore, in spite of the richness of its OB star formation, the RMC would probably have been classified by Mooney and Solomon as being devoid of OB star formation, probably because there are no *embedded* O or early B stars. Furthermore, if the Rosette OB association had fewer O and B stars, its far-infrared surface brightness and luminosity would likely be even smaller. It is quite possible, therefore, that all of the clouds classified as IR-quiet are Rosette-like objects or similar sources with even fewer O and B stars.

A similar problem exists if one tries to use HII regions to trace the star formation in inner Galaxy GMCs as was done by Myers *et al.* (1986). They used the Altenhoff *et al.* (1970, 1978) 5 GHz surveys to find HII regions associated with molecular clouds, and found a few GMCs (a smaller percentage than Mooney and Solomon) with very low star formation efficiencies based on the absence of detectable 5 GHz flux. However, the Rosette Nebula is known to have an emission measure of about 3000 cm^{-6} pc (Bottinelli and Gouguenheim 1964), well below the detection limit at which HII regions are unambiguously identified in the Altenhoff survey. Other solar vicinity OB associations (*e.g.* the Mon OB1 association which contains the well known young cluster NGC 2264) have HII regions with even lower emission measures. Low level IR or radiocontinuum emission may simply signal that the conditions in inner Galaxy molecular clouds without giant HII regions are like those in the Rosette Nebula, and not due to the absence of O stars, or star formation in general. Such situations can arise when the action of O stars destroys the dust, when the angle subtended by the molecular cloud at the ionizing stars is small, or both.

Lockman (1990) has considered the detectability of the Rosette Nebula in the inner Galaxy in a recent review. He concludes that the Rosette Nebula would not have been detected in *any* discrete source survey of the inner Galaxy to date, including his own sensitive recombination line survey (1989). Consequently, the abundance of low surface brightness HII regions in the inner Galaxy is unknown, and may account for the clouds identified by Mooney and Solomon and by Myers *et al.* that have little apparent star formation.

There may therefore be no discrepancy between the fraction of GMCs that form stars in the solar vicinity (virtually all) and that found in the inner Galaxy. The Mooney and Solomon clouds not only may have star formation efficiencies similar to that of other GMCs in the Galaxy, but they could contain substantial OB star formation and still be IR quiet according to their definition. On the other hand, some of the clouds in their sample may have low star formation rates, but much more sensitive observations are necessary to determine this. The Maddalena-Thaddeus cloud remains the only known molecular cloud where the star formation rate within the cloud is known to be a small fraction of the rate observed in local OB associations. It is possible with directed observations to set better limits on the stellar content of individual inner Galaxy molecular clouds than are currently available, and it may be possible to find clouds with star formation rates significantly smaller than those found in local GMCs. On the other hand, if the inner Galaxy clouds form stars with the same efficiency as local GMCs (Mooney and Solomon 1988), we might expect that finding GMCs with significantly depressed star formation efficiencies inside the solar circle will prove difficult.

4 ANGULAR MOMENTUM

4.1 Rotation vs. Shear

That GMCs rotate has long been inferred from the velocity gradients observed across them (*e.g.* Kutner *et al.* 1977). The Kutner *et al.* observations of the Orion cloud show a continuous velocity gradient suggestive of solid body rotation. However, the velocity gradients may also be due to shear resulting from the momentum deposited by energetic stellar winds. Bally (1989) has argued, for example, that the large scale velocity gradient in the L1641 cloud in Orion results from the interaction of the molecular gas with the energetic stellar winds from the O stars in the association. Could it be that the large scale gradients of all GMCs in which they are found are the result of similar stellar wind activity?

Table 3 gives the values of the largest known velocity gradients for entire GMCs.

TABLE 3. Measurable Velocity Gradients for GMCs

Cloud	Gradient $km\,s^{-1}\,pc^{-1}$	Reference
Rosette	-0.18	Blitz and Thaddeus (1980)
Mon R1	$\sim$ -0.2	Blitz (1978)
W3	0.30	Thronson *et al.* (1985)
Orion	-0.10	Kutner *et al.* (1977)

The negative sign indicates that the inferred angular velocity is antiparallel to the angular velocity vector of the galactic disk. Other velocity gradients published for GMCs are significantly smaller in absolute value than the values quoted above. Note that the observed gradients listed in Table 3 are much larger than the galactic shear due to differential rotation. In each case, the most recent burst of star formation has taken place at or near the edge of a prolate molecular cloud, and, with the exception of the W3 cloud, the gradient is in a direction roughly parallel to the galactic plane along the long axis of the cloud. In the W3 cloud, the gradient is nearly perpendicular to the plane of the Galaxy in a layer argued by Lada *et al.* (1978) and Thronson *et al.* (1985) to be swept up by the Cas OB6 association. It appears likely that the velocity gradient in this cloud is related to the sweeping action of the stellar association. Can an OB association induce a shear in a GMC along the *long* axis of the cloud if the stars are situated at one end? Shear thus induced can in principle produce a positive or negative velocity gradient depending on the orientation of the cloud with respect to the line-of-sight. In what follows, we consider only the three clouds with gradients roughly parallel to the plane; we omit W3.

If GMC velocity gradients were due to the acceleration of cloud material by stellar winds, then the geometry of the complexes must have undergone a considerable rearrangement. The material most distant from the accelerating stars must have at one time been closest to them. The timescale required to reach the present state is $\sim 1 - 2 \times 10^7$ y. The material must have remained at all times molecular, because there is no evidence in any of the clouds that the gas has condensed from a previously atomic or ionized state. Nevertheless, we know that HII regions tend to ionize and evaporate the molecular gas rather than simply to accelerate it. A good example is the molecular gas most closely associated with the Rosette Nebula. The mean density of the optical nebula is nearly equal to the mean density of the molecular cloud suggesting that the stars have simply caused a phase change in the pre-existing gas. Also, there is remnant molecular gas seen in projection against the optical nebula, the well known optical globules, and these have been shown to be evaporating (Herbig 1974, Schneps *et al.* 1980). Furthermore, in all cases the clouds are prolate with the long axis pointing away from the direction of the accelerating stars. It is hard to understand how it would be possible to accelerate such a large mass of neutral material so that the the acceleration is linear rather than shell-like.

Perhaps the most serious objection to a wind induced gradient for the GMCs comes from the conservation of linear momentum. If the clouds in Table 3 have a mean mass of $\sim 10^5$ $M_\odot$, the linear momentum imparted to the clouds has a typical value of $\sim 5 \times 10^5$ $M_\odot$ km s^{-1}. This implies, from Newton's third law, that gas has been accelerated in the opposite direction with the same momentum. Orion has a bubble of ionized gas expanding in a direction away form the molecular cloud (Reynolds and Ogden 1979), but the mean radial velocity of the ionized gas has the wrong sign to conserve the linear momentum. The ionized gas, therefore, should have no relation to the velocity gradient in the cloud. Furthermore, no such "counterjets" of material are evident in the vicinity of any of the other two GMCs in either atomic, molecular or ionized gas. Moreover, if the mass of any such gas is small compared to that of the GMC itself, it would be accelerated to velocities that would stand out clearly in spectral line surveys of such gas in the galactic plane.

It therefore seems reasonable to conclude that the velocity gradients observed in CO are due to rotation rather than shear, and the remainder of this section proceeds from this conclusion.

4.2 Conservation of Angular Momentum

The angular momentum of a GMC should reflect the angular momentum of the interstellar medium from which the cloud formed, thus the specific angular momentum of a GMC (that is, the total angular momentum per unit mass of a cloud complex) is an important parameter for understanding its history. In spite of this, little is known about the angular momentum of GMCs. Mestel (1966) considered the problem of the angular momentum of a planet condensing from a protoplanetary disk; this work is applicable in part to the problem of GMC formation from the disk of the Galaxy. In the case of a GMC that condenses from the general interstellar medium, however, it is possible, perhaps even likely that magnetic fields play a role (e.g. Mouschovias, Shu and Woodward 1974), a possibility not considered by Mestel and which adds some complexity to the problem. Let us, in any event, consider how the specific angular momentum of a GMC compares to that of the material from which a cloud formed.

Initial Angular Momentum States. The specific angular momentum, J/M, of the general ISM with which one wants to compare that of a GMC depends in detail on how a cloud forms and the angular momentum distribution of the disk of the galaxy in which it forms. We might consider, for example, that a cloud forms when a disklike region becomes gravitationally unstable and collapses to form the cloud. In this case, we expect a result similar to that considered analytically by Mestel (1966) who showed that for a cloud that has condensed from a disk with a flat or rising rotation curve, the angular momentum is always prograde in an inertial frame of reference. If the rotation curve is falling, then the rotation can be either prograde or retrograde depending on the details of the cloud formation. Furthermore, it is the form of the rotation law *locally* that determines the initial angular momentum of the cloud.

We wish to calculate $1/2R^2\Omega$, the specific angular momentum for the interstellar medium from which a local GMC forms. The quantity R is the size of the region that has collapsed to form a GMC. Let us assume first that the cloud is formed from the condensation of atomic gas only. We now ask from what radius must a cloud contract to form a cloud with a mass of 2×10^5 $M_\odot$? Let us assume that the geometry of the initial configuration of the condensing gas is that of a cylinder with a diameter equal to that of its height. This assumption will not have a large effect on the final results unless the initial geometry is very different in one direction from the others. The surface density of atomic gas in the vicinity of the Sun is 5 $M_\odot$ pc^{-2} (Henderson, Jackson and Kerr (1982). The local effective scale height of the atomic gas is 200 pc (Falgarone and Lequeux 1973), thus the midplane density of atomic gas is 0.5 cm^{-3}. The radius of a cylinder that would contract to form a typical GMC in the solar vicinity is therefore 140 pc.

If the GMCs formed from initially *molecular* gas, R would increase by only about 35% because the surface density of molecular gas locally is ~ 2 $M_\odot$ pc^{-2} (Dame *et al.* 1987). If the GMCs form from a combination of atomic and molecular gas, R will of course have some intermediate value. The dependence of R on the initial geometry is weak because it scales as the 1/3 power of the initial mass. We may then take the value of 140 pc for R to be a reasonable lower limit, but even considering the uncertainties of geometry and the inclusion of molecular gas, it is probably not in error by more than about 50%.

The initial value of Ω depends on the details of the collapse (or equivalently accretion) of the gas that forms the GMC. Consider what happens for collapse along a line of constant radius in the limit that all of the collapsing gas lies at a single distance from the galactic center. This needle-like collapse has zero angular velocity in the rotating frame centered on the LSR (which I will call the LSR frame), and is +0.025 $km\,s^{-1}$ pc^{-1} in the inertial frame. Consider now a similar collapse which takes place in the perpendicular direction, along a line of galactic radius. In the LSR frame, Ω is determined by the large scale shear in the Galaxy. Assuming a flat rotation curve, the value of Ω in the solar neighborhood is -0.025 $km\,s^{-1}\,pc^{-1}$; in the inertial frame, Ω is zero. We find therefore that the value of the initial angular momentum depends on the details of the collapse. The angular momentum of the gas that initially forms a GMC can have a range of values, therefore, from 0 to 250 $km\,s^{-1}$ pc (or 0 to 7.7 $\times 10^{25}\,cm^2\,s^{-1}$) assuming that the GMC formed in a location where the rotation curve is flat. Allowing for the uncertainty in R, this value can be higher by about a factor of 2.

Final Angular Momentum States. Consider a typical GMC in the solar vicinity with a mass of $2 \times 10^5\ M_\odot$. Such clouds have a typical dimension of 45 pc (Blitz 1978), a maximum dimension of about 100 pc and tend to be elongated along the Galactic plane (Stark and Blitz 1978; Blitz 1980). If the true shape of the clouds is that of a cigar, the specific angular momentum is $1/3R^2\Omega$; if it is a disk, then the specific angular momentum is $1/2R^2\Omega$. As stated above, the values given in Table 3 are the largest published velocity gradients for entire GMCs. The velocity gradient of a *typical* GMC is less than half that of the L1641 cloud, and published maps make it difficult to discern gradients smaller than about 0.02 $km\,s^{-1}\,pc^{-1}$. In a survey of the literature, Blitz (1980) found that only half of the GMCs known at that time had a measurable velocity gradient. Thus, if the velocity gradient is entirely due to rotation, Ω has an extreme value for GMCs of 0.3 $km\,s^{-1}\,pc^{-1}$ in the solar vicinity, but has a typical value somewhat less than 0.05 $km\,s^{-1}\,pc^{-1}$. The specific angular momentum of a typical GMC is therefore somewhat less than about $\pm$ 15 pc $km\,s^{-1}$ or $5 \times 10^{24}\,cm^2\,s^{-1}$, for a cloud with a typical diameter of 45 pc. The specific angular momentum for the four clouds listed in Table 3 is shown below in Table 4.

Table 4. Angular Momentum of Four Local GMCs

Cloud	$1/2R^2\Omega$ $km\,s^{-1}$ pc
Rosette	-47
Mon R1	-45
W3	27
Orion	-74

For the clouds with measurable gradients, the sense of the rotation is *retrograde* with respect to the galactic rotation in a rotating coordinate frame of reference centered on the local standard of rest (LSR). That is, the three clouds (other than W3) have higher velocities on the low longitude side of the cloud. To transform to an inertial frame of reference we must add the angular velocity of the LSR, Ω_0, to the observed angular velocities; about 0.025 $km\,s^{-1}\,pc^{-1}$. This value is small compared to the measured angular velocites for the GMCs in the table above. Thus, the sense of the angular momenta of the three clouds with negative gradients is retrograde even in an inertial frame of reference. The GMCs without measurable gradients have angular momenta consistent with zero in the inertial reference frame. It was noted above that most of the mass in GMCs typically resides in dense clumps which may themselves be spinning. However, because of the R^2 dependence of the angular momentum, and because clumps do not, as a general rule, show large velocity gradients (although there are many specific counterexamples), the angular momenta of the clumps are not likely to contribute significantly to the total angular momentum of a GMC.

Comparison. Because the clouds listed in Table 3 are all strongly *counter-rotating*; that is, in the inertial frame, these clouds cannot have conserved angular momentum from the initial states calculated above. Mestel (1966) pointed out that retrograde rotation can occur if the disk from which the clouds form has a falling rotation curve. The Oort A constant expected in the solar vicinity for a flat rotation curve is 12.5 $\mathrm{km\,s^{-1}\,kpc^{-1}}$. However, almost all measurements of this quantity suggest that the value is 15 $\mathrm{km\,s^{-1}\,kpc^{-1}}$ (Kerr and Lynden-Bell 1986), a value which implies that locally the rotation curve is falling. Such local variations are expected in the vicinity of a spiral arm. The value of the Oort A constant does not affect the collapse of a cloud along a line of constant radius. However, for the case of radial collapse, Ω has a value of -0.005 $\mathrm{km\,s^{-1}\,pc^{-1}}$ in the inertial frame. The corresponding specific angular momentum is -50 $\mathrm{km\,s^{-1}}$ pc, but may be somewhat higher if the initial geometry is different from the one considered above.

Thus all of the clouds with large gradients have angular momenta within the values expected from collapse from the diffuse interstellar medium. However, the counter-rotating clouds put a severe constraint on how the clouds can have formed. That is, they must have formed where the rotation curve is locally falling, and they must have collapsed very nearly along a line of Galactic radius. Becuase this constraint is so severe, it is important to establish unequivocally that the large velocity gradients are due to rotation and not to shear. In this way, statistical studies of the angular velocities of many GMCs can provide important information about how the clouds formed.

Now, consider the *typical* GMCs. These have angular momneta which are only a small fraction of the allowable range. Does this mean that angular momentum is shed at an early stage in the collapse, or does it tell us that the initial angular momentum is frequently large, but in most cases is shed at an early stage? Understanding this question will likely tell us much about how GMCs form and evolve.

It is important to recognize that the observed values of J/M quoted above are reasonably well determined. For a molecular cloud, the velocity differences across a cloud can be measured to an accuracy of about 1 $\mathrm{km\,s^{-1}}$ (measurement errors are considerably smaller). Large velocity gradients are therefore measurable to an accuracy of 10-20%. The major source of uncertainty is probably the distance to a cloud, which, for clouds like Orion and the Rosette are probably as low as about 20%. For the ISM, the largest uncertainties are the value of the Oort A constant, which is probably known to an accuracy of 20%, and the midplane density of the atomic hydrogen gas, which translates into the distance R out to which the ISM must be collected to form a GMC. The uncertainty in the density is probably less than 50%, but in any event, the distance R depends only on the 1/3 power of the midplane density, and thus J/M on the the 2/3 power. The major uncertainty for the specific angular momentum of the ISM is the actual process by which GMCs are formed. For example, if the efficiency of transformation of the ISM into molecular material is low, then R may be bigger, and the resulting J/M may be larger.

Studies of the angular momentum distribution among the GMCs simply do not exist at present in spite of the importance that they have for understanding how the molecular clouds form from the interstellar medium. It is unknown, for example, whether there are any examples of clouds with measurable prograde rotation in the LSR frame. Even the results quoted above come partly from unpublished analyses of molecular clouds and from personal communications. There is a wealth of information, however, in the large scale surveys of the CO distribution in the Galaxy. For example, there have been hundreds of GMCs that have been catalogued (Solomon, *et al.* 1987; Scoville *et al.* 1987) in the inner Galaxy. It would be an easy task to determine to what degree GMCs which exhibit measurable velocity gradients will have the sense of the angular momentum antiparallel to the angular momentum of the Galaxy, yet this has never been done. Nor has a detailed study of the angular momentum of GMCs as a function of Galactic radius been done. Many other questions about the distribution of angular momentum are answerable with existing data.

REFERENCES

Adler, D.S, 1988, Ph.D. Dissertation, University of Virginia.

Adler, D.S., and Roberts, W.W., 1992, *Ap. J.*, in press.

Altenhoff, W., Downes, D., Goad, L., Maxwell, A., and Rinehart, R., 1970, *Astr. Ap. Suppl.*, **1**, 319.

Altenhoff, W., Downes, D., Pauls, T., and Schraml, J., 1978, *Astr. Ap. Suppl.*, **35**, 23.

Bally, J., Langer, W.D., Stark, A.A., and Wilson, R.W., 1987, *Ap. J. (Letters)*, **312**, L45.

Bally, J., 1989, in *Low Mass Star Formation and Pre-Main Sequence Objects*, ESO

Workshop Proceedings No. 33, B. Reipurth, ed., ESO: Munich, p.1.
Bash, F.N., Green, E., and Peters, W.L., 1977, *Ap. J.*, **217**, 464.
Bertoldi, F., and McKee, C.F., 1992, *Ap. J.*, *in press.*
Blitz, L., 1978, Ph.D. Dissertation, Columbia University.
Blitz, L., 1980, in *Giant Molecular Clouds in the Galaxy*, Solomon and Edmunds, eds., Pergammon:Oxford, p.1.
Blitz, L. and Shu, F.H., 1980, *Ap. J.*, **238**, 148.
Blitz, L. and Thaddeus, P., 1980, *Ap. J.*, **241**, 676.
Blitz, L., 1987a, in *Millimetre and Submillimetre Astronomy*, Wolstencroft and Burton, eds., (Kluwer:Dordrecht), p.269.
Blitz, L., 1987b, in *Physical Processes in Interstellar Clouds*, Morfill and Scholer, eds., (Reidel:Dordrecht), p.35.
Bloemen, J.B.G.M., *et al.* , 1984 *Astr. Ap*, **139**, 37.
Bloemen, J.B.G.M., *et al.* , 1986, *Astr. Ap*, **154**, 25.
Bottinelli, L., and Gouguenheim, L., 1964, *Ann. d'Ap.*, **27**, 685.
Boulanger, F., and Perault, M., 1988, *Ap. J.*, , **330**, 964.
Clemens, D.P., 1985, *Ap. J.*, **295**, 402.
Cohen, R.S., Cong, H-I., Dame, T.M., and Thaddeus, P., 1980, *Ap. J. (Letters)*, **239**, L53.
Dame, T.M., Elmegreen, B.G., Cohen, R.S., and Thaddeus, P., 1986, *Ap. J.*, **305**, 892.
Dame, T.M., *et al.* , 1987, *Ap. J.*, **322**, 706.
de Geus, E., 1988, Ph.D. Dissertation, Leiden University.
de Geus, E., Bronfman, L., and Thaddeus, P., 1990, *Astr. Ap*, **231**, 137.
Digel, S., Bally, J., and Thaddeus, P., 1990, *Ap. J. (Letters)*, **357**, L29.
Elmegreen, B.G., and Lada, C.J. and Dickinson, D.F., 1979, *Ap. J.*, **230**, 415.
Falgarone, E., and Lequeux, J., 1973 *Astron. Ap.*, **25**, 253.
Greene, T.P., and Young, E.T., 1989, *Ap. J.*, **339**, 258.
Henderson, A.P., Jackson, P.D., and Kerr, F.J., 1982, *Ap. J.*, **263**, 116.
Herbig, G.H., 1974, *Pub. A.S.P.*, , **86**, 604.
Issa, M., MacLaren, I., and Wolfendale. A.W., 1990, *Ap. J.*, **352**, 132.
Kerr, F.J., and Lynden-Bell, D., 1986, *M.N.R.A.S.*, **221**, 1023.
Kutner, M.L., Tucker, K.D., Chin, G., and Thaddeus, P., 1977, *Ap. J.*, **215**, 521.
Lacey, C.G., and Fall, S.M., 1985, *Ap. J.*, **290**, 154.
Lada, C.J., Elmegreen, B.G., Cong, H-I., nad Thaddeus, P., 1978, *Ap. J. (Letters)*, **226**, L39.
Lada, C.J., and Wilking, B.A., 1984, *Ap. J.*, , **287**, 610.
Leisewitz, D., Bash, F.N., and Thaddeus, P., 1989, *Ap. J. Suppl.*, **70**, 731.
Liszt, H.S., and Burton, W.B., 1983, in *Kinematics, Dynamics, and Structure of the Milky Way*, W.L. Shuter, ed., (Reidel:Dordrecht), p.135.

Liszt, H.S., Burton, W.B., and Xiang, D.,1981, *Astron. Ap.*, **140**, 303.
Lockman, J., 1989, *Ap. J. Suppl.*, **71**, 469.
Lockman, J., 1990 in *Radio Recombination Lines: 25 years of Investigation*, Gordon and Sorochenko, eds., Kluwer: Dordrecht, p.225.
Maddalena, R., and Thaddeus, P., 1985, *Ap. J.*, **294**, 231.
Magnani, L., Blitz, L., and Mundy, L., 1985, *Ap. J.*, **295**, 402.
Maloney, P., 1988, *Ap. J.*, **334**, 761.
Maloney, P., 1990, *Ap. J. (Letters)*, **348**, L9.
McKee, C.F., 1989, *Ap. J.*, **345**, 782.
Mead, K.M., and Kutner, M.L., 1988, *Ap. J.*, **330**, 399.
Mestel, L., 1966, *M.N.R.A.S.*, **131**, 307.
Mooney, T.J., and Solomon, P.M., 1988, *Ap. J. (Letters)*, **334**, L51.
Mouschovias, T., Shu, F.H., and Woodward, 1974, *Astr. Ap*, **33**, 73.
Myers, P.C., Dame, T.M., Thaddeus, P., Cohen, R.S., Silverberg, R.F., Dwek, E., and Hauser, M.G., 1986, *Ap. J.*, **301**, 398.
Reynolds, R.J., and Ogden, P.M., 1979, *Ap. J.*, **229**, 942.
Sanders, D.B., Scoville, N.Z., and Solomon, P.M., 1985, *Ap. J.*, **289**, 323.
Schneps, M.H., Ho, P.T.P., and Barrett, A., 1980,*Ap. J.*, **240**, 84.
Scoville, N.Z., Yun, M.S., Clemens, D.P., Sanders, D.B., and Waller, W.H., 1987, *Ap. J. Suppl.*, **63**, 821.
Sodrowski, T.J., Dwek, E., Hauser, M.G., and Kerr, F.J., 1987, *Ap. J.*, **322**, 101.
Solomon, P.M., Rivolo, A.R., Barrett, J., and Yahil, A., 1987, *Ap. J.*, **319**, 730.
Stark, A.A., and Blitz, L., 1978, *Ap. J. (Letters)*, **225**, L15.
Stark, A.A., 1983, in *Kinematics, Dynamics, and Structure of the Milky Way*, W.L. Shuter, ed., (Reidel:Dordrecht), p.127.
Stark, A.A., 1984, *Ap. J.*, **281**, 624.
Thronson, H.A., Lada, C.J., and Hewagama, T, 1985, *Ap. J.*, **297**, 662.
Ungerechts, H. and Thaddeus, P., 1987, *Ap. J. Suppl.*, **63**, 645.
Wilking, B.A., Lada, C.J., and Young, E.T., 1989, *Ap. J.*, **340**, 823.
Zuckerman, B. and Palmer P., 1974, *Ann. Rev. Astron. Ap.*, **12**, 279.

Propagating Star Formation

B.G. ELMEGREEN

IBM Research Division, T.J. Watson Research Center

P.O. Box 281 Yorktown Heights, NY 10598 USA

ABSTRACT: Triggering mechanisms for star formation on small, intermediate, and large scales are discussed from both an observation and theoretical point of view. On a small scale, globules that are surrounded by high pressure can be induced to collapse into stars because of a dynamical instability if the compression follows motion along the magnetic field lines, but for compression perpendicular to the field lines, the triggering mechanism is primarily one of enhanced magnetic diffusion. On an intermediate scale, as in an OB association several million years old, triggering follows the collapse of swept-up neutral gas after a time $\sim 0.5/(G\rho)^{1/2}$ for ambient cloud density ρ. This collapse produces dense new cores in the layer which may or may not be unstable to further collapse, but in either case, the normal processes of dissipation and magnetic diffusion lead to rapid cluster formation in these cores. On a large scale, as in the vicinity of OB associations several $\times 10^7$ years old, star formation occurs along the periphery of a swept-up shell as the dense gas collapses into whole new molecular clouds. The time scale for this triggering is again $\sim 0.4/(G\rho)^{1/2}$ for ambient density ρ. Large scale triggering also requires $Q < 1$ for disk stability parameter Q in order for the collapse time to be less than the galactic shear time.

1 INTRODUCTION

This paper is a brief review of the observations and theory of pressure-induced star formation. The physical mechanisms involved are separated into three basic categories depending on the relative size of the compressed and the compressing regions. When the compressed region is much smaller than the compressing gas, as in the case of a small dark globule embedded in an extended HII region, or a small diffuse cloud inside a supernova cavity, then the process of triggered star formation is related to the direct squeezing of a somewhat fixed, or slightly ablated, mass of neutral material. When the compressed gas has a size that is comparable to both the size of the high pressure region and the size of the high pressure source, as in the case of a giant molecular cloud exposed to the compressing and sweeping action of an old HII region and extended OB subgroup, then the process of triggered star formation is nearly a lateral collection of neutral material at the edge of the expanding HII region, followed by the self-gravitational collapse of this material into new molecular cores. When the compressed gas is much larger than the compressing gas, as when the ambient neutral medium is swept into a shell by the expansion of a hot bubble around an OB association, then the process of triggered star formation is related to the gravitational collapse of the shell, which forms whole new molecular clouds. A direct squeezing of pre-existing clouds may be important in this latter case too.

The physical processes for each of these cases are not well modelled or understood yet, but we can grasp the essential physics from simple theoretical considerations. The next three sections review representative observations of these processes and discuss the available theory.

2 SMALL SCALE TRIGGERED STAR FORMATION: GLOBULE SQUEEZING

2.1 Observations

A recent survey of bright-rimmed globules in HII regions led Sugitani, Fukui, and Ogura (1991) to identify 44 with IRAS sources that are likely examples of implosion-induced star formation. These globules have the same range of size and mass as isolated Bok globules, but they have luminosity-to-mass ratios exceeding those of isolated globules by a factor of 100. This excess luminosity led Sugitani et al. to suggest that intermediate mass stars form there, and they estimated that possibly 15 percent of such stars around the Galaxy form in this way.

The shapes of the globules in their survey ranged from slightly curved to comet-like, which is similar to the sequence of shapes expected for the time evolution of a dense globule or small cloud core that is compressed and swept back by an adjacent HII region (e.g., Sandford, Whitaker, and Klein 1982). For the models, star formation is assumed to occur in the last stages of the compression, during the comet-like phase when the globule has become very dense, but in the observations, young stars appear even in the slightly curved globules, which look morphologically young (see Figure 1). This raises the question of whether star formation was actually triggered by the HII region. The young stars could have formed in the small cores independently of the HII region, and then the HII region shaped the peripheral gas afterwards. Alternatively, the collapse of the dense core to a star could be much faster than the peripheral restructuring, in which case star formation could be triggered even in morphologically young bright rims.

The relative importance of external pressure and internal gravity on the boundedness of a cloud can be assessed from the dimensionless pressure $P/G\sigma^2$, for external pres-

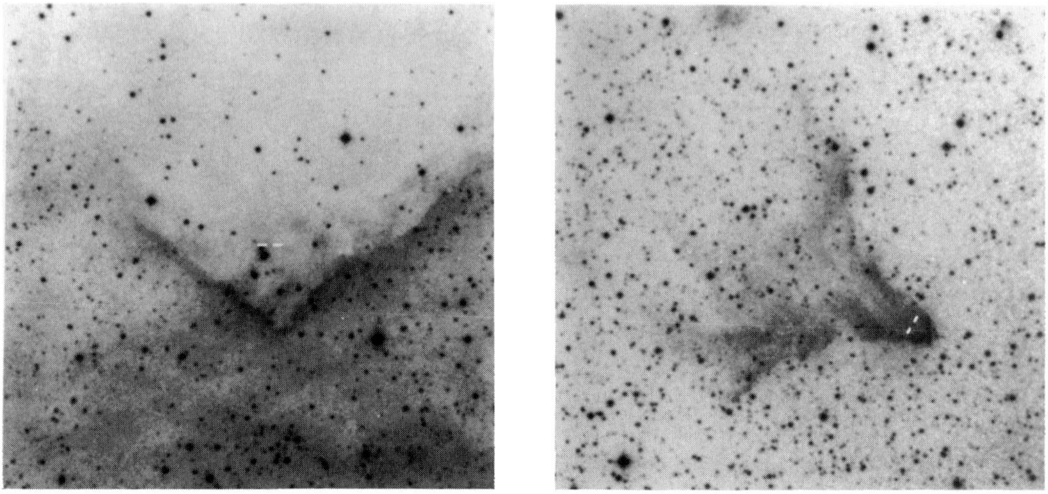

Figure 1 - Two bright-rimmed globules from Sugitani et al. (1991), showing IRAS point sources at the positions of the tic marks.

sure P and internal mass column density σ. The utility of this parameter for understanding cloud structure and formation was discussed extensively by Elmegreen (1991a). Diffuse clouds have $P/G\sigma^2 >> 1$ and self-gravitating clouds have $P/G\sigma^2 <<$ 1. Here the same dimensionless parameter arises from the ratio of the gravitational free-fall time in the unshocked cloud, $(G\rho_0)^{-0.5}$, to the time for a shock front to propagate to the cloud center, which is approximately $R/(P/\rho_0)^{0.5}$; P is the external pressure that drives the shock into the cloud at a velocity of approximately $(P/\rho_0)^{0.5}$ for initial internal density ρ_0. The ratio of these two time scales is $(P/G\sigma^2)^{0.5}$ for initial column density $\sigma \sim R\rho_0$. Because this ratio is larger than 1 for diffuse clouds at the ambient pressure, it must be much larger than 1 for diffuse clouds in a high pressure, and this implies that compression to stellar densities can be sped up greatly by the shock. But for an initially self-gravitating piece of a molecular cloud, the initial value of $P/G\sigma^2$ is probably less than 1 and then the excess pressure from the ionized gas might not be much more than the pressure of the previous molecular environment, which is already large. In this case, the external pressure would not speed up the collapse much and the importance of compression for triggering star formation would be less obvious.

The study by Sugitani et al. illustrates how common it is for dense globules with IRAS sources to be surrounded by ionized gas – they counted approximately two star-forming globules per HII region at any one time. This observation is consistent with the recent finding that molecular clouds have highly convoluted boundaries, with protruding bulbs and hollow cavities everywhere in a fractal-like structure (Dickman et al. 1990; Falgarone, Phillips and Walker 1991). Thus it appears that young HII regions expand into the cavities and surround the bulbs, compressing these protrusions and ionizing their gas. The smallest bulbs are probably completely ionized, but the largest can presumably shield themselves long enough to make a star.

The compression of diffuse clouds inside supernova cavities is also likely, but it is more difficult to observe. Brand et al. (1983) discuss the evidence for triggered star formation in a cometary globule in the Gum nebula, and a catalogue of other cometary globules is in Hawarden and Brand (1976). The fraction of stars that form this way is unknown, but it is not likely to be high because the total diffuse cloud mass is generally much smaller than the total mass in giant molecular clouds. Models of the shock-cloud interaction were made by Rozyczka and Tenorio-Tagle (1987) and Mac Low et al. (this conference). These models concentrate mostly on the shock structure and on turbulent disruption of cloudy material leading to the formation of a comet tail; they are not yet at the point where the possibility of triggered star formation can be assessed in any detail (as discussed in another context by Nelson at this conference).

2.2 Theory

The instability condition for the collapse of a uniform, pressure-bounded, isothermal sphere of mass M and sound speed c is (Ebert 1955; Bonner 1956)

$$P > P_{max} = \frac{1.4c^8}{G^3 M^2}.$$

When the external pressure exceeds P_{max}, the internal pressure can no longer support the cloud and it must collapse. In the 1950's, pressure-induced star formation on a small scale was generally considered to be governed by this instability condition: individual clouds were thought to be squeezed from stable to unstable configurations (Dibai 1958).

The presence of a magnetic field changes this picture. For a magnetic sphere, there is another collapse condition in addition to a critical boundary pressure. The need for an additional condition can be seen most easily from the magnetic virial equation:

$$4\pi R^3 \left(P + \chi \frac{B^2}{8\pi} \right) = \frac{3MkT}{\mu} - \frac{3GM^2}{5R},$$

where χ depends on the field distribution and boundary condition and equals 2/3 for the case where $B \propto R^{-3}$ outside the cloud (Spitzer 1978). With flux freezing, $B \propto 1/R^2$, and the magnetic and gravity terms both scale as $1/R$. This means that if the magnetic energy is sufficient to prevent collapse at some initial time, then it will always prevent collapse regardless of P, so there is no critical pressure.

Instability in a magnetic cloud requires two conditions (Mouschovias and Spitzer 1976). The first is that the field must be sufficiently weak that gravity dominates the magnetic pressure, i.e.,

$$\frac{B_0^3}{G^{3/2}\rho_0^2 M} < 280$$

for initial field strength B_0 and initial density ρ_0 before contraction along field lines. This field strength limit is equivalent to the requirement that the cloud mass exceed a critical magnetic mass:

$$M > M_{mag} = \frac{B_0^3}{280 G^{3/2}\rho_0^2}.$$

Secondly, the external pressure must be sufficiently large to overcome the internal gas pressure (with help from gravity):

$$P > \frac{1.89 c^8}{G^3 M^2 \left\{1 - (M_{mag}/M)^{2/3}\right\}^3}$$

where $M > M_{mag}$. Note that the first condition does not depend on the pressure and the second condition does not depend on the magnetic field strength once the first condition is satisfied by a wide margin, $M >> M_{mag}$.

How do these two conditions apply to star formation in compressed globules? The theory assumes that a uniform cloud with B_0, ρ_0, and external pressure P is given, and then determines if this cloud can contract to a stable equilibrium. For the HII region - globule problem, we imagine that compression and ionization act on the boundary of a cloud piece, giving the initial condition for the collapse. Then the HII region pressure is P in these equations, and the field strength and density in the squeezed globule are B_0 and ρ_0. Note that this field strength and density are not the values before the compression (which could be much smaller), but rather the values after the compression, at the beginning of the collapse.

Now we can rewrite ρ_0 and B_0 in terms of P: let $\rho_0 = \alpha P/c^2$ and $B_0^2/8\pi = \beta P$ for scaling parameters α and β. Then the two conditions for collapse are (1) that the mass must satisfy

$$M > M_{mag} = 0.011 M_{\odot} \frac{c_{0.1}^4 \beta^{3/2}}{P_6^{1/2} \alpha^2}$$

and (2) that the pressure condition for collapse, which can be written in terms of a similar mass threshold, must satisfy

$$M > M_p = 0.034 M_\odot \frac{c_{0.1}^4}{P_6^{1/2}} \left(1 - (M_{mag}/M)^{2/3}\right)^{-3/2}.$$

Note that if the field and thermal pressures in the globule are comparable to the external pressure, $\alpha \sim 1, \beta \sim 1$, then the two conditions are approximately the same. Here $c_{0.1}$ is c in units of $0.1\mathrm{kms}^{-1}$ and P_6 is P in units of $10^6 \mathrm{k_B cm^{-3} K}$ for Boltzmann constant $\mathrm{k_B}$.

Now consider what happens when the compression is perpendicular to the ambient magnetic field in the cloud. Then β increases because of the compression and α decreases (compared to values before the compression). So perpendicular compression can increase M_{mag} and make the cloud piece more stable. This implies that compression perpendicular to the field makes it harder for a cloud piece to overcome its previous magnetic resistance to collapse.

When the compression is parallel to the ambient field, $\alpha \sim 1$ and β decreases, in which case M_{mag} decreases. Then an initially stable cloud piece ($M < M_{mag}$) can become unstable ($M > M_{mag}$) because M_{mag} decreases as a result of the compression. Collapse follows in this nearly field-free case if the mass exceeds the usual Bonner-Ebert mass given by the pressure condition. Thus, compression parallel to the field makes it easier for a cloud piece to overcome its previous magnetic resistance to collapse. This is the same triggering mechanism as in the model by Elmegreen and Lada (1977), as reviewed in Elmegreen (1985).

What about magnetic diffusion? The diffusion rate is given by

$$\tau_D \sim \frac{\nu_{in}}{\omega_A^2} \sim \frac{n_i \sigma_{in} c}{k^2 {v_A}^2} \propto \frac{n_i \rho R^2}{B^2} \propto \frac{\rho^{3/2} R^2}{B^2}$$

for ion-neutral collision rate ν_{in} and cross section σ_{in}, ion density $n_i \propto \rho^{1/2}$, Alfven frequency and speed ω_A and v_A, and wavenumber $k \sim 1/R$. Compared to the gravitational free fall time, this scales as

$$\frac{\tau_D}{\tau_{ff}} \propto \left(\frac{\rho R}{B}\right)^2.$$

When there is a compression of some column of gas parallel to B, ρR is constant in the column and B is constant, so τ_D/τ_{ff} is constant and the compression does nothing to speed up diffusion in relative terms. When there is compression perpendicular to the field, ρ/B is constant but R decreases, so τ_D/τ_{ff} can decrease, leading to more rapid diffusion and eventual collapse.

In summary, compression with a magnetic field is very different than compression without a field. Pressure alone does not trigger dynamical collapse when a magnetic field is present. For compression predominantly parallel to the field, collapse is triggered because M_{mag} decreases during the compression (so gravity overcomes magnetism), and not because $P > P_{crit}$, which is relatively easy to satisfy once $M >> M_{mag}$. For compression predominantly perpendicular to the field, dynamical collapse is inhibited because of the increased field strength, but the field diffuses out of the gas more rapidly. Then enhanced diffusion is what triggers collapse, by lowering M_{mag} to a value less than M.

The stability criterion for a *rotating*, magnetic cloud was given by Tomisaka et al. (1989). If we write the second condition above as the function $M_p(P, c, M_{mag})$, then the critical mass with rotation is found by a numerical solution of the equilibrium states to be

$$M_{cr}^2 = M_p^2 + \left(\frac{4.8cj}{G}\right)^2$$

for specific angular momentum j. This implies (Tomisaka et al.) that cloud collapse can be initiated by a reduction in magnetic flux (diffusion), a reduction in angular momentum, or an increase in boundary pressure. In the first case, M_{mag} decreases, so M_p decreases and M_{cr} decreases, at which point M can find itself greater than M_{cr} and collapse will follow. In the second case, M_{cr} decreases directly. In the third case, M_p decreases and so M_{cr} decreases (for $M > M_{mag}$ already).

The compression of a rotating cloud has the same parallel and perpendicular distinction as the non-rotating cloud discussed above: i.e., compression parallel to the field triggers collapse by decreasing M_{mag} and compression perpendicular to the field triggers collapse by increasing the rate of diffusion. In addition, compression parallel to the field decreases j by preserving the angular rotation rate as the density increases (the loss of angular momentum occurs along the field lines). Then the decrease in both M_{mag} and j can trigger the collapse for the parallel case.

3 INTERMEDIATE-SCALE TRIGGERING: OB ASSOCIATIONS

3.1 Observations

If we view star formation in OB associations on a large scale, measuring perhaps tens of parsecs rather than tenths of a parsec as in the previous section, then another process of triggering becomes apparent. An example is shown in Figure 2. This is a diagram of the OB association and molecular cloud complex in Cepheus OB3.

Approximately 8 million years ago, a cluster began to form in the far side of the cloud at a velocity equal to the cloud's velocity, which is about -10kms^{-1}. During the next 4 million years this cluster apparently pushed on the cloud and gave the adjacent portion of it an excess velocity of about 5kms^{-1} in our direction. This pushing formed a cluster that is now 4 million years old, with a velocity of about -15kms^{-1}, which is still the velocity of the intermixed molecular gas. The rest of the cloud is at the original velocity of -10kms^{-1}.

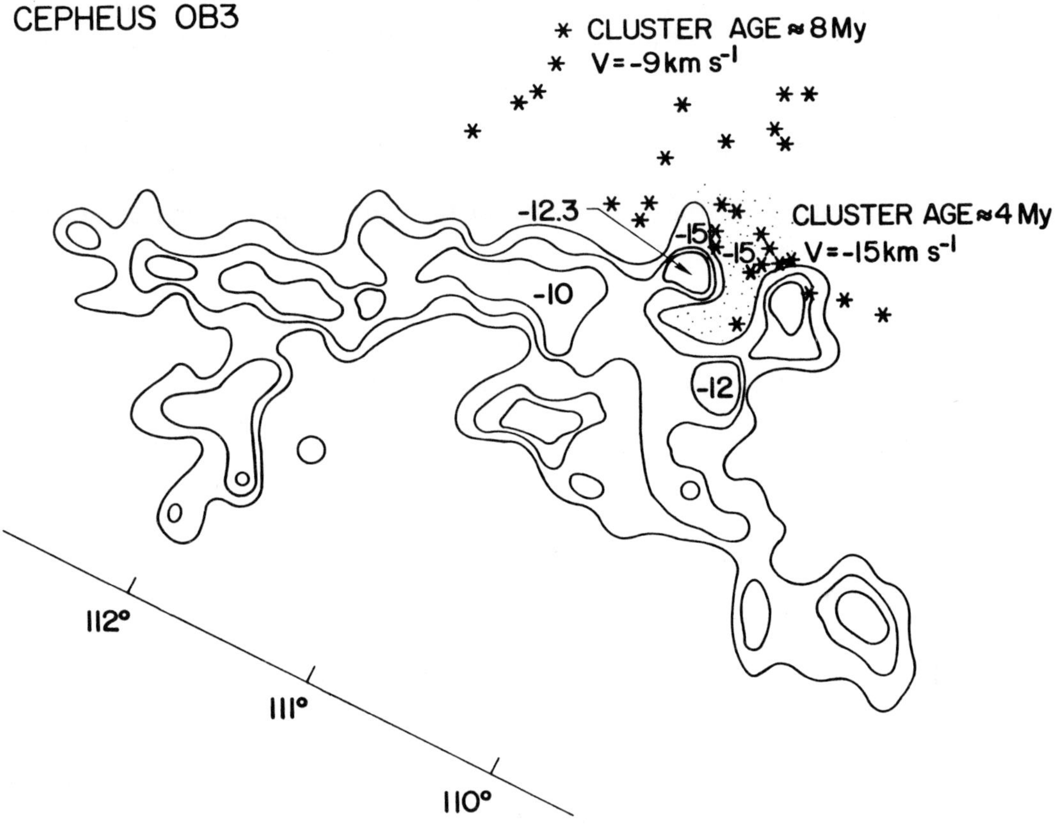

Figure 2 - CO map with stars and velocities of the Cepheus OB3 association (from Elmegreen and Wang 1988, with CO data from Sargent 1979).

Many other old OB associations show a similar morphology: there is an old cluster of stars adjacent to a molecular cloud, an abrupt jump in the cloud's velocity and density at the HII region boundary, and a young cluster of stars (e.g., Lada et al. 1991) in the dense moving gas. The embedded stars are also seen to share the peculiar motion of the dense boundary gas (Wilking et al. 1984; Thronson et al. 1985), which indicates that they formed *after* the gas was pushed. Extensive CO mapping in the

Orion region (Bally et al. 1987) shows in addition a comet-like tail of gas streaming away from the dense cloud core that is adjacent to the HII region.

These observations suggest that intermediate-scale triggering forms whole clusters in single cloud cores of mass $10^3 - 10^4 M_\odot$, that it has a relatively long time delay (several My) after the formation of the older cluster, and that there are relative motions of a few $\mathrm{km s}^{-1}$ between the older cluster and the younger cluster. The gas that forms the younger cluster is at the same velocity as the embedded stars. The relative velocities are also consistent with the expected shock speeds, considering the HII region pressures and the neutral cloud densities, and considering the expanded size of the HII region and the cluster age difference. Thus the mechanism for intermediate scale triggering differs from the mechanism for small scale triggering, not only because of the larger mass involved, but also because the intermediate-scale process includes substantial cloud motions at the periphery of the HII region. There may be a tendency to direct this motion along the prevailing magnetic field lines.

Theory suggests that there *should* be a physical difference between small scale and intermediate scale triggering because the expansion of an HII region changes character as it ages. A young HII region spreads quickly because of the high pressure and strong radiation field, and many of the globules near the edge can be engulfed by ionized gas and either completely ionized or squeezed into stars. The general rate of ionization keeps up with the expansion rate too, so most of the shocked neutral gas becomes ionized unless it is in a dense globule. But an old HII region expands more slowly and the ionization rate of neutral material at the edge falls far below the accretion rate of neutral material at the expansion front. This causes a thick neutral layer to accumulate near the edge of an old HII region.

In the classical theory, the total mass engulfed by the expansion increases as $\rho_0 R^3$ and the ionized mass increases as $\rho_i R^3$ for ionized density ρ_i. Because $\rho_i^2 R^3$ is constant for a fixed radiation source, ρ_i decreases as $R^{-3/2}$. Thus the fraction of the pressurized mass that is neutral and forced to move with the expansion increases with time as

$$\frac{M_{neutral}}{M_{total}} = 1 - \left(\frac{R}{R_0}\right)^{-3/2},$$

where R_0 is the initial HII region size; typically $R/R_0 > 10$ at late times.

This result implies that an old HII region forces the cloud clumps to move sideways because of neutral clump-clump collisions and not because of the rocket effect with direct ionization. Then they join the expanding neutral layer without ever being exposed to ionizing radiation. It also implies that the mechanism of triggering at the edge of an old HII region is not radiation-driven implosion of pre-existing globules, but the forced motion of these globules into a dense neutral layer or shell. This layer presumably forms stars in the same way as other large parts of the cloud do, by collapsing gravitationally into massive cloud cores. But the layer can collapse much faster than more remote parts of the cloud (once it becomes unstable) because of its higher density.

3.2 Theory

We would like to understand the stability properties of the compressed and moving neutral layer that forms at the boundary between an old expanding HII region and a molecular cloud. For some situations we can probably ignore the clumpiness of this layer (but see Elmegreen 1988). The only calculations available at this time are for plane parallel layers without magnetic fields, so we follow this analysis here.

The collapse rate ω of an infinitely extended isothermal layer is given by the dispersion relation

$$\omega^2 = -k^2c^2 + 2\pi G\sigma k$$

for wavenumber k and mass column density σ. Vishniac (1983) showed that this equation also applies to a pressure-confined layer provided one side of it is behind a *steady* shock front. Then the total collapse time t_{coll} follows from the equations

$$\sigma = \rho_0 v_s t_{coll}$$

$$v_s = (\rho_s c^2/\rho_0)^{1/2}$$

$$\omega_{max} = \frac{\pi G\sigma}{c} = 1/t_{coll}$$

which give $t_{coll} = (\pi G\{\rho_0\rho_s\}^{1/2})^{-1/2} = (\pi G\rho_0 M)^{-1/2}$ for postshock and preshock densities ρ_s and ρ_0, and Mach number M. This total collapse process is faster than the gravitational time scale in the unshocked portion of the cloud because of the Mach number dependence. We refer to this as the short time scale; it is relevant to shocks at *constant* speed and to certain special geometries for a decelerating layer (Elmegreen 1989).

Any collapse on this short time scale leads to condensations that are *unbound* to the layer and emerge from the front of it if the layer decelerates. For layer deceleration proportional to v_s/t, the ratio of the relative acceleration of a dense object out of the front of the layer to the gravitational acceleration toward the layer is approximately equal to the Mach number, which exceeds one. This implies that newly formed stars do not stay in the layer if they form on the short time scale, and it may explain the observation of moving O stars with bow shocks (as observed by van Buren et al. 1990).

The collapse of a *decelerating* layer behind a shock front should be different than the collapse of a layer with constant speed because condensations that begin to form in a decelerating layer migrate forward into the shock as the layer slows down. Then they are subject to an excess of erosion and a divergence of the postshock flow at the front, which breaks them apart. If gravity is ignored, then the result is a kinematic instability (Vishniac 1983) leading to oscillations and fragmentation. When gravity is included, the result is a change in the dispersion relation given above to one with a primary resistance to collapse from the divergent flows at the front, and not from internal pressure (Elmegreen 1989). In the decelerating layer around an old HII region, erosion at the front surface becomes so strong that density perturbations cannot collapse until they become so massive that their escape velocity is comparable to the shock speed.

A numerical and analytical calculation of the dispersion relation in this case gives (Elmegreen 1989)

$$\omega^2 = -\epsilon k^2 v_s^2 + 2\pi G\sigma k$$

where now c is replaced by the shock speed v_s; typically $\epsilon \sim 0.5$. Then the time scale for the gravitational collapse of the layer comes from the maximum growth rate $\omega_{max} = \pi G\sigma/(\epsilon^{0.5} v_s) = 1/t_{coll}$ and from the equations $\sigma = \rho_0 v_s t_{coll}$ and $v_s = (\rho_s c^2/\rho_0)^{1/2}$ as before. This gives a new collapse condition

$$t_{coll} \sim \frac{0.5}{(G\rho_0)^{1/2}} = \frac{40My}{n(H)^{1/2}},$$

which is independent of ρ_s or M. This time scale is essentially the gravitational time in the unshocked part of the cloud. According to the theory of Elmegreen and Lada (1977), it accounts for the delay between generations of star formation in OB associations. In this model, the shock does not really affect the overall gravitational collapse time in the cloud because the unshocked cloud is probably somewhat stable anyway. The rate limiting process during star formation is the dissipation of energy, not self-gravitational collapse. The shock removes a large fraction of this supporting turbulent energy (and it removes angular momentum when the shock motion is parallel to the field), and then the gas in the layer can collapse faster than it would have collapsed without the compression.

The detailed calculation (Elmegreen 1989) indicates that the perturbed column density of the layer increases at the rate $4(G\rho_0)^{1/2}$, but because of the deceleration, the pressure decreases with time and so the density does not increase as fast as the column density. The density takes about twice as long as the column density to increase, and this gives the more relevant time scale for star formation, which is $0.5/(G\rho_0)^{1/2}$, the same as the value given by the simple argument above.

This result suggests that the time delay between generations of OB stars in an association depends mostly on the average density of the molecular cloud. If we use the molecular cloud correlations (Solomon et al. 1987), which suggest that the average density is related to the mass as $\rho = 2 \times 10^{-19}(M/M_\odot)^{-1/2}\mathrm{gcm}^{-3}$, then the time scale is related to the cloud mass by the equation

$$t_{coll} = 2.4 M_5^{1/4} \mathrm{My}$$

for a mass in units of $10^5 M_\odot$. This is essentially the observed time scale separating the generations if OB stars in an association (e.g., Blaauw 1964), for which $M_5 \sim 2$ to 5.

4 LARGE-SCALE TRIGGERING: GIANT SHELLS AND GALACTIC SHEAR

4.1 Observations

When the age of an OB association exceeds about 10 million years, the continuous high pressure from HII regions, winds, and supernovae can lead to a large-scale expansion into the surrounding low-density interstellar medium. This expansion produces a giant cavity surrounded by a shell or cylinder through the galactic plane (Bruhweiler et al. 1980; McCray and Kafatos 1987). If the ambient density is sufficiently high, and the local galactic shear and Coriolis forces sufficiently low (see below), then the accumulated neutral gas can collapse along the periphery of the shell and produce new molecular clouds. Because the average mass column density in the galactic plane near the Sun is about $13 M_\odot pc^{-2}$ (Gilmore, this conference) and higher in the inner part of the galaxy, the total amount of accumulated mass in an expanded region of radius R is at least $4 \times 10^5 (R/100pc)^2 M_\odot$, which is enough for several new cloud complexes. Star formation in these new clouds should begin spontaneously when their internal turbulent and magnetic energies dissipate, and then the shell will appear as a ring or ellipse of molecular clouds, HII regions, and star formation surrounding the old OB association. The time scale for this ring to collapse and form stars is several tens of millions of years, so the old OB association in the center may be dispersed and difficult to recognize when the next generation of star formation begins.

The Large Magellanic clouds and several other nearby galaxies contain examples of rings of HII regions surrounding older clusters. Figure 3 shows one from the LMC discussed by Meaburn et al. (1989). A list of regions with propagating star formation on this scale was given by Elmegreen (1987). Magellanic irregular galaxies tend to display this morphology more often than galaxies the size of the Milky Way because there is less shear in the smaller galaxies; then the rings are easier to recognize. A poster paper showing such regions in the small galaxy NGC 2366 was given at this conference by Bomans and Hopp.

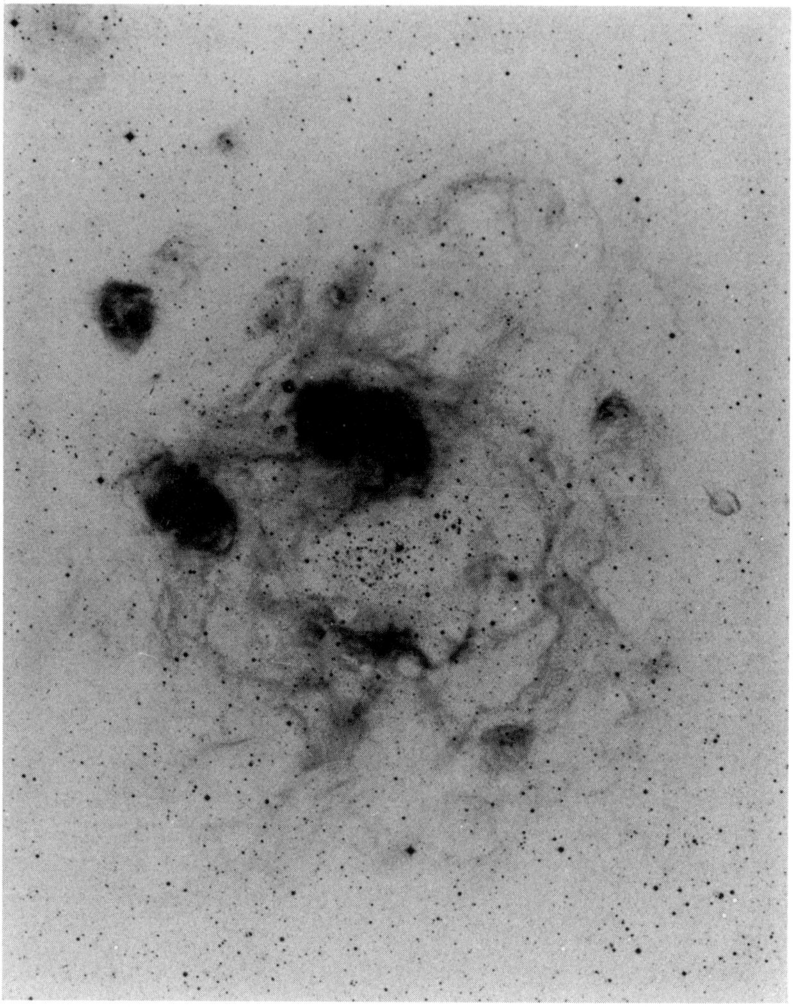

Figure 3 - A large region of star formation in the LMC with an old dispersed cluster in the center and HII regions with apparently younger age around the periphery (from Meaburn et al. 1989).

Another possible example of long range propagating star formation is the Lindblad ring, with the Orion, Perseus and Sco-Cen associations representing collapsed condensations along the periphery, and other gas such as the great rift also participating in the general systematic expansion (Elmegreen 1982). At the center of the ring is the old Cas-Tau OB association, found by Blaauw (1984). A different fit to the Lindblad ring, including also the Lac OB I association, was given by Olano (1982). Figure 4 shows a diagram of the features in this region, with LSR velocities indicated.

The velocity of the Sco-Cen association is from Petrie (1962). The ring appears as a nearly continuous distribution of positive-velocity HI (Lindblad 1967) and CO (Cohen et al. 1980; Bronfman et al. 1989) extending to high galactic latitude. A fit to the expansion based on a purely kinematic model for ballistic particles (Lindblad, et al. 1973) accurately reproduces the distribution of gas at known distances from the sun (compare the dashed line in the figure, which is the kinematic model, to the observed gas distribution). This seems to imply that much of the expansion is in fact kinematic and not driven by persistent high pressures from the previous OB association, nor resisted by drag from gas accumulation. This is consistent with the loss of pressure resulting from a blowout perpendicular to the plane, and to the loss of pressure that should result from the decay in luminosity of the Cas-Tau association. The appearance of ballistic motion might also indicate that much of the material in the ring is from the dense cloud formerly associated with the old association.

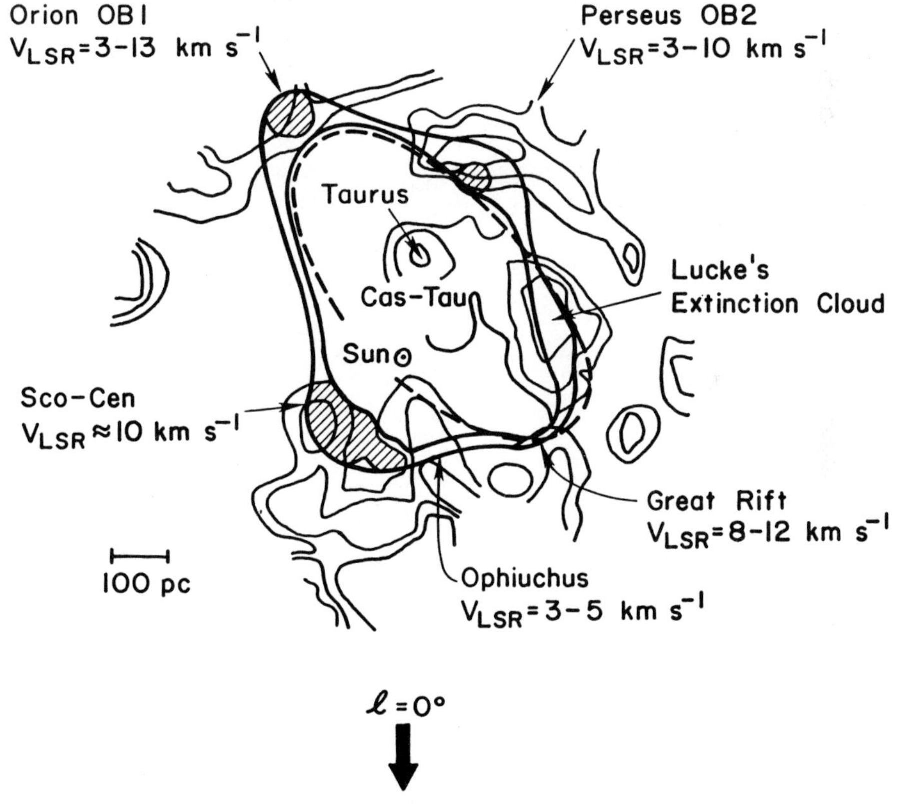

Figure 4 - Schematic diagram of local clouds and young clusters, showing Lindblad's free expansion solution (dashed), known extinction and molecular clouds (light lines), OB associations (hatched), and a representative shell fit (heavy lines) centered on the old Cas-Tau association. The expansion speed of each part is shown.

In the next section, the time scale for the collapse of an expanding structure is estimated in the manner of Section 3.2, and a constraint on the galactic value of Q is derived.

4.2 Theory

The mechanism for the propagation of star formation on a large scale appears to be the gravitational collapse of material in the swept-up shell surrounding an old OB association or other pressure source. This collapse leads to new cloud complexes and then star formation begins spontaneously in these new clouds, just as it does in other clouds formed by different mechanisms. Star formation would also seem to be triggered when the high pressure engulfs existing clouds and squeezes them (e.g., Dopita et al. 1985). We concentrate on the first mechanism here because the second is similar to that discussed in Section 2.2.

Considering the discussion in Section 3.2, the growth rate for gravitational collapse inside an expanding, decelerating shell is given by the approximate dispersion relation

$$\omega^2 = -\epsilon k^2 v_s^2 + 2\pi G\sigma k - \omega_{exp}^2$$

where ω_{exp} is the expansion rate $\sim v_s/R$. This relation has a peak growth rate

$$\omega_{peak} = \frac{\pi G\sigma}{\epsilon^{0.5} v_s}\left(1 - Q_{exp}^2\right)^{1/2}$$

where

$$Q_{exp} = \frac{\epsilon^{0.5} v_s \omega_{exp}}{\pi G\sigma} \sim \frac{\epsilon^{0.5}\omega_{exp}^2 R}{\pi G\sigma} \sim \frac{3\epsilon^{0.5}\omega_{exp}^2}{\pi G\rho_0} < 1$$

or

$$t > \frac{0.58\epsilon^{1/4}}{(G\rho_0)^{1/2}}$$

for $\omega_{exp} = v_s/R = 0.6/t$ for the expansion from a steady wind or other steady power source (Castor et al. 1975). Thus the growth occurs after the gravitational free fall time in the ambient medium.

Now we write the peak growth rate as as

$$\omega_{peak}^2 = \frac{1}{t_{coll}^2} = \left(\frac{\pi G\sigma}{\epsilon^{0.5} v_s}\right)^2\left(1 - Q_{exp}^2\right) = \left(\frac{\omega_{exp}}{Q_{exp}}\right)^2\left(1 - Q_{exp}^2\right)$$

from which we get

$$1 = \left(\frac{0.6}{Q_{exp}}\right)^2\left(1 - Q_{exp}^2\right)$$

and $Q_{exp} = 0.51$ at the time of collapse. This implies that

$$t_{coll} = \frac{0.8\epsilon^{1/4}}{(G\rho_0)^{1/2}}$$

for ambient density ρ_0. Now we need a value of ϵ. For a plane parallel layer that decelerates as in an OB association, we took $\epsilon \sim 0.5$ from a detailed calculation (Elmegreen 1989). There is no analogous calculation for a decelerating ring or shell in a galactic environment, with rotation, shear, Coriolis forces, etc., but we suspect that ϵ will be smaller than 0.5 at the late time implied by t_{coll}. This is because a dense perturbation will not move directly through the front of the shell with the Galactic Coriolis force acting on it. It should instead have a significant velocity component parallel to the surface of the shell. This sideways motion should become significant when t_{coll} is comparable to the inverse of the Galactic epicyclic frequency κ. To account for this effect, we take $\epsilon = 0.1$, and then write

$$t_{coll} = \frac{0.4}{(G\rho_0)^{1/2}}.$$

This is a sensible result for the local average density of about 1.5 atoms cm^{-3}, because then $t_{coll} \sim 26$ million years, which is reasonable for the Lindblad ring and the star formation around it.

This result has an interesting implication for triggered star formation on a galactic scale. Recall that Quirk (1972) and Kennicutt (1989) found a Q threshold for star formation in galaxies,

$$Q = \frac{\kappa c}{\pi G \sigma_g} < 1.4$$

for mass column density in the galactic plane σ_g and epicyclic frequency κ. The usual explanation for this is that it results from spontaneous gravitational instabilities in the disk. But it can also result from propagating star formation if the gravitational collapse of the swept-up shell is required to occur before shear and Coriolis forces distort the shell too much.

To demonstrate this point, we use a result from Palous, Franco, and Tenorio-Tagle (1990), that an expanding shell begins to stall because of the Coriolis force when $t \sim 0.25 t_{ref}$, which gives $t_{stall} \sim 1.25/\kappa$. The criterion for star formation is that the collapse time has to be less than the stall time:

$$t_{coll} = \frac{0.4}{(G\rho_0)^{1/2}} < \frac{1.25}{\kappa} = 0.25 t_{ref}.$$

To convert this to an equation with Q, we have to write ρ_0 in terms of σ_g and c, which involves the scale height of the gas. We use the relation

$$P = \rho_0 c^2 (1 + \alpha + \beta) = \frac{\pi}{2} G \sigma_g \sigma_T$$

for midplane gas pressure P, relative magnetic and cosmic ray pressures α and β, and total gas+star column density σ_T in the gas layer. More discussion of this pressure relation is in Elmegreen (1991b). As a result,

$$t_{coll} = \frac{0.4}{(G\rho_0)^{1/2}} = \frac{1.0c}{\pi G\sigma_g}\left(\frac{(1+\alpha+\beta)\sigma_g}{\sigma_T}\right)^{1/2}.$$

Then we find that $t_{coll} < 1.25/\kappa$ if

$$Q = \frac{\kappa c}{\pi G\sigma_g} < \frac{1.25}{(1+\alpha+\beta)^{1/2}}\left(\frac{\sigma_T}{\sigma_g}\right)^{1/2} \sim 1.25$$

for α and β equal to 0.5 and $\sigma_T = 2\sigma_g$. This result implies that there is a Q threshold for large-scale propagating star formation that is similar to the observed Q threshold for all star formation.

REFERENCES

Bally, J., Langer, W.D., Stark, A.A. & Wilson, R.W. 1987, ApJ (Letters), 312, L45.

Blaauw, A. 1964, Ann.Rev.Astr.Ap., 2, 213.

Blaauw, A. 1984, Irish AJ, 16, 141.

Bonner, W.B. 1956, MNRAS, 116, 351.

Brand, P.W.J., Hawarden, T.G., Longmore, A.J., Williams, P.M. & Caldwell, J.A.R. 1983, MNRAS, 203, 215.

Bronfman, L., Alvarez, H., Cohen, R.S. & Thaddeus, P. 1989, ApJ Suppl., 71, 481.

Bruhweiler, F.C., Gull, T.R., Kafatos, M. & Sofia, S. 1980, ApJ (Letters), 238, L27.

Castor, J., McCray, R. & Weaver, R. 1975, ApJ (Letters), 200, L107.

Cohen, R.S., Cong, H., Dame, T.M. & Thaddeus, P. 1980, ApJ (Letters), 239, L53.

Dopita, M.A., Mathewson, D.S. & Ford, V.L. 1985, ApJ, 297, 599.

Dibai, E.A. 1958, Sov.Astron.J., 2, 429.

Dickman, R.L., Horvath, M.A. & Margulis, M. 1990, ApJ, 365, 586.

Ebert, R. 1955, Z. Astrophys., 37, 222.

Elmegreen, B.G. 1982, in Submillimetre-Wave Astronomy, ed. J. Beckman & J. Phillips, (Cambridge: Cambridge University Press), p.1.

Elmegreen, B.G. 1985, in Birth and Infancy of Stars, ed. R. Lucas, A. Omont & R. Stora, (Amsterdam: Elsevier Science Publishers B.V.), p. 215.

Elmegreen, B.G. 1987 in IAU Symposium No. 115, Star Forming Regions, ed. M. Peimbert & J. Jugaku, (Dordrecht:Reidel), p. 457.

Elmegreen, B.G. 1988, ApJ, 326, 616.

Elmegreen, B.G. 1989, ApJ, 340, 786.

Elmegreen, B.G. 1991a, in Protostars & Planets III, ed. E.S. Levy & M.S. Matthews, University of Arizona Press, in press.

Elmegreen, B.G. 1991b, ApJ, in press for September 1.

Elmegreen, B.G. & Lada, C.J. 1977, ApJ, 214, 725.

Elmegreen, B.G. & Wang, M. 1988 in Molecular Clouds in the Milky Way and External Galaxies, ed. R.Dickman, R. Snell and J. Young, (Berlin: Springer-Verlag), p. 240.

Falgarone, E., Phillips, T. & Walker, C.K. 1991, ApJ, in press.

Hawarden, T.G. & Brand, P.W.J.L. 1976, MNRAS, 175, 19p.

Kennicutt, R.C. 1989, ApJ, 344, 685.

Lada, E.A., DePoy, D.L., Evans, N.J. III. & Gatley, I. 1991, ApJ, 371, 171.

Lindblad, P.O. 1967, BAN, 19, 34.

Lindblad, P.O., Grope, K., Sandqvist, Aa. & Schober, J. 1973, A&A, 24, 309.

MacLow, M.M., Klein, R.I., McKee, C.F., Stone, J.A. & Norman, M.L. 1991, this conference.

McCray, R. & Kafatos, M. 1987, ApJ, 317, 190.

Meaburn, J., Solomos, N., Laspias, V. & Goudis, C. 1989, A&A, 225, 497.

Mouschovias, T.C. & Spitzer, L.Jr. 1976, ApJ 210, 326.

Olano, C.A. 1982, A&A, 112, 1965.

Palous, J., Franco, J. & Tenorio-Tagle, G. 1990, A&A, 227, 175.

Petrie, R.M. 1962, MNRAS, 123, 501.

Quirk, W.J. 1972, ApJ (Letters), 176, L9.

Rozycka, M. & Tenorio-Tagle, G. 1987, A&A, 176, 329.

Sandford, M.T.II, Whitaker, R.W. & Klein, R.I. 1982, ApJ 260, 183.

Sargent, A.I. 1979, ApJ, 233, 163.

Solomon, P.M., Rivolo, A.R., Barrett, J. & Yahil, A. 1987, ApJ, 319, 730.

Spitzer, L., Jr. 1978, Physical Processes in the Interstellar Medium, (New York: Wiley Interscience).

Sugitani, K., Fukui, Y. & Ogura, K. 1991, ApJ Suppl., in press.

Thronson, H.A., Lada, C.J. & Hewagama, T. 1985, ApJ, 297, 662.

Tomisaka, K., Ikeuchi, S. & Nakamura, T. 1989, ApJ, 341, 220.

van Buren, D., Mac Low, M.M., Wood, D.O.S. & Churchwell, E. 1990, ApJ, 353, 570.

Vishniac, E.T. 1983, ApJ, 274, 152.

Wilking, B.A., Harvey, P.M., Lada, C.J., Joy, M. & Doering, C.R. 1984, ApJ, 279. 291.

Is the Cloud Population of M31 Different from That of the Milky Way ?

John M. Dickey

University of Minnesota

and

Elias Brinks

National Radio Astronomy Observatory

ABSTRACT

An important research topic for the 1990's is to understand the diversity of interstellar media among spiral galaxies. As a start, we have already sufficiently detailed knowledge of the phases of the interstellar gas in M31 that we can discern both differences and similarities between it and the Milky Way. Overall, the molecular cloud phase is underabundant by at least a factor of three relative to the Milky Way, whereas the atomic gas mass is somewhat larger. Looking in more detail, the individual molecular and atomic clouds in M31 look remarkably similar to their Milky Way counterparts. A major difference which has recently been detected in the atomic gas of the two galaxies is that the peak brightness temperature of the λ21-cm line is 50% higher in M31 than in the Milky Way. This has been interpreted as a fundamental difference in the temperature of the cloud population, but we show that it could also be a natural result of the difference in our perspective from outside the disk of M31 but inside the cloud layer in the Milky Way.

1. AGGREGATE PROPERTIES

Whether the interstellar media of M31 and the Milky Way are similar or dissimilar depends on the size scale on which we look to compare them. On large scales the most striking difference is the distribution of gas in the inner disk. M31 does not have a ring rich in molecular gas at radii 4 to 5 kpc as the Milky Way has; instead the inner disk is quite empty of all forms of interstellar matter. This emptiness makes the total molecular mass of M31 much lower than that of the Milky Way, by at least a factor of three (see Table). Outside of radius 6 kpc M31 is as rich in gas as the Milky Way. As in the Milky Way at the solar circle and beyond, the gas in M31 is primarily atomic. In the outer regions M31 has at least as much gas as the Milky Way, and the disk apparently extends further than ours. Thus its overall atomic mass is probably higher than that of the Milky Way, in spite of the empty inner disk. How M31's inner disk came to be empty to a much larger radius than the Milky Way's, and whether the interstellar cloud population is different, either as a cause or a result of this large scale difference, is not completely understood. More extensive reviews of the large scale molecular and atomic gas distributions in M31 can be found in Dame *et al.* (1991), Koper *et al.* (1991), Young and Scoville (1991), Knapp (1990) and Brinks (1990).

Table : Overall Atomic and Molecular Gas Masses (10^9 $M_\odot$)

	M31	Milky Way
H I mass :	4.6 Braun and Walterbos, 1991	3.5 Henderson *et al.*, 1982
H_2 mass :	0.3 Dame *et al.*, 1991	1.2 Bloemen *et al.*, 1986

The puzzle of why M31 and the Milky Way should be so different in their molecular masses but so similar in their atomic gas masses invites us to look more closely at the molecular and atomic clouds in the two galaxy disks. To understand how the processes of agglomeration, fragmentation, condensation, and collapse of clouds determine the ultimate mixture of atomic and molecular species requires more information about the spectrum of cloud masses and column densities. For the Milky Way we have detailed observational data on all kinds of clouds, at least in the solar neighborhood, and M31 is close enough that it is possible, although difficult, to obtain comparable information. The purpose of this contribution is to contrast the properties of the clouds, particularly the atomic clouds, in the two systems in hopes of understanding how these differences might explain the difference in aggregate gas properties of the two galaxies.

2. GIANT MOLECULAR CLOUDS

Looking on the scale of giant molecular clouds, 50 to 100 pc, the individual atomic and molecular clouds in M31 look very much like those in the Milky Way (Lada *et al.* 1988, Vogel *et al.* 1987, Wilson and Scoville 1989, Israel 1991). Sizes and linewidths measured by aperture synthesis instruments match those seen in Milky Way molecular surveys (Young 1990). As in the Milky Way, molecular clouds contain large amounts of atomic gas as envelopes and substrate (Wannier *et al.* 1983). Thus it seems that although the total molecular content of M31 is low compared to the Milky Way, and low compared to other Sb's, this difference is not reflected in the structure of individual molecular clouds.

Beyond these few giant molecular clouds which have been mapped in CO with aperture synthesis telescopes, we know much less about the population of interstellar clouds of all kinds and masses in M31 than we know about this population in the Milky Way. For the Milky Way, inventories of the larger molecular clouds have been taken using single dish CO surveys (reviewed by Scoville and Sanders 1987, Elmegreen 1987) which cover masses from a few times 10^4 to a few times 10^6 $M_\odot$. The slope, α, of the number density spectrum ($\alpha = d \log\rho(m) / d \log m$, where $\rho(m)\, dm$ is the number of clouds per unit volume in the mass range m to m+dm) is in the range -1.6 to -1.8 for molecular clouds in this mass range (Tereby *et al.* 1986, Solomon *et al.* 1987, Ferrini *et al.* 1990). We do not yet know whether this cloud mass spectrum is universal. Calibration of molecular cloud masses is still uncertain, depending on the controversial "size-linewidth relation" (Maloney 1990). In spite of these uncertainties, it is clear that the bulk of the molecular mass is in a relatively small number of the largest clouds, because α is definitely greater than -2, at least in the Milky Way.

3. DEFINING CLOUDS

A more controversial issue is how to define and measure "clouds" in emission surveys. Drawing boundaries in position and velocity space is often a subjective process; even when an algorithm is used to do this consistently and completely the results always depend on the beam size of the telescope which produced the survey. For example, clouds catalogued from the Columbia survey (Cohen *et al.*, 1986) made with an 8.3' telescope beam, correspond to aggregates of clouds in the FCRAO survey (Scoville *et al.*, 1987), which was made with at 50" beam. Similar effects are seen in H I, *eg.* when comparing high velocity clouds found in the Bell Labs catalog (Stark *et al.*, 1991) to those found in the Westerbork/Dwingeloo survey (Wakker 1990).

The problem is that the structure of the interstellar gas is difficult to describe mathematically; it covers a very broad range of sizes from AU to kpc scales, and the limited spatial dynamic range of emission surveys filters out all but a very narrow range of structure sizes (Scalo, 1990, Houlahan and Scalo, 1990). So, although we can say that certain clouds exist in M31 which are similar to certain Milky Way giant molecular clouds, to compare inventories of the cloud populations in the two systems using properly corresponding survey data is not possible yet. We need a more robust way to measure the cloud population in order to tell whether there is a difference between the cloud mass spectrum in M31 and the Milky Way.

4. CLOUDS AS ABSORPTION LINES

Historically, much of our early understanding of the structure of interstellar matter came from optical absorption line surveys (Adams 1949, Münch 1965). More recently absorption surveys have also been made in the radio by observing H I at λ21-cm, which has the advantage that both emission and absorption can be measured (reviewed by Dickey and Lockman 1990, Kulkarni and Heiles 1988). Absorption surveys are geometrically different from emission surveys, each absorption spectrum samples only the very small solid angle subtended by the bright background source. The background sources are typically randomly distributed, so absorption surveys give only a statistical sample of gas properties along lines of sight which are essentially delta functions spatially. In contrast, emission surveys give a complete sample of all the gas, but with spatial structure smoothed by convolution with the beam size. Comparing ρ(M) determined from the two kinds of surveys gives consistent results, but assumptions about cloud cross-sections are used which oversimplify the question of cloud geometry (Dickey and Garwood, 1989).

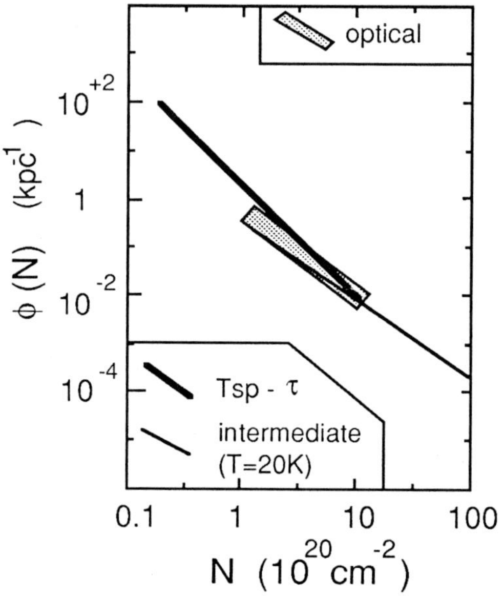

Figure 1. The abundance of clouds (ie. interstellar absorption lines at 21-cm or optical wavelengths) per unit line of sight length, as a function of column density. For the λ21-cm lines two assumptions are used to translate optical depth to column density depending on the typical cloud temperature (see Dickey and Garwood, 1989).

The absorption data are most naturally interpreted in terms of a spectrum of the number of clouds Φ(N) of various column densities N per unit line of sight path length. Here the definition of "clouds" is simplified by counting absorption lines of different optical depths; typically these are distinct, separate, and nearly Gaussian shaped in velocity. Whether these separate, individual spectral features correspond to separate, individual spatial structures is not clear, but it is an easy assumption to make. With this and a few other assumptions needed to convert optical depth into column density (Dickey and Garwood, 1989, Hobbs, 1974), we obtain $\Phi(N)$ as shown on figure 1. There is good agreement between optical and radio results for column density N of about 10^{20} cm^{-2}. Different assumptions give slopes for $\Phi(N)$ from -1.6 to -2.2; these translate into slopes for $\rho(M)$ of $-1.9 > \alpha > -2.1$. Thus for diffuse clouds there is roughly equal contribution to the total mass of gas in the cool phase from each logarithmic interval of cloud mass. Whether this is a coincidence, the effect of some profound astrophysical law of cloud evolution, or the result of a pernicious, unsuspected observational selection effect is still not known.

For M31 we do not yet know much about the form of the spectra $\Phi(N)$ and $\rho(M)$, but with modern optical and radio telescopes it is possible to study absorption through the disks of nearby spirals, and so we may hope to determine whether this "cloud spectrum" varies from galaxy to galaxy, or has a universal form. Preliminary attempts at this have been made by Dickey and Brinks (1988). The column densities probed by atomic lines seen in absorption in the radio, optical, and uv are typically more than 100 times lower than those of the molecular clouds discussed above; the atomic clouds are much more diffuse and lower in mass than the clouds counted in most molecular surveys at low latitudes. "Diffuse molecular clouds" seen at high latitudes, like the "cirrus" seen in the far-IR, may constitute an intermediate population.

5. ATOMIC CLOUDS AND INTERCLOUD GAS

The molecular gas is confined to regions of high density, and high column density, if only because the molecules must be shielded from photodissociating radiation, so it occupies only a tiny fraction of the volume of a galaxy disk, a few tenths of a percent in the Milky Way and even less in M31. The atomic gas occupies a much larger fraction of the volume, perhaps as much as 50%, but the **cool** atomic gas (60 to 75 K), which is the primary constituent of the diffuse clouds, fills only 3% to 5%. The rest of this paper will concentrate on this cool atomic phase, which is intermediate between the molecular clouds and the warm (~6000 K) atomic gas in temperature, density, filling factor, and perhaps also in evolutionary sequence. M31 has somewhat less cool H I than the Milky Way, about 20% to 25% ; it may also be that the temperature of the cool phase in M31 is warmer than in the Milky Way. The objective here is to understand how the thermal properties of the atomic gas differ between the two galaxies, in search of clues as to why the molecular gas mass is so different in spite of the atomic masses being about the same.

5.1 Survey Results for the Distribution of Gas Perpendicular to the Plane

We know quite a bit about the distribution of the warm and cool H I in the Milky Way. The average distribution perpendicular to the galactic plane, n(z), has both a narrow component and a broad component. The first is roughly Gaussian, with dispersion 90 pc and midplane density $n(0) = 0.4\ cm^{-3}$. The broad component is best modelled as the sum of a Gaussian with dispersion 225 pc plus an exponential with scale height 430 pc, having midplane densities 0.11 and $0.06\ cm^{-3}$ respectively (Lockman and Gehman, 1991). M31 has midplane density of about $0.6\ cm^{-3}$ at radius corresponding to the solar circle, and z dispersion of 120 pc, so it is very similar to the Milky Way in the z-distribution of its H I (Brinks and Burton, 1984). In the solar neighborhood we can identify the 90 pc H I component with the clouds, which consist of cool gas which is seen in absorption plus some associated warm gas. The components with the much larger scale heights are dominated by warm gas. This identification is based on absorption surveys, which show that the opacity of the H I also is distributed roughly as a Gaussian with dispersion about 90 pc. Since the opacity of the λ21-cm line is proportional to gas density divided by temperature the warm component does not show up in absorption.

To compare the results of λ21-cm surveys of M31 with those of the Milky Way it is convenient to work in the units of the observations. In the optically thin limit the column density N is given by the velocity integral of the spectrum of emission brightness temperature, $T_B(v)$:

$$\frac{\int T_B(v)\,dv}{[\mathrm{K\ km\ s^{-1}}]} = \frac{N}{[\mathrm{cm^{-2}}]} \times \frac{1}{1.823 \times 10^{18}} = 1.69 \times \frac{<n>}{[\mathrm{cm^{-3}}]} \times \frac{\Delta\ell}{[\mathrm{pc}]}$$

where $\Delta\ell$ is the path length along the line of sight contributing to the observed spectrum. The corresponding integral of the optical depth spectrum, $\tau(v)$, measured in absorption, gives the opacity :

$$\frac{\int \tau(v)\,dv}{[\mathrm{km\ s^{-1}}]} = \frac{N}{T_{spin}}\left[\frac{\mathrm{K}}{\mathrm{cm^{-2}}}\right] \times \frac{1}{1.823 \times 10^{18}} = \frac{<\kappa>}{[\mathrm{km\ s^{-1}\ pc^{-1}}]}\,\frac{\Delta\ell}{[\mathrm{pc}]}$$

where opacity has units $km\ s^{-1}\ pc^{-1}$ because it is defined as :

$$<\kappa> \equiv \frac{\int \tau(v)\,dv}{\Delta\ell}$$

In the above formulae the mean density and opacity, <n> and <κ>, are line of sight averages, so for example a 1 kpc path length at midplane would give $1.0 \times 10^3\ K\ km\ s^{-1}$ for the integral of the emission spectrum, assuming small optical depth. Dividing emission by absorption gives a number with the units of temperature, which we call the "spin temperature", T_{spin}. This is in general a function of velocity, but we can compute an average value by dividing the spectrum integrals, as :

$$\frac{\int T_B(v)\,dv}{\int \tau(v)\,dv} = \langle T_{spin} \rangle$$

This does not necessarily correspond to a physical temperature, since it is a blend of much warmer and much cooler gas, but still it is useful for understanding the thermal state of the H I.

5.2 The Significance of the Opacity, $\langle\kappa\rangle$

The mean opacity, $\langle\kappa\rangle$, is a fundamental quantity for understanding not only the absorption in the λ21-cm line, but the brightness temperature as well. In the solar neighborhood at midplane this has the value $\langle\kappa\rangle \cong 8\times10^{-3}$ km s^{-1} pc^{-1} (Garwood and Dickey, 1988), so for a Gaussian z-distribution with dispersion 90 pc we should find $\int\tau(v)\,dv \cong 0.9$ km s^{-1} for an absorption spectrum toward $|b| = 90^o$. In fact there is hardly any absorption toward either galactic pole, but this value is in good agreement with $\langle|\sin b| \times \int\tau(v)dv\rangle$ averaged over many lines of sight at intermediate latitudes, ie. it is an average for the interstellar gas within about 1 kpc of the sun. The corresponding average column density perpendicular to the disk, $\langle|\sin b| \times \int T_B(v)dv\rangle$ is 180 K km s^{-1} = 105 cm^{-2} (see Kulkarni and Heiles, 1987). The ratio of these gives a mean spin temperature for lines of sight passing entirely through the disk, $\langle T_{spin}\rangle$ = 200 K. This tells us the global mixture of warm and cool phases, if we assume a temperature for the cool, absorbing phase. Taking 60 K for the cool H I temperature, T_c, we find the fraction of H I (by mass) in the cool phase, f_c, to be :

$$f_c = \frac{N_{cool}}{N_{warm} + N_{cool}} = \frac{T_c}{\langle T_{spin} \rangle} = 30\%.$$

We would get different numbers for $\langle T_{spin}\rangle$ and f_c if we considered only a line of sight which stays entirely in the disk at z = 0; taking the ratio of $\langle n\rangle$ to $\langle\kappa\rangle$ gives $\langle T_{spin}\rangle$ = 120 K, and $f_c \cong 50\%$. The cloud phase is a larger fraction of the total H I at the midplane than it is integrated over all z because the cool gas is more tightly confined to low z than the warm gas.

The opacity, $\langle\kappa\rangle$, has the same units as a velocity gradient, $dv/d\mathcal{L}$, and it is useful to compare its numerical value to the velocity gradient of the disk in order to understand better the results of low latitude surveys of λ21-cm emission. Absorption lines at λ21-cm typically have widths of 2 to 4 km s^{-1}, much narrower than emission lines because they correspond exclusively to cool gas. A typical line of sight at low latitudes has many such absorption lines spread over velocity by the differential rotation of the galaxy (see Burton 1988). If the velocity gradient is high, ie. a short line of sight distance projects onto a broad velocity range of the spectrum, then the gas will be on average optically thin, since the absorption lines will be spread over a wide bandwidth. On the other hand, if the magnitude of the velocity gradient is less than $<\kappa>$ then the gas will be on average optically thick. In this case absorption lines, most of which are individually of low optical depth, will overlap so as to make the gas optically thick on average over the velocity range for which $dv/d\ell$ is less than $<\kappa>$.

5.3 The Peak Brightness Temperature

Since the earliest surveys of λ21-cm emission at low latitudes it has been clear that the brightness temperature of the Milky Way disk saturates at a value of about 125 K (Oort, 1956, Kerr and Westerhout, 1965). That this is a saturation process is evident from the fact that many emission profiles are flat-topped over a broad range of velocities in the inner

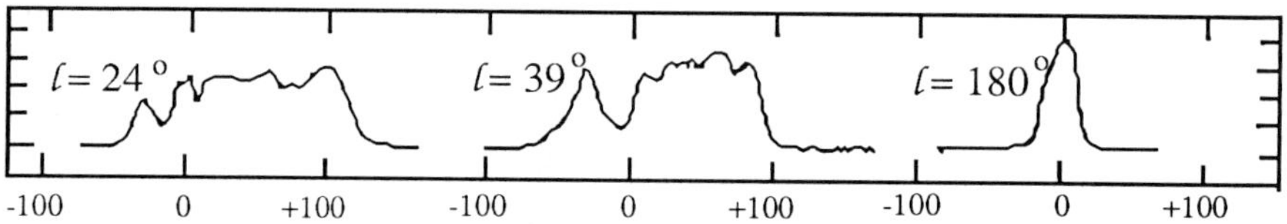

Figure 2. Spectra of λ21-cm emission at latitude b=0°. Two typical inner galaxy spectra are shown on the left, on the right is a spectrum toward the anticenter. The inner galaxy emission (0<v<+100 km s^{-1} on the first quadrant spectra) is saturated, although because the telescope beam does not resolve the cool disk fully, the peak antenna temperature is less than the true peak brightness temperature of 125 to 130 K.

galaxy, and in the outer galaxy toward the anticenter where the velocity gradient is small. This is illustrated on Figure 2 by two spectra drawn from the Berkeley survey (Weaver and Williams, 1973). A simple explanation for this flattening (inherent in the results of Baker and Burton, 1975) is that it corresponds to regions where the velocity gradient is less than $<\kappa>$, so that the gas is optically thick on average. The value of the brightness temperature at saturation, $T_{B,max}$, is not simply the temperature of the cool gas, but it is the sum of this cool phase temperature, T_C, **plus** the brightness temperature contributed by the warm, optically thin gas **in front of** the absorbing clouds. As long as $<\kappa> > dv/d\ell$, $T_{B,max}$ will not show a distinct variation with longitude or velocity, which is at least roughly the case in both the inner Milky Way or in M31.

A simple example of the radiative transfer determining the value of the peak brightness temperature is shown on figure 3. For a given velocity interval Δv in the spectrum, corresponding by galactic rotation to a given line of sight interval $\Delta\ell$, there may be a sequence of optically thin, warm gas regions interspersed with cool, partially optically thick H I clouds. The brightness temperature is the sum of the column densities of the warm regions, attenuated by the foreground optical depth, plus the kinetic temperatures of the

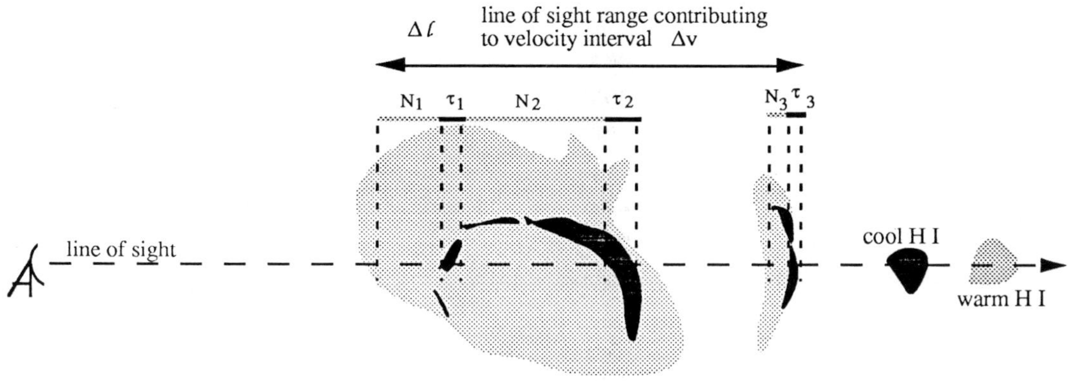

Figure 3. Peak brightness temperature example.

cool regions, each times $(1 - e^{-\tau})$. Assuming all the cool clouds to have kinetic temperatures T_c, gives, for three clouds with intervening warm gas :

$$T_B = N_1 + \exp(-\tau_1) \times [\, N_2 + \exp(-\tau_2) \times N_3 \,] \;+\; T_c \times \{1 - \exp[-(\tau_1+\tau_2+\tau_3)]\}$$

with N_i the warm gas column densities between successive cool clouds, as shown on figure 3. If $<\kappa>$ is greater than the velocity gradient, $dv/d\ell$, then the sum of the optical depths is greater than one, and in the limit of many short intervals N_i and τ_i the brightness temperature saturates at :

$$T_{B,max} = \frac{1.69 <n_w>}{<\kappa>} \left[\frac{(\mathrm{km\ s^{-1}\ pc^{-1}})}{\mathrm{cm^{-3}}} \right] + T_c$$

where $<n_w>$ is the density of the warm phase only, which is half the total at midplane, ie. $<n_w> \cong 0.3\ \mathrm{cm^{-3}}$. Thus the warm phase contributes 60 to 65 K to $T_{B,max}$ and the cool phase contributes another 60 K or so at midplane, which explains the observed value of $T_{B,max}$ of 125 to 130 K at low latitudes.

5.4 Comparison to M31

Looking down into the disk from above gives quite a different perspective; in particular the fact that the warm gas is distributed much more widely in z than the cool clouds causes the peak brightness temperature to be considerably higher. If we could look down at the Milky

Way disk our line of sight would first traverse the warm, optically thin gas, collecting a total column density of 1.3 x 10^{20} cm^{-2} or 72 K km s^{-1} before even reaching the upper edge of the cool cloud layer ($\sigma_z \cong$ 90 pc). Looking straight down the brightness temperature of the λ21-cm line would not saturate, since the Milky Way is on average optically thin seen face-on. This is because the effective path length through the cool layer ($\sqrt{2\pi}\ \sigma_z$ = 226 pc) is less than the mean free path, $\delta\ell$. The mean free path is the distance through the cloud layer needed to fill with optically thick gas the velocity range δv corresponding to the random motion of the cloud ensemble. In the Milky Way, as in most spiral galaxies, $\delta v \cong 6$ km s^{-1}, so :

$$\delta\ell = \frac{\delta v}{\langle\kappa\rangle} = 750 \text{ pc}$$

thus only if the galaxy were seen inclined by more than about 70° to the line of sight would the brightness temperature saturate, even if $\langle\kappa\rangle > dv/d\ell$. The value we would expect for $T_{B,max}$ for the Milky Way seen from outside at inclination of, say, 77°, is the value derived above **plus** the "overburden" of warm gas at high z which must be traversed by the line of sight before reaching the optically thick layer. Seen at inclination 77° this amounts to 320 K km s^{-1} of brightness temperature (N = 5.8x10^{20} cm^{-2}), which will be spread over a velocity range δv plus whatever differential rotation is present along the high-z part of the line of sight, typically a total of 8 to 10 km s^{-1}. So, seen from outside, the peak brightness temperature will be raised by 30 to 40 K from about 130 K seen from inside to about 170 K.

We are fortunate to have M31 nearby, inclined at 77°, with roughly the same H I properties as the Milky Way, so that we can measure the peak brightness temperature of the λ21-cm line seen from outside. It is only with the VLA survey of Braun (1990) that sufficiently high spatial and velocity resolution is available to measure $T_{B,max}$ without suffering from observational blending effects. There seems to be some longitude effect in the peak brightness temperatures, but they are indeed in the range 160 K to 180 K, as expected if M31 is similar to the Milky Way in its mixture of cool and warm H I. It is also possible that the temperature of the clouds themselves, T_c, is higher by 30 to 50 K, as suggested by Braun and Walterbos (1989, 1991), or that the mixture of warm and cool phases is different. Either of these differences should show up as a difference in $\langle\kappa\rangle$, which is directly observable through absorption surveys of M31. Preliminary results seemed to show that $\langle\kappa\rangle$ was **lower** in M31 than in the Milky Way (Dickey and Brinks, 1988), but more complete surveys show that $\langle\kappa\rangle$ is close to the Milky Way value (Braun and Walterbos, 1991, Dickey and Brinks, 1992).

6. SUMMARY

Whether the mixture of warm and cool phases of the atomic gas in M31 is the same as in the Milky Way, whether the cloud phase temperature is the same, and whether $<\kappa>$ is the same are crucial questions for understanding the overall differences in the interstellar media of the two galaxies. If these are different, as suggested by Braun and Walterbos (1991), it may well be the result of a difference in the interstellar pressure, with M31 having a pressure of about half that in the Milky Way. The pressure is determined on a large scale by the gravitational potential of the disk and on smaller scales by the formation rate of massive stars. In most of the M31 disk the star formation rate is lower than in the Milky Way. Since massive star formation is proportional to the density of molecular clouds (to at least the first power) this difference is clearly linked to the relative absence of the molecular phase in M31. The deeper question is how this different mixture of atomic and molecular phases is linked to the difference in pressures, if any. In order to properly understand the differences among spiral galaxies we need to understand the feedback in the loop connecting star formation, the atomic gas, and the molecular clouds. Studies of the interstellar gas in nearby galaxies offer contrasts which show how the different pieces can fit together.

This research is supported in part by NSF grant 87-22990 to the University of Minnesota. The National Radio Astronomy Observatory is Operated by Associated Universities, Inc. under agreement with the National Science Foundation.

7. REFERENCES

Adams, W.S., 1949, *Ap. J.,* **109**, 354.
Baker, P.L. and Burton, W.B., 1975, *Ap. J.*, **198**, 281.
Bloemen, J.B.G.M., *et al.*, 1986, *Astron. Astrophys.*, **154**, 25.
Braun, R., 1990, *Ap. J. Suppl.*, **72**, 755 and 761.
Braun, R. and Walterbos, R.A.M., 1989, in *The Interstellar Medium in External Galaxies : The Contributed Papers*, eds. D. Hollenbach and H.A. Thronson, (NASA Conf. Proc. 3084), p. 269.
Braun, R. and Walterbos, R.A.M., 1991, *Ap. J.*, in press.
Brinks, E., 1990, in *The Interstellar Medium in Galaxies*, eds. H.A. Thronson and J.M.Shull, (Dordrecht : Kluwer), p. 39.
Brinks, E. and Burton, W.B., 1984, *Astron. Astrophys.*, **141**, 195.
Burton, W.B., 1988, in *Galactic and Extragalactic Radio Astronomy*, 2nd ed., eds. G.L. Verschuur and K. Kellerman, (New York : Springer-Verlag) p. 295.
Cohen, R.S., Dame, T.M., and Thaddeus, P., 1986, *Ap. J. Suppl.*, **60**, 695.
Dame, T.M., *et al.*, 1987, *Ap. J.*, **322**, 706.

Dame, T.M., Koper, E., Israel, F.P., and Thaddeus, P., 1991, in IAU Symp. #146 : *Dynamics of Galaxies and their Molecular Cloud Distributions*, eds. F. Combes and F. Casoli, (Dordrecht : Kluwer), p. 23.

Dickey, J.M. and Brinks, E., 1988, *M.N.R.A.S.*, **233**, 781.

Dickey, J.M. and Brinks, E., 1992, in preparation.

Dickey, J.M. and Garwood, R.W., 1989, *Ap. J.*, **341**, 201.

Dickey, J.M. and Lockman, F.J., 1990, *Ann. Rev. Astr. Astrop.*, **28**, 215.

Elmegreen, B.G., 1987, in *Interstellar Processes*, eds. D.J. Hollenbach and H.A.Thronson, (Dordrecht : Reidel), p. 259.

Ferrini, F., Palla, F., and Penco, U., 1990, *Astron. Astrophys.*, **231**, 391.

Garwood, R.W. and Dickey, J.M., 1988, *Ap. J.*, **338**, 841.

Henderson, A.P., Jackson, P.D. and Kerr, F.J., 1982, *Ap. J.*, **263**, 116.

Hobbs, L.M., 1974, *Ap. J.*, **191**, 395.

Houlahan, P. and Scalo, J.M., 1990, *Ap. J. Suppl.*, **72**, 133.

Israel, F.P., 1991, in IAU Symp. #146 : *Dynamics of Galaxies and their Molecular Cloud Distributions*, eds. F. Combes and F. Casoli, (Dordrecht : Kluwer), p. 13.

Kerr, F.J. and Westerhout, G., 1965, in Stars and Stellar Systems Vol. 5 : *Galactic Structure*, eds. A. Blaauw and M. Schmidt, (Chicago : Univ. of Chicago Press), p. 167.

Knapp, G.R., 1990, in *The Interstellar Medium in Galaxies*, eds. H.A. Thronson and J.M.Shull, (Dordrecht : Kluwer), p. 3.

Koper, E., Dame, T.M., Israel, F.P., and Thaddeus, P., 1991, in IAU Symp. #146 : *Dynamics of Galaxies and their Molecular Cloud Distributions*, eds. F. Combes and F. Casoli, (Dordrecht : Kluwer), p. 27.

Kulkarni, S.R. and Heiles, C., 1987, in *Interstellar Processes*, eds. D.J. Hollenbach and H.A.Thronson, (Dordrecht : Reidel), p. 87.

Kulkarni, S.R. and Heiles, C., 1988, in *Galactic and Extragalactic Radio Astronomy*, 2nd ed., eds. G.L. Verschuur and K. Kellerman, (New York : Springer-Verlag) p. 95.

Lada, C.J., *et al.*, 1988, *Ap. J.*, **328**, 143.

Lockman, F.J. and Gehman, C., 1991, *Ap. J.*, in press.

Maloney, P., 1990, *Ap. J. Lett.*, **348**, L9.

Münch, G., 1965, in Stars and Stellar Systems Vol. 5 : *Galactic Structure*, eds. A. Blaauw and M. Schmidt, (Chicago : Univ. of Chicago Press), p. 203.

Oort, J.H., 1955, in *Vistas in Astronomy*, ed. A. Beer, **1**, 607 (London : Pergamon).

Scalo, J.M., 1990, in *Physical Processes in Fragmentation and Star Formation*, eds. R. Capuzzo-Dolcetta, C. Chiosi, and A. DeFazio, (Dordrecht : Kluwer), p. 151.

Scoville, N.Z. and Sanders, D.B., 1987, in *Interstellar Processes*, eds. D.J. Hollenbach and H.A.Thronson, (Dordrecht : Reidel), p.21.

Scoville, N.Z., *et al.*, 1987, *Ap. J. Suppl.*, **63**, 821.

Solomon, P.M., Rivolo, A.R., Barrett, J., and Yahil, A., 1987, *Ap. J.*, **319**, 730.

Stark, A.A., *et al.*, 1991, *Ap. J. Suppl.,* in press.
Tereby, S., Fich, M., Blitz, L., and Henkel, C., 1986, *Ap. J.*, **308**, 357.
Vogel, S.N., Boulanger, F., and Ball, R., 1987, *Ap. J. Lett.*, **321**, L145.
Wakker, B., 1990, Ph. D. Thesis, Rijksuniversiteit te Groningen.
Wannier, P.G., Lichten, S.M., and Morris, M., 1983, *Ap. J.*, **168**, 727.
Weaver, H. and Williams, D.R.W., 1973, *Astron. Astrophys. Suppl.*, **8**, 1.
Wilson, C. and Scoville, N.Z., 1989, *Ap. J.*, **347**, 743.
Young, J.S., 1990, in *The Interstellar Medium in Galaxies*, eds. H.A. Thronson and J.M.Shull, (Dordrecht : Kluwer), p. 67.
Young, J.S. and Scoville, N.Z., 1991, *Ann. Rev. Astron. Astrophys.*, **29**, in press.

Numerical Simulations of Star Formation

PONGRACIC, H., CHAPMAN, S.J., DAVIES, J.R., DISNEY, M.J., NELSON, A.H.* , & WHITWORTH, A.P.

* reviewer

Physics & Astronomy Dept., University of Wales College of Cardiff, Cardiff, U.K.

INTRODUCTION

It is generally accepted that understanding star formation is one of the most pressing problems of astrophysics today, and a solution to it would have far reaching implications for galaxy evolution and cosmology in general. It is almost self-evident that gravity is the principal agency that pulls interstellar material together to form a star. However there are many other competing physical processes involved, and the precise details of when, how and why stars form still eludes us. On the other hand the subject has been developing very rapidly in recent years, driven by the wealth of observational data coming out of modern telescopes in various wavelength bands. To back these observations up and to help us unravel the complexities of intensity maps, spectra etc., we need good theoretical models of the various processes and objects involved.

Mathematically of course the problem is highly non-linear and geometrically complex, so that analytical techniques are of little value; we need reliable and powerful computer simulations to make any kind of realistic attack on the problem. What I hope to convince you of today is that such reliable and powerful techniques are now becoming available.

SIMULATIONS OF VIOLENT HIGH-MASS STAR FORMATION TO-DATE

Numerical simulations of star formation can be divided into two types (1) those which follow the evolution of an already formed protostar, often in spherical or cylindrical symmetry, and (2) those which consider the transformation of interstellar matter into protostellar matter, without the protostar being explicitly present in the initial conditions. Calculations of type (1) have recently been reviewed by Boss [1990] and I will not consider them further here. Type (2) can be further sub-divided into (a) quiescent formation of low mass stars, when, for instance the support of magnetic stresses is removed by ambipolar or turbulent diffusion of the magnetic field out of a cloud condensation, and (b) Violent formation of high mass stars, in which collision of clouds or cloudlets of supersonically moving gas, either with themselves or with a large scale shock, leads to the fragmentation and collapse of protostars. We will discuss here only calculations of the type 2(b). We will first review previous work in this field, and then go on to describe some recent work on the collision of interstellar clouds being carried out at Cardiff.

One of the first simulations of the collision of two interstellar clouds was due to Stone [1970]. He considered the head-on collision of two 350 $M_{\odot}$ spherical clouds with cylindrical symmetry about the collision axis. The clouds were 1 pc in radius, at an initial temperature of 25° C and had a collision mach number of 60. As with all these calculations the gas forms a dense shocked layer in the collision interface between the two clouds, and the high pressure here causes radial outflow away from the collision axis. Stone used an energy equation to obtain the pressure, which turned out to be very nearly equivalent to a polytropic equation of state with pressure proportional to $\rho^{0.86}$ (ρ being gas density) i.e. the cooling terms dominated at high density, so that the gas cools as it compresses. This turns out to be very important for the character of the solution in later work, but in Stone's case it did not stop the whole gas ensemble bouncing with the clouds re-expanding back the way that they had came. This was probably due to the stiff equation of state, the high Mach number, and possibly the lack of resolution implied in his 10×10 grid. He concluded that "the tremendous compressions which occur do not, in general, increase the likelihood of star formation in cloud collisions". The "in general" here was fortuitous since subsequent calculations have shown that it is indeed possible for collision to lead to gravitational collapse and star formation.

The next calculation worthy of note was that due to Woodward [1976] in which he considers (again in 2-d with cylindrical symmetry about the collision axis) the collision of a spherical gas cloud with a large scale shock. The gas cloud had 500 $M_{\odot}$, a radius of 15 pc, a temperature of 140° K, and the collision had a Mach number of 25. The pressure was given by a cooling equation of state (i.e. softer than isothermal). The increased pressure behind the shock causes a compression wave to propagate into the cloud, and the ram pressure on the front face enhances the compression there. The whole cloud flattens along the collision axis, while at the same time the surface of the cloud goes unstable to a Rayleigh-Taylor type instability on the leading face, and a Kelvin-Helmholtz instability on the sides where the shocked gas is streaming past the cloud in the cloud's rest frame. This calculation was very high resolution for its day with 96×188 grid points in the outer shocked gas, and 60×60 grid points in the cloud. The end result of his calculation is a highly flattened and distorted pancake of gas with high density condensations in the fingers of gas created by the instabilities. Although this calculation was non-self gravitating, on the basis of simple estimates

Woodward was able to conclude that some of the condensations were gravitationally bound, but that the gas as a whole was not collapsing. On the former point he was probably correct, but on the latter, probably wrong, as we shall see later.

Jumping 10 years Hunter et al [1986] carried out the next significant simulation of colliding interstellar gas. This time two cylindrical gas streams were projected at each other with a Mach number of 5, the radius of the cylinders was 0.2 pc, the gas temperature 100° K, and the gas density was 10^{-21} g cm^{-3} . The pressure was obtained from an energy equation with enhanced cooling at higher density. Hunter et al obtained significant fragmentation of the shock interface, due to a Rayleigh-Taylor type instability, leading to the formation of dense clumps of gravitationally bound gas (actually rings from the symmetry). They also obtained noticable central condensation toward the collision axis due to general gravitational collapse. The grid used was 25×50 and the maximum density increase over the initial density was by a factor of 500. An important conclusion of Hunter et al's work was that to obtain fragmentation leading to condensation (i.e. to avoid simple bounce in the collision) the effective equation of state has to be softer than isothermal, i.e. cooling.

One great drawback of these early calculations was that they were grid based and restricted to 2-dimensions, due to the limitations of the computers of the day. The problem really has to be tackled in three-dimensions, and the tools to do this were developed in the eighties. One of these is the use of Smoothed Particle Hydrodynamics (SPH) to simulate the gas flow and pressure forces [Lucy, 1977; Ginghold & Monaghan, 1977]. In this model the gas is represented by an ensemble of extended particles which interact by short range forces to mimic pressure. The model is geometrically very flexible, and adapts to three dimensions with minimal effort. SPH was in fact developed partly in the context of Star Formation simulations, and the first use of it in the context of violent high mass star formation was by Lattanzio & Henriksen [1988]. they performed a whole series of calculations of the collision of two 8000 $M_\odot$clouds at a variety of impact parameters, including cloud rotation with various orientations of the cloud spin and orbital angular momentum. Their clouds had a radius of 12.25 pc and a temperature of 74° K, and the Mach numbers of the collisions varied form 3 to 10. When collapse occurred they generally found that the two clouds coallesced to a central collapsing object, with no signs of fragmentation. They obtained one collapsing cloud even when the impact parameter was equal to the cloud radii; this is at variance with our results in which we obtain two collapsing clouds for such large impact parameters. Possibly the difference is due to lack of numerical resolution since they calculated the gravity using a grid based method. The lack of fragmenation was due partly to this also, but partly also to the artificial viscosity which they used in their SPH model, although they argue that the physical viscosity due to turbulence may be comparable to the viscosity that they used. Also they used an isothermal equation of state, which tends to inhibit fragmentation. One thing that this paper highlighted was that central collapse toward the centre of mass is indeed an important feature of these collisions. The largest density enhancement over the initial density which they obtained at the centres of the collapsed objects was a factor of 500.

The flexibility of the SPH model enabled more complicated geometries to be considered, and Monaghan and Varnas [1988] simulated the interaction of an ensemble of 48 cloudlets moving randomly in an interstellar cloud. The total mass of the cloud was 3350 $M_{\odot}$, and the total radius was 5 pc (radius of each cloudlet $\sim$ 0.5pc). The temperature of the gas was 80° K, and the r.m.s. velocity of the cloudlets varied from Mach 1 to 3. An energy equation with cooling was used to obtain the pressure. The cloud collapses as a whole, with merging of individual cloudlets. In the case of Mach 2 random velocities, 8 final sub-clouds were formed, mostly rotationally supported against further collapse. A density increase by a factor of 10^3 was obtained at the centres of the most condensed sub-clouds. These results are the ones which most resemble ours, but we obtain much more highly condensed (by a factor of 10^7) rotationally supported objects, probably because of higher spatial resolution. Monaghan & Varnas calulate the gravity on a fixed 58×58×58 grid, whereas we use a method which is not restricted by the resolution of a grid.

A final calculation worth mentioning is that of Nagasawa and Miyama [1987], who use SPH and a multipole expansion technique for the gravity, which avoids the restrictions of a fixed grid. They considered the head-on collision of two clouds at Mach 3; the calculation is scaled so that the mass of each cloud is 0.56 times the equilibrium mass of the Bonnor-Ebert solutions use for the initial cloud state. The end result is central collapse toward the centre of mass, and they terminate the calculation when the central density reaches 10^4 times the initial density. They use both an isothermal equation of state, and a cooling equation of state with pressure proportional to $\rho^{0.73}$. In the former case they obtain no fragmentation, while in the latter they do (but still with general central collapse), as expected form the original Hunter et al calculation.

In summary the aims of these previous calculations has been (i) to convert interstellar matter by some collision scenario into protostellar matter, and to set up the initial conditions for protostellar collapse, and (ii) to investigate in a rudimentary way the initial mass function and the statistics of binaries. While the calculations have fallen short of success in achieving these aims, they have nevertheless made considerable progress and this has been due not simply to increasingly powerful computers, but also to increasingly powerful numerical models. The work we have been engaged in at Cardiff takes this a stage further, by introducing the Tree-Code method for solving for gravity [Barnes & Hut 1986, Hernquist 1987], which is geometrically very flexible and yields very much increased resolution for this problem. The result is a notable advance in the numerical simulation of protostellar formation.

SIMULATIONS OF CLOUD-CLOUD COLLISIONS AT CARDIFF

i) NUMERICAL MODEL

The method used in the simulation at Cardiff is a combination of Tree-Code Gravity (TCG), and a modified form of SPH in which we are able to resolve very compact dense structures by allowing the particle size to vary with position.

In TCG, the force on any one particle is obtained by grouping distant particles in clusters. There is a hierarchy of clusters which is stored in the computer in a tree structure. We are grateful to Lars Hernquist for the use of his Fortran version of TCG, which we have modified to include SPH. In SPH, the gas is represented by an ensemble of spherically-symmetric particles. The mass of an individual particle is smoothly distributed about its centre according to a kernel function, which is centrally peaked and finite in extent. The outer radius of the kernel is 2h. At any point in the computational domain, the net density, pressure, velocity, etc. are obtained by summing contributions from all the particles within a distance 2h. The contributions are weighted by the kernel function. The number N of particles within a radius 2h must be much greater than one if the code is to give an acceptable approximation to hydrodynamics. Typically $N_k \sim 50$ is sufficient. The high spatial resolution of our simulation derives from the fact that we allow h to be a local function of particle density. We vary h so that the number of particles within a kernel radius is statistically the same everywhere, i.e. $h \propto n^{-1/3}$, where n is the particle number density.

In the simulations described here we use 4000 particles. The equations of motion are integrated using a time-centred scheme, with positions and velocities being evaluated at alternate half time-steps. The value of the time-step is taken to be

$$\Delta t = 0.3\ min \left\{ \frac{h}{\alpha_{max}}, \frac{h}{v_{max}}, \left(\frac{h}{\dot{v}_{max}}\right)^{\frac{1}{2}} \right\}$$

where α and v are the local isothermal sound speed and bulk velocity respectively. The time-step becomes very small at the end of the simulation, due to the high density in the protostar.

ii) PHYSICAL CONDITIONS

The first simulation starts with two equal, uniform-density, uniform-temperature, spherical clouds approaching one another head-on along the x-axis at a supersonic relative speed, 1.62 km s^{-1} (Mach 4.3). Figure 1a shows the particle positions at the beginning of the simulation (t=0) projected onto the (x,y)-plane. The clouds initially have mass $M_0 = 75\ M_{\odot}$(each), radius $R_0 = 1$ pc, and density $\rho_0 = 1.2 \times 10^{-21}$ gm cm^{-3} . We have also performed calculations for the same cloud parameters with non zero impact parameter.

The initial clouds are globally stable against contraction, but since they have uniform density they are not locally in equilibrium. Left to its own devices, a single cloud relaxes towards a centrally concentrated equilibrium structure on a sound-crossing time-scale ($\sim 5.3 \times 10^6$ years). In our simulation, this is prevented from happening because the clouds collide with one another on a much shorter time-scale.

The pressure is given by a cooling equation of state of the form

$$p = \alpha_0^2 \rho_0^{2/3} \rho^{1/3} + \alpha_1^2 \rho \tag{1}$$

where $\alpha_0 = 0.57$ km s^{-1} , and $\alpha_1 = 0.18$ km s^{-1} . Hence, in the initial clouds the sound speed is $\alpha \simeq 0.375$ km s^{-1} (corresponding to $\sim 100^\circ$ K, if we adopt cosmic abundances and assume the hydrogen to be mainly molecular), whilst in the limit of very large density $\alpha \rightarrow \alpha_1 = 0.15$ km s^{-1} (corresponding to $\sim$ 10 K).

Implicit in equation (1) is the assumption that thermal equilibrium is achieved instantaneously, and that cooling radiation escapes freely. Equation (1) reproduces adequately the broad variation of sound speed with density revealed by detailed calculations of the thermal and chemical balance of optically thin interstellar gas. Order of magnitude estimates indicate that the time-scale on which thermal equilibrium is achieved is much less than the dynamical time-scale, and that trapping of cooling radiation is just starting to become significant at the end of the simulation reported here.

iii) RESULTS

For the head-on (b = 0) calculation it is convenient to divide the evolution into two phases. In the first phase, $0 < t < t_1$ (where $t_1 \simeq 1.45 \times 10^6$ years), the clouds collide and produce a dense layer, with expansion in the outer parts similar to that of Stone. At the end of this first phase the density in the layer has reached a value $\sim 10^3 \rho_0$. In the second phase, $t_1 < t < t_2$ (where $t_2 \simeq 1.6 \times 10^6$ years), the protostar condenses out of this layer. When the protostar reaches a density of $\sim 10^{10}\ \rho_0$ at $t = t_2$, its collapse is halted by rotation.

Figures 1b-d show particle positions projected onto the (x,y)-plane at times $t \simeq 0.96 \times 10^6$ years, $t = t_1$ and $t = t_2$. Although the protostar condenses out between t_1 and t_2 , it is buried deep within the densest part of the layer and cannot be resolved on the scale of Figure 1d.

In Figures 2a-d, the density at each particle is plotted against x. The times are the same as those illustrated in Figures 1a-d. The formation of a very compact, very dense protostar is evident from Figure 2d. Figure 3 shows how the maximum density varies with time. At the end of the calculation the rate of increase of density in the centre of the protostar slows down dramatically as rotation becomes important. In Figures 4 and 5 we zoom in on the protostar at the end of the simulation ($t \simeq t_2$). Figures 4a-c show the density at each particle plotted against x, y and z. We see that the protostar has a well defined centrally peaked density profile and is $\sim 10^6$ times denser than its immediate surroundings. Because of the rotation, the extent of the protostar in the z direction is about one third its extent in the x and y directions.

Figures 5a-c show velocity vectors of the particles in the protostar projected onto the (y,z), (z,x) and (x,y)-planes. We see that the protostar rotates, approximately as a solid body, about an axis which by chance is nearly aligned with the z-axis. Particles outside the protostar accrete onto it. At this instant, the protostar is expanding, having just bounced; but this is a transient episode and contraction - albeit quasistatic contraction - is soon resumed.

At this stage the protostar has mass $\sim 16.5\ M_\odot$and rotation period $\sim 10^3$ years. Its extent in the x and y directions is ~ 400 AU and its extent in the z direction is ~ 80 AU. It is displaced from the centre of symmetry of the overall collision by $\Delta x \sim 2500$ AU, $\Delta y \sim 2400$ AU, and $\Delta z \sim 900$ AU. The subsequent evolution of the protostar will depend on the efficiency with which magnetic stresses, which are not modelled here, remove angular momentum from the protostar.

When we examine the various energy components we find that at $t = t_2$ the gravitational potential energy has reached a very large (negative) value, and most of the binding energy has been radiated away during the final collapse phase. Even though the gravitational potential energy has increased by two orders of magnitude, the total energy (including radiated energy) is conserved to within a few percent.

For impact parameter $b = R_0$ two collapsed objects are formed near the centres of the two respective clouds, after an elongated structure is formed during the collision which compresses the two centres (see Fig 6).Figure 7 shows the density profiles of these two objects. The higher density one has attained rotational support, while the

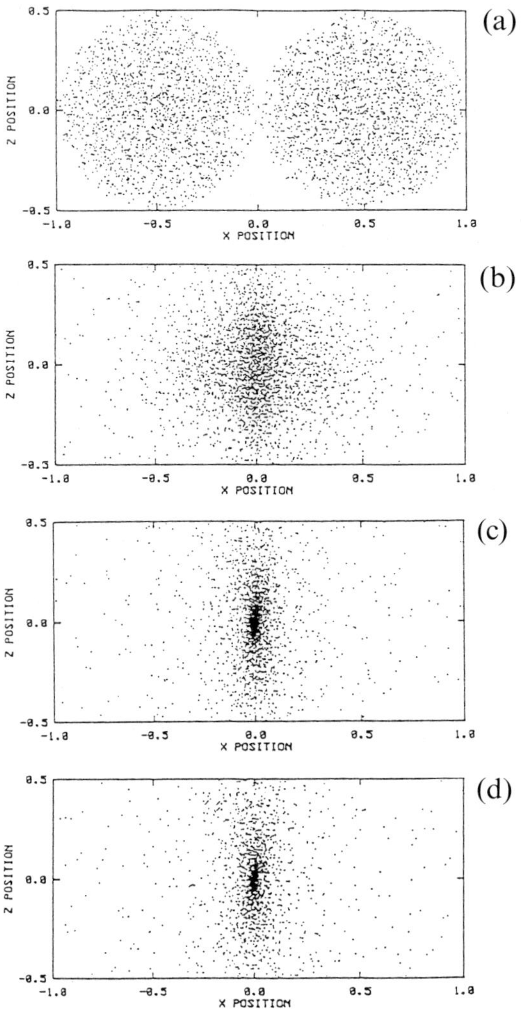

Figure 1. Head-on (b=0) collision. Particle positions projected onto the (x,y)-plane at times (a) $t = 0$; (b) $t = 0.96\times10^6$ years; (c) $t = t_1 = 1.45\times10^6$ years; (d) $t = t_2 = 1.60\times10^6$ years (end of simulation). The length unit is 2 pc.

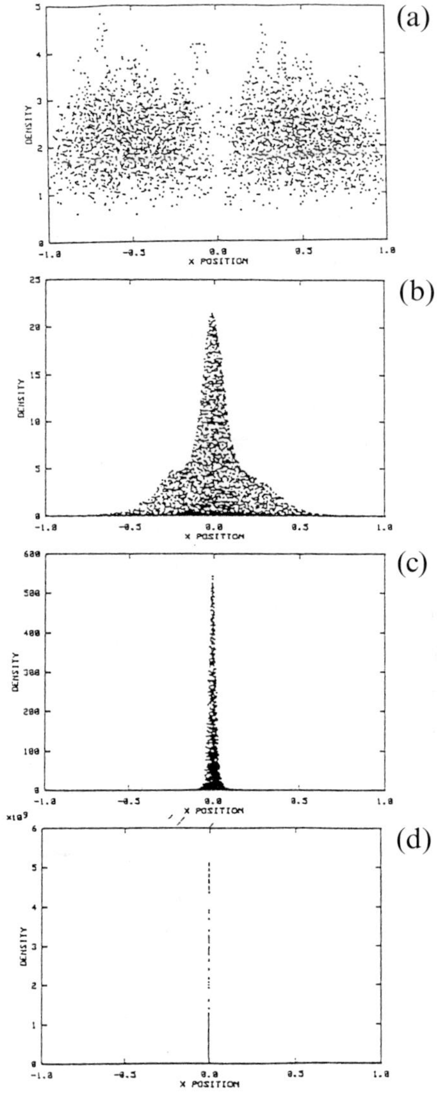

Figure 2. The density at each of the particles plotted against its x-coordinate, at the same times portrayed in Figure 1. The density unit is 6.3×10^{-22} gm cm^{-3}, which corresponds to a molecular hydrogen number density of ~ 200 cm^{-3}.

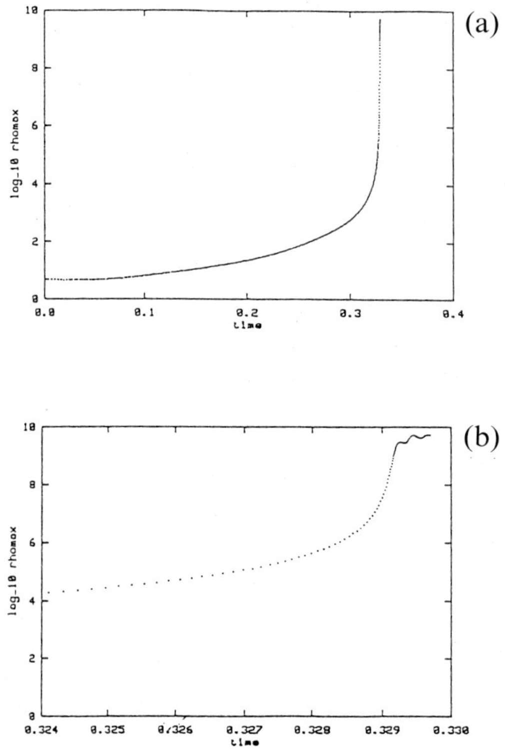

Figure 3. The maximum density as a function of time. (a) full time span of the calculation (b) last ~ 2% of the time. The time unit is 4.9 X 10^8 years.

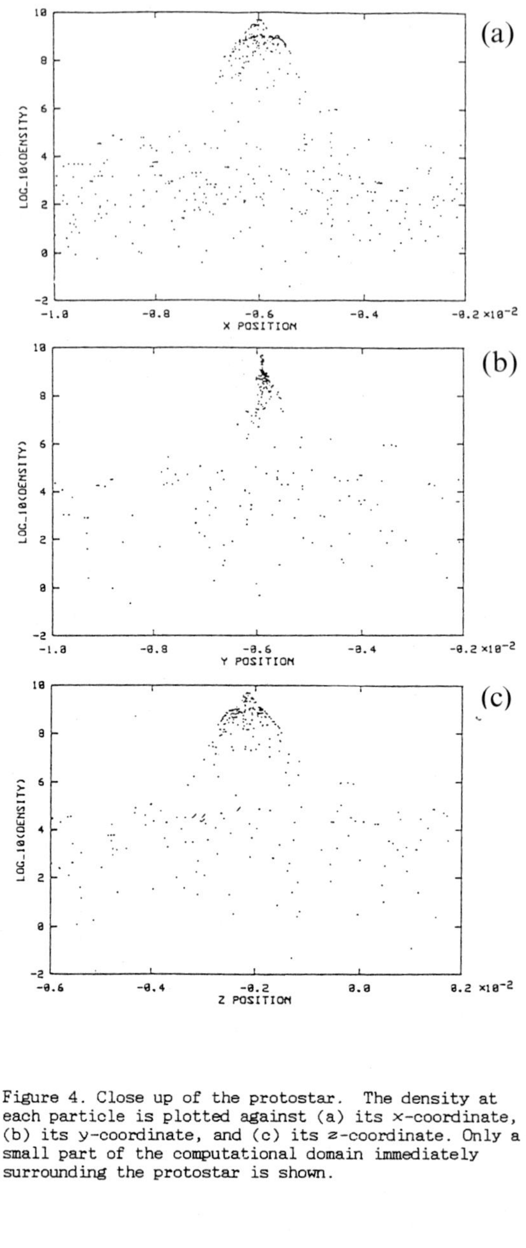

Figure 4. Close up of the protostar. The density at each particle is plotted against (a) its x-coordinate, (b) its y-coordinate, and (c) its z-coordinate. Only a small part of the computational domain immediately surrounding the protostar is shown.

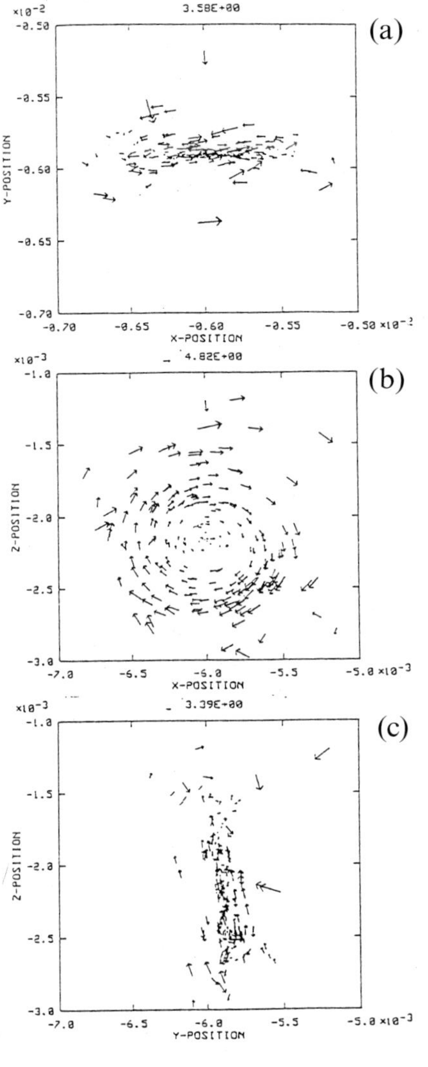

Figure 5. Here the particles near the protostar are shown projected onto the x–y, x–z, and y–z planes. The particles are represented by arrows, with the back of the arrow at the particle position, and the size and direction of the arrow giving the velocity of the particle in the corresponding plane. The largest velocities shown are ~ 7 km s^{-1}.

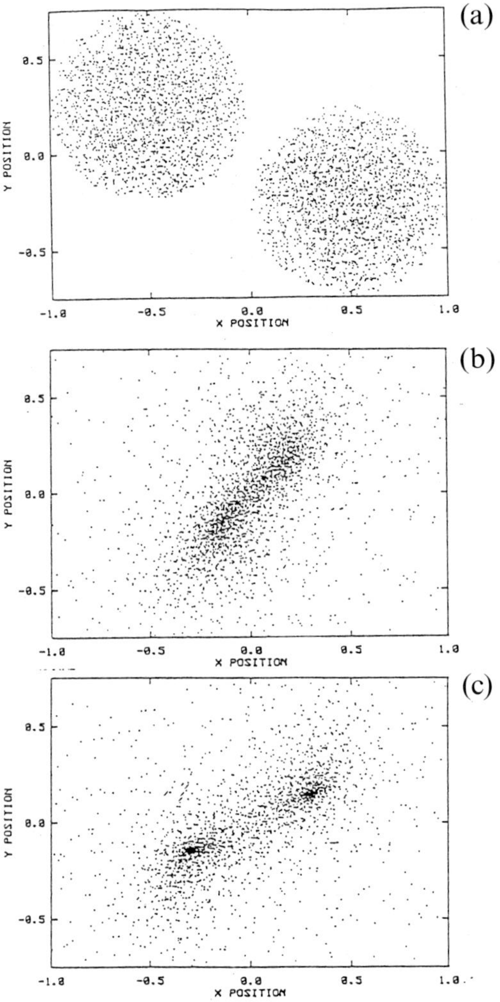

Figure 6. b = R_o collision. Particle positions projected onto the x–y plane at times (a) $t = 0$, (b) $t = 1.52 \times 10^6$ years, and (c) $t = 1.99 \times 10^6$ years.

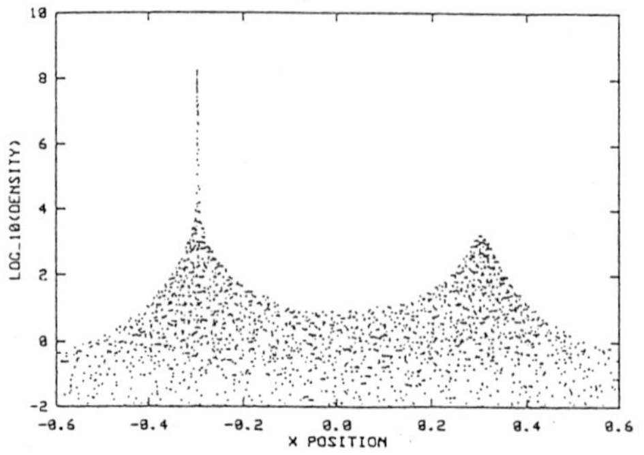

Figure 7. b = R_o collision. The density at each particle in the range $-0.6 < x < 0.6$ plotted against its x-coordinate.

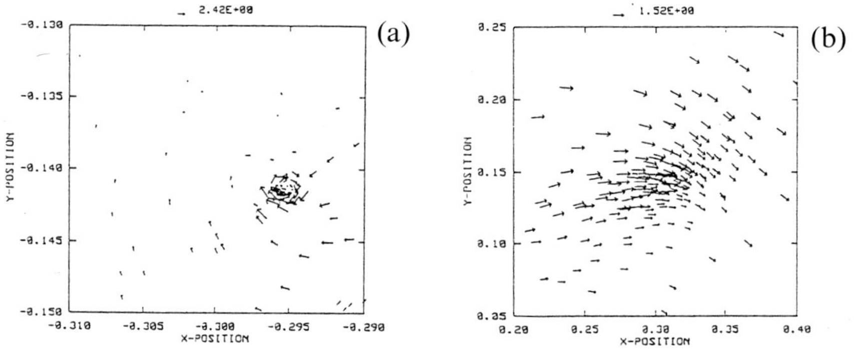

Figure 8. b = R_o collision. Close up of the two protostars. Particle positions and velocities are shown as in Figure 5.

lower density one is still collapsing at the end of the calculation. This can be seen from the velocity structure of the two in Figure 8. The separation of the two objects is 1 pc, and they are not bound. These results contrast with previous non-zero b calculations which were unable to form double condensations. The main reasons for the difference here are the cooling equation of state and the greater accuracy of TCG for the collapsing objects.

CONCLUSIONS

The protostars formed in these simulations behave in a manner remarkably like that expected of a real protostar, collapsing to a high-density rotating disc, radiating away gravitational binding energy and continuing to grow by accretion. The dimensions of the protostar in the $b \simeq 0$ cases are comparable to those expected for protoplanetary discs. This simulation demonstrates that the combination of Tree Code Gravity and Smoothed Particle Hydrodynamics, with varying particle size, provides a very powerful tool for studying star formation. The enormous dynamic ranges of density and spatial resolution which can be obtained mean that the evolution of protostars can be accurately modelled to high density, and their formation can be identified unambiguously.

REFERENCES

Barnes, J. & Hut, P. Nature 324, 446 (1986). bb Boss, A.P. in Interstellar Processes, eds. Hollenbach, D.J., & Thronson Jnr., H.A. (pub D.Reidel), 321 (1990)
Gingold, R.A. & Monaghan, J.J. Mon. Not. roy. astron. Soc. 181, 375 (1977)
Hernquist, L. Astrophys. J. Suppl. Ser. 64, 715 (1987).
Hunter, J.H., Jnr., Sandford, M.T., II, Whitaker, R.W. & Klein, R.I. Astrophys. J. 305, 309 (1986).
Lattanzio, J.C., & R.N.Henriksen Mon. Not. roy. astron. Soc. 232, 565 (1988).
Lucy, L.B. Astron. J. 82, 1013-1024 (1977).
Monaghan, J.J. & Varnas, S.R. Mon. Not. roy. astron Soc. 231, 515 (1988).
Nagasawa, M. & Miyama, S.M. Prog. theor. Phys. 78, 1250 (1987).
Stone, M.E. Astrophys. J. 159, 293 (1970).
Woodward, P.R. Ap.J. 207,484 (1976).

ACKNOWLEDGEMENTS

We thank Dr. M. Noguchi for useful discussions regarding the Cardiff calculations. These were carried out on the CRAY XMP at the Rutherford Appleton Laboratory.

THE EVOLUTION OF CLUMPY GAS IN GALAXIES

H.W. Yorke[1], R. Spurzem[2], R. Kunze[2]

[1]Institut für Astron. und Astrophys., Am Hubland, DW-8700 Würzburg
[2]Institut für Theor. Physik und Sternwarte, Olshausenstr. 40, DW-2300 Kiel 1

ABSTRACT

The numerical treatment of a clumpy interstellar medium is reviewed. Emphasis is placed on the ansatz of a multicomponent mixture of clouds and an ambient gas, for which the clouds are treated stellar dynamically and the gas by the hydrodynamic equations — supplimented by "source terms" describing the interactions between the components. These source terms and their numerical treatment are briefly discussed. As an example of the use of a multicomponent numerical model the expansion of a supernova remnant in a 'homogeneously clumpy' medium has been calculated under the assumption of spherical symmetry. The effects of various interaction terms and the sensitivity of the results on the particular choice of poorly known parameters are discussed. It is shown that the evaporation of small clouds in particular causes the largest deviations in the SNR evolution as compared to the non-clumpy approach.

1.0 Introduction: The Clumpy ISM

The interstellar medium (ISM) is a complex mixture of atomic, molecular, and ionized gas and dust grains permeated by cosmic rays and magnetic fields. It is known to be highly inhomogeneous in density (from $n \lesssim 10^{-2}$ cm^{-3} to $n \gtrsim 10^{6}$ cm^{-3}) and temperature (from $T \lesssim 10$ K to $T \gtrsim 10^{6}$ K) over a large range of scale sizes. We are accustomed to think of the ISM as a mixture of several co-existing phases, which correspond to three distinct temperature regimes (see Fig. 1). The hot ($T \approx 10^{6}$ K), tenuous gas results from supernova explosions. Although thermally instable, the longevity of this phase is due to its inability to cool effectively. At somewhat lower temperatures ($T \approx 10^{5}$ K) the thermal excitation of ions – especially OVI – by electrons increases dramatically and the likelihood of observing gas in this quickly cooling temperature range is correspondingly low. The warm ($T \approx 10^{4}$ K) gas and cool ($T \approx 10^{2}$ K) gas can be in thermal equilibrium (heating by cosmic rays and UV starlight = cooling), while in pressure equilibrium with each other and with the hot phase, corresponding to the line $nT \approx 10^{3}$ K cm^{-3} in Fig. 1. The warm and the cool phases are thermally stable over a limited range of pressure (Field *et al.* 1969), which is consistent with that observed in the ISM.

Cox & Smith (1974) were first to point out that supernova explosions occur often enough to overlap and conceivably form a network of hot "tunnels" which they estimate to fill about 10% of the volume in the disk of our Galaxy. Since then, the mass fractions and volume filling factors of each of the three phases has been the subject of much debate and can still be considered an open question. According to the "three phase model" of McKee & Ostriker (1977), most of the volume in the galactic plane is filled by hot gas. Embedded in this gas are clouds (cool neutral medium = CNM) surrounded

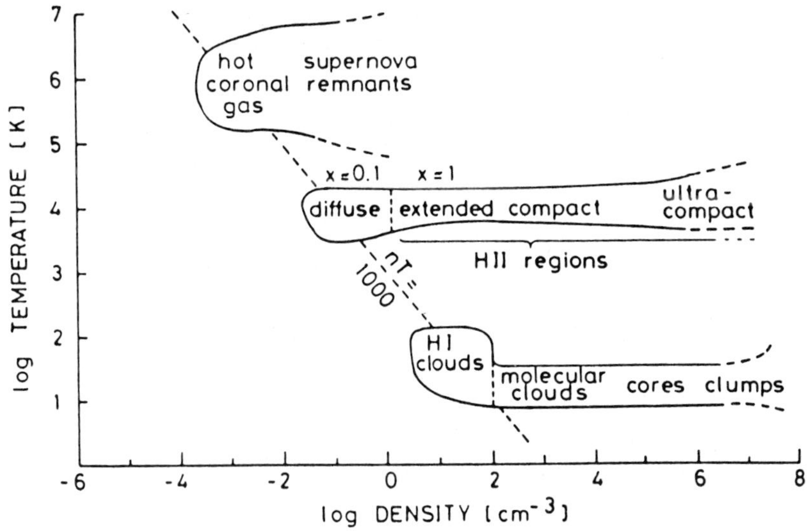

Figure 1. *Phases of the interstellar medium (schematic) in the* $\log n$ *vs.* $\log T$ *plane (from Yorke 1987). Whereas the components lying along the line* $nT \approx 10^3$ K cm^{-3} *could be in "quiescent" pressure equilibrium with each other, the other components are governed by other forces, such as magnetic fields, internal turbulence, or gravity, and/or they are in a state of violent motion – collapse or expansion.*

by a warm halo with both neutral (WNM) and ionized (WIM) parts. Crucial to this picture is a large volume filling factor for the hot gas ($\sim 70\%$). Cox (1989), on the other hand, has argued that the volume filling factor must be small – perhaps as low as 10% (Cox & Slavin 1990). Indeed, it is likely that the mass and volume ratios could vary greatly from galaxy to galaxy and even within a galaxy (*e.g.* in spiral arms as opposed to interarm or nuclear regions). The interested reader is referred to Tenorio-Tagle & Bodenheimer (1988), McKee (1990), and Spitzer (1990) for more complete discussions.

2.0 Clumpy Gas Dynamics

The ISM plays a critical role in the formation and evolution of galaxies and – due to its fundamental importance for star formation – their overall morphological properties. It is therefore not surprizing that much effort has been invested in studying ISM dynamics. Theoretically, a number of approaches have been adopted, including phenomenological models designed to explain the observed size scales and velocity dispersion, direct numerical solutions of the hydrodynamical equations – sometimes modified to take "turbulence" into account, and a statistical treatment of turbulence with assumed closure relations.

Due to limited computer resources, particularly memory and CPU time, the discretization of the partial differential equations of hydrodynamics is restricted to $\lesssim 10^6$ grid cells and $\lesssim 10^6$ time steps in an explicit calculation. For a 2D calculation this generally translates to a spatial resolution of $\Delta l \gtrsim 10^{-3}$ L, where L is the overall length scale of the problem. The size of the time step δt is restricted by the Courant-Friedrichs-Lewy (CFL) condition ($\delta t \lesssim \Delta l/u$, where u is the relevant velocity scale – either sound velocity, Alfvèn speed or the velocity of the gas itself), and this in turn limits the overall evolution time $\lesssim 10^6$ δt which can be considered for a particular problem.

Is a numerical treatment of the gas dynamical equations valid when considering the evolution of the ISM? In order to answer this question, we must ask ourselves:

1. Does the volume $(\Delta l)^3$ contain a sufficient number of particles to be treated statistically?
2. Is $\Delta l >> \lambda_{\rm coll}$ and $\delta t >> \tau_{\rm coll}$, so that the velocity distribution of the particles is sufficiently Maxwellian? ($\lambda_{\rm coll}$ is the mean free path and $\tau_{\rm coll}$ the mean time between collisions.)
3. Is $\Delta l << \rho/|\nabla\rho|$ and $\delta t << \rho/|\dot{\rho}|$?

For most astrophysical applications the answer to the first questions is 'yes', the answer to the second is 'yes, except in shocks' (which must be treated differently), whereas the answer to the third question is 'no', *i.e.* we generally deal with a clumpy, rapidly fluctuating medium. It is important to realize that *inhomogenities can significantly modify the global evolution* and that *the usual hydrodynamic equations are thus not valid for the ISM in general.* What can we do to rectify this situation?

— We can attempt to choose Δl (and δt) sufficiently small, so that the inequalities in question 3. are fulfilled.
— Fluctuations on the sub-grid level can be treated as an additional (turbulent-) pressure. Equations are required to estimate the evolution of turbulent energy density. Such equations can be derived from phenomenological considerations or through a more rigorous statistical treatment of turbulence.
— The clumpy ISM can be interpreted as a system of "clouds" and a system of stars moving in an ambient intercloud medium. This mixture is then treated as a multicomponent medium with complex interaction terms.

After briefly discussing the first two approaches, we will consider in greater detail the evolution of a multicomponent medium.

2.1 Spatially resolved hydrodynamics

In the first approach we try to spatially resolve the smallest clumps. Norman *et al.* (1988), for example, used a program called ZEUS to calculate the passage of an adiabatic shock through a clumpy medium permeated by a weak uniform magnetic field. The clumps were assumed to be randomly distributed rods extending infinitely in the y-direction (it was a 2D calculation), each with a gaussian density distribution in the xz-plane. The maximum initial density contrast was 20; temperature was chosen for pressure equilibrium. The magnetic field was initially oriented in the x-direction, parallel to the Mach 10 planar shock. Their results show the corrugation of the shock front as it encounters the clumps, the development of horseshoe-shaped clumps behind the shock (cloud-crushing), and vertical stretching of magnetic field lines.

As a second example of magnetohydrodynamics with spatially resolved clumps, MacLow *et al.* (1992 – this volume) have calculated the interaction of a shock front with an individual cloud, showing in greater detail the phenomenon of cloud-crushing.

The most obvious disadvantage of this approach is the limited global scales ($L \sim 10^3\ \Delta l$) which can be treated. This shortcoming is especially apparent, when one realizes that clouds themselves have a clumpy substructure (*e.g.* Blitz 1990) and (sub-) clumpiness may persist down to scales significantly below 1 pc.

2.2 "Turbulent" Hydrodynamics

It would be impossible to do justice to the complex topic of turbulence in such a short review. Actually, there is a vast amount of literature on the subject, which can be summarized by the statement that we do not really understand interstellar turbulence (see *e.g.* the review by Scalo 1987). In the simplest numerical approach, an additional isotropic pressure term $P_{\rm turb}$ is included in the equation of motion. The hydrodynamic equations must then be supplimented by equations describing the evolution of $P_{\rm turb}$, often a formitable task.

Consider, for example, the equation of momentum conservation for an incompressible ($\rho =$ const.) fluid:

$$0 = \rho \frac{\partial \mathbf{v}}{\partial t} + \rho(\mathbf{v} \cdot \nabla)\mathbf{v} + \nabla P - \rho\nu\nabla^2\mathbf{v} \quad ,$$

where $\mathbf{v}$, ρ, P, and ν are the fluid velocity, density, pressure, and kinematic viscosity. The incompressibility condition reduces the mass conservation equation to $\nabla \cdot \mathbf{v} = 0$. Replacing the fluid velocity by a time averaged part $\mathbf{u}$ and a rapidly fluctuating part $\mathbf{u}'$, *i.e.* $\mathbf{v} \rightarrow \mathbf{u} + \mathbf{u}'$ with $<\mathbf{u}'> = 0$, we can write the time average of the momentum equation in the form:

$$0 = \rho \frac{\partial u_i}{\partial t} + \rho u_k \frac{\partial u_i}{\partial x_k} + \frac{\partial P}{\partial x_i} + \frac{\partial \rho <u'_i u'_k>}{\partial x_k} - \rho\nu \frac{\partial^2 u_i}{\partial x_k \partial x_k}$$

where we have expressed the equation, component by component, in a Cartesian coordinate system, using implied summation over repeated indices. Note that the last term in this equation $\rho<u'_i u'_k>$ has the appearance of an additional non-isotropic pressure, which we identify with the turbulent pressure. This equation demonstrates that you cannot calculate the mean flow without knowledge of the fluctuations. A similar equation for P' can also be derived, which we omit for simplicity.

By subtracting the mean flow equations from the original unaveraged equations, we obtain equations describing the evolution of the fluctuations, *i.e.* for the momentum equation considered here:

$$0 = \frac{\partial u'_i}{\partial t} + \frac{\partial u'_i u'_k}{\partial x_k} + \frac{1}{\rho}\frac{\partial P'}{\partial x_i} - \nu \frac{\partial^2 u'_i}{\partial x_k \partial x_k} + u'_k \frac{\partial u_i}{\partial x_k} + u_k \frac{\partial u'_i}{\partial x_k} - \frac{\partial <u'_i u'_k>}{\partial x_k}$$

The interpretation of each term as a physical effect can be found in Deissler (1984). From this 'fluctuation equation' we can determine the rate of change of $<u'_i u'_j>$ at a single point (one-point correlations) by taking the time average of $u'_j \partial u'_i/\partial t + u'_i \partial u'_j/\partial t$:

$$0 = \frac{\partial <u'_i u'_j>}{\partial t} + <u'_j u'_k>\frac{\partial u_i}{\partial x_k} + <u'_i u'_k>\frac{\partial u_j}{\partial x_k} + u_k \frac{\partial <u'_i u'_j>}{\partial x_k} + \frac{\partial <u'_i u'_j u'_k>}{\partial x_k}$$
$$+ \frac{1}{\rho}[<u'_i \frac{\partial P'}{\partial x_j}> + <u'_j \frac{\partial P'}{\partial x_i}>] - \nu \frac{\partial^2 <u'_i u'_j>}{\partial x_k \partial x_k} + 2\nu < \frac{\partial u'_i}{\partial x_k}\frac{\partial u'_j}{\partial x_k}> \quad .$$

The purpose of this exercise was to demonstrate explicitly the appearance of a large number of couplings which arise from the nonlinear terms in the original fluid equations. In particular, we see third order correlations $<u'_i u'_j u'_k>$ in the equation for the time evolution of the second order correlations. An even more complicated expression can be derived for $\partial <u'_i u'_j u'_k>/\partial t$, but it contains the fourth-order correlation $<u'_i u'_j u'_k u'_l>$. To complicate the situation even further, we note that the one-point correlation equations tell us nothing about the transfer of turbulence between scales. In order to derive an equation for the two-point correlation function, the fluctuation equation is written at two different positions, they are similarly combined and they similarly contain higher order correlations. Fluid turbulence is a spatial *and* temporal problem. In order to obtain a solvable system of equations with this approach, assumptions must be made. Existing methods of treating interstellar turbulence differ in how they handle the 'closure problem', *i.e.* in how higher-order correlations can be expressed as a function of lower-order correlations (see *e.g.* Scalo 1987 and Deissler 1984).

2.3 Multicomponent models

Another theoretical approach to treat clumpiness in the ISM is to consider a system of discrete clouds embedded in a gaseous medium. The clouds can interact among themselves and with the gas and stars. The various different multicomponent approaches differ in the manner in which the cloud component is treated and the number of and sophistication with which the interaction terms are treated. The cloud system could, for example, be treated as an N-body system of finite spheres or "cloud fluid" equations similar to those of hydrodynamics can be derived (Scalo & Struck-Marcell 1984).

Consider the Boltzmann equation

$$\frac{\partial f^{(s)}}{\partial t} + (\mathbf{v} \cdot \nabla_{\mathbf{r}}) f^{(s)} + (\mathbf{a} \cdot \nabla_{\mathbf{v}}) f^{(s)} = \left(\frac{\delta f^{(s)}}{\delta t}\right)_C$$

for the time evolution of the one-particle distribution function $f^{(s)} = f^{(s)}(\mathbf{r}, \mathbf{v}, t)$ of particle type 's', where $\mathbf{a} = d\mathbf{v}/dt$ is the acceleration and $(\delta f^{(s)}/\delta t)_C$ is the collision term, describing both the sudden jump of particles from one part of phase space to another and source (sink) terms due to production (destruction) of particles. Defining velocity moments by $\langle lmn \rangle = \int v_x^l v_y^m v_z^n f^{(s)} dv_x dv_y dv_z$, multiplying the Boltzmann equation by velocity components to various powers, and integrating over velocity space, we obtain a sequence of velocity moment equations for *e.g.* Cartesian coordinates:

$$\frac{\partial}{\partial t}\langle 000\rangle + \frac{\partial}{\partial x}\langle 100\rangle + \frac{\partial}{\partial y}\langle 010\rangle + \frac{\partial}{\partial z}\langle 001\rangle = \cdots$$
$$\frac{\partial}{\partial t}\langle 100\rangle + \frac{\partial}{\partial x}\langle 200\rangle + \frac{\partial}{\partial y}\langle 110\rangle + \frac{\partial}{\partial z}\langle 101\rangle + \cdots$$
$$\vdots$$

By defining $\rho_{(s)} = \langle 000 \rangle$, $\rho_{(s)} u_{(s)x} = \langle 100 \rangle$, $\rho_{(s)}(\sigma^2_{(s)xx} + u^2_{(s)x}) = \langle 200 \rangle$, ..., the first few equations in the series:

$$\frac{\partial \rho_{(s)}}{\partial t} + \frac{\partial \rho_{(s)} u_{(s)j}}{\partial x_j} = \sum_k \left(\frac{\delta \rho_{(s)}}{\delta t} \right)_k$$

$$\frac{\partial \rho_{(s)} u_{(s)x}}{\partial t} + \frac{\partial \rho_{(s)} u_{(s)x} u_{(s)j}}{\partial x_j} + \frac{\partial \rho_{(s)} \sigma^2_{(s)xx}}{\partial x} + \cdots = \sum_k \left(\frac{\delta \rho_{(s)} u_{(s)x}}{\delta t} \right)_k$$

$$\vdots$$

resemble those of hydrodynamics, except (1) they include source (sink) terms, (2) they allow anisotropic velocity distributions (anisotropic pressure), and (3) the moment expansion must be truncated at some point and supplimented by closure relations, which express the highest order moments included as a function of the other moments. Indeed, the usual hydrodynamical equations can be recovered from the moment equations by assuming (1) no sources or sinks (= conservation of mass, momentum, energy), (2) isotropic pressure, and (3) an equation of state (= closure relation).

Considering a mixture of clouds, stars and an ambient gas, the types of 'source' terms which must be included in the above equations are :

a) phase transitions
  — cloud formation out of the ambient medium and its destruction via electron conduction (evaporation; *e.g.* Begelman & McKee 1990, McKee & Begelman 1990), via the champagne effect (photo-evaporation; *e.g.* Yorke 1986), or by an inverse thermal instability (*e.g.* Biermann *et al.* 1972), say, during expansion after cloud-cloud collisons;
  — star formation (this produces a stellar component out of the cloud component)
  — stellar mass loss from aging stars

b) dynamical interactions amoung the components, such as
  — drag forces on clouds or stars moving relative to the ambient gas (dynamic friction)
  — cloud-cloud collisions (this also reshuffles the distribution of cloud masses)
  — cloud-crushing (interaction of a cloud with an intercloud shock wave)

In addition, the internal properties of each component can vary with time, *i.e.* stars age, the clouds' temperature, pressure and radius may vary, the medium can evolve chemically, *etc.*

In the following we will consider – as a specific example of a multicomponent calculation – the spherically symmetric (1D) expansion of a supernova remnant into a clumpy medium. We refer also to the 2D calculations presented by Kunze *et al.* (1992 – this volume) of the evolution of a superbubble in a clumpy, stratified galactic disk.

3.0 The Expansion of a Supernova Remnant into a Clumpy Medium

The problem of the expansion of a supernova remnant into a homogeneously clumpy medium has been considered by a number of authors (see *e.g.* Cowie *et al.* 1981 and the discussion by McKee 1988 and Wolff & Durisen 1987). Previously in multicomponent calculations, the cold component has been treated as an ensemble of clouds of equal mass and radius. Because, however, the scales of inhomogenities in the ISM vary by several orders of magnitude, one can expect the detailed form of the mass spectrum to be of importance.

We consider the evolution of a supernova remnant in a medium containing clouds with a mass spectrum $n(m) \propto m^{-\alpha}$ with $\alpha = 3/2$ and $\alpha = 2$, using a numerical multicomponent (6 cloud, 1 gas) model. Each cloud component has its own mean mass and radius, which are calculated simultaneously with the dynamics. The following physical effects have been taken into account: cloud evaporation by electron thermal conduction, cloud acceleration by ram pressure, cloud crushing due to the blast wave passing the cloud, and cloud formation due to thermal instabilities in the cooling remnant. The explosion itself has been highly idealized. Thermal and kinetic energy totaling 10^{51} erg are deposited within a small volume of the multicomponent ISM. The chemical composition is assumed to be solar; the effects of gamma ray acceleration, magnetic pressure and cosmic rays are not explicitly included.

3.1 The dynamical equations

In a volume V filled with gas and an ensemble of clouds which fill a volume $V_{\rm cl}$, we have to distinguish between the average values of ambient gas density ρ and pressure P over V and the intrinsic values $\tilde{\rho}$, $\tilde{P}$ within the volume $V - V_{\rm cl}$. Defining the volume filling factor $f = V_{\rm cl}/V$ of clouds, we note that $\rho = \tilde{\rho}\,(1-f)$ and $P = \tilde{P}\,(1-f)$, and the hydrodynamic equations for the gas in spherical symmetry become (Spurzem *et al.* 1991):

$$\frac{\partial \rho}{\partial t} + \frac{1}{r^2}\frac{\partial}{\partial r}(r^2 \rho u) = X$$

$$\frac{\partial \rho u}{\partial t} + \frac{1}{r^2}\frac{\partial}{\partial r}(r^2 \rho u^2) + \frac{\partial P}{\partial r} + \rho\frac{GM_r}{r^2} + \frac{P}{1-f}\frac{\partial f}{\partial r} + \mu_1 = Y$$

$$\frac{\partial P}{\partial t} + \frac{1}{r^2}\frac{\partial}{\partial r}(r^2 P u) + \frac{2}{3}\frac{P}{r^2}\frac{\partial}{\partial r}(r^2 u) - \frac{2}{3}\frac{P}{1-f}\left(\frac{\partial f}{\partial t} + u\frac{\partial f}{\partial r}\right) + \mu_2 + \frac{2}{3}\frac{1}{1-f}(\Lambda - \Gamma) = Z$$

where M_r is the total mass (gas + stars + clouds) contained within a sphere of radius r; μ_1 and μ_2 are terms associated with artificial viscosity (Tscharnuter & Winkler 1979); Λ (Γ) is the net interstellar cooling (heating) rate, known functions of $\tilde{\rho}$ and $\tilde{P}$; and X, Y, and Z are the source terms arising from phase transitions and dynamical interactions.

Each cloud component is described by a system of moment equations up to second order:

$$\frac{\partial \rho_{(s)}}{\partial t} + \frac{1}{r^2}\frac{\partial}{\partial r}(r^2 \rho_{(s)} u_{(s)}) = X_{(s)}$$

$$\frac{\partial \rho_{(s)} u_{(s)}}{\partial t} + \frac{1}{r^2}\frac{\partial}{\partial r}(r^2 \rho_{(s)} u_{(s)}^2) + \rho_{(s)}\frac{GM_r}{r^2} + \frac{\partial P_{r(s)}}{\partial r} + \frac{2P_{r(s)} - P_{t(s)}}{r} = Y_{(s)}$$

$$\frac{\partial P_{r(s)}}{\partial t} + \frac{1}{r^2}\frac{\partial}{\partial r}(r^2 P_{r(s)} u_{(s)}) + 2P_{r(s)}\frac{\partial u_{(s)}}{\partial r} = Z_{r(s)}$$

$$\frac{\partial P_{t(s)}}{\partial t} + \frac{1}{r^2}\frac{\partial}{\partial r}(r^2 P_{t(s)} u_{(s)}) + 2P_{t(s)}\frac{u_{(s)}}{r} = Z_{t(s)}$$

where we have denoted the radial and tangential components of the anisotropic pressure by $P_{r(s)} = \rho_{(s)}\sigma^2_{r(s)}$ and $P_{t(s)} = \rho_{(s)}\sigma^2_{t(s)}$. Closure relations for the higher order moments are included implicitly as dissipational effects on the right hand sides of the last two equations.

The above system of equations, 3 gas dynamical and $4 \times S$ moment equations (S is the number of stellar dynamical components: clouds + stars) and a 1D Poisson equation for gravitational acceleration GM_r/r^2 — supplimented by initial conditions and $4 \times (S+1)$ boundary conditions — are discretized using backward time differences and solved implicitly with a Newton-Raphson-Henyey procedure. The detailed form of the source term equations are given in Spurzem *et al.* (1991).

3.2 Ram pressure drag

As a single example of how an interaction term is incorporated into all equations, we consider the ram pressure of gas streaming relative to the system of clouds 's'. Consider first a solid sphere of radius R moving through the gas. We expect a decelerating force $\mathbf{F}_\mathrm{D} = -\pi R^2 \tilde{\rho} v \mathbf{v}_\mathrm{rel}$ where $\mathbf{v}_\mathrm{rel}$ is the velocity difference between the sphere and the gas. v is approximately the sound speed c_S for subsonic flow ($|\mathbf{v}_\mathrm{rel}| << c_S$). For supersonic flow ($|\mathbf{v}_\mathrm{rel}| >> c_S$) we expect $v \approx |\mathbf{v}_\mathrm{rel}|$. By analogy we can derive a decelerating force on spherical clouds:

$$\left(\frac{\delta \rho_{(s)} u_{(s)}}{\delta t}\right)_\mathrm{drag} = -C_D\, \tilde{\rho}\, \rho_{(s)} \frac{\pi R^2_{(s)}}{m_{(s)}} \sigma^\mathrm{rel}_{(s)} \cdot (u_{(s)} - u)$$

with $\sigma^\mathrm{rel}_{(s)} = \left(\sigma^2_{r(s)} + \sigma^2_{t(s)} + (u^2_{(s)} - u)^2\right)^{1/2}$. We take the dimensionless parameter $C_D = 1$ as in Cowie *et al.* (1981). McKee (1988) discussed C_D in greater detail and concluded that C_D varies between 0.25 and 1 (using the cloud's cross section rather than $\pi R^2_{(s)}$), depending of the cloud geometry and the ratio $|\mathbf{v}_\mathrm{rel}|/c_S$.

With this expression for the decelerating force on each cloud system, the other source terms follow from conservation arguments:

$$\left(\frac{\delta \rho u}{\delta t}\right)_\mathrm{drag} = -\sum_{s=1}^{S} \left(\frac{\delta \rho_{(s)} u_{(s)}}{\delta t}\right)_\mathrm{drag}$$

$$\left(\frac{\delta P_{r(s)}}{\delta t}\right)_\mathrm{drag} = -C_D\, \tilde{\rho}\, \rho_{(s)} \frac{\pi R^2_{(s)}}{m_{(s)}} \sigma^\mathrm{rel}_{(s)} \cdot \sigma^2_{r(s)}$$

$$\left(\frac{\delta P_{t(s)}}{\delta t}\right)_\mathrm{drag} = -C_D\, \tilde{\rho}\, \rho_{(s)} \frac{\pi R^2_{(s)}}{m_{(s)}} \sigma^\mathrm{rel}_{(s)} \cdot \sigma^2_{t(s)}$$

$$\left(\frac{\delta P}{\delta t}\right)_\mathrm{drag} = \frac{1}{3} \sum_{s=1}^{S} C_D\, \tilde{\rho}\, \rho_{(s)} \frac{\pi R^2_{(s)}}{m_{(s)}} \sigma^\mathrm{rel}_{(s)} \cdot \left(\sigma^2_{r(s)} + \sigma^2_{t(s)} + 2(u_{(s)} - u)^2\right)$$

Table 1. *Parameters of cases considered of supernova explosions in a clumpy medium. C_C is a dimensionless parameter associated with cloud crushing (not discussed here). α is the mass spectral index $n(m) \propto m^{-\alpha}$. Other symbols are discussed in text. Not included in the table are cases 5 and 9, for which only a single cloud component was assumed with $m = 10\ M_\odot$ and $25\ M_\odot$, respectively.*

$\alpha = 2$	$\alpha = 1.5$	features
1	11	$C_D = 1$, $C_C = 1$, $m_{(s)}/\mathrm{M}_\odot = 1, 2, 5, 10, 20, 50$
2	12	without evaporation
3	13	$C_D = 0.1$
4	14	$C_C = 0.1$
6	16	$m_{(s)}/\mathrm{M}_\odot = 1.5, 7.5, 35$
7	17	$m_{(s)}/\mathrm{M}_\odot = 2, 4, 10, 20, 40, 100$
8	18	$T_C = 50$ K, $n_C = 40\ \mathrm{cm}^{-3}$ initially

3.3 The numerical problem: initial conditions and boundary conditions

In order to test the various effects described above to the "classical case" of a supernova remnant expanding into a non-clumpy medium, we have conducted a series of calculations in which the "non-classical" effects have been turned off or on. For the "classical" comparison case, we have deposited 4 $\mathrm{M}_\odot$ within $r_1 = 2$ pc, setting its expansion velocity to $v(r) = v_0\, r/r_1$, where v_0 is choosen so that $E_{\mathrm{kin}} = 10^{51}$ erg. The partially ionized ($x = 0.2$) non-clumpy medium had a density $n_0 = 0.1\ \mathrm{cm}^{-3}$ and temperature $T = 5000$ K. The outer boundary conditions — all dependent variables remained constant at the outermost radius ($r_N = 1000$ pc) — did not influence the flow, because the blast wave never got that far. The numerical grid was divided into 200 concentric shells, spaced equidistantly in $\log r$ from $\log r_1$ to $\log r_N$.

For the clumpy cases calculated the following modifications were necessary. The intercloud density was set to $\tilde{n}_{IC} = 0.1\ \mathrm{cm}^{-3}$. Six cloud components were considered with masses $m_{(s)}\ /\ \mathrm{M}_\odot = 1, 2, 5, 10, 20$, and 50 (for most cases), although other combinations were also tried (see Table 1). The mass ratio of cloud to intercloud medium in the volume considered was $\sum_s m_{(s)}/M_{IC} = 10$. The internal cloud parameters $n_C = 20\ \mathrm{cm}^{-3}$ and $T_C = 100$ K resulted in an initial volume filling factor $f \approx 5\%$ (except cases 8,18), assuming pressure balance with the partially ionized intercloud medium.

3.4 Numerical results and discussion

The evolution of the supernova remnant is significantly modified by the presence of evaporating clouds. The gas interior to the expanding blast wave is cooler and denser (see Fig. 2) for our "standard" case 1, as compared to the classical cloudless case. Indeed, the density of the intercloud medium changes by a factor of only ~ 3 between the hot ($T \simeq 2 \times 10^6$ K) interior and the warm ($T \simeq 10^4$ K) exterior. The corresponding density contrast for the classical case is about an order of magnitude greater. Surely this will have important observational consequences, especially for the X-ray emission. Other differences between these two cases are: the blast wave moved faster, it traveled

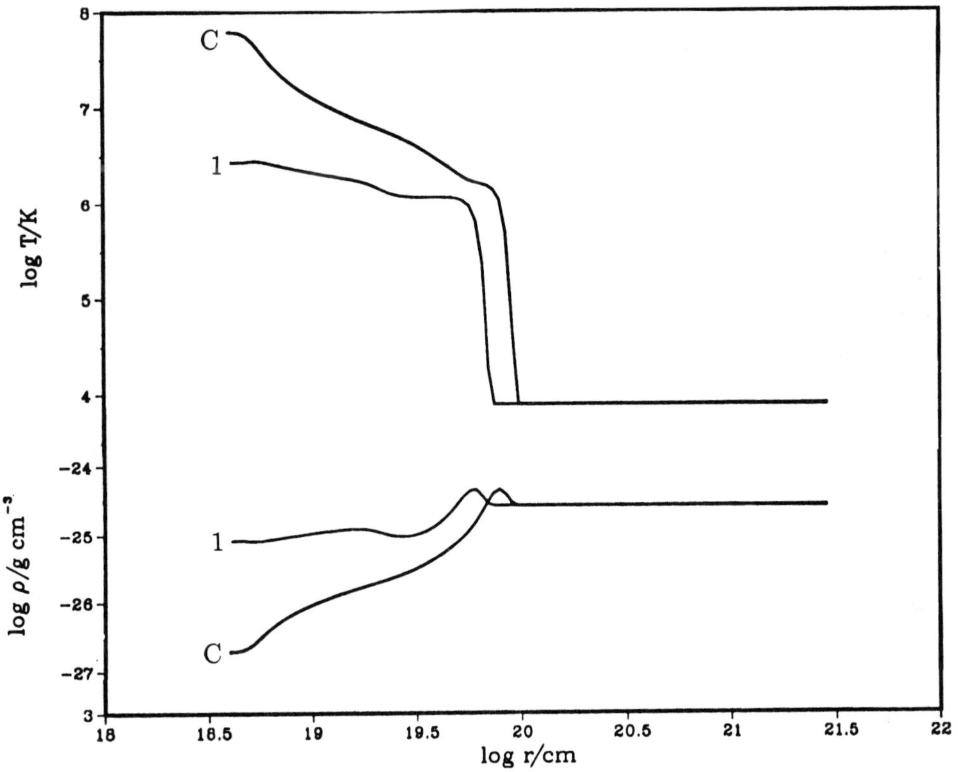

Figure 2. *Time evolution of density (bottom curves) and temperature (top curves) for the "classical" case (C) and case 1 (see Table 1).*

further and the cooling phase began later during the classical comparison calculation.

The evolution of radius $R_{\rm cloud}$ and mass $M_{\rm cloud}$ of the cloud component which initially started at 1 $M_\odot$ is shown in Fig. 3 as a function of both position and time. The position of the blast wave can be followed as a function of time in both diagrams by the locus of points at which clouds begin to evaporate or they are crushed. Not shown is the velocity evolution of this and the other cloud components. Initially at rest, the clouds were accelerated to velocities of $\sim$ 10 km s^{-1} – close to the center expansion velocities exceeding 20 km s^{-1} were attained. Newly formed clouds can be expected to move at the velocity of the expanding material – $\sim$ 100 km s^{-1} in early phases (Kunze *et al.* 1992 – this volume), but cloud formation was not extremely important for the cases presented here.

Finally, we wish to summarize the results of the other comparison calculations. The greatest deviation from the "classical" non-cloudy case was attained when evaporation was greatest. Case 11, for example, with a cloud mass distribution $n(m) \propto m^{-3/2}$ rather than $n(m) \propto m^{-2}$ (case 1), had fewer low mass clouds, and evaporation of cloud material was less effective. Thus, the evolution of case 11 was more similar to the evolution of

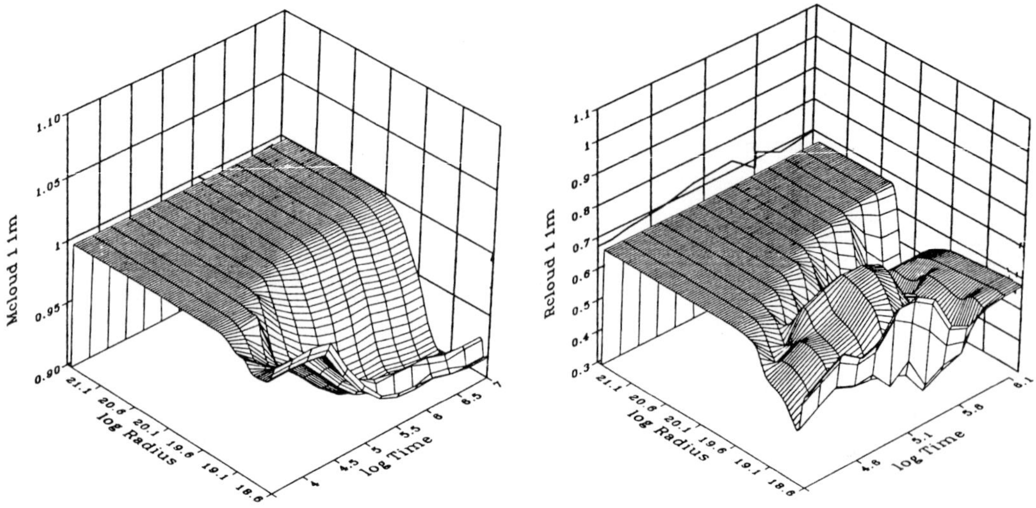

Figure 3. *Cloud mass (left) and cloud radius (right) as a function of evolutionary age and position for the smallest cloud component in case 1, initially at 1 $M_{\odot}$.*

the non-evaporating cases 2 & 12, or to the evolution with larger clouds, cases 5 & 9 (see Table 1). Decreasing dynamical friction (cases 3 & 13) had little effect on the intercloud medium's and the blast wave's time evolution, but the amount of kinetic energy deposited into bulk cloud motion was significantly reduced. Reducing crushing (cases 4 & 14) modified the overall evolution by enhancing evaporation, because the intrinsic cloud density remained lower than for cases 1 & 11. Lowering the initial cloud temperature (cases 8 & 18) had the opposite effect. The intrinsic density was higher and less cloud material was evaporated.

The present calculations, described in detail by Spurzem *et al.* 1991, demonstrate the feasibility of numerical multicomponent calculations for the evolution of clumpy media. There is still much work to be done in the framework of this model – especially the "mesaphysics" of cloud evolution (cloud-cloud collisions and cloud crushing) should be improved. It would be very useful to compare these results to clumpy hydrodynamical calculations with an entirely different approach.

ACKNOWLEDGEMENTS

We acknowledge support through DFG grant Yo 5/5-2. Numerical calculations were performed at the HLRZ, Jülich and at the LRZ in Munich.

REFERENCES

Begelman, M.C., McKee, C.F.: 1990, *Astrophys. J.* **358**, 375.

Biermann, P., Kippenhahn, K., Tscharnuter, W., Yorke, H.W.: 1972, *Astron. Astrophys.* **19**, 113.

Blitz, L.: 1990, in *The Evolution of the interstellar Medium*, ed. L. Blitz (San Francisco: Publ. Astron. Soc. Pac.), p. 273.

Cowie, L.L., McKee, C.F., Ostriker, J.P.: 1981, *Astrophys. J.* **247**, 908.

Cox, D.P.: 1989, in *Structure and Dynamics of the Interstellar Medium, IAU Coll. No. 120*, eds. G. Tenorio-Tagle, M. Moles, J. Melnick (Dordrecht: Reidel), p. 500.

Cox, D.P., Smith, B.W.: 1974, *Astrophys. J. Lett.* **189**, L105.

Cox, D.P., Slavin, J.D.: 1990, in *Extreme Ultraviolet Astronomy*, eds. R.F. Malina, S. Bowyer (New York: Pergamon), p. 268.

Deissler, R.G.: 1984, *Rev. Mod. Phys.* **56**, 223.

Field, G.B., Goldsmith, D.W., Habing, H.J.: 1969, *Astrophys. J. Lett.* **155**, L149.

Kunze, R., Yorke, H.W., Spurzem, R.: 1992, this volume.

MacLow, M.-M., Klein, R.I., McKee, C.F., Stone, J.M., Norman, M.L.: 1992, this volume.

McKee, C.F.: 1988, in *Supernova Remnants and the Interstellar Medium*, eds. R.S. Roger, T.L. Landecker (Cambridge: Cambridge Univ. Press), p. 205.

McKee, C.F.: 1990, in *The Evolution of the Interstellar Medium*, ed. L. Blitz (San Francisco: Publ. Astron. Soc. Pac.), p. 3.

McKee, C.F., Begelman, M.C.: 1990, *Astrophys. J.* **358**, 392.

McKee, C.F., Ostriker, J.P.: 1977, *Astrophys. J.* **218**, 148.

Norman, M.L., Dickel, J.R., Livio, M., Chu, Y.-H.: 1988, in *Supernova Remnants and the Interstellar Medium*, eds. R.S. Roger, T.L. Landecker (Cambridge: Cambridge Univ. Press), p. 223.

Scalo, J.M.: 1987, in *Interstellar Processes*, eds. D.J. Hollenbach, H.A. Thronson, Jr. (Dortrecht: Reidel), p. 349.

Scalo, J.M., Struck-Marcell, C.: 1984, *Astrophys. J.* **276**, 60.

Spitzer, L., Jr.: 1990, *Ann. Rev. Astron. Astrophys.* **28**, 71.

Spurzem, R., Kunze, R., Yorke, H.W.: 1991, to be subm. to *Astron. Astrophys.*

Tenorio-Tagle, G., Bodenheimer, P.: 1988, *Ann. Rev. Astron. Astrophys.* **26**, 145.

Tscharnuter, W.M.; Winkler, K.H.: 1979, *Comp. Phys. Comm.*, **18**, 171.

Wolff, M.T., Durisen, R.H.: 1987, *Mon. Not. Roy. Astron. Soc.* **224**, 701.

Yorke, H.W.: 1986, *Ann. Rev. Astron. Astrophys.* **24**, 48.

Yorke, H.W.: 1987, in *Radiation in Moving Gaseous Media, Saas-Fee Lecture Series No. 18*, eds. Y. Chmielewski, T. Lanz, (Sauverny-Versoix: Geneva Obs.), p. 193.

State of the Interstellar Medium, Phase Transitions, and Galactic Shocks

L.S. MAROCHNIK[1], A.A. SUCHKOV[2]

[1] Space Research Institute of the USSR Academy of Sciences, Profsoyuznaya 84/32, Moscow, 117810, USSR
[2] Rostov State University, Zorge 5, Rostov-on-Don, USSR

ABSTRACT

This review highlights some recent developments in cosmic gas dynamics. A controversy regarding the steady-state and supernova-dominated models of the ISM is discussed. It is concluded that the steady-state model seems to be the more compatible with current observations. We also discuss the self-regulated ISM which involves a negative feedback between the ISM and star formation. We argue that magnetic fields in the ISM are plausibly subjected to turbulent diffusion which strongly reduces their effect upon the dynamics of the ISM. The review presents an illustration of how gas flows through the density waves produce large-scale galactic shocks that generate small scale gas clouds. It also gives an account of two new types of gas flows through the density waves, namely an accretion wave and a multiphase flow.

1 INTRODUCTION

During the last decade there has been a tremendous advance in all-wavelength observations of cosmic gas in and outside galaxies. A wealth of new phenomena involving various dynamical processes in rarefied gas has been discovered, which has stimulated rapid development of new areas in cosmic gas dynamics. Also the classical problems of the interstellar gas dynamics have not been left intact. An example is provided by the theory of the interstellar medium where the widely accepted supernova-dominated model of McKee and Ostriker (1977) has been questioned as not being compatible with the data using a new generation of radio astronomy tools (cf. Heiles, 1987).

The aim of the present review is mainly to provide a short account of recent developments in cosmic gas dynamics. We shall discuss the controversy between the two-phase steady-state model and the three-phase supernova-dominated model of the ISM, as well as some other approaches to the description of the ISM. Another topic will be gas flows in density waves of spiral galaxies. We shall consider the formation of galactic shocks and phase transitions occurring in them. Some other types of gas flows developing in density waves will also be reviewed, including accretion waves, multiphase flows, and the flows with contact discontinuites. We shall also make some comments on the role played by magnetic fields in the dynamics of the ISM.

2 THREE-PHASE SUPERNOVA-DOMINATED MODEL VERSUS TWO-PHASE STEADY-STATE MODEL

The outstanding work of McKee and Ostriker (1977) entirely revolutionized the understanding of the ISM and strongly influenced all subsequent advances in ISM studies.

The main goal of their work was to present a new comprehensive model of the ISM based on ideas essentially different from those involved in the two-phase steady-state model of Pikel'ner (1967) and Field *et al.* (1969).

Two major developments in ISM physics motivated the model. First, measurements of the soft X-ray background suggested the presence of a large amount of very rarefied gas at temperatures $\sim 10^5 - 10^6\ K$, far above those predicted in the model for the high-temperature phase. Second, there was a growing realization that supernova explosions occurring in the galactic disk would strongly influence the state of the ISM, by making in it large, long-lived cavities filled with an extremely tenuous gas at temperatures of about $10^5 - 10^6\ K$. This gas was expected to be the source of the s X-ray emission; also it could be responsible for OIV absorbtion lines observed in spectra of many distant stars.

The main physical process involved in the McKee and Ostriker model is the interaction of a supernova blast wave with various gaseous components of the ISM. As the blast wave propagates, it leaves behind a hot, rarefied gas and preexisting cloud which, after being compressed at the shock front, undergo heating and ionization by the ultraviolet and soft X-ray radiation. As a result, the small clouds evaporate, while the larger ones remain as dense, cold entities enveloped by extended warm, ionized or neutral gas evaporated from the cloud edges. In this way three phases arise: the hot, warm, and cold ones. The typical density and temperature of the hot component was estimated to be $\sim 3 \times 10^{-2} cm^{-3}$ and $5 \times 10^5\ K$, respectively.

Given the supernova rate S, it is possible to estimate the fraction of volume occupied by the hot phase (which is not present at all in the "standard" two-phase model). With $S \sim 10^{-13}\ pc^{-3} yr^{-1}$, McKee and Ostriker found the volume filling factor to be 0.7-0.8, which implied that the major part of the galactic disk is pervaded by this hot gas. This very surprising result led to the conclusion that: "The standard two-phase model proposed by Field, Goldsmith, and Habing (1969) for which (T_0, n_0) $(10^4\ K, 10^{-1}\ cm^{-3})$ cannot be maintained if our estimate of supernova rate S is reasonably accurate." The filling factors of the warm and cold components represented by the two-phase clouds were estimated as 0.2 and 0.02, respectively, with the major fraction of the cloud mass being contained in the cold cores of the clouds.

The model of McKee and Ostriker explained OIV and X-ray observations and accounted for the cloud velocity dispersion; regarded as points in favour of the model were its predictions for pressure, density, and ionization of clouds with their warm envelopes (cf. Tenorio-Tagle and Bodenheimer, 1988). The supernova model was favoured by many authors also due to its obvious unavoidability from the physical point of view once conventional rate and energetics of Type II supernova has been adopted.

However, the subsequent progress in radio observations of the interstellar hydrogen, especially of its small-scale structure in nearby spiral galaxies, raised severe doubts of the applicability of the model to the actual ISM. As argued by Heiles (1987), spiral galaxies show a very smooth HI distribution over a major part of their disks, which is quite inconsistent with the prediction of the supernova-dominated model. Also, in our Galaxy the warm-phase gas occupies about a half of the disk volume, as opposed to $\sim 20\%$ in the model. Heiles (1987) indicated a few other inconsistencies; in particular, he stated that both a small number of hot supercavities in the galactic disk and an actually too low soft X-ray background are not compatible with a large filling factor for the hot phase in our Galaxy. As a result, he preferred the two-phase model of Pikel'ner (1967) and Field *et al.* (1969).

The dominance of the hot phase was questioned also on several other grounds. For example, Turner (1981) noted that "A hot tenious medium cannot be compressed by, nor develop, a SDW /spiral density wave/ shock when it passes through the spiral gravitational field. Nevertheless, there is observational evidence for a density enhancement in the HI (factor of ~ 10) in the SDW shocks in external galaxies".

A study of Richter and Rosa (1988) seems to provide an explanation for the discrepancy between the supernova-dominated model and observations. These authors have found evidence (also suggested by Heiles (1987)), that the supernova rate has been strongly overestimated for the majority of galaxies (and just as strongly underestimated for a small number of remaining galaxies which produce the bulk of all supernova detected over years of supernova patrolling). This result is consistent with the evidence that the star formation rate, and hence the supernova rate is very uneven in the Galaxy, with short bursts providing the bulk of stars and enrichment.

The bursts are separated by long periods of low star formation, hence low supernova rate.

Consequently the main state of gas in spiral galaxies still appears to be an HI - dominated disk; perhaps only for rather short times, when a burst of star formation is occurring, will the state of the interstellar gas correspond to that described by the supernova-dominated model.

3 SELF-REGULATED ISM

The later development of the supernova-dominated model proceeded mainly along the lines which considered the ISM as a global galactic phenomenon governed in a self-regulated manner by star formation. This self-regulation isolates a set of timescales, from $\sim 10^7$ yr (the time scale for a local burst of star formation) to $\geq 10^9$ yr (the time scale for inhibited star formation between the global bursts). As a result, the ISM is envisioned as a system with a complex behaviour that, nevertheless, bears features of temporal regularity. Essential for the understanding of self-regulation is the interaction of the gaseous disk, where star formation and large-scale shocks driven by a global density wave occur, with the gaseous halo, which may be site of cloud formation.

The idea is that multiple supernova explosions in a stellar cluster and association produce a hot, expanding superbubble whose size is comparable to the disk thickness. Such a superbubble breaks through the disk into the halo, drastically influencing the resulting effect of supernova upon the ISM. Due to the vertical density gradient, a superbubble originating in the mid-plane of the disk develops as a very narrow cylinder rather than as a sphere , and breaks out into the halo. As a result, the bulk of the energy and mass of the hot gas runs through this kind of "chimney" into the halo. Even more pronounced is the escape of mass and energy from the disk in case of superbubbles originating above the midplane. In this case there must form in the gaseous disk a cavern opening into the halo which produces a cumulative effect on the outflowing material. Also in this case, both the hot-gas mass and its energy rapidly escape from the region of supernova activity. In both cases, one may expect a substantial reduction of the effect of the large energy release in star-forming regions on the state of the gaseous disk.

At the same time, the above process results in circulation of the gas from the disk into the halo and back again, with the gaseous matter experiencing drastic changes over the cycle. The following scenario is envisaged. A star-forming region which is connected with a region of high-density, low-temperature gas (say, a giant molecular cloud) ultimately produces a hot superbubble, with the major fraction of the gas being ionized and heated to $\sim 10^6$ K. The hot gas sweeps out the disk into the halo where it gradually cools. After sufficient cooling, it may form cold clouds through thermal instability; the clouds fall down into the disk again, replenishing it with fresh material for further star formation. This is the widely-known "galactic fountain" model which is frequently invoked to describe disk-halo connections regulated by active star formation (see, e.g. Shapiro and Field, 1976).

Li and Ikeuchi (1989) and Norman and Ikeuchi (1989), emphasizing a duct-like structure of a superbubble developing in a plane-stratified medium in the gaseous disk-halo system, proposed a new model of the ISM, namely the chimney model. The model describes mass circulation between the disk and halo via superbubbles as discussed above. By considering the formation rate and lifetime of superbubbles, they found that thousands of "chimneys", that is duct-like "superbubbles" elongated perpendicular to the disk, must be steadily "smoking" in the Galaxy; the mass circulation rate has been estimated to be 1.6 $M_\odot$ yr^{-1}.

An interesting implication for the global star formation rate is provided by the model in terms of the enhanced supernova rate expected in young galaxies or starburst galaxies. Li and Ikeuchi (1989) found that the temporal behaviour of the ISM in this case reveals a periodicity; they suggested that it may indicate recurrent bursts of star formation.

To model a full self-regulation of the ISM, one has to include negative feedback between the ISM and star formation. The idea of a global, negative feedback was central to the theory of active phases of the evolution of the Galaxy (Marsakov and Suchkov, 1976; Suchkov and Shchekinov, 1979; Suchkov, 1981). The theory predicted the formation of discrete galactic components in a few major star formation events, each being cut off by the cumulative effect from a multitude of supernova appearing during a star-formation burst, and new star formation events occurring only after a delay long enough to ensure transfer of the galactic gas back into predominantly dense, cold clouds. In its most extreme form, the active phases in galaxies are suggested to affect drastically the global dynamics of the galactic gas by developing a strong galactic wind, with much gaseous mass being even expelled from a galaxy. This is the scenario envisioned in the "hot" model of galaxy formation by Berman and Suchkov (1991). The "hot" model predicts the disks of spiral galaxies to form only a few Gyr after the formation of halos: this is the time required for the gas to again become capable of producing stars following the preceding active phase. Self-regulated cycles in the galactic disk, including disk-halo circulation, have been discussed by Tenorio-Tagle and Bodenheimer (1988). They combined the model suggested by Larson with the picture of superbubbles to produce the following scenario. As envisioned by Larson (1988), star formation beginning in a massive molecular cloud is self-limited by the ionization caused by the intense ultraviolet radiation of newly-born OB stars. The fraction of mass that has time to enter into stars has been estimated by Larson as $\sim 5\ \%$; the rest of the gas becomes ionized and streams away from the molecular cloud at a speed of about 30 km s^{-1}; this streaming of ionized gas is called the "champagne flow". Stellar winds and supernova explosions may also contribute to dispelling the mass of the original cloud; however, these factors mainly operate at later stages, when condensation of gas into stars has already been stopped by ionization. New star formation may begin only after $\sim 10^8\ yr$; this is about the time needed to transfer the ionized gas back into the molecular clouds, with gravitational instability, galactic shocks, and cloud coagulation being among the prime operating mechanisms. Tenorio-Tagle and Bodenheimer supplement this picture by introducing the formation of a superbubble at the late stages of star-formation; the superbubble breaks through and discharges the overpressure into the halo over a time scale of $\sim 10^7\ yr$, i.e., long before the last supernova are expected to explode (the respective time scale is $5 \times 10^7\ yr$). A new star-formation burst in this region requires agglomeration of the swept-out gas to form new giant molecular clouds. This would take $\sim 10^8\ yr$, which is thus the time scale of periodicity of star-formation bursts.

It has to be remarked that this time scale was obtained earlier by Suchkov and Shchekinov (1979), in a model which assumed that a major mechanism to cut off star formation is the overall cosmic ray background and or X-ray background contributed by the multitude of supernova remnants appearing during a star-formation burst. The time scale under discussion was derived along lines similar to those discussed above; it was emphasized that the time is insufficient to account for the long periods of inhibited star formation that made the major galactic component distinctly separate. Consequently, the model of Suchkov and Shchekinov, which considered the global self-regulation by the ISM of the star formation, and that of Larson can be pertinent only to a short-scale periodicity in the star-formation rate. Moreover, they can be pertinent only to local star-formation events rather than to the global ones responsible for the galactic components, star-forming complexes of $\sim 1\ kpc$ in size, which have been advocated over many years by Efremov (1989), as being distinct entities in our Galaxy as well as in nearby other galaxies, may be the signature of these local events.

The discussion may be summarized as follows. There is a set of self-regulating cycles that govern star formation and ISM on a variety of time scales and length scales. The most significant ones are the global cycles incorporating the bulk of the Galaxy, in which most of the gas becomes hot and is removed from the overall star-forming region for $\sim 10^9\ Gyr$. These cycles are believed to be responsible for the formation of galactic components. The "second-order" cycles see to encompass regions of about 1 kpc in size and operate on a time scale of about 10^8 yr; they are related to superbubbles and star forming complexes. Finally, there are probably more rapid "third order" cycles operating on a length scale of about 100 pc which do not involve superbubbles and disk-halo mass interchange, their operational mode may be of the kind proposed by Larson (1988). It is evident that all three types of cycles are specified primarily by the volumes involved (beyond the halo extension, inside the halo extension, inside the disk) and by the respective time scales.

4 ROLE OF MAGNETIC FIELDS

The most important dynamical property of the interstellar gas lies in its enormous compressibility caused by radiative energy losses. It allows the gas to develop very large density contrasts; this is believed to occur within the galactic disks when cold dense clouds are formed from a tenuous warm gas. The corresponding physics is rather well understood. But a lucid physical picture for the dynamical behaviour of such a gas becomes immediately blurred once magnetic fields are involved. Magnetic fields, being "frozen" into the gas, and behaving as an elastic medium, resist efforts to compress the gas across the force lines. Since the interstellar magnetic field possesses a highly entangled component (alongside with a regular one), the corresponding magnetic pressure in fact acts isotropically (on scales larger than the homogeneity scale of the field). As far as the magnetic pressure within the galactic disk is comparable to the thermal pressure, the medium as a whole, i.e. gas plus field, is expected to behave essentially adiabatically, its dynamics being governed mainly by the interstellar field. But in this case, the strong compression necessary to produce a dense cloud in a rarefied medium is impossible. Gas contraction along the force lines of a presumably regular magnetic field can, however, hardly be accepted as a universal mechanism for cloud formation, because the real interstellar field does contain a strong random component. Nonetheless, the clouds are formed somehow in gaseous disks of galaxies.

Similarly, a regular component of a large-scale magnetic field directed along a spiral arm, being "frozen-in", must prevent the interstellar gas from forming along the arm a highly compressed region known as a galactic shock; yet such shocks are observed.

All this taken together suggests that magnetic fields can efficiently be carried away from a contracting gaseous element.

A mechanism for efficient transport of the embedded magnetic field in cosmic gas bodies has been proposed more than two decades ago: namely as a rapid diffusion of the field due to turbulence (cf. Parker, 1973). The ISM in the galactic disk is known to be turbulent over a very large range of scales, extending down to subparsec scales. Therefore, the enhanced turbulent diffusion of the field is expected to operate on the scale of the galactic disk as a whole (Parker, 1972), and down to the scales of the smallest clouds.

In case of strong turbulence, the turbulent diffusion is believed to occur on a time scale of the order of the time required for the sound to travel through a given gaseous element. If so, on scales larger than this time, the embedded magnetic fields would not be able to prevent this element from contracting to high density. On the other hand, the same sound travel time characterizes the rate at which a gaseous element can change its density due to a dynamical process. This implies that the capability of a rarefied, radiatively cooling gas to develop large density contrasts would not be drastically changed by allowing for magnetic fields if the gas is turbulent: the fields would only slow this process, rather than totally inhibit it.

These considerations suggest that ignoring magnetic fields in describing the dynamical behaviour of the ISM is in certain respects much more adequate than involving the approximation of fully "frozen-in" fields. The latter approach is hardly capable of accounting for the most important processes in the ISM, namely formation of dense clouds and strong galactic shocks, while the former one predicts and excellently describes them.

The following discussion, which does not explicitly take into account interstellar magnetic fields, may thus be regarded as based on the conviction that the fields can very efficiently withdraw from any contracting gas volume due to turbulent diffusion.

5 GALACTIC SHOCKS AND CLOUD FORMATION

As discussed in Section 3, there are currently good reasons to consider the ISM along the lines of the two-phase model rather than of the three-phase model. Let us consider an initially circular motion of gas in a galactic disk where a gravitational field of a stellar density wave gradually arises. In case of a tightly wound, two-armed spiral pattern rotating at a speed Ω_p, the equations governing the motion are as follows (cf. Roberts, 1969); Marochnik *et al.*, 1983):

$$\frac{du}{dt} - 2\Omega(v - v_0) = -\frac{1}{\rho}\frac{\partial p}{\partial x} - \frac{\partial \phi_S}{\partial x},$$

$$\frac{dv}{dt} + \frac{\kappa^2}{2\Omega}(u - u_0) = 0,$$

$$\frac{dx}{dt} = u,$$

$$\frac{d\rho}{dt} + \rho\frac{\partial u}{\partial x} = 0,$$

$$\frac{3}{2}\frac{d}{dt}\frac{p}{\rho} + \frac{p}{\rho}\frac{\partial u}{\partial x} = -L\left(\rho, T, S_i\right),$$

$$\phi_S = A(t)\,\cos\left(\frac{2x}{r\sin i}\right),$$

where u and v are velocity components directed perpendicular and tangentially to the spiral arm, respectively; x is the spiral coordinate orthogonal to the direction along the arm; $L(\rho, T, S_i)$ is the rate of internal energy variation due to radiation cooling and cosmic ray heating; ϕ_S is the gravitational potential of the density wave, $A(t)$ is its amplitude; and κ is the epicyclic frequency. The coordinate x is related to the polar galactocentric coordinates, r and θ, as

$$x = r\left[\cos i \ln(r/r_0) + (\theta - \Omega_p t)\sin i\right]$$

where i is the pitch angle of the arm. The reference frame rotates at a speed Ω_p, and

$$u_0 = r(\Omega - \Omega_p)\sin i, \quad v_0 = r(\Omega - \Omega_p)\cos i,$$

are the velocity components of the background circular motion (the subscript "$_0$" refers to the initial values of gas parameters).

Marochnik *et al.* (1983) have numerically integrated the above equations for a set of models differing in n_0, Ω_p, A, r, Ω, κ, and S_i . To illustrate the most conspicuous features of the solutions, we present here the one corresponding to $n_0 = 0.05\ cm^{-3}$, $\Omega_p = 23\ km\ s^{-1}\ kpc^{-1}$, $r = 5\ kpc$, $A = -265\ km^2\ s^{-2}$ (which is a stationary value after $t \simeq 3\times10^8$ years), $i = 7^o$, $\Omega = 46\ km\ s^{-1}\ kpc^{-1}$, and $\kappa = 73\ km\ s^{-1}\ kpc^{-1}$.

Fig. 1 shows density profiles across a spiral arm at three different moments of time. It can be seen that the gas pressure and density grow in the potential well of the density wave, and a galactic shock gradually forms. When the density at the shock front reaches a value corresponding to the upper critical point of the equilibrium curve, a certain part of the gaseous mass rapidly cools and contracts, with two contract discontinuities being formed at the boundaries; the density and temperature jump at the boundaries by about two orders of magnitudes, the pressure varies much less, and the velocity remains initially almost continuous. The cooled mass, displayed as a density peak in the upper panel of Fig. 1, is carried away as a whole out of the shock front and moved down-stream with the main flow. It is very natural to interprete this process as a formation of a cool, dense cloud from initially warm gas.

Formation of the cloud reduces the pressure and density at the shock front, and some time is required for these parameters to reach the upper critical values again. After that, the process is repeated and the second cloud is formed, then the third one, and so on (see the lower panel of Fig. 1). So the solution obtained suggests that a galactic shock can really be the "machine" which "manufactures" clouds in galaxies.

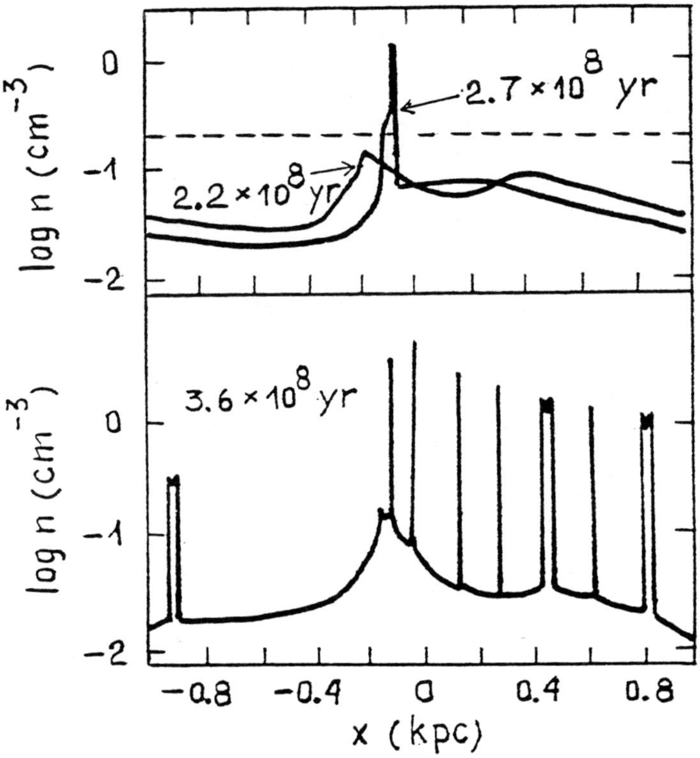

Fig. 1. Development of density profiles across a spiral arm. Indicated is the time elapsed after the gravity of the density wave is turned on. The zero point on the X-axis corresponds to the center of the gravitational well of the density wave. The dashed line indicates the critical density corresponding to the upper critical point of the equilibrium curve in the pressure-density plane. Note the formation of a density peak on the shock front as the latter reaches the upper critical point.

The density peaks seen in Fig. 1 are in fact unresolved in terms of their true spatial structure and dimension on the scales of Fig. 1. Their true size can be seen in Fig. 2, where which shows on a much smaller scale the structure of a "cloud" corresponding to one of the peaks from Fig. 1.

We illustrate this with "clouds" with rather specific behaviour. As seen in Fig. 1, the gas flow exhibits essentially non-stationary behaviour, with gas parameters significantly varying both in the clouds and the ambient warm gas. In case of the cloud shown in Fig. 2, the density in the cloud centre at some moment has dropped below the value corresponding to the lower critical point of the equilibrium curve. As a result, the temperature rapidly jumps up here, producing a rapid pressure increase. This leads to a shock propagating outwards. The cloud inflates, forming a bubble of hot gas inside a dense, cold shell.

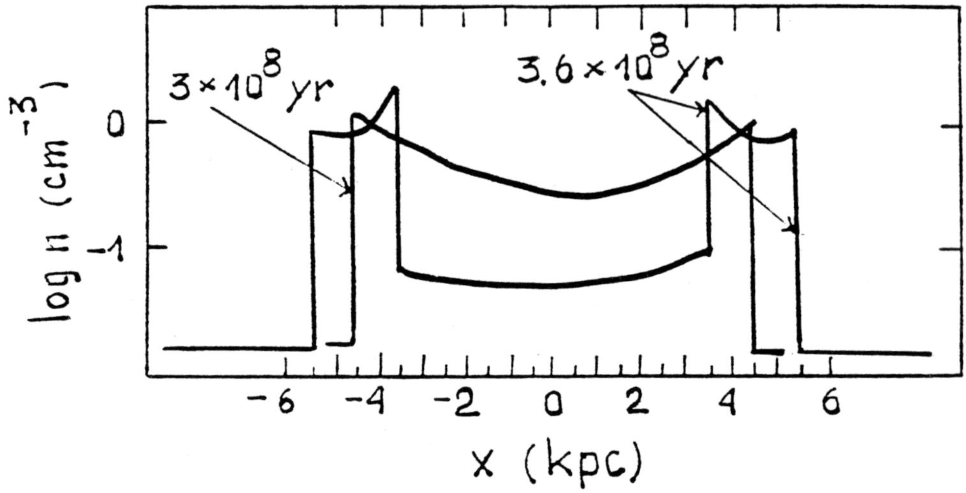

Fig. 2. Density profiles across the first density peak formed as shown in Fig. 1. The peak is interpreted as a cloud formed on the front of the galactic shock. The cloud, inflated because of the central density, overshoots above the critical density, n_{cr}^{up}. The inflation forms a shock on the inner boundary of the dense shell; the outer boundary represents a contact discontinuity slowly propagating into ambient warm gas.

It is important to note that formation of separate density peaks requires rotation, or, to be more exact, the action of the Coriolis force. The above equations, but with $\Omega = 0$, yield a density profile with an infinitely growing peak on the shock front; in this case, the "cloud" cannot leave the gravity potential well (Timonin *et al.*, 1986). This result may be pertinent to the general problem of cloud formation.

6 THE ACCRETION WAVE

A pattern completely different from the one discussed in Section 5 arises for the same parameters but for an initial combination of values which corresponds to the cold phase. A wave of unusually large amplitude ($n/n_0 \sim 100$) arises in this case in the center of the density wave potential well (see Fig. 3). Large gas masses are delayed for a long time in a narrow region of the wave front, the gas velocity decreases to extremely low values, and, moreover, there are cases when the velocity beyond the front generally changes sign so that gas will flow into the region of the front from both directions. This type of gas flow is called an accretion wave (Marochnik et al. 1983).

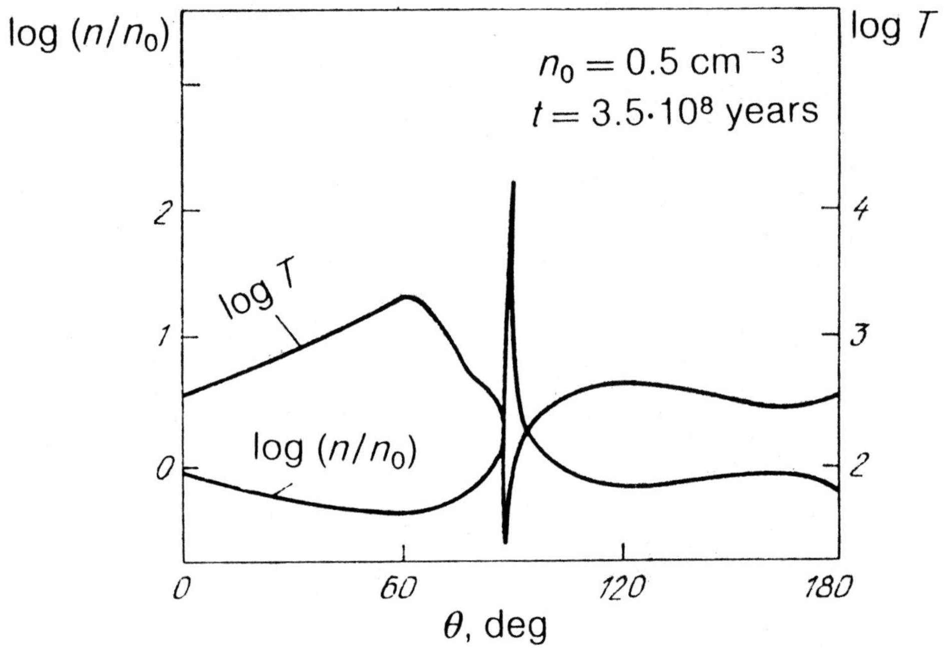

Fig.3. An accretion wave. The wave becomes quasi-steady- state after a time $t \simeq 3 \times 10^8$ years

An accretion wave, with the gas being extremely compressed and cooled in it, seems to be a highly favoured place in the Galaxy for H_2 to form, for giant molecular clouds to arise, and for intense star formation to operate. Strong compression also involves powerful enhancement of gamma radiation and synchrotron radio emission. These are just the effects we observe in the Great Molecular Ring at 5 kpc from the center of the Galaxy. Therefore it is very tempting to suggest that the Great Molecular Ring is due to an accretion wave. Perhaps such rings exist in other galaxies as well. They could be detected as annular regions of strongly enhanced synchrotron and CO emission. Appropriate observations are needed.

7 MULTIPHASE FLOW

Calculation of this same variant, but with the value R = 10 kcp instead of R = 5 kpc, revealed one more type of motion in the gas. The change of R is reflected most significantly in the unperturbed velocity of the gas with respect to the spiral potential: it decreases strongly and becomes comparable with the initial speed of sound (it was approximately six times faster than the speed of sound in the previous example). This circumstance radically changes the nature of the gas motion in comparison with the previous case: instead of an accretion wave, a flow is developed here that has been called a three-phase flow (Marochnik et al. 1983). The flow characterizes the situation. The gas density near the center of the potential well at first increases smoothly, by approximately a factor of two. Far from the center, the density decreases by approximately the same factor, but here this decrease is accompanied by a transition of the density from $n > n_{2cr}$ to $n < n_{1cr}$ (see Efremov *et al.* 1989 for details). At $\mathrm{n} < \mathrm{n}_{1cr}$, the gas is very rapidly heated to temperatures $T > T_{1cr}$. As a result, an enormous region arises in which the density is only half its initial value, whereas the temperature increases to about 30 times its initial value. In essence, two phases arise with a sharp, abrupt temperature change at their boundary; the density is discontinuous at the phase separation boundary, and generally changes only slightly in space.

An abrupt pressure change obviously arises at the phase boundary. Then it compresses the layers of the original cold phase adjacent to the boundary, and as a result a narrow region of strongly compressed, cold gas (a cold cloud) is formed. The abrupt temperature change at this boundary increases, but the abrupt pressure change decreases due to the formation of an abrupt density change. There are now already three strongly differing states, i.e. three "phases", with abrupt parameter changes at their boundaries. First, there is a phase where the parameters are close to their original values. Second, there is a phase with a high temperature ($T \sim 10^4\ K$) and a fairly high density (there is an abrupt presure change at its boundary). Third, there is a phase in which the density is almost an order of magnitude higher than its original value at practically the original temperature (there are abrupt changes of density and pressure at its boundary; see Fig. 4).

The hot phase is near the upper critical point; therefore fluctuations of density in it caused by dynamical processes in turn lead to phase transitions. Those elements of volume, in which the density turns out to be higher than the critical density, are rapidly cooled, forming new cold phase clouds.

In conclusion, we note that one can divide the structural elements arising during the flow of interstellar gas in the field of a spiral wave into two types. First, there are strongly nonlinear waves, for example, galactic shock and accretion waves. Here the characteristic regions of parameter change (let us say a shock-wave front) move through the gas; each element of a mass of gas enters into and exits from such a region. Second, there are regions of different phases that are connected with given elements of gas and which move together with them (cold-phase clouds and intercloud gas). "Clouds" which are generated at a shock wave front belong to this second group in our examples.

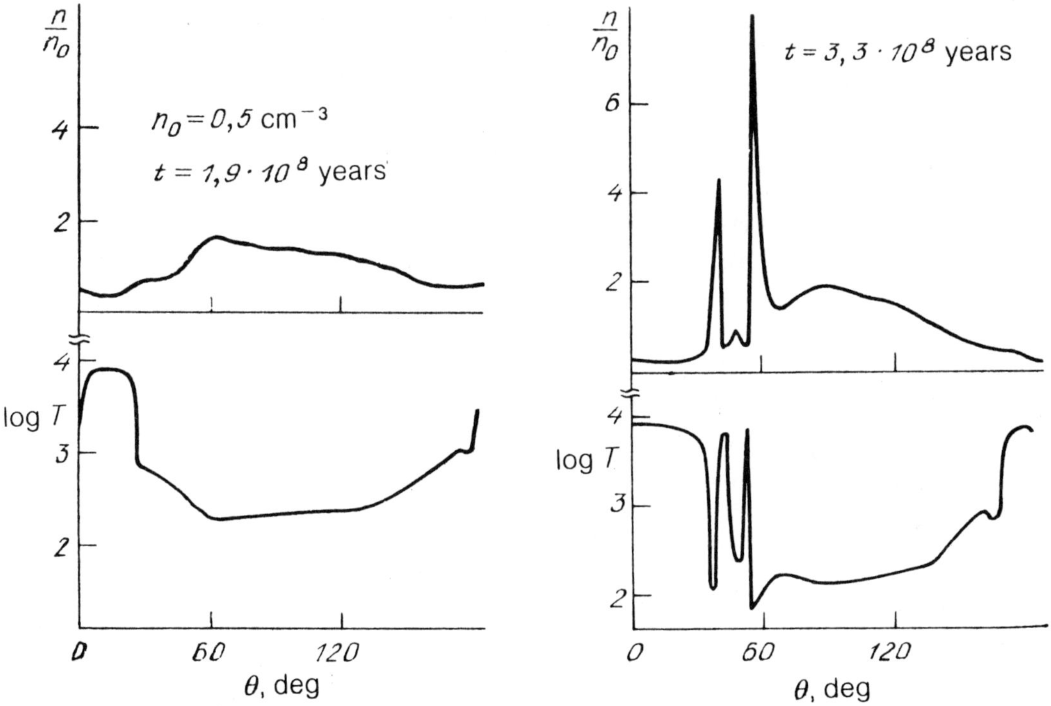

Fig. 4. The development of three-phase flow. Third-phase cold cloud, with densities much less than the density of the original cold phase, arise after a time $t = 2 \times 10^8$ *years*.

REFERENCES

Berman, V.G. and Suchkov, A.A., 1991, submitted to Astrophys. Space Sci.

Efremov, Yu.N., 1989, Centers of Star Formation in Galaxies (in Russian), Nauka, M.

Efremov, Yu.N., Korchagin, V.I., Marochnik, L.C. and Suchkov, A.A., 1989, Soviet Physics Uspekhi 32 (4), 310.

Field, G.B., Goldsmith, D.W., and Habing, H.J., 1969, Ap. J. Lett. 155, L149.

Heiles, C., 1987, Ap. J. 315, 555.

Larson, R.B., 1988, in "Galactic and Extragalactic Star Formation", eds. R.E. Pudritz, M. Fish, Dordrecht, Reidel.

Li, F. and Ikeuchi, S., 1989, Publ. Astron. Soc. Japan 41, 221.

Marochnik, L.S., Berman, V.G., Mishurov, Yu.N. and Suchkov, A.A. 1983, Astrophys. Space Sci. 89, 177.

Marsakov, V.K. and Suchkov, A.A., 1976, Pisma Astron. Zh. 2, 381

McKee, C.F. and Ostriker, J.P., 1977, Astrophys. J. 218, 148.

Norman, C.A. and Ikeuchi, S., 1989, Astrophys. J. 345, N 1.

Parker, S.H., 1973, Astrophys. Space Sci. 22, 279.

Pikel'ner, S.B., 1976, Astron. Zh. (Russ.) 44, 1915.

Richter, O.-G., Rosa, M., 1988, Astron. Astrophys. 286, 219.

Shapiro, P.R. and Field, G.B., 1975, Astrophys. J. 205, 762.

Suchkov, A.A., 1981, Astrophys. Space Sci., 77, 3.

Suchkov, A.A., 1988, Astron. Zh. 65, 1 (Sov. Astron. 32, 1).

Suchkov, A.A., Berman, V.G. and Mishurov, Yu.N., 1978, Astron. Zh. 64, 708 (Sov. Astron., 31, N4).

Suchkov, A.A. and Shchekinov, Yu.A., 1979, Soviet Astronomy 23, 665.

Turner, B.E., 1981, In: The phases of the interstellar medium E. J.M.Dickey, Proc. Workshop, NRAO, Green Bank, p. 69.

Tenorio-Tagle, G. and Bodenheimer, P., 1988, Ann. Rev. Astron. Ap. 26, 145.

Gravitationally Induced Giant Molecular Cloud Formation Inside Superclouds

IGOR G. KOLESNIK

Main Astronomical Observatory, Ukrainian Academy of Sciences, 252127 Kiev, Goloseevo, USSR

Giant Molecular Clouds (GMCs) and star forming regions tend to occupy the central parts of superclouds, the largest and most massive structures in galaxies (Efremov, 1989; Elmegreen and Elmegreen, 1983). Masses of about $10^7 M_\odot$ and sizes of 2 - 2.5 kpc are typical for superclouds. They are formed by gravitational instability of galactic disks (Elmegren, 1987). Evidently, the processes of superclouds and the formation of GMCs are closely connected.

Kolesnik (1987, 1991) showed that the thermal properties of the interstellar gas in a supercloud are responsible for the GMC condensation in it. The main points of this process are the foolowing: The interstellar gas in galactic disks is thermally unstable; it decays to the warm ($T_1 = 10^4 K$) or the cold ($T_0 = 10^2 K$) phases depending on whether or not the partical density exceeds some critical level n_{LD}. Inside the supercloud, selfgravity in the homogeneous tenuous medium causes the density profile to decrease outside, but near the centre the density can grow to this critical value n_{LD}. As a result, the gas cools and collects into a cold dense core. This process can be called gravitationally induced cold core formation. It was found that this process leads to cold core formation with parameters and properties consistent with those of GMCs.

Two conditions that are necessary and sufficient for GMC formation inside the supercloud must be fulfilled:

$$n_{core} > n_{LD}, \qquad \text{and} \qquad N_{env} > N_1.$$

The first condition determines the fraction of the supercloud's mass that can be transformed into a cold core. The second one makes the molecules formation in it possible, because the cold core becomes molecular only if the central parts of the supercloud have a column density N_{env} sufficient for shielding the core from destroying external radiation. The minimal shielding column density N_1 is equal to about $5 \times 10^{20} cm^{-2}$.

The condition $n_{core} \geq n_{LD}$ is very sensitive to the supercloud's mass M_{SC}. It is only fulfilled if $M_{SC} \leq 5 \times 10^6 M_{\odot}$. This explains why star forming regions are connected with the largest structures in galaxies. Chemical composition and radiation fields that determine properties of interstellar matter and values of the n_{ND} are not very different in different places. This makes the similarity of the star complexes and superclouds in different galaxies and in different places of the same galaxy understandable.

The mass of the cold core M_C is strongly dependent on the external pressure P_{ext} of the gas in the place where superclouds are formed. In typ ical conditions $M_C \approx 10^5 - 10^6 M_{\odot}$ the equilibrium of the core is maintained by turbulent pressure with turbulent velocities. $v_t < c_1 \approx 7\ km\ s^{-1}$, where c_1 is the sound velocity in the warm gas of temperature T_1. The Jeans mass at c_1 is about $2 \times 10^6 M_{\odot}$. If the external pressur e is high enough to form the core having $M_C > 2 \times 10^6 M_{\odot}$ overall collapse and global star formation occurs. In this way, jumps of P_{ext} can stimulate the bursts of star forming activity.

REFERENCES

Efremov, Yu. N., 1989, Stellar Complexes, Harvard, London
Elmegreen, B.G., Elmegreen, D.M., 1983, MNRAS, 203, 31
Elmegreen, B.G., 1987, ApJ, 312, 626
Kolesnik, I.G., 1987, Kinematics and physics of celestial bodies, 3, 47
Kolesnik, I.G., 1991, AA, 243, 239

Heat and Radiation Driven Implosion of Interstellar Clouds

ILYA KOVALENKO[1], YURI SHCHEKINOV[2]

[1]University of Volgograd, Russia

[2]Institute of Physics, University of Rostov, Russia

1 INTRODUCTION

One of the most interesting processes in the interstellar medium is evaporation of HI clouds driven by heat transfer. This process determines, among other things, the ionic composition and global energy budget of interstellar gas (McKee and Ostriker, 1977), and the dynamics of SN remnants (Cowie, McKee and Ostriker, 1981). Usually, the study of cloud evaporation is conducted in a 'thermal-wave' approximation, which involves neglecting hydrodynamical motions. In this communication we present some results of the calculations of the hydrodynamical picture of evaporation of clouds in spherical symmetry. The detailed description will be published elsewhere (Kovalenko and Shchekinov, 1991).

2 THE EXCESS OF PRESSURE AT THE CLOUD'S EDGE

Consider a cloud being initially in *pressure equilibrium* with surrounded intercloud gas. At $t \geq 0$ this cloud is subjected to heating by thermal conductivity or UV with $\lambda > 912$ Å (or cosmic rays) absorbed by the optically thick outer layer of the cloud. In both cases the excess of pressure arises at the absorbing layer and it excites a simultaneous flow of gas towards, and outwards from, the center of cloud (Figure 1).

3 IMPLOSION OF CLOUD

Thus, evaporation of a cloud embedded in hot intercloud gas, or exposed to the absorbing heating radiation is accompanied by contraction of cloud. The fate of the cloud depends on the ratio of the cloud's mass to the Jeans mass M_J for unperturbed cloud parameters and intensity of energy transfer:

-in the models with initial cloud parameters $n_o = 10^2 cm^{-3}, T_o = 10^2 K$, and intensity of heating UV radiation at 1000 Å equal to $U_\nu = 3\ 10^{-8}$ photon $cm^{-2}s^{-1}Hz^{-1}$, all

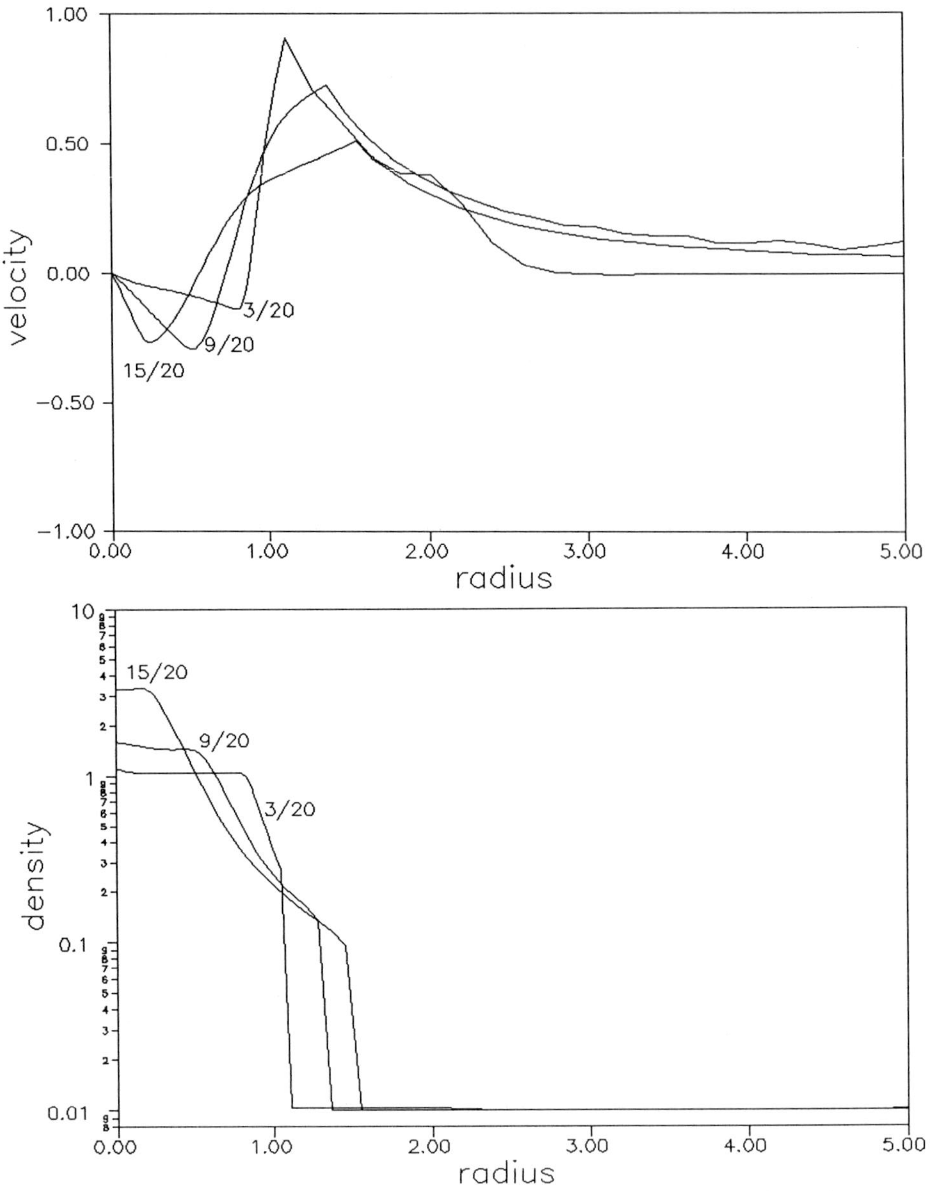

Fig.1. Initial dynamics of cloud implosion by UV heating radiation $U_\nu = 3\ 10^{-8}$ photons $cm^{-2}s^{-1}Hz^{-1}$ at $\lambda = 1000$ Å: velocity in units of sound speed (top), and density in units of initial cloud density (bottom). Cloud parameters are $n_o = 10^2 cm^{-3}, T_o = 10^2 K, R_{cl} = 2.66pc$, where R_{cl} is the radius of the cloud. The extinction cross-section $\sigma_{ex} = 2\ 10^{-21} cm^2$. Curves are labelled by time interval from the beginning in units of hydrodynamical time R_{cl}/v_{sound}.

clouds with mass larger than about $M_{cr} = 0.05\ M_J$ are subjected finally to the self-gravitating contraction. M_{cr} decreases as U_ν increases;

-for clouds with initial parameters $n_o = 1 cm^{-3}$, $T_o = 10^4 K$, embedded in the hot coronal gas with $n_h = 10^{-3} cm^{-3}$, $T_h = 10^6 K$, $M_{cr} \simeq 10^{-2} M_J$.

REFERENCES

McKee, C.F., and Ostriker, J., 1977, Astrophys.J., 218, 148

Cowie, L.L., McKee, C.F., and Ostriker, J., 1981, Astrophys.J., 247, 908

Kovalenko,I., and Shchekinov, Yu., 1991, Astron. and Astrophys. Transactions, in press

The Fractal Appearance of the Closest Molecular Cloud L1457

THOMAS ZIMMERMANN, JÜRGEN STUTZKI, GISBERT WINNEWISSER

I.Physikalisches Institut Universität Köln,Zülpicher Str.77,5000 Köln 41,Germany

Star Formation in Clumpy Clouds. In the last years clumpiness has proven to be the most striking characteristic of molecular clouds observed at all spatial scales. However, until now little is known about the physical processes producing and sustaining the clumpy and turbulent structure, which probably is crucial for star formation. Investigation of the clumpy molecular gas structures on all scale sizes is one important goal of present day astrophysics.

The Closest Molecular Cloud. The closest known molecular cloud L1457 (or: MBM12) is particulary well suited for studying molecular cloud structure, dynamics, and excitation conditions down to very small linear scales. It is located in Aries, about 20° west of the Taurus dark cloud complex (Ungerechts and Thaddeus 1987), at a distance of 65 pc (Hobbs, Blitz, and Magnani 1986). L1457 shows a clumpy structure (see Figs. below) and a large velocity dispersion which is derived from the present line components seen over a range of more than 10 km s^{-1}.

Observations. We have extensively studied L1457 with the KOSMA 3 m telescope in ^{12}CO, ^{13}CO, and $C^{18}O$ (1–0), ^{12}CO and ^{13}CO (2–1), and ^{12}CO (3–2) with angular resolutions of 4′, 2′, and 1′ respectively (Zimmermann and Ungerechts 1988, 1990 (Fig.2); Zimmermann and Winnewisser 1991 (Figs.3,4)). These observations reveal the clumpy structure of the molecular gas in L1457 down to scales of 0.02 pc corresponding to 1′ at 65 pc. Now we are extending our investigation to the HI and infrared emission (Corneliussen and Zimmermann 1991, Fig.1) of L1457, and we will study the clumpy structure with the IRAM 30 m telescope down to spatial scales of 800 AU.

Clumpiness and Fractal Appearance. To understand which forces (gravitation, turbulence, magnetic fields) cause the clumpy structures of molecular clouds at different spatial scales, it is important to quantitatively characterize the observed structure and properties of the gas. We use a clump fitting procedure (Stutzki and Güsten 1990) to determine the clump properties, e.g. size distribution and mass spectrum. An analysis of the observed maps by means of fractal geometry allows to study whether the clumps form a hierarchy and the structures show self-similarity over scales of several orders of magnitude (Scalo 1990; Falgarone 1990); first results yield a fractal dimension of $D_H = 1.4$ in the CO maps of L1457.

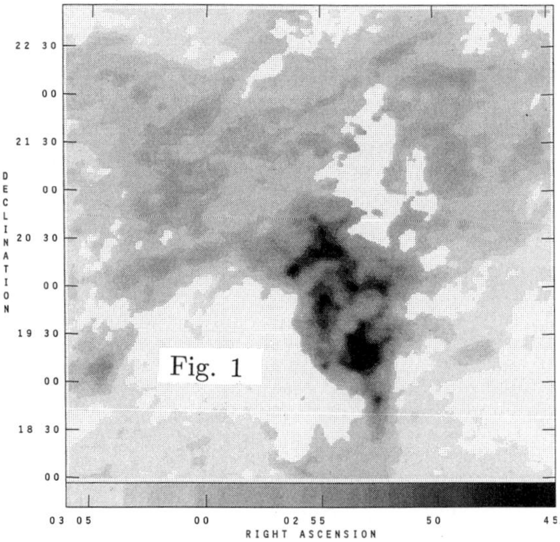

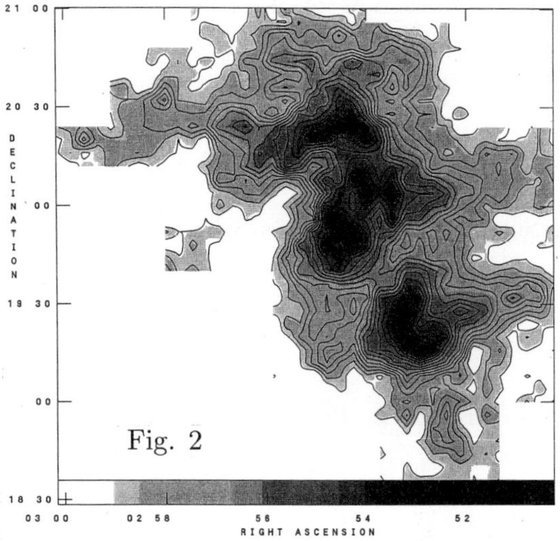

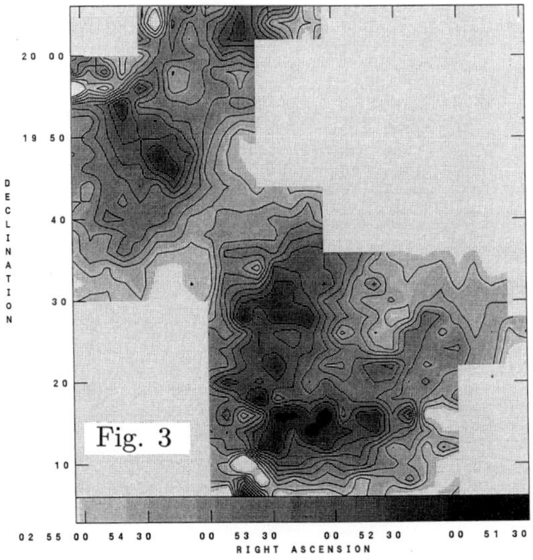

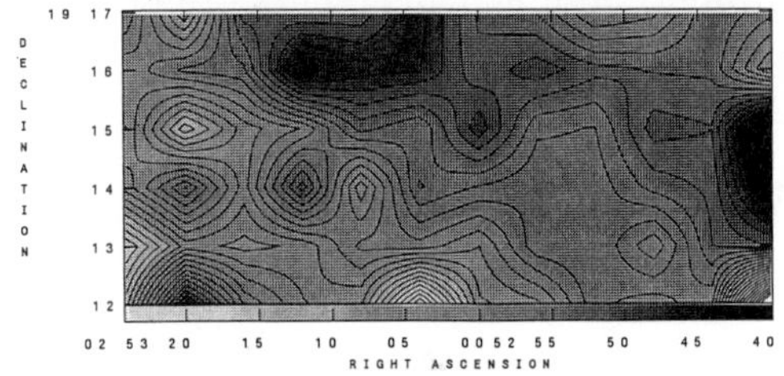

Fig. 4

Fig.1 Colour corrected ($\alpha = 1$) 100 μm IRAS flux density of L1457, calculated from the colour temperature obtained by 60 μm and 100 μm GEISHA sky flux maps.
Fig.2 Integrated main beam brightness temperature (T_{mb}) of ^{12}CO (1-0) in L1457, observed on a 4' grid.
Fig.3 Integrated T_{mb} of ^{12}CO (2-1) in L1457, observed on a 2' grid.
Fig.4 Integrated T_{mb} of ^{12}CO (3-2) in L1457, observed on a 1' grid.

REFERENCES

Corneliussen & Zimmermann: 1991, *in prep.*

Falgarone: 1990, IAU Symposium No.120, Tenorio-Tagle, Moles, & Melnick (Eds.)

Hobbs, Blitz, & Magnani: 1986, *Astrophys. J.* **306**, L109

Scalo: 1990, in: *Physical Processes in Fragmentation and Star Formation*, Capuzzo-Dolcetta, Chioso, & Di Fazio (Eds.)

Stutzki & Güsten, 1990: *Astrophys. J.* **356**, 513

Ungerechts & Thaddeus: 1987, *Astrophys. J. Suppl.* **63**, 645

Zimmermann & Ungerechts: 1988, in: *The Physics and Chemistry of Interstellar Molecular Cluods*, Winnewisser & Armstrong (Eds.), p. 216

Zimmermann & Ungerechts: 1990, *Astron. Astroph.*, **238**, 337

Zimmermann & Winnewisser: 1991, *in prep.*

Multiline Studies in Orion

CARSTEN KRAMER, JÜRGEN STUTZKI, GISBERT WINNEWISSER

I.Physikalisches Institut, Universität Köln,Zülpicher Str.77,5000 Köln 41,Germany

The Orion giant molecular cloud, at a distance of about 450 pc, is the nearest region showing recent massive star formation. It belongs to the best investigated star formation regions (see e.g. Genzel and Stutzki, 1989).
All observations described here were done with the KOSMA 3m radiotelescope and its 230 and 345 GHz Schottky receivers. Backend was an acousto optical spectrometer with 270 kHz resolution bandwidth. See Winnewisser et al. (1986, 1989) for further details.

CO(3-2) observations in Orion A
We observed Orion A in the $^{12}CO(3\text{–}2)$ transition. So far more than 1500 spectra were taken within a 0.5 deg^2 region on a $1'$ grid near the BN-object. The lineprofile at IRc2 exhibits high-velocity wings which extend over 125 km s^{-1} above the 100 mK rms noise level. Figure 1 shows a map of integrated CO-intensities. The map comprises the HII regions M42, M43 and further to the north NGC 1977. The data will be used for a comparison with $^{12}CO(1\text{-}0)$ and (2-1) data published by Castets et al. (1990). An analysis of Gierens et al. (1991) leads to the conclusion that the molecular cloud must be clumpy on scales of roughly 1 pc. These clumps are subject to exterior UV-radiation resulting in photodissociation regions throughout the whole volume of the cloud.

CO(2–1), ^{13}CO(2–1), and CO(3–2) observations in Orion B
We also observed the central 0.7 deg^2 region of the giant molecular cloud Orion B (L 1630) in $^{12}CO(2\text{–}1)$, $^{13}CO(2\text{–}1)$, and $^{12}CO(3\text{–}2)$. The CO (2–1) and (3–2) maps were done on a $2'$ and $1'$ grid respectively. All three maps were centered on the HII region NGC 2024. We find a correlation between the $^{13}CO(2\text{–}1)$ emission and the thick north-south dust lane in front of NGC 2024 as well as with the bright reflection nebula NGC 2023. The aim of these investigations is to gain knowledge about the large scale structure and about excitation conditions in Orion B. It is planned to analyse our (2–1) and (3–2) data together with (1–0) data from Bally et al. (1989) using various radiation transfer models (Kramer et al., 1991).

References
Bally, J., Stark, A.A., Wilson, R.W., Langer, W.D.: 1989, in *The Physics and Chemistry of Interstellar Molecular Clouds*, G. Winnewisser, J.T. Armstrong (publ.),

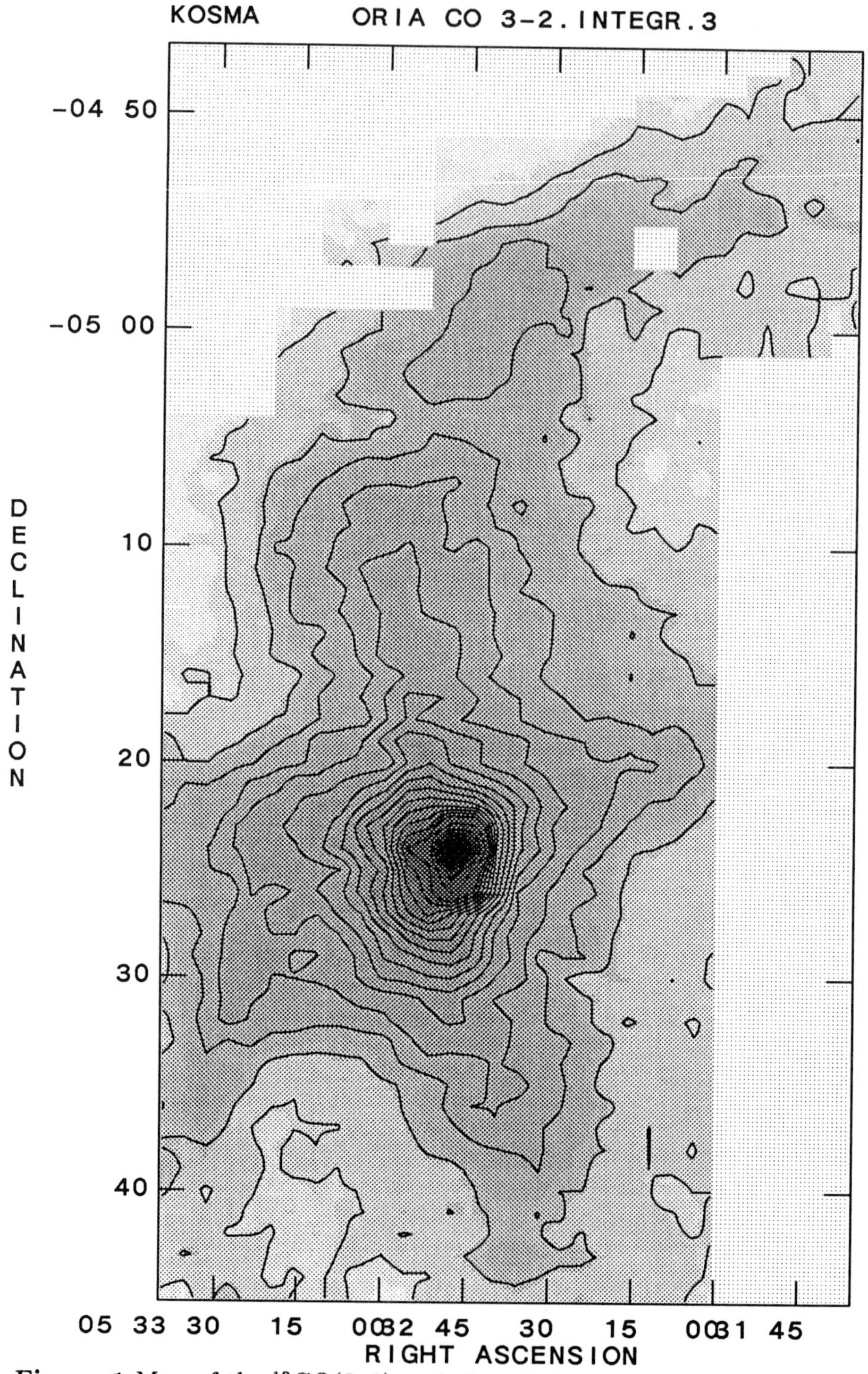

Figure 1 Map of the $^{12}CO(3–2)$ emission of the central part of Orion A.

Springer, Heidelberg
Castets, A., Duvert, G., Dutrey, A., Bally, J., Langer, W.D., Wilson, R.W.: 1990, A&A **234**, 469
Genzel, R., Stutzki, J.: 1989, Annu. Rev. A&A **27**, 41
Gierens, K., Stutzki, J., Winnewisser, G.: 1991, in preparation
Kramer, C., Stutzki, J., Winnewisser, G.: 1991, in preparation
Winnewisser, G.,Bester, M., Ewald, R., Hilberath, W., Jacobs, K., Krotz-Vogel, W., Miller, M., Olberg, M., Pauls, T., Pofahl, E., Rau, G., Schieder, R., Stubbusch, H., Vowinkel, B., Wieners, C., Zensen, W.: 1986, A&A **167**, 207
Winnewisser, G., Zimmermann, P., Hernichel, J., Miller, M., Schieder, R., Ungerechts, H.: 1989, A&A **230**, 248

HI self-absorption in a molecular cloud

C. FELDT, H.J. WENDKER

Hamburger Sternwarte

1. THE OBSERVED CLOUD

Names: North America-Pelican Nebulae = W80 = NGC7000/IC5070 = DR27 = S117 = L935
Distance: $500\,pc$ (Wendker et al.(1983), Strayzis et al.(1989))
Mass: $6 \cdot 10^4\,M_{sol}$ (Bally and Scoville (1980))
Diameter: $20\,pc$
Position: $\alpha, \delta\,(1950) = 20^h\,54^m,\,43^\circ\,30\,',\, l = 84.5^\circ,\, b = -1^\circ$

2. OBSERVATIONS (SEE FIG. 1)

HI: DRAO synthesis telescope + Effelberg $100\,m$-RT, aperture synthesis including short baselines, resolutions: $2\,'$, $1.32\,km/s$
CO($J = 1 - 0$): KOSMA $3\,m$-RT, resolutions: $3.8\,'$, $0.5\,km/s$
^{13}CO($J = 2 - 1$): KOSMA $3\,m$-RT, resolutions: $3.0\,'$, $0.3\,km/s$
H_2CO($J = 1 - 1$): Effelsberg $100\,m$-RT, resolutions: $3.0\,'$, $0.5\,km/s$

3. EVIDENCES AND CONCLUSIONS

The l-b-v correspondence of HI self-absorption with the molecular emission of CO, ^{13}CO and H_2CO:

We conclude that there is cold ($T_s < 40\,K$) atomic hydrogen mixed with the molecular hydrogen (traced by the above molecules).

Origin of the cold HI gas

The ratio N_{HI}:N_{H_2} is between 0.2 and $\simeq$ 1% depending on the assumed *expected HI emission profile.* Within the errors the HI abundance might be explained by H_2 dissociated by low energy cosmic rays (de Jong (1972), de Jong et al.(1980)). Nonetheless we suspect that part of the cold HI is still a residue of the maternal gaseous HI cloud. Therefore the cloud is young and has not reached chemical equilibrium. Although its mass clearly exceeds the Jeans mass no star formation is yet evident.

HI self-absorption as a sensitive tool
For lines of sight in the galactic plane high HI background temperatures (here $\simeq 80\,K$) are provided. From the self-absorption lines the HI spin temperature has to be $T_s < 40\,K$. We estimate it to be $10 - 20\,K$ because the lines are not saturated. Volume densities are $n_{HI} \simeq 10\,cm^{-3}$. Note that if $T_s \simeq 10\,K$ even $n_{HI} = 1\,cm^{-3}$ (!) would be detectable here.

HI self-absorption as an independent thermometer
From the depth of the line an independent estimate of the kinetic temperature within the dense regions of the dark cloud is possible.

HI self-absorption as a *high density tracer*
The best correspondence of the HI self-absorption regions is found with the H_2CO distribution which itself is a tracer for high densities. Moving to lower densities (and higher temperatures) the HI would not be detectable any more in self-absorption. On the outer layers of the dark cloud the temperature would be high enough to let the HI show up again as an envelope of HI emission against the background as found by Wannier et al.(1983).

4. REFERENCES

Bally,J., Scoville,N.Z.: 1980, *Astrophys. J.* **239**, 121

de Jong,T.: 1972, *Astr. Astrophys.* **20**, 263

de Jong,T., Dalgarno,A., Boland,W.: 1980, *Astr. Astrophys.* **91**, 68

Strayzis,V., Goldberg,E.P., Meistas,E., Vansevicius,V.: 1989, *Astr. Astrophys.* **222**, 82

Wannier,P.G., Lichten,S.M., Morris,M.: 1983, *Astrophys. J.* **268**, 727

Wendker,H.J., Benz,D., Baars,J.W.M.: 1983, *Astr. Astrophys.* **124**, 116

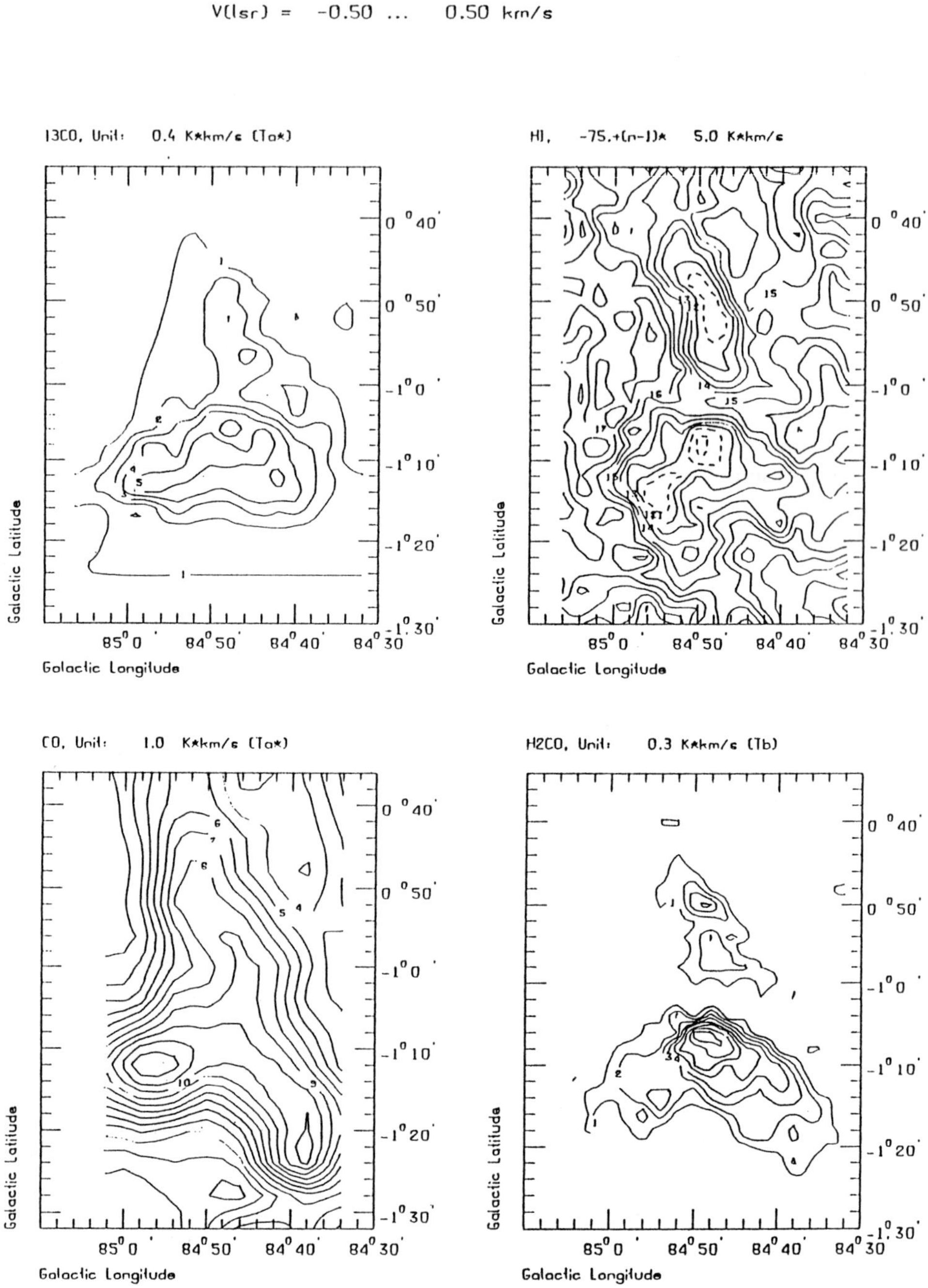

Fig. 1: l-b maps of the CO (lower left), H_2CO (lower right), ^{13}CO (upper left) and HI (upper right) line temperature integrated over a narrow velocity interval.

High Resolution IR Images of the Young Cluster GM-24

M. TAPIA[1], M. ROTH[2] AND J.A. LOPEZ[1]

[1]Instituto de Astronomia, UNAM, Apartado Postal 877, Ensenada, 22830, México.
[2] Las Campanas Observatory of the CIW

GM-24 (17^h 13.7^m;$-36°$ $17'.8$) is a "cometary nebula" associated with a dense molecular cloud. Previous work has established that the nebula is part of a "blister" *HII* region located in the "hot spot" of a larger CO cloud and that star formation has recently been (and possibly still is) active in this region. We have obtained near-infrared *JHKL* and $Br\gamma$, high spatial resolution ($\sim$ 1") images with the 1.5m and 4m telescopes of the Cerro Tololo Interamerican Observatory. These were complemented by an IRSPEC $2.1 - 2.2\,\mu$m spectrum taken with the 3.6m telescope at ESO. The main conclusions from our analysis are:

Some 30 stars form an embedded IR cluster in an area that could not be resolved by IRAS and is listed as one entry in the *Point Source Catalogue.* Previous single-detector scans identified only four sources inmersed in extended emission.

The Br_γ emission is spatially coincident with 2 cm radio-continuum emission. The measured Br_γ and radio - continuum fluxes allow an estimation of the extinction ($A_V \geq 50$).

Warm dust outlines the boundaries of the *blister HII* region, possibly a result of compression heating caused by the expansion of the ionized region.

The total luminosity of the region is $L_{tot} \simeq 1.5 \times 10^5 L_\odot$, a value which is consistent with the presence of 4–6 B0 *ZAMS* stars; this result is in agreement with previous radio determinations. At present a determination of the luminosity function is unrealistic due to the lack of high spatial resolution $10 - 20\,\mu$m array observations.

We can estimate the efficiency of star formation to be either $\sim$ 20% or $\sim$ 9% assuming that the faint end of the observed stellar population (at K) is formed either by obscured *ZAMS* B-type stars or low-mass pre-main sequence objects respectively.

The details of this work have appeared in *Astronomy and Astrophysics* **242**, 388.

The Interaction by Dynamical Friction of Embedded Young Stellar Clusters and the ISM

A. Just, B.M. Deiss
Institut für Theoretische Physik, J.W. Goethe-Universität, Frankfurt

Open stellar clusters are born in dense regions of the interstellar medium. It is likely that a large fraction of gas is not converted into stars. In this case it is not a priori clear whether the cluster remains gravitationally bound after removing the gas. The energetic state of the stripped cluster depends strongly on the interaction with the gas in the early phase of evolution.

Most present day models (see e.g. Verschueren & David 1989) neglect friction forces between the stars and the ISM. In these scenarios some energy input (e.g. continously by stellar winds, or suddenly by nearby HII-regions or supernovae) is responsible for the thermal evolution of the gas. The young cluster evolves via the interaction with the total gravitational potential.

In dense molecular gas ($n_{H_2} > 10^5 cm^{-3}$) friction forces are very effective even if the stars have no extended envelope. Dynamical friction due to the individual gravitational fields of the stars leads to short deceleration times. Adopting the friction coefficient of a homogeneous gas in thermal equilibrium (Deiss et al. 1990) the braking timescale of moderately supersonic stars ($v \approx 1 km/s$) is of the order of $10^3 - 10^4 yr$. A cooling time short compared to the free fall time is assumed. The corresponding cross section is $10\pi R_A^2$ for supersonic motion. $R_A = 2GM/v^2$ is the accretion radius of a star with mass M and velocity v . This cross section agrees with that of dynamical friction in a collisionless gas (Chandrasekhar 1960). For subsonic motion dynamical friction is strongly supressed by pressure forces. If the extension of the stellar envelopes (of stellar winds e.g.) exceeds R_A, friction by ram pressure is more effective yielding even shorter timescales.

The corresponding lengthscale for dynamical coupling of the stars and the ISM is of the order of $10^{-3} - 10^{-1} pc$, which is smaller than the mean distance of the stars. Thus stellar motion creates turbulence in the ISM with a scale length of the order of the mean distance of the stars. The turbulence in return reduces the mean friction force due to statistically enhanced relative motion between the individual stars and the ISM.

Neglecting the statistical reduction factor and averaging over a Gaussian velocity distribution yields an energy loss timescale of about $10^5 - 10^6 yr$ for the stellar cluster. Since this timescale exceeds well the free fall time, virialization of the stellar component and a quasihydrostatic state of the ISM can be assumed. In the hydrostatic equation turbulent pressure may be added to the thermal pressure. But for the stellar dynamics the turbulent motion of the gas acts quite differently than thermal motions because the turbulent motion is macroscopic.

In this model the energetic evolution of stars and ISM is controlled by energy transport and loss of the gas component. The evolution of the young cluster in that phase of strong coupling must be analized by detailed models. Two limiting cases can be discussed in advance. If the stellar component virializes mainly in its own gravitational field, energy loss by friction leads to contraction and heating of the cluster. On the other hand, if the gas potential dominates, the stellar cluster will cool down with less contraction. In the first case gravitational binding of the cluster increases, whereas in the second case the binding energy of the isolated cluster depends strongly on the potential well of the gas component.

The above discussion shows that the friction forces of the stars and the ISM can essentially influence the final energetic state of the young stellar cluster when the molecular gas is removed.

Chandrasekhar,S.:1960 Principles of Stellar Dynamics, Dover Pub., Inc., New York
Deiss,B.M.,Just,A.,Kegel,W.H.:1990 Astron.Astrophys. **240**,123
Verschueren,W., and David,M.:1989 Astron.Astrophys. **219**,105

The dynamics of the subsystem of open clusters in the two-component model of our Galaxy

S.A.KUTUZOV[1], L.P.OSSIPKOV[1], E.M.NEZHINSKIJ[2]

[1]Leningrad State University, Russia

[2]Institute for Theoretical Astronomy, Sanct Petersburg, Russia

INTRODUCTION

This study is a continuation of work initiated by Prof. Barkhatova; it is based on the catalogue of 69 open clusters of Barkhatova et al., 1987. There are two purposes to our investigation. The first one is to find galactic orbits of clusters, which is a problem (see review by Ossipkov,1981) demanding knowledge of the galactic potential. Our second purpose is to check the existence of non-random groups (*complexes*) of open clusters, as described by Barkhatova et al.,1988.

THE MODEL OF THE GALACTIC POTENTIAL

We have used the two-component model of gravitational potential of our Galaxy proposed by Kutuzov and Ossipkov 1989. The first component represents the disk and bulge and corresponds to the Kuzmin - Malasidze potential in the galactic plane. The second component represents the nucleus and halo and corresponds to the Kuzmin - Veltmann 'limiting' model. The heliocentric distance, $R_\odot$, the circular velocity, $V_\odot$, the Oort and Kuzmin parameters A and C, and the total mass, M, have the following values:

$$R_\odot = 8.23\ kpc, \quad V_\odot = 226\ km\ s^{-1}, \quad A = 13.4\ km\ s^{-1}\ kpc^{-1},$$

$$C = 75\ km\ s^{-1}\ kpc^{-1}, \quad M = 4.4 \times 10^{11} M_\odot.$$

ORBITS OF OPEN CLUSTERS

The data necessary for orbit finding have been taken from the catalogue of Barkhatova et al., 1987. We have used two distance scales: the one by Barkhatova and the one by Hagen. The latter leads to larger vertical velocities. All orbits are found to be box ones: the boxes are trapezium-like, practically rectangles (even for such old clusters as NGC 188 and NGC 2420). The dimensionless elements of orbits practically coincide with the ones we found earlier for the one-component galact ic model. The initial values of the adiabatic invariant of vertical motion, J, (Ossipkov,1976) have been calculated and the conservation of J along the orbits have been checked. We have found that if $J < 1$ $km\ s^{-1}\ kpc^{-1}$ then changes of J do not exceed 4%.

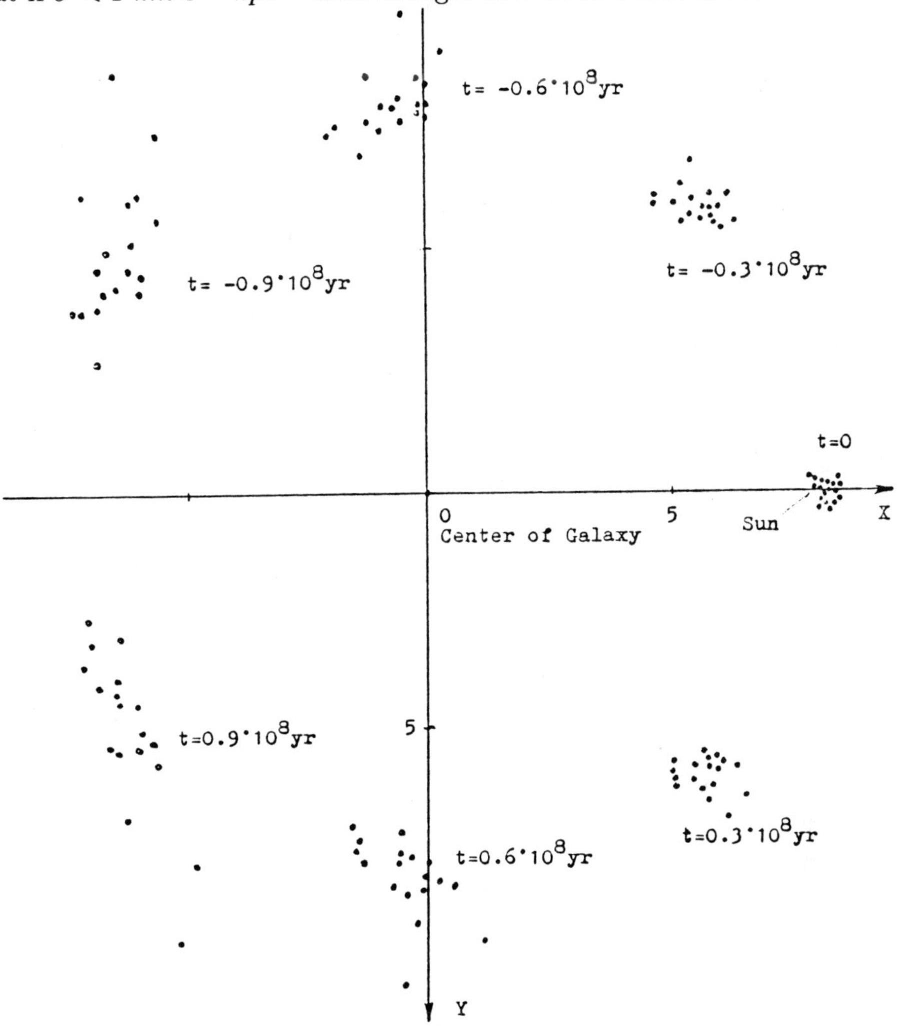

Fig.1.

We found no correlation between orbit elements and age and metallicity of the clusters. For instance, the old clusters NGC 188, 2420, 7789 have the following value of eccentricities: 0.079, 0.296 and 0.059.

COMPLEXES OF OPEN CLUSTERS

Barkhatova et al. (1988) supposed the existence of 11 complexes of open clusters.

Complex I (Per) includes NGC 889, 884 Tr 1, and others, Complex II (corresponding to the Local System) includes Plei, α Per, NGC 752, and others. According to the statistical analysis by Barkhatova et al. (1989), their mean diameter is 400 *pc* in the galactic plane and 100 *pc* in the vertical direction; the mean velocity dispersion is 100 $km^2\ s^{-2}$.

DYNAMICAL EVOLUTION OF COMPLEXES

We found that at the present time the complexes are the most compact (in agreement with our previous calculations, Barkhatova et al., 1986). As an example, we show the dynamical evolution of complex II in Fig.1. Several explanations for this phenomenon could be considered:
A. There are no real complexes as physical systems: but there is no doubt in the reality of such a system as Complex I.
B. The complexes are real but their velocities are fictitious due to errors of proper motions. However, using proper motions from different sources leads to the same result.
C. The masses of the complexes are much larger than the sum of the masses of clusters. In reality the large number of GMC's in the region of complexes makes such a possibility seem likely.

We estimated the minimal masses, M_c, of complexes from the Hill condition of stability and found $M_c \sim 10^6 - 10^7 M_\odot$ (Nezhinskij, Ossipkov, 1987). Using a generalization of the virial theorem (with allowance for galactic attraction), it is possible to find that the virial mass of the Complex I is about $5 \times 10^7 M_\odot$, and that of Complex II is $10^7 M_\odot$ (Barkhatova, Ossipkov 1987).

REFERENCES

Barkhatova, K.A., Kutuzov, S.A., Ossipkov, L.P., 1987, Astron. Zh. 64, 956.
Barkhatova, K.A., Kutuzov, S.A., Ossipkov, L.P., 1988, Kin. fiz. neb. tel. 4, 64.
Barkhatova, K.A., Nezhinskij, E.M., Ossipkov, L.P., 1986, Astron. Tzirk. 1472, 7.
Barkhatova, K.A., Ossipkov, L.P., 1987, Astron. Tzirk. 1519, 1.
Barkhatova, K.A., Ossipkov, L.P., Kutuzov, S.A., 1989, Astron. Zh. 66, 1154.
Kutuzov, S.A., Ossipkov, L.P., 1989, Astron.,Zh.,66, 965.
Nezhinskij, E.M., Ossipkov, L.P., 1987, Astron. Tzirk. 1479, 1.
Ossipkov, L.P., 1976, Pisma v Astron.,Zh.,2, 367.
Ossipkov, L.P., 1981, Dvizh. isk. i estestv. nebesn. te, Sverdlovsk, 18.

The Mass Function For Faint Stars

Pavel Kroupa[1] and Christopher A. Tout[2]

[1]Institute of Astronomy, Madingley Road, Cambridge CB3 0HA, England
[2]Lick Observatory, University of California, Santa Cruz, CA 95064, U.S.A.

ABSTRACT
We describe a model of the distribution of low-mass stars by luminosity in which the effects of unresolved binaries and structure in the mass–luminosity relation are included. We find that one mass function can account for the luminosity function observed locally and at distances of $\approx 100\,\mathrm{pc}$, even though the shapes of these luminosity functions differ significantly if most 'stars' on the sky are, in fact, binary systems with uncorrelated masses. The same mass function is also consistent with the (sparse) data on open clusters. These conclusions, if confirmed, form important constraints on star-formation theories.

1. INTRODUCTION

The number of main-sequence stars increases steeply with decreasing mass. However, the least massive stars are very difficult to observe because they are exceedingly faint. Little is known about them and their contribution to the mass of the Galactic disc is difficult to establish. The study of the distribution of stars in the Galactic disc and in young open clusters is very important for constraining star formation theories.

The volume density of stars per unit mass interval, or the stellar mass function, can only be deduced indirectly from the stellar luminosity function, which is the observed number density of stars per unit magnitude interval. These two functions are related through the slope of the mass–luminosity relation. Although this relation remains uncertain at the faint end, points of inflexion, which result from effects of molecules and ions in the outer layers of the stars, are instrumental in understanding the structure seen in the luminosity function (Kroupa, Tout & Gilmore 1990, hereafter KTG).

2. THE DATA

The stars within 5.2 pc of the sun have well determined absolute luminosities and are considered to form a complete sample for all magnitudes (Dahn, Liebert & Harrington 1986 and Henry & McCarthy 1990). For brighter magnitudes than $M_V \approx 10$ the sample of stars selected from the Gliese Catalogue is complete to within 20 pc (Wielen, Jahreiss & Krüger 1983). These nearby stars are well studied and most binaries have been identified. Fig. 1 shows the luminosity function for this sample and demonstrates that the faint end remains poorly constrained. The statistical weight at the faint magnitudes is improved using photographic surveys which reach out to distances of $\approx 100\,\mathrm{pc}$ (see for example Stobie, Ishida & Peacock 1989). Distance estimates are made using the method of photometric parallax which relies on an absolute magnitude–colour relation which is poorly defined for the reddest stars.

This together with the large proportion of binaries among stars (Halbwachs 1986 and Duquennoy & Mayor 1991) which remain unresolved in such surveys makes the conversion of photographically derived luminosity functions to a mass function problematic. Fig. 1 shows typical examples of the results of such surveys and we see that a significant discrepancy exists for $M_V > 12$ in that the local sample of stars appears to contain a significantly larger space density of faint stars than the Galactic field.

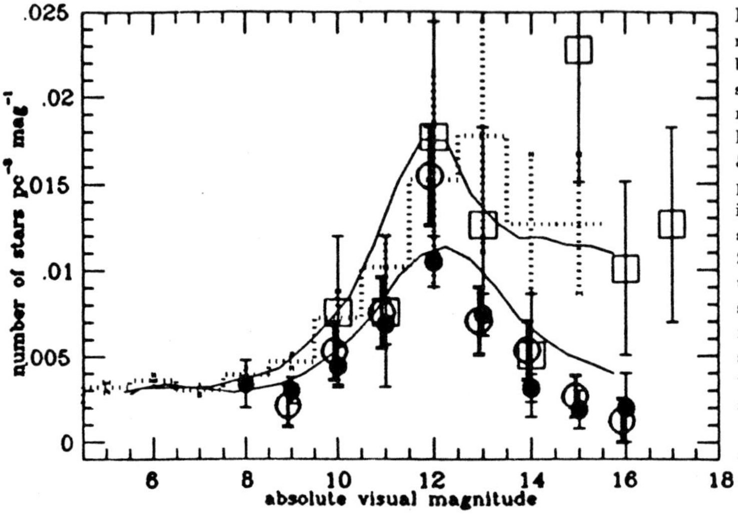

FIG. 1.—The empirical local luminosity function based on data compiled by Wielen, Jahreiss & Krüger (1983) is shown as a histogram. Open squares represent the determination by Dahn, Liebert & Harrington (1986) and Henry & McCarthy (1990). Two examples of photographically determined luminosity functions are shown as open and solid circles (Reid & Gilmore 1982 and Stobie, Ishida & Peacock 1989, respectively). Malmquist and Galactic disc scale-height corrections have been removed. The upper curve depicts our single-star model and the lower curve the system luminosity function. Both are derived from the same mass function (Eqn. 1). Errorbars denote Poisson uncertainties.

3. THE MODEL

This discrepancy may be resolved if we take into account the structure in the mass–luminosity relation discussed by KTG and the large proportion of binary systems among all 'stars' on the sky (Kroupa, Tout & Gilmore 1991). We assume that the mass function, $\xi(m)$, is a double power-law from which we choose stellar masses at random. Assuming the masses of binary components to be uncorrelated we generate a population of systems of which a fraction, f, consists of unresolved binaries in which both components are main sequence stars. The luminosities are obtained from the mass–luminosity relation derived by KTG. We compute model single-star and system luminosity functions by taking into account the effects of a metallicity dispersion, age dispersion and measurement errors as well as an exponential Galactic disc with scale-height h (Kroupa & Tout 1991a).

4. RESULTS

The best agreement is obtained as f approaches unity. The number of stars per pc^3 in the mass interval m to $m + dm$, $\xi(m)dm$, can be represented by

$$\xi(m) = n \begin{cases} (m/0.5)^{-\alpha_1}, & \text{if } 0.08 \le m \le 0.5 \\ (m/0.5)^{-2.0}, & \text{if } 0.5 \le m \le 1. \end{cases} \tag{1}$$

where, locally, $n \approx 0.1$ stars pc^{-3}, $\alpha_1 \approx 1.1$ and m is the mass of the star in solar units. The corresponding single-star and system luminosity functions are compared to the empirical data in Fig. 1. The mass in stars less massive than the sun comprises 30–40% of the Oort limit if that is taken to be $0.1\,M_\odot \mathrm{pc}^{-3}$. Correct normalization to the data is achieved with an exponential galactic-disc scale height of $h \approx 215$ pc, *i.e.* overall the solar neighbourhood appears somewhat over dense in stars with the generally accepted value $h \approx 325$ pc.

Applying this model to young open clusters we find consistency with the observed luminosity functions in the Hyades and Praesepe clusters (Kroupa & Tout 1991b). As an example we compare our model with $f = 0.8$ with the empirical Praesepe luminosity function in Fig. 2.

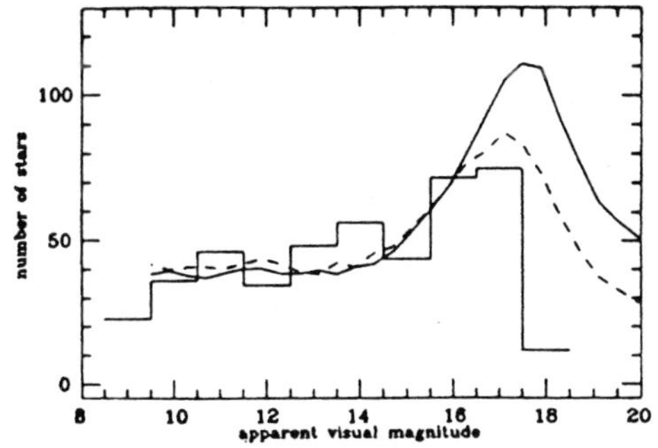

FIG. 2.—The observed Praesepe luminosity function (histogram, Jones & Stauffer 1991) is compared to our model (solid curve) assuming a population of 80% binaries, a cluster age of 4×10^8 years, distance of 159 pc, single chemical composition and the mass function given by Eqn. 1 with $\alpha_1 = 1.1$. The dashed curve represents a model with $\alpha_1 = 0.6$. The observations are incomplete in the faintest bin and are probably incomplete in the $V = 17.01$ bin.

5. CONCLUSIONS

Our results suggest that the same mass function can account for the luminosity functions observed locally, at $\approx$ 100 pc distance and within young clusters if the effects of a large proportion of binary stars with uncorrelated component masses and the structure in the mass–luminosity relation are taken into account. The data on open clusters remains sparse and we must be cautious with this conclusion and await further observational data. The hypothesis that the stellar masses in binary systems are uncorrelated is also controversial but recent studies appear to support it (Halbwachs 1987 and Duquennoy & Mayor 1991).

If the above picture is true then stars less massive than 1 $M_\odot$ contribute no more than 40% to the Oort limit. The implication for star-formation is that most stars form in binary systems, that the component masses are uncorrelated and that the same spectrum of masses is formed irrespective of the initial conditions in molecular clouds in the Galactic disc.

We would like to thank Gerard Gilmore and Neill Reid for stimulating discussions. PK thanks Trinity College, Cambridge, for a grant from the Rouse Ball Research Fund and the University of Cambridge for an Isaac Newton Studentship.

REFERENCES

Dahn, C. C., Liebert, J., & Harrington, R. S. 1986, *A. J.*, **91**, 621.
Duquennoy, A. & Mayor, M. 1991, *A. J.*, in press.
Halbwachs, J. L. 1986, *Astron. Astrophys.*, **168**, 161.
Halbwachs, J. L. 1987, *Astron. Astrophys.*, **183**, 234.
Henry, T. J. & McCarthy, D. W. 1990, *Ap. J.*, **350**, 334.
Jones, B. F. & Stauffer, J. R. 1991, preprint.
Kroupa, P. & Tout, C. A. 1991a, in preparation.
Kroupa, P. & Tout, C. A. 1991b, in preparation.
Kroupa, P., Tout, C. A., & Gilmore, G. 1990, *M. N. R. A. S.*, **244**, 76.
Kroupa, P., Tout, C. A., & Gilmore, G. 1991, *M. N. R. A. S.*, in press.
Reid, N. & Gilmore, G. 1982, *M. N. R. A. S.*, **201**, 73.
Stobie, R. S., Ishida, K., & Peacock, J. A. 1989, *M. N. R. A. S.*, **238**, 709.
Wielen, R., Jahreiss, H., & Krüger, R. 1983, in *The Nearby Stars and Stellar Luminosity Function, IAU Colloq. No. 76*, eds Davis Philip, A. G., and Upgren, A. R., Davis Press, New York, p.163.

The KOSMA 3m Radiotelescope

N. SCHNEIDER, J. STUTZKI, G. WINNEWISSER

I. Institute of Physics, University of Cologne, Germany

The KOSMA (Kölner Observatorium für Submillimeter und Millimeter Astronomie) 3m radiotelescope on Gornergrat near Zermatt in Switzerland was installed in 1985 by the I. Physikalisches Institut, University of Cologne. Since early 1986 the observatory has been in successful operation. It is supported by the Deutsche Forschungsgemeinschaft (DFG) through grant SFB 301 as well as by special funding from the Land Nordrhein–Westfalen. The observatory is administered by the Hochalpine Forschungsstationen Jungfraujoch und Gornergrat, Bern/Switzerland. It is used by staff astronomers of Cologne as well as by visitors from other institutes.

The 3m telescope has a resolution of $2'$ at 230 GHz and $1'$ at 345 GHz. The pointing accuracy is smaller than $10''$ and the surface accuracy of the dish is 30 μm rms. The observatory is equipped for the the observation of interstellar and atmospheric molecular lines in the millimeter and submillimeter range :

• The **frontends** are 3 receivers for the frequency ranges 130-170 GHz (SIS-mixer with a receiver noise temperature T_R of 50-100 K), 220-250 GHz (Schottky-mixer with T_R of 400 K) and 345 GHz (Schottky-mixer with T_R of 380 K). The future developments are largely concentrating on SIS junction production in the cologne laboratory, which recently started operation and higher frequency Schottky receivers.

• The **backends** are 2 acousto optical spectrometers (AOS). The medium (high) resolution spectrometer has a channel spacing of 166.5 (31.6) kHz, a total bandwidth of 287 (64.7) MHz and 1724 (2048) channels. A broad band AOS (1.2 GHz bandwidth) with variable resolution is presently under construction and will be available in 1992.

According to the good climatic conditions at an altitude of 3200 m (10134 ft), the telescope is ideal for radioastronomical observations at high frequencies. Weather statistics from the last two winter seasons show that the zenith transparency at 350 GHz is >0.60 (τ=0.50) for 60 % of the time. The summer period shows 230 GHz opacities between 0.13 and 0.75 with an average value of 0.3 during 70% of the total available time. Extrapolation to higher frequencies predicts 25 % of the time in winter to be useful for obsevations in the higher atmospheric windows.

The astronomical projects concentrate on three points :

- Investigation of the structure and dynamics of galactic molecular clouds

- Frequency surveys and chemistry of molecular clouds

- Large scale structure of the Galaxy and external galaxies

Furthermore molecular tracers in the atmosphere, mainly CO and O_3, are frequently observed.

Part III

EVOLUTION OF GALAXIES

Density Wave Resonances in Galaxies

D.M. ELMEGREEN AND B.G. ELMEGREEN

Vassar College Observatory, Poughkeepsie, NY 12601

and IBM T.J. Watson Research Center, Yorktown Heights, NY 10598

ABSTRACT: Optical features and computer enhancement of two-arm and three-arm symmetries in spiral galaxies have been used to determine the positions of spiral wave resonances and to show that most two-arm spirals probably have inner wave reflection barriers. The three-arm spirals are apparently driven by asymmetries in the two-arm spirals. We conclude that most two-arm spirals are self-sustaining wave modes and that the three-arm components remove energy from the two arms. Interacting galaxies with only two arms have not had enough time for the three-arm component to become prominent.

1 INTRODUCTION

Resonances in spiral galaxies can be recognized by the locations of key optical features that are best revealed by enhanced and specially processed images. By isolating the m=2 and m=3 symmetric components of a galaxy, we can estimate the extent of spiral density waves and wave modes. We have studied grand design galaxies with two long, continuous and symmetric arms, such as M81, multiple arm galaxies with apparent inner symmetry but lots of outer arms, such as M101, and flocculent galaxies with short random spiral arm segments, such as NGC 5055. In this paper, we examine density wave theories in the context of observations of multiple arm galaxies.

Most classification systems are based on the appearance of galaxies in the blue band, since blue images are commonly available. For understanding spiral structure, it is often more useful to consider color images, since the distributions of young and old disk stars are revealed in the blue and near-infrared passbands, respectively. Over the past decade, we have photographed galaxies in B and I bands, digitized the plates by scanning with a microdensitometer with a pixel resolution of about 2", and stored the resultant 512x512 images for further processing. The average of the B and I bands is used to produce a G image, then the three images are photographed on computer through B, G, and R filters and combined to make a composite color image. To aid in studying the morphology of randomly inclined galaxies, the color images are then deprojected by stretching pixels along the minor axis by a factor equal to the inverse cosine of the inclination. Next, the effective dynamic range of each image is increased by removing an azimuthally averaged radial profile, in which a boxcar average is made over several adjacent pixels, and the rms variation at each radius is normalized to a constant value. In this way, spiral structure can be discerned with greater contrast over a larger part of the disk. Many more features become apparent in the enhanced images, such as arm symmetry, the endpoints of the spiral arms, spurs and gaps.

We showed previously that these and other optical features can be identified with specific resonance locations (Elmegreen, Elmegreen and Seiden 1989; Elmegreen and Elmegreen 1990). In order to estimate where these locations occur according to the density wave theory, recall that a star in the disk undergoes radial and azimuthal oscillations that lead to epicyclic motion in a co-moving reference frame. The epicyclic frequency, κ, depends on the local angular rotation rate, Ω , and its derivative:

$$\kappa = 2\Omega\left(1+\frac{d\ln\Omega}{2d\ln R}\right)^{1/2}.$$

Resonances occur when the difference between the angular rotation and the pattern speed, $(\Omega - \Omega_p)$, equals κ /m for mode m, which is such that a star completes m/n epicycles in the time it takes to go from one arm to the next in a galaxy with n symmetric arms. The outer Lindblad resonance (OLR) occurs with m=−2, the inner Lindblad resonance (ILR) has m=2, the inner 4:1 resonance is at m=4, and other inner and outer resonances follow similarly. At resonance radii, stellar orbits absorb wave energy since their natural epicyclic orbits are synchronous with the density wave perturbations.

This paper discusses the research that determined the relationship between theoretical resonances and distinctive optical features, so that an enhanced image of a galaxy can be used to reveal its kinematics.

2 DATA ANALYSIS

2.1 Resonances from rotation curves vs. observations

Rotation curves can be fit in the optical parts of a galaxy (beyond the inner regions) reasonably well by a power law in which V(R) is proportional to R^α. The ratios of all resonances are specified by the rotation curve; thus, if the location of one resonance is known (or assumed), all of the others can be derived. Consider M81; its rotation curve was measured by Visser (1980), from which we derive a slope $\alpha = -0.2$. The ratio of the inner 4:1 resonance to the OLR is given by

$$\frac{R(\text{inner}4:1)}{R(\text{OLR})} = \frac{\left(1 - \frac{1}{2}\left[\frac{1+\alpha}{2}\right]^{1/2}\right)^{\frac{1}{1-\alpha}}}{\left(1 + \left[\frac{1+\alpha}{2}\right]^{1/2}\right)^{\frac{1}{1-\alpha}}},$$

so we expect R(4:1)/R(OLR) = 0.49. This ratio agrees with what we obtain by using the precise rotation curve and not just the power law fit. If we assume that the OLR is at the optical ends of the spiral arms, then we know where the 4:1 resonance occurs: it happens to be where there are symmetric spurs in between the main arms and where there are gaps in the spiral arms. Our goal is to determine whether there are sufficiently many distinctive features present in an individual galaxy to give a self-consistent fit for the resonances (which then also specifies the pattern speed).

Qualitatively, the spurs and arm gaps in M81 can be understood by considering stellar orbits at the inner 4:1 resonance: a star completes 2 epicycles in an arm-to-arm time, so there is a loop in the orbit exactly midway between the two arms, which is where the spurs are observed (see Artymowicz and Lubow 1991). A schematic of wave modes in M81, in which a wave bounces at an inner reflection point, shows that the arms are unusually bright where the inward and outward-moving waves constructively interfere (which is at the radius of corotation here), and the arms have a gap where the waves destructively interfere (which is near the 4:1 resonance here). We believe that this alternating symmetric pattern of bright and dark along the arms is the first direct evidence for spiral wave modes in a galaxy.

Lin and Lowe (private communication) have used Visser's rotation curve to create a modal model for M81, and compared it with our data for the m=2 component. Fourier transforms of the models and our data show that the model peaks and gaps occur in exactly the same locations as in the data. These features are sometimes easier to see in an unwrapped galaxy image, in which the spiral arms are plotted as logarithm of the radius, log R, versus the azimuthal angle, θ. Spiral arms usually are approximately straight on these plots, and the tangent of the pitch angle is given by the slope of the arms measured from the vertical. In M81, the main arms clearly have gaps and a phase shift where the arms continue, at about 0.4 R_{25}, where R_{25} is the radius at which the surface brightness is 25 mag/square arcsecond.

Rotation curves are difficult to obtain for the low inclination galaxies whose spiral structure is most discernible. When rotation curves are not available, we use the universal rotation curve fit by Persic and Salucci (1991), which depends on the absolute magnitude of a galaxy. We convert their fit to a power law representation for the outer parts of galaxies, between about 1 and 3 scale lengths, by the equation

$$\alpha = \frac{\ln\left(\frac{2.2+\Delta}{2.2-1.2\Delta}\right)}{\ln 3.2},$$

where

$$\Delta = 0.12 + 0.096(M_B + 21.5).$$

As before, the expected ratios of resonances can be determined.

The spiral galaxy NGC 4321 is another grand design galaxy that we previously fit to resonances (Elmegreen, Elmegreen and Seiden 1989). It has a prominent ridge of star formation along the spiral arms in the inner regions, which is common in other galaxies as well. It also has arm gaps and interarm spurs. If we again assume that the outer arms end at the OLR, then corotation is at the end of the star formation ridge and the inner gaps and spurs correspond to the inner 4:1 resonance, based on Persic and Salucci's (1991) rotation curve and the previous resonance ratio equations. Other locations for the resonances were tried and rejected when key optical features missed the resonances.

2.2 Symmetrized images

Global density waves create a symmetric effect in a galaxy. In order to examine the overall symmetry of spiral arms, we developed a method for producing computer symmetrized images of galaxies. To search for m=2 symmetry, we take an enhanced image I and perform the operation

$$S_2 = I - [I(\theta) - I(\theta + \pi)]_t \,,$$

where t stands for truncation to 0 for negative numbers; the leftover asymmetric material can be represented by $A_2 = I - S_2$. The S_2 representation plots all even symmetric features, including the four, six, etc. arm structures. To search for m=3 symmetries, we make an image of the intensities

$$S_3 = 2I - [I(\theta) - I(\theta + 2\pi/3)]_t - [I(\theta) - I(\theta - 2\pi/3)]_t \,.$$

In this representation, the brightness of three equally spaced features represents the strength of the component; i.e., features that occur on all three sides of an equilateral triangle are twice as bright as features that occur on only two sides of such a triangle. The surprising result we found is that many multiple arm and grand design galaxies contain substantial m=3 structure in addition to m=2 structure. This m=3 structure is present on the original images, of course, but it is masked by the two-arm component. In NGC 4321, the inner and outer extent of the m=3 structure is exactly the range and location expected for the 3:1 resonances, based on the other resonance radii that were independently determined. This result suggests that analysis of the m=3 structure may provide additional resonance features that are useful for fitting more complicated multiple arm galaxies. These images also show that multiple arm galaxies can be understood as the superposition of m=2 and m=3 structure.

We have successfully applied this analysis to 18 spiral galaxies. For example, consider the barred spiral galaxy NGC 613, shown in Figure 1. In this figure and the ones that follow, the top left image is the enhanced image where the original color image has been deprojected and the azimuthally averaged radial profile subtracted. The white bars on the right and left represent R_{25} and 100 pixels, respectively. The image on the middle left is the m=2 (S_2) image. The middle image on the right is the A_2 image, showing the asymmetries. The top right picture is the S_3 structure. The bottom left image repeats the middle left image except that circles indicating resonances have been marked. The outermost one in all cases is the best fit for the outer Lindblad resonance. The middle circle marks the corotation resonance, and the innermost circle marks the inner 4:1 resonance. The bottom right image repeats the m=3 image, with circles delineating the inner and outer 3:1 resonances for the same pattern speed as the 2:1 resonances. At first glance,the enhanced image of NGC 613 appears to show two sets of symmetric arms. The S_2 image shows that only one set of arms is m=2 symmetric; the S_3 image shows that there is substantial m=3 structure. Thus, the multiple structure of this galaxy is almost entirely due to a superposition of m=2 and m=3 structures. For NGC 613, we place the OLR where the m=2 structure ends, at 0.9 R_{25}. Then the bar extends to corotation, with star formation between the inner 4:1 resonance and corotation. The m=3 structure shows an inner triangle. The radius of a circle inscribed within the innermost m=3 triangular structure is the same as the radius of the m=3 resonance, based on the m=2 pattern speed. The outermost m=3 arms at 0.76 R_{25} are at the outer m=3 resonance.

The multiple arm galaxy NGC 5457, shown in Figure 2, has a fascinating arm structure. On a conventional enhanced image, it appears symmetric in the central regions and lopsided and multiple-armed in the outer. The S_2 image shows four distinct spiral arms in the central regions, while the S_3 image shows three distinct spiral arms. Suppose that the OLR is where the m=2 symmetric arms end, at $0.75R_{25}$. Some HII regions appear beyond this m=2 symmetry limit, since gas spirals can extend beyond the OLR. Then the three-arm structure has limits bracketed by the inner and outer 3:1 resonances. At corotation, the ridge of bright inner star formation ends. At the inner 4:1 resonance, the spiral structure has four distinct, equally spaced arms, which is analogous to having two main arms with two spurs midway between the arms. Thus, the rather complicated structure in this galaxy also can be understood as a simple superposition of the m=2 and m=3 spirals, with strong spurs at the inner 4:1

resonance.

Another interesting multiple arm spiral is NGC 628, shown in Figure 3. Its two-arm structure is very broad, and it has three distinct arms in the S_3 image. Letting the OLR be at the end of the m=2 component places corotation at the end of the bright star-forming ridge along the main arms, and the inner 4:1 resonance at the symmetric inner spurs. The m=3 component ranges from the inner to the outer 3:1 resonances.

We find that the inner m=2 component for most spiral galaxies does not extend to the ILR but generally ends closer to the 4:1 resonance, while the inner m=3 component usually extends exactly to the inner 3:1 resonance. This result suggests that the m=2 component is a wave mode, with a reflection at an inner high Q barrier or bar that is outside the absorbing ILR, whereas the m=3 component may be a driven wave rather than a mode. Possibly the two-arm mode is driving the three-arm wave.

In NGC 5194 and NGC 3031, which both have a well-developed m=2 structure, the m=3 structure is much weaker than in the other galaxies in this sample. NGC 5194 shows an inner triangle for the m=3 component, without substantial m=3 arms. NGC 3031 shows a circular m=3 component, also without strong arms. These are the only two galaxies in our sample which are likely to be strongly interacting with another galaxy, and this fact may shed further light on the way in which the m=3 structure develops: if the m=2 structure is young in these galaxies, then the m=3 driven structure may not have had time to develop.

3 SUMMARY

The results of our work (given in more detail in Elmegreen, Elmegreen and Montenegro 1991), may be summarized as follows: we have identified 5 distinct optical features that occur at key resonance locations. At the OLR, the main symmetric m=2 structure ends abruptly. Usually this radius is near R_{25}. At the outer 3:1 resonance, the m=3 structure, when present, ends. At the corotation resonance, there is often a ridge of star formation in the main inner arms that terminates. Sometimes there is a circle of interarm star formation (as in NGC 4321), and at corotation strong dust lanes become noticeably weaker (sometimes switching from the concave to the convex sides of the main arms). At the inner 4:1 resonance, sometimes there are spurs located midway between the main spiral arms. Occasionally, there is m=4 structure in the form of four equally spaced spiral arms. At the inner 3:1 resonance, the m=3 structure generally begins.

The m=2 structure does not go to its ILR, whereas the m=3 structure generally does. This result suggests that the m=2 structure reflects in the inner regions and so is a wave mode (as evidenced also by gaps from interference patterns of inward and outward-moving waves) rather than just a wave. The m=3 termination at the inner 3:1 resonance suggests, on the other hand, that it is a driven wave and not a mode. Since the m=3 structure extends from the inner and outer 3:1 resonances determined from the location of the m=2 structure, the two components have the same pattern speed. We suggest that the m=3 waves are driven by asymmetries in the the m=2 spiral pattern (i.e., the m=1 component from the asymmetry interacts with the m=2 component to produce the m=3 component).

For interacting galaxies with strong m=2 but weak m=3, as in NGC 3031 and NGC 5194, we suggest that the m=2 spirals were recently triggered by companions, and have not had time to develop prominent driven three-arm structure. After sufficient time, we expect that the m=3 component will grow in these galaxies and then they will appear multiple-armed because of the superposition of the two spirals.

D.M.E. is grateful for travel grants from the American Astronomical Society and Vassar College.

REFERENCES

Artymowicz, P., and Lubow, S. 1991, preprint.

Elmegreen, B.G. and Elmegreen, D.M. 1990, Astrophysical Journal, 355, 52.

Elmegreen, B.G., Elmegreen, D.M. and Montenegro, L., 1991, Astrophysical Journal, in press.

Elmegreen, B.G., Elmegreen, D.M. and Seiden, P.S., 1989, Astrophysical Journal, 343, 602.

Persic, M. and Salucci, P. 1991, Astrophysical Journal, 368, L60.

Visser, H.C.D. 1980, Astronomy and Astrophysics, 88, 149.

FIGURE 1: NGC 613 - These photos show NGC 613. The top left photo is the enhanced composite of the B and I images, deprojected and with the azimuthally averaged radial profile removed. The middle left photo is the S_2 image, and the bottom left photo is the S_2 image with the OLR, corotation and inner 4:1 resonances indicated by circles. The top right photo is the S_3 image, the middle right is the A_2 image, and the bottom right photo is the S_3 image with circles indicating outer and inner 3:1 resonances based on the same pattern speed as the S_2 image.

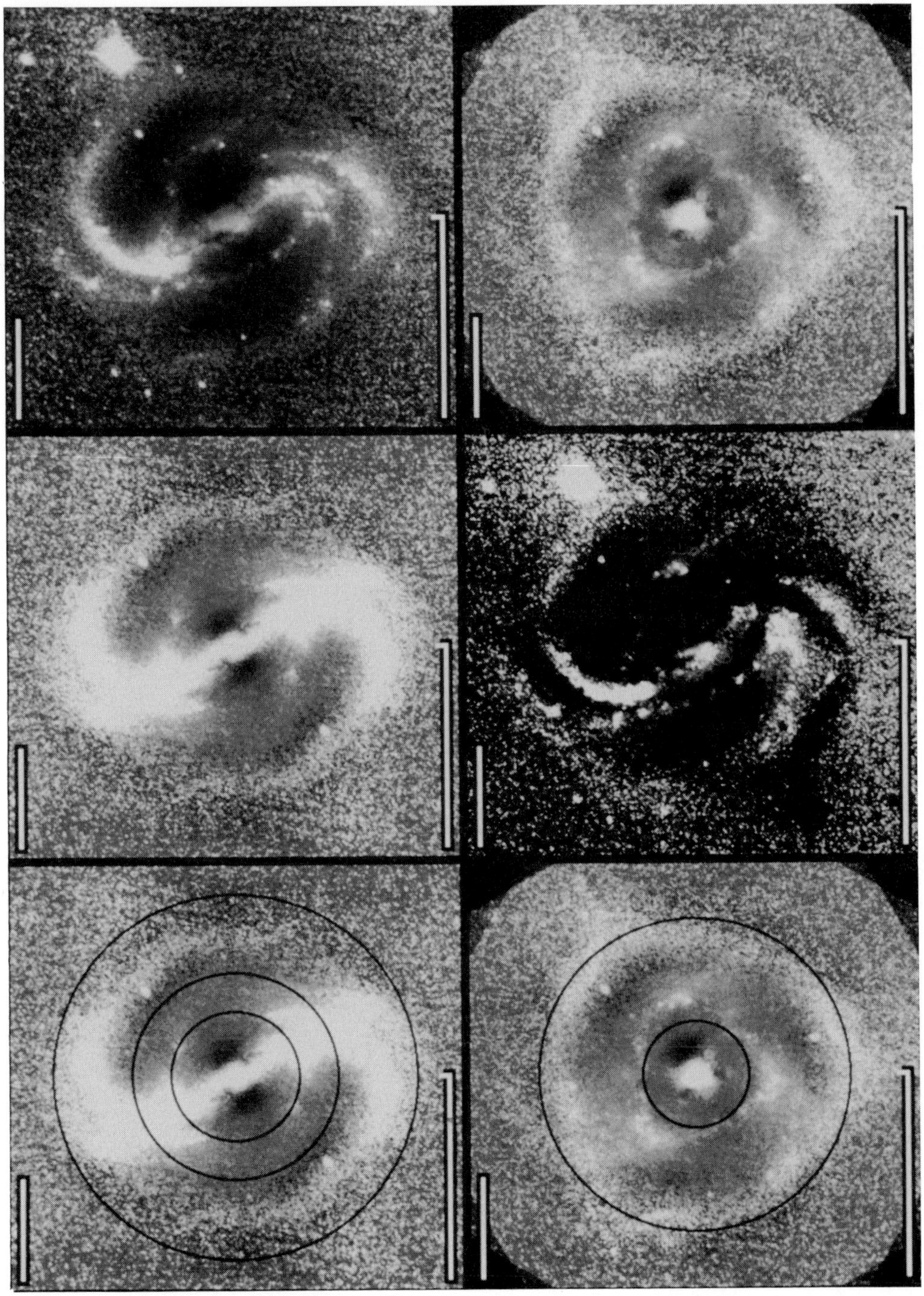

FIGURE 1.

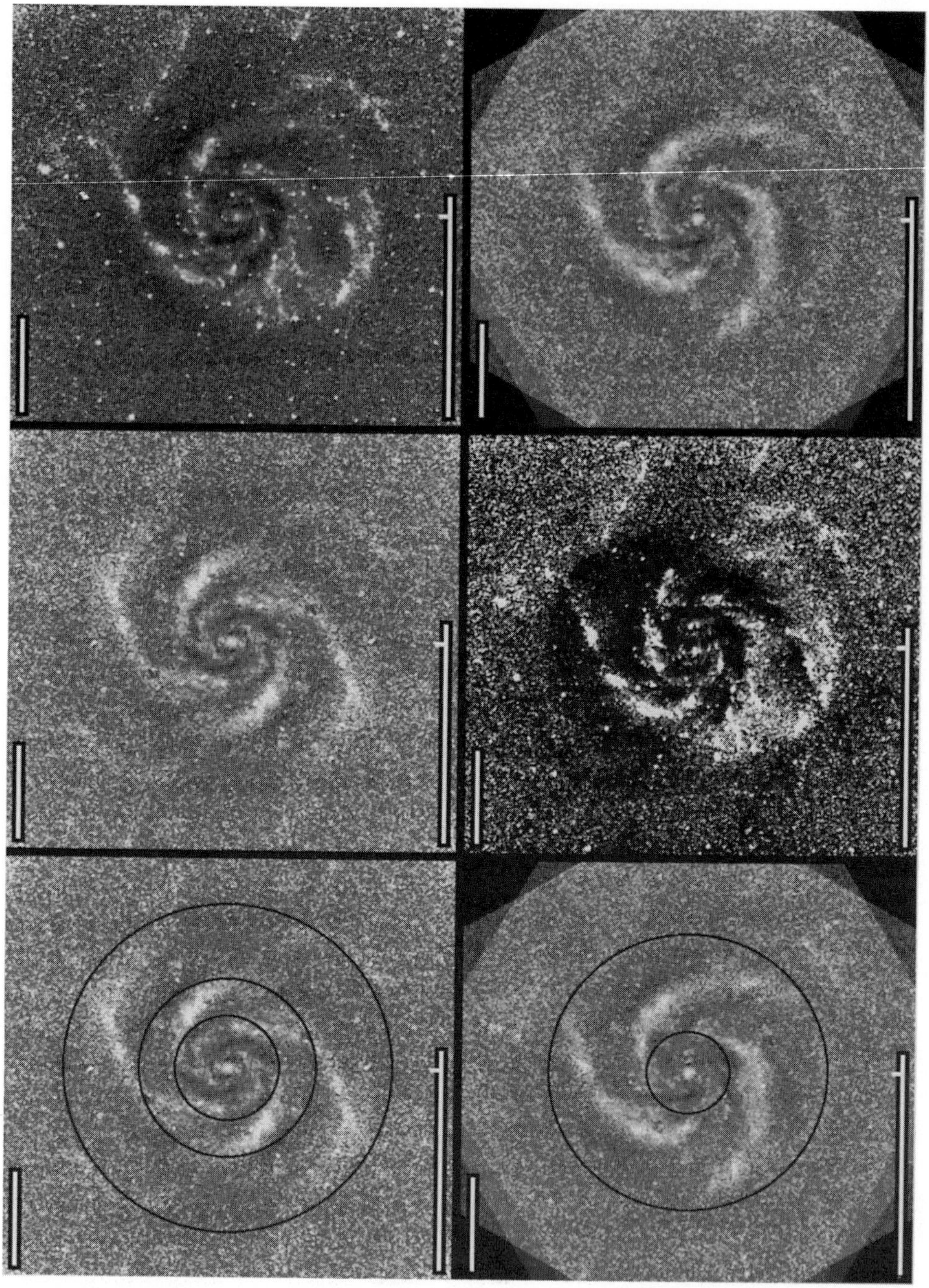

FIGURE 2: NGC 5457 These photos show the blue images of NGC 5457, with the same convention as for Figure 1.

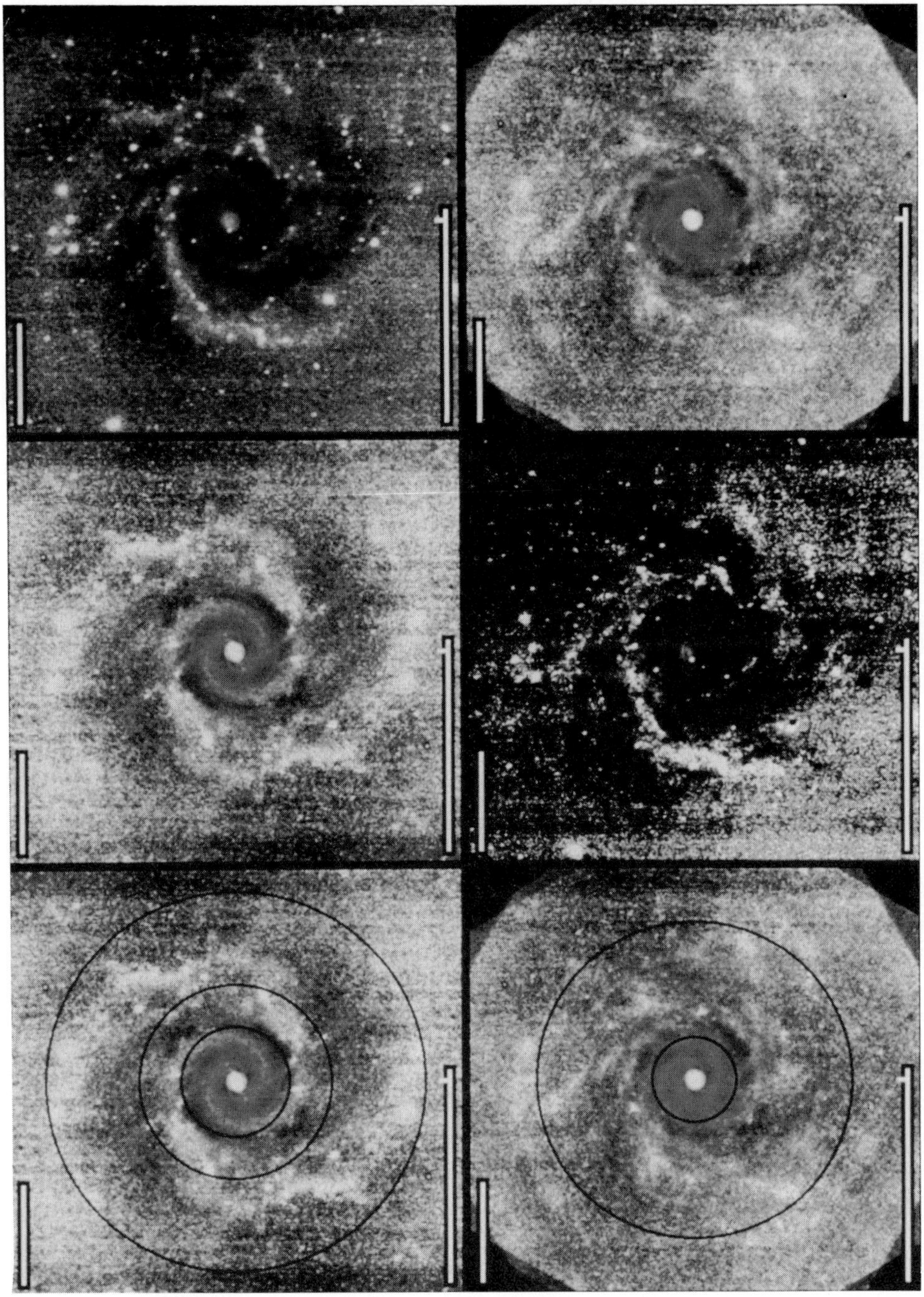

FIGURE 3: NGC 628 These photos are of NGC 628, with the same convention as in the previous figures.

Formation of the Inner Ring in the Grand Design Galaxies

CHI YUAN

Department of Physics, City College of the University of New York, New York, NY 10031 and National Tsing Hua University, Hsin Chu, Taiwan, ROC

ABSTRACT Ring structures are often found in the grand design galaxies, forming the inner boundary to the spiral region. The prototype of these is NGC 5364. These rings, in the past, have often been referred as the inner Lindblad resonance without further elaboration. In this report, we examine the properties of wave propagation near the resonance and interpret the ring structure as a star formation region which is occupies by a pair of tightly wound nonlinear density waves of the interstellar gas, resonantly excited by the exterior grand design spirals.

1 INTRODUCTION

Ring structures are often observed in spiral galaxies, with or without a bar. A large class of them are also seen in early-type S0 galaxies (Buta 1990). In this report, we do not intend to address the problem of ringed galaxies in general, but rather focus on the origin of rings for a small class of galaxies. These galaxies all have well defined two-arm spirals and usually referred in the literature as the grand design galaxies. Their rings, if any, form the inner boundary to the spiral region. The most famous example of these is NGC 5364. Other examples are NGC 3344, NGC 4725. Less well known is M 81, which, according to the recent observations (Kaufman et al 1989; Kaufman 1991), also has a ring of HI and UV excess at 2000Å from which the stellar arms and gas arms emerge.

Often the formation of such a ring is attributed to the inner Lindblad resonance (ILR) without further explanation (e.g. Rots 1975). Judging from the fact that the Lindblad resonances play a role of absorbing incoming waves (Mark 1971), it is rather difficult to imagine that they can simutaneously excite the region to produce a ring. Furthermore, because of the existence of the central bulge, short trailing waves are usually reflected before they even reach the Lindblad resonances (Lin and Lau 1979). This makes the idea to identify the ILR as a ring more difficult to accept. In this report, we shall propose a mechanism, based on the density wave theory, to explain the formation of rings in the grand design galaxies. The idea and its physical implications are described in the next following sections.

2 RESONANCE EXCITATION

The disk galaxy under consideration consists of both the stellar component and the gaseous component. The grand design spirals are mainly made of stars. The gas in the same region is essentially driven by the density waves of the stars. According to the linear density waves, short trailing density waves may be viewed as to originate from the corotation (CR) and propagate inward toward the ILR. Because the existence of a central bulge of stars, the disk mass of the stellar component diminishes quickly toward the center. This leads to a drastic increase of the stability parameter, Q, and hence causes the Q-barrier of the stars (QBS) to position itself in front of the IRL. Thus, before the short trailing waves have a chance to reach the ILR, they are reflected as long trailing waves and propagate outward toward the corotation again to complete a cycle (e.g. Lin and Lau 1979). Therefore, the grand design spirals, in principle, can never go beyond the QBS, and reach the ILR. The gravitational force associated with these waves, however, can act on the ILR from a distance, as an external periodic perturber which resonantly excites the gas there to form a ring-like structure.

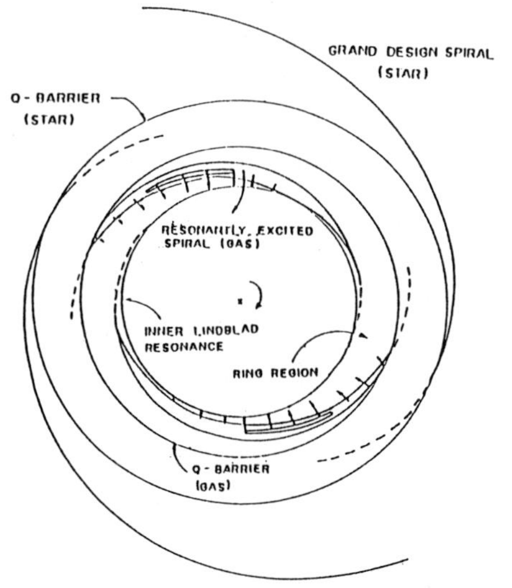

Figure 1 Schematical drawing of resonance excitation of spiral waves

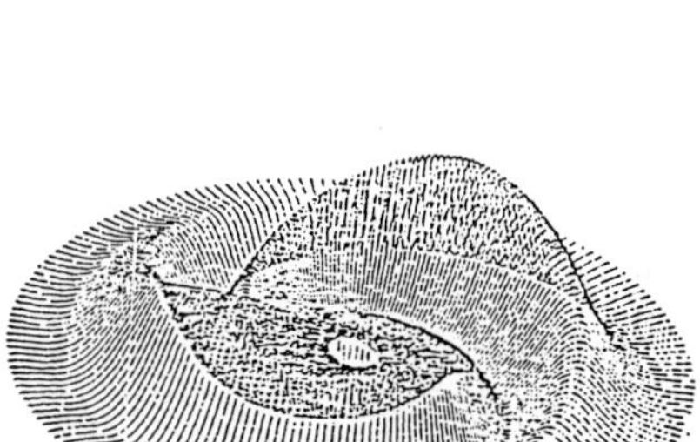

Figure 2 Surface density of a pair of resonantly excited waves in the gas

The resonance excitation of density waves by an periodic perturber is a subject receiving considerable attention lately. The theory has been successfully applied to the waves in the Saturn's rings (Goldriech and Tremaine 1979; Shu, Yuan and Lissauer 1985). It is also applied to the gaseous component of the Milky Way to explain the 3-kpc Arm phenomenon (Yuan 1984). According to the theory, the gas particles at the ILR in response to a periodic forcing will be organized collectively into long trailing waves which propagate outward. These waves, however, cannot go very far before they are turned back at the Q-barrier of the gas (QBG) as short trailing waves, propagating inward. They too cannot go very far before they are attenuated by the viscosity of the interstellar gas. The relative locations of the QBS, the QBG and the ILR are shown in Figure 1.

The short trailing waves, also shown in Figure 1, can easily develope into waves of large amplitude, even within the short distance before their total attenuation. The large-amplitude waves are likely evolving into shock waves if the gas is polytropic (so far we only treat the isothermal gas). Therefore, we expect that star formation takes place there as the interstellar clouds sweep through the high compression regions or shock formation regions. Unlike the exterior grand design regions, the angular speed of the interstellar clouds is large relative to the pattern speed of the spiral waves (same as that outside). The newly formed stars would have enough time to move out of the spiral arms and populate the entire annular region occupied by the spiral waves of the gas. This annular region is expected to be narrow, since the viscosity is very effective in damping the waves and limit the short trailing waves to a short distance inward. Thus, we expect to observe a ring structure of optical prominence in some of the grand design galaxies.

An numerical example of a pair of short trailing waves resonantly excited by a doubly periodic forcing is shown in Figure 2 to simulate the ring formation in a grand design galaxy. Depicted is the surface density of the waves under a moderate damping for an isothermal gas. The perturbation field at the ILR is 5% of the mean field there. Details of the model and calculations will be published elsewhere (Yuan 1991).

3 THE LOCATION OF THE INNER LINDBLAD RESONSCE

Direct observations indicate that the rings are physically connected to the exterior grand design spirals. In fact, they form the inner boundary to the spiral region. But, the ILR is not necessarily located inside the ring. The situation is somewhat complicated. First, we note that ILR, QBG and QBS always come in the following order, counting radially outward. The grand design spirals are density waves mainly of stars. Thus, they cannot go beyond the QBS, although some residuals of the waves may leak inside of it. If the QBG is close to the QBS , then the ring which is supposed to lie immediately inside the QBG will appear physically connected to the grand design spirals. This, we believe, is the case for NGC 5364. As for the ILR, it may lie next to the QBG inside the ring, or it may lie further inward outside the ring. That does not affect our results.

Similar situation also holds for M 81. Yet the location of the ILR is an issue of controversy (Kaufman et al 1989 and Visser 1980) . If we interpret the ring of HI and UV excess to be the manifestation of nonlinear short waves reflected at the QBG, then like the previous case we have a situation that QBS and QBG are close to each other. The ILR, therefore, may be located either at 2.5 kpc (Visser) or at 4 kpc (Kaufman et al). As long as it lies inside the QBG, the result does not change. The points we want to stress, however, are that the ring need not coincide with the ILR and the grand design spirals should emerge from the QBS, not from the ILR.

On the other hand, the QBG may lie closely to the ILR and both may lie at a distance sufficiently away from the QBS. Then, we would expect to see an inner ring detached from the grand design region. This, however, is not obvious in the observation. There are several possible explanations: (1) Such a situation rarely happens in nature. The QBG is always situated close to the QBS. (2) The strength of the perturbing field is too weak at such a distance to excite nonlinear density waves. (3) There may not be enough gas in the disk to participate in the waves. The last point may be the explanation why many of the grand design galaxies do not have a ring, since the gas component often tends to diminish quickly toward the central region for those galaxies.

4 CONCLUDING REMARKS

In this report, we try to clarify the different roles played by the ILR, the QBG and the QBS in relation to the formation of ring structure in the grand design galaxies. We suggest that the ring may be formed out of a pair of tightly wound nonlinear spiral density waves, which are resonantly excited in the gaseous disk by the gravitational field associated with the grand design spirals.

We have pointed out that the location of the ring depends sensitively on the location of the QBG, but is rather independent of the location of the ILR. Since the Q value for the gas varies appreciably with the basic model of the galaxies, many situations may happen. The ring may appear physically connected to the grand design spiral region, as in NGC 5364. It may be a detached inner ring. It may not exist at all if there is not enough gas in the inner disk. It seems important to look into the basic models of the grand design galaxies including those of the gas and its correlation with or without the rings for the grand design galaxies. This will help to establish statistics to check our approach.

ACKNOWLEDGEMENTS

I wish to thank Professor C. C. Lin and Dr. Bruce Elmegreen for useful discussions. The work is supported by the National Research Council of R.O.C. under Grand NSC80-0208-m007-02.

REFERENCES

Buta, R. (1986), *Ap.J.*, **351**, 62.
Goldreich, P. and Tremaine, S. (1979), *Ap.J.*, **233**, 857.
Kaufman, M. (1990), private communication.
Kaufman, M., Bash, F. N., Hine, B., Rots, A. H., Elmegreen, D. M. and Hodge, P. ...W. (1989), *Ap.J.*, **345**, 674.
Lin, C. C. and Lau, Y. Y. (1979), *Studies Appl. Math.* **60**, 97.
Mark, J. W.-K. (1971), *Proc. Nat. Acad. Sci.*, **68**, 2095.
Shu, F. H., Yuan, C. and Lissauer, J. J. (1985), *Ap.J.*, **291**, 356.
Rots, A. H. (1975), *Astron. Astrophy.*, **45**, 43.
Visser, H. C. D. (1980), *Ap.J.*, **88**, 149.
Yuan, C. (1984), *Ap.J.*, **281**, 600.
Yuan, C. (1991), in preparation.

GAS CLOUD DYNAMICS IN SPIRAL GALAXIES

F. Combes[(1)], S. Garcia-Burillo[(2)], M. Gerin[(1)]
[(1)]Dep. Radioastronomie Millimétrique, DEMIRM
Observatoire de Meudon, F-92 190 Meudon, FRANCE
and Ecole Normale Supérieure, 24 Rue L'homond, F-75 005 Paris
[(2)] IRAM, 300 Rue de la Piscine, F-38 406 St-Martin d'Hères

The role of large-scale dynamics is fundamental in the gas behaviour in a spiral galaxy. Star formation is observed to be triggered in spiral arms, and it has long been proposed that density waves can alter the star formation rate significantly, by gas compression and shock waves (Roberts 1969). But may be this factor is not the most important (Elmegreen & Elmegreen 1986): star formation is not globally enhanced in grand-design galaxies. What might be of larger impact on gas behaviour are the gravitational torques exerted by a non-axisymmetric potential on the gaseous spiral (Combes 1988). These torques can drive large radial transports of gas in short time-scales, and can re-structure the galaxy entirely. These torques might be responsible for starbursts, in gathering large amounts of gas in a small area, generally towards the center.

When there exists a stationnary density wave, like in a barred galaxy, the pattern speed of the perturbation is well defined, as are the resonances between the gas motions and the pattern speed. The Lindblad resonances then determine the limits of action of the gravitational torques: outside corotation, the gas is driven outwards, towards the OLR, and inside corotation, inwards to the ILR (Schwarz 1984, Combes & Gerin 1985). In this manner, gaseous rings can be formed, in time-scales depending on the perturbation strength. These structures might be formed more transiently, in spiral galaxies without a stationnary pattern.

To better understand gas streaming in spiral galaxies, we therefore must know where are the principal resonances, and get more insight in the nature of gas orbits, to interprete the dynamical perturbations in the observed velocity field. To answer these questions, we have undertaken simulations of the gas streaming in known potentials, deduced from red images of the studied galaxy. The simulations are non self-gravitating, but the gravitational potentials are as close as possible from the actual ones.

I- Gas dynamics in M51

To illustrate the method, let us describe the case of Messier 51, where a wealth of observational data are available. We are concerned here only with the dynamics of the molecular component, which is confined in the inner 2 or 3 arcmin, corresponding to the inner coherent two-armed spiral structure. The gaseous disk extends much farther, and in particular a new HI arm has been discovered extending beyond the companion (Rots et al 1989). However, there seems to be a discontinuity in the inner spiral pattern, and the external arms.

Adopted potential:

A near-infrared image of the galaxy is the best tracer of the total mass: it traces the old stellar population, without perturbing dust features. We used a CCD 2μ image from Gatley (priv. comm.) for the whole central region. However, farther from the center, the signal-to-noise was not enough, and we used a red plate, digitized with the MAMA automatic machine in Paris. The resulting image is displayed in fig. 1, where the essential feature is the central bar, already discovered by Pierce (1986), Zaritsky & Lo (1986).

From this de-projected surface brightness, we can deduce the surface-density, assuming an M/L ratio at each radius. To determine this M/L ratio, we first computed the potential, by an FFT method, and deduced the rotation curve which is then fitted to the observed one. The required M/L ratio was not varied more than a factor 2 for all the radial range.

Since we don't know the z-distribution of the mass, we assumed a constant-thickness stellar disk (1kpc), to extend the potential in the third dimension. Since the molecular gas layer is very thin (50-100pc), we can assume that the x-y dependences of the forces have a cylindrical geometry.

Numerical method

The ensemble of molecular clouds were simulated as ballistic particles suffering collisions, but not as a fluid since the equilibrium time by collisions is longer than the dynamical time, to cross spiral arms for instance (Casoli & Combes 1982). They were distributed according to a mass spectrum, between 10^3 to 10^6 M_O. We call GMC clouds more massive then 3 10^5 M_O. Small clouds grow by collisions, and when they reach the GMC's phase, they have a finite life-time with rtespect to dispersion by star-formation, of 4 10^7yrs. Then all the GMC's mass is re-injected into the interstellar medium, in the form of small clouds.

Figure 1: Deprojected surface brightness adopted for M51, used to obtain the gravitational potential.

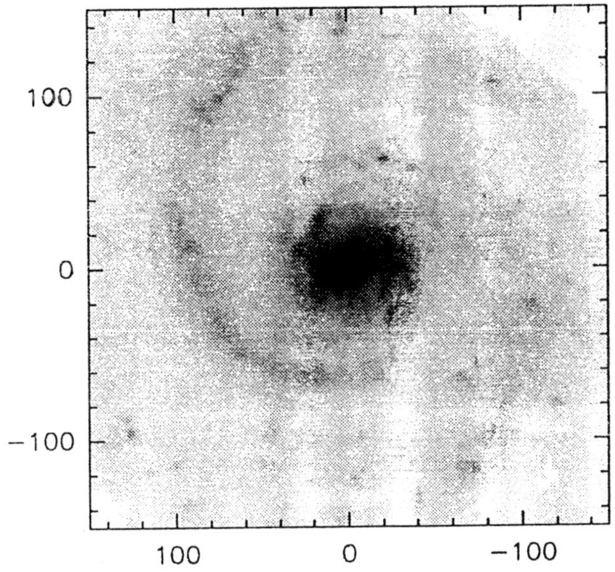

Figure 2: Results of 4 simulations carried out with different pattern speed Ω_p; a) 40km/s/kpc; b) 20km/s/kpc; c) 10km/s/kpc; d) 27km/s/kpc.

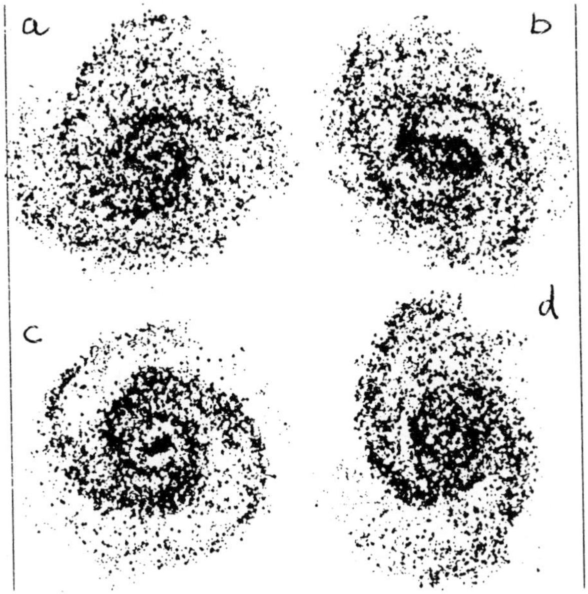

Energy is also re-injected into the medium, since these small clouds are launched with a velocity dispersion of 10km/s. This energy compensates for that dissipated in collisions. This contributes to cancel out the effect of viscosity.

The collision grid (resolution of the simulations) has a cell of 130pc. The number of clouds simulated varies between 2 to 4 10^4Mo. The gas mass simulated was 10^9 M_o, although the total molecular mass detected in M51 is 10 times higher.

Clouds are launched in the axisymmetric part of the potential, with a circular velocity. The non-axisymmetric part is introduced progressively, and reaches its maximum value at the present time. The perturbing potential comes also in part from the tidal interaction of the companion. The orbit of NGC 5195 is computed beforehand with a 3-body code, taking into account the true M51 potential, but not including dynamical friction. In the assumed scenario, the tidal interaction perturbs essentially the outer parts of M51, and the perturbations are then swing-amplified to generate the coherent spiral pattern in the inner parts, that is in our model introduced with the same time-scale as the orbital period.

Fitting parameters

The main parameter used to fit the gaseous distribution observed in M51 is the pattern speed Ω_p of the perturbation. We assumed that there exists an average of the pattern speeds, constant in radius, even though it is likely that there exist a wave packet. Also, we assume only one wave system, which is justified by the smooth variation of the Ω-κ/2 curve. In early-type galaxies, with high central mass concentrations, there could exist two systems of waves, the pattern speed being higher in the center (Sellwood & Sparke 1988).

The gas clouds suffer only a few collisions in one revolution. Therefore, they can be considered as ballistic particles following particle orbits between collisions. For the gas to be trapped in the imposed perturbation, it is essential that the precession rate Ω-κ/2 of the particles is matched to Ω_p, at least in average. This is shown in fig. 2, where several trials are displayed with different pattern speeds Ω_p. The observed spiral structure, and inner morphology is only obtained for the fitted value Ω_p = 27km/s/kpc. We have also slightly varied the inner potential (M/L ratio), to better obtain the observed rotation curve a posteriori. Indeed, the observed velocities in the center are not circular due to the oval potential; the dynamical minor axis is not perpendicular to the major axis. The tilt is towards the west side, indicating that the gas runs in orbits perpendicular to the bar in the

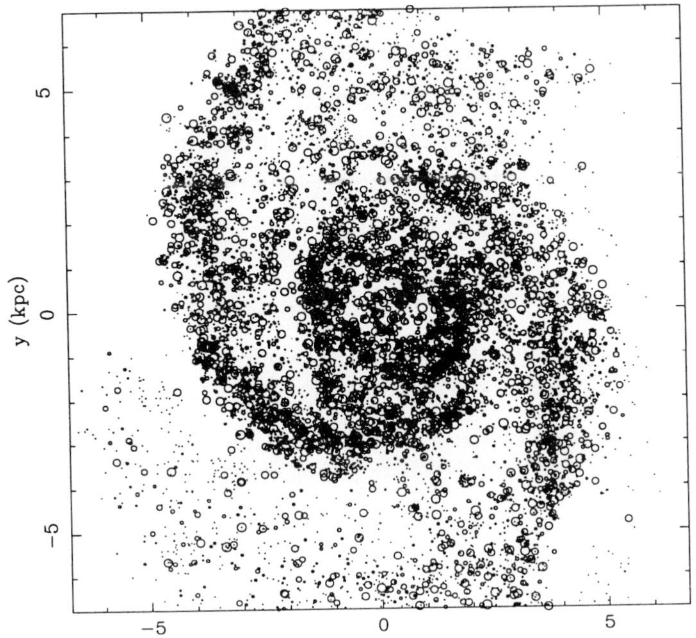

Figure 3: a) Particle plots of the best model; clouds are plotted with their real size. b) Velocities in the rotating frame; the arrow at the top right represents 100km/s.

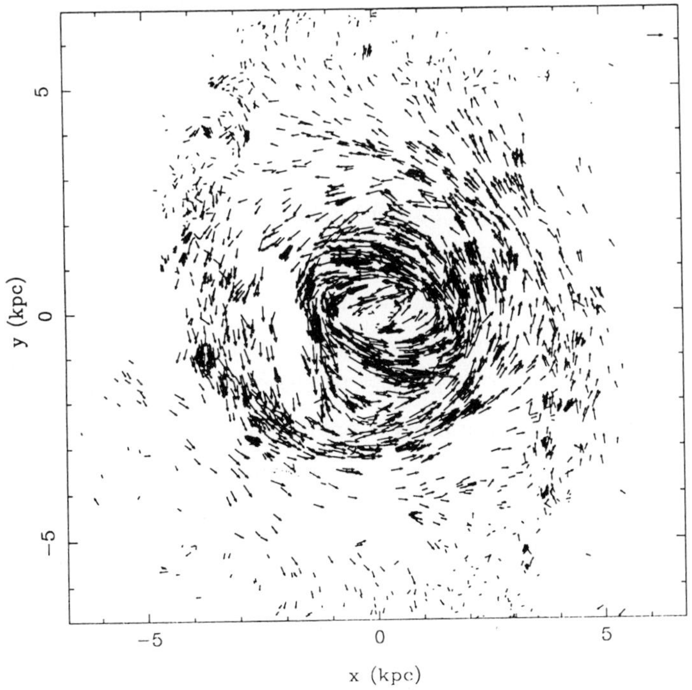

inner ring. We have also found that structure in the model.

The result for the inner morphology is that the gas is accumulating in a ring, corresponding to the beginning of the spiral pattern. This ring corresponds to the inner Lindblad resonance of the pattern. There exists orbits parallel to the bar in the center, but in the ring, the gas is more frequently in orbits perpendicular to the bar. The ring encircles the oval distortion (fig. 3).

Observation of the model

To better compare with the molecular component data from Garcia-Burillo & Guélin, obtained in CO(2-1) with the IRAM 30m telescope, we "observe" the model and built a data-cube from the convolution of the expected emission of all clouds by the telescope beam (12").

Each cloud is assumed to obey the relationships observed in the Milky Way (e.g. Solomon et al 1987): internal velocity dispersion $\sigma_V \propto size^{0.5}$, and size $\propto M^{0.5}$, implying the relationship between mass and internal velocity dispersion: σ_V(km/s) = 0.23 $M(M_o)^{0.25}$; for the mass range in our modeled clouds, σ_V varies between 1.3 to 7.4km/s. These relations imply that the H_2 column density is the same for all clouds, and every cloud contributed an emission proportional to its mass in the beam (proportional to its area). This is true as long as clouds **do not overlap** in the same line-of-sight at a given velocity, which is the usual assumption for extragalactic clouds: CO emission is then considered as a good tracer of the H_2 mass, although each cloud is individually optically thick.

On fig. 3 however, where we have plotted the clouds with their real sizes, it is obvious that blending of clouds occur, at least in the arms. We have therefore devised a model to take into account this overlapping, without computing an actual radiative transfer model, too sophisticated here. Within each velocity channel (of 3km/s wide), we take into account all material seen in front of a given cloud, and consider that it is absorbed proportionally to exp($-\tau$), with τ being proportional to the column density of the material. We take into account the area overlapped between the two clouds, according to their position and relative sizes. This procedure leads to some saturation of the spectra in the arms, and contributes to broaden the resulting spectra.

Comparison between the observed and modeled distribution

Fig. 4 displays the CO(2-1) map from Garcia-Burillo & Guélin (1991): the essential features are the inner ring, from where start the spiral structure; the isovelocity curves

Figure 4: Observed CO map (grey scale from Garcia-Burillo & Guélin 1991) with isovelocity contours superposed.

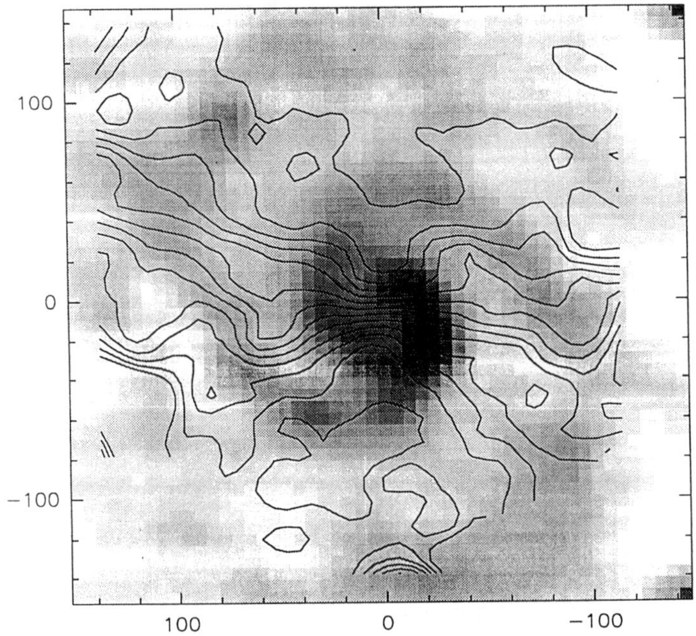

Figure 5: Same as fig. 4, but for the model.

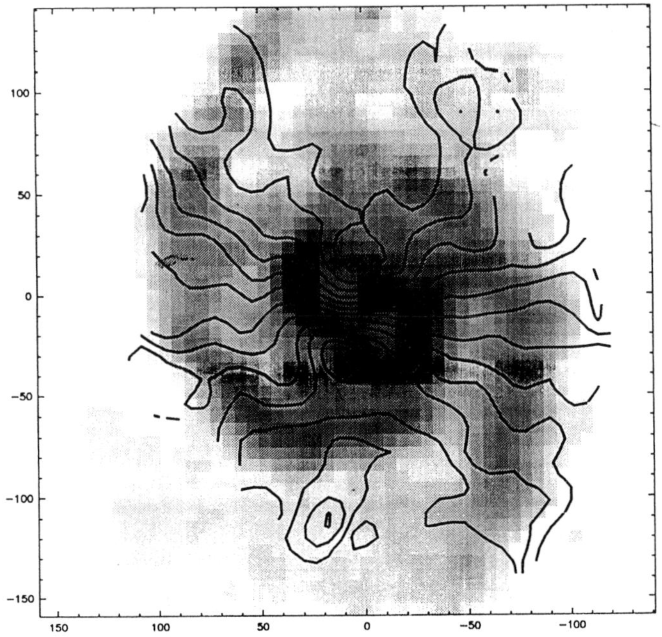

reveal strong streaming motions across the arms, up to 20km/s projected on the sky, i.e. up to 60km/s in the M51 plane.

Fig. 5 displays the corresponding figures for the model. The inner ring and spiral structure are reproduced correctly. The isovelocity curves reveal the same order of magnitude for the streaming motions. The arm-interarm contrast is also of the same order of magnitude as in the observations (fig. 6). The contrast is rather low in the center, because of a background of emission (the rotation period is of the same order of magnitude as the time to cross an arm). But in the outer arms, the emission goes to zero in the interarm, and the contrast is higher than 10.

The large linewidths of the spectra observed in M51 are the only feature that is hard to reproduce in the model. The FWHP widths are surprisingly large in the observations, they can reach 60km/s in the southern arm. Of course, the linewidth increases significantly towards the center, but this can be interpreted in terms of large velocity gradients in the beam, due to the rising rotation curve, and non-circular motions. In the disk however, and due to the small inclination of M51 (i=20°), the V-gradient is rather small and cannot explain the linewidths; even the streaming motions, which are resolved out, and anyway are in average of the order of 10km/s in projection. The linewidths show a pronounced maximum in the arms (fig. 7).

Broadening the linewidths

A possible solution to the problem is to take into account the **blending** of clouds in the arms. The ensemble of clouds is now "optically thick" globally, and spectra are broadened by saturation. With the order of magnitude of the number of clouds that are indeed observed (10^{10} M_o of H_2 is observed in M51), we have succeeded to double the line-spectra in the arms, from 20km/s to 40km/s to FWHP. The consequence is that CO emission is no longer a good tracer of the H_2 mass. In the arms, we can miss a significant fraction of the mass (factor 4), but globally over the galaxy, this is only a factor 2.

If this hypothesis is right, the surface filling factor within each velocity channel must be of the order of 1 for the envelopes of GMC. Since the main beam temperatures observed are about 0.8K, this means that the excitation temperatures are not above 4K. This is quite low, but compatible with the low CO(2-1)/CO(1-0) line ratio, in average 0.5, suggesting that the CO(2-1) line is subthermally excited. The ^{13}CO line is also in agreement with this hypothesis, since the ^{13}CO linewidths are observed narrower than the ^{12}CO ones.

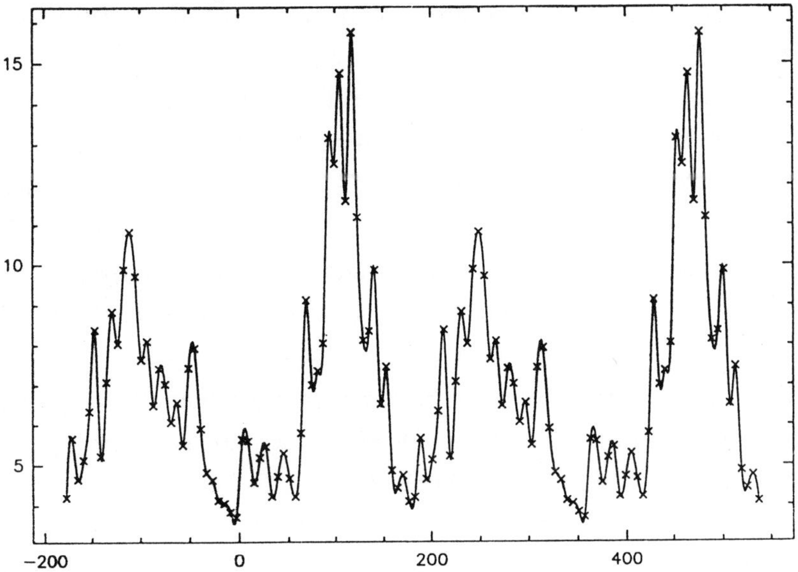

Figure 6: Azimutal profiles of the intensity, displaying the arm-interarm contrast for a) the observations, b) the model.

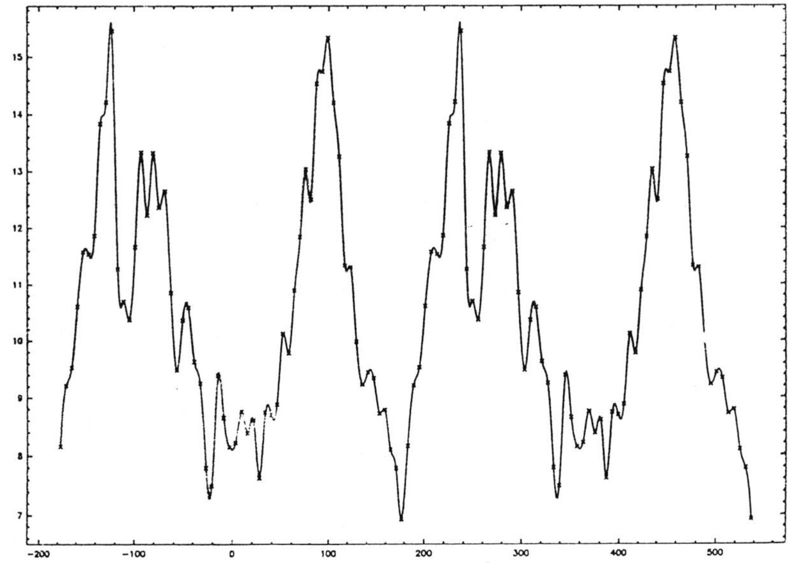

Another way out is to invoke large-scale turbulence. If spectra are broadened by such a large amount of turbulence, cloud blending is somewhat avoided in the same time. Small-scale turbulence, generated by stellar activity (supernovae, stellar winds, HII regions, etc...) can contribute at most with 10km/s, and is not sufficient here.

The large-scale turbulence needed can only come from the rotation itself, through the shear due to *differential rotation* (Fleck 1981). The order of magnitude of the predicted turbulence is somewhat low, i.e. 2A = 20km/s to 50km/s maximum, where A is the Oort constant= $1/2\ [\Omega - d(\Omega r)/dr]$. The maximum velocity dispersion measured is 25km/s, and since it is a 1D measurement, we deduce $\sigma = 25\sqrt{3} = 43$km/s.

Part of the dispersion could also come from gravitational instabilities; in the center, the gas may have reached self-gravitation in the thin layer where it is confined. If the dispersion is lower than the critical one to be stable (Toomre criterion, 1964), then instabilities will heat the gas. The critical radial dispersion can be computed from the observed H_2 surface density μ(gas), and from the epicyclic frequency κ given by the rotation curve: $\sigma(\text{crit}) = 3.36\ G\ \mu/\kappa$. This value is always lower than 10km/s, all over the inner disk. Since what we observe is essentially a z-velocity dispersion, since the galaxy is almost face-on, one must compute in fact the critical dispersion to avoid bending instabilities. This has been computed by Araki (1985) and found to be only 0.3 σ(crit) in the plane. This is not sufficient to account for the large observed linewidths. It is to be noted however that the potential in the z-direction is not known, and has been assumed in the model by analogy with galaxies seen edge-on; a thick disk in M51 would help to solve the problem.

If we assume that large-scale turbulence of the amount of V_d = 43km/s is present at a scale d=500pc (the beam-size), and if we assumed a developped turbulence, then we can estimate the decay-time by energy flow towards the lower scales. The dissipation rate per unit mass is of the order of $\varepsilon = V_d^3/d$ (Landau & Lifshitz 1954). The energy per unit mass is $V_{rot}^2/2 = (200\text{km/s})^2/2$, and the decay time-scale $\tau = d\ V_{rot}^2/2V_d^3 = 10^8$ yrs. Let us recall that in our own galaxy, where V_d = 10km/s, $\tau = 10^{10}$yrs, and there is no problem to maintain the gaseous disk over a Hubble time. In M51, the large-scale turbulence would drive all the gaseous disk towards the center in the time-scale of a revolution period.

II- Gas dynamics in NGC 4736

We have applied the same method to investigate the gas dynamics in the RSab galaxy NGC 4736. This galaxy is quite peculiar, revealing an inner ring at a radius of 50" = 1.5kpc, and an outer ring at 7.8 kpc, but in between no very clear and coherent spiral structure. The

Figure 7: Contours of the velocity dispersion of the observed CO map, superposed on the intensity map (grey scale).

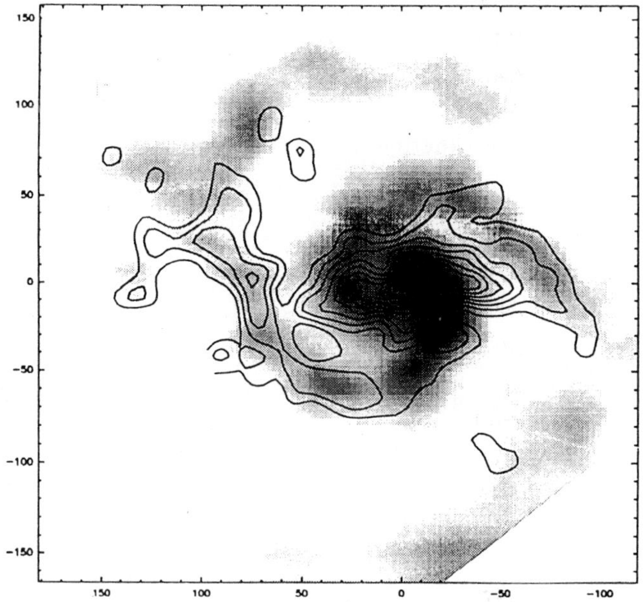

Figure 8: Intermediate picture of the simulations for NGC 4736 (T=0.4 Gyr), showing the contrasted inner ring; the outer ring is not yet formed.

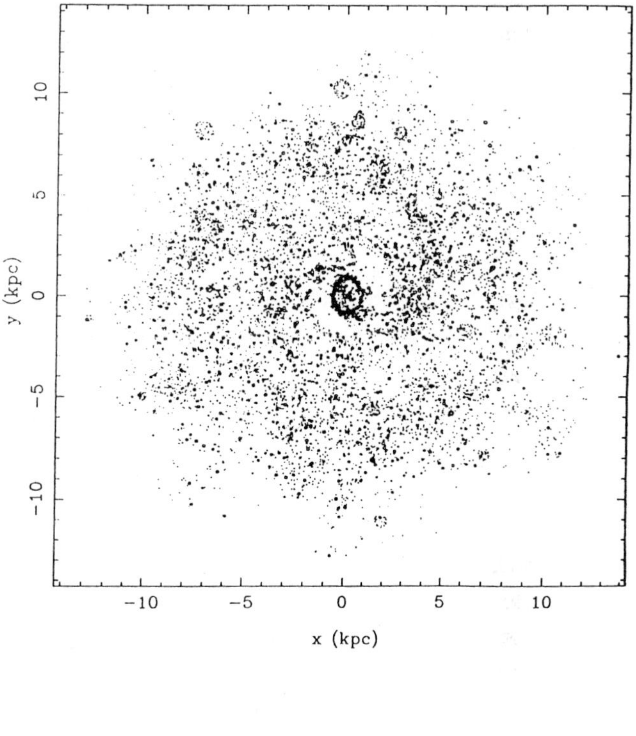

inner and outer rings are traced by HI gas (Bosma et al 1977), and molecular gas is prominent in the center and inner ring (Gerin et al 1990). The galaxy is not strongly barred, but possesses an oval distortion. The presence of two rings makes of this galaxy a good test for theoretical models of ring formation. If the outer ring corresponds to the outer Lindblad resonance of the oval perturbation, a precise knowledge of the rotation curve then yields the angular velocity of the pattern; one can then check whether there exists and inner Lindblad resonance, and whether it corresponds to the Ha-HI-CO inner ring.

In fact, there seems to exist two ILRs in the galaxy, since the pattern speed Ω_p is quite low, and the mass highly concentrated towards the center. To obtain the stellar surface density and derive the potential, we used the luminosity profile of Kent (1987), assuming that the galaxy is axisymmetric for radii $R<500$pc, and 5.5kpc $<R<13$kpc, and oval with an axis ratio of 0.8 in between.

The time-scale for the formation of the inner ring by the gravitational torques due to the oval perturbation is very short: the ring can be seen as soon as 350 Myr (2 periods of rotation of the oval). The formation time-scale for the outer ring is relatively longer (2Gyr = 13 periods) and corresponds to the dynamical time at these radii. The inner ring is quite thin, while the outer ring is much wider. This could be explained since there is no gravitational torques confining the outer ring in its outer boundary. Also we do not take into account self-gravity (fig. 8).

Conclusion

From gas simulations and fit to the observed distribution, we have found that the main optical and molecular spiral structure of M51 develops between the inner Lindblad resonance and corotation. The inner oval distortion ends up just at ILR. Inside the molecular "ring", the gas orbits are parallel to the oval distortion, but they rotate as soon as the ring begins. Streaming motions are reproduced in the gas with the right order of magnitude. The very large observed linewidths cannot be reproduced with the model, unless blending of clouds is taken into account and CO spectra are broadened by opacity effects. If the linewidths are due to large-scale turbulence, a very short life-time ($\approx 10^8$ yrs) of the gaseous disk can be predicted.

The method has been used to check ring formation in the galaxy NGC 4736. The simulations show that even a very weak oval perturbation is able to form these rings by its gravitational torques on the gas. However, the formation time-scales for the inner and outer rings are very different, and the inner ring is much more contrasted.

We have carried out gas simulations in other galaxies, and in particular in a strongly barred galaxy NGC 7479. In this case, the pattern speed was compatible with the corotation at the end of the bar, as is suggested from the orbit theory (Contopoulos & Papayanopoulos 1980). Is there a qualitative or a quantitative difference between the inner oval distortions frequently observed in grand-design spiral galaxies, and the strong bar components? It is possible that the former have not enough strength in their potential to trap orbits as far as the corotation; alternatively, oval distortions could form with a higher speed than that of the spiral, and would slow down in the course of trapping orbits, and gaining strength. Gas simulations suggest that the two kinds of bar perturbations differ qualitatively at least concerning their role in the gas orbits and resonances.

References

Araki S. (1985) PhD thesis, Massachussetts Institute of Technology
Bosma A., van der Huslt J.M., Sullivan W.T. (1977) A&A **57**, 373
Casoli F., Combes F. (1982) A&A **110**, 287
Combes F. (1988) in "Galactic and Extragalactic Star Formation", ed. R. Pudritz & M. Fich, p. 475
Combes F., Gerin M. (1985) A&A **150**, 327
Contopoulos G., Papayanopoulos T. (1980) A&A **92**, 33
Elmegreen B.G., Elmegreen D.M. (1986) Ap.J. **311**, 554
Fleck R.C. (1981) Ap.J. **246**, L151
Garcia-Burillo S., Guélin M. (1991) in "Dynamics of Galaxies and their Molecular Cloud Distributions", ed. F. Combes & F. Casoli, Kluwer, IAU 146, p. 67
Gerin M., Casoli F., Combes F. (1991) A&A in press
Kent S.M. (1987) A.J. **93**, 816
Landau L.D., Lifshitz E.M. (1954) Hydrodynamica, ed. Mir, Moscow
Pierce M.J. (1986)A.J. **92**, 285
Roberts W.W. (1969) Ap.J. **158**, 123
Rots A.H., Bosma A., van der Hulst J.M., Athanassoula E., Crane P.C. (1990) A.J. **100**, 387
Schwarz M.P. (1984) M.N.R.A.S. **209**, 93
Solomon P.M., Rivolo A.R., Barrett J.W., Yahil A. (1987) Ap.J. **319**, 730
Sellwood J.A., Sparke L.S. (1988) M.N.R.A.S. **231**, 25P
Toomre A. (1964) Ap.J. **139**, 1217
Zaritsky D., Lo K.Y. (1986) Ap.J. **303**, 66

Multi–Phase Models for the Evolution of Galaxies

F. FERRINI

Dipartimento di Fisica, Università di Pisa, piazza Torricelli 2, 56100 Pisa, Italy

1 INTRODUCTION

The luminous history of galaxies is determined by the number of stars born as function of the mass at any instant of time, that is by the combination of the Star Formation Rate with the Initial Mass Function. In turn, the knowledge of these quantities allows to determine the stellar population history and consequently the Supernovae rates, the nucleosynthetic enrichment and then the abundances of elements. The conversion of gas into stars, which proceeds, as established both by observations and theory, via the molecular clouds phase, is the fundamental process in the life–cycle of galaxies. The restitution of material from stars to gas is important as well: it changes the initial conditions of the new born stars, and influences the processes of clouds and stars formation. Self–regulation and propagation in the star formation cannot be disregarded. Furthermore, galaxies are not isolated systems: Supernovae events and winds, as well as tidal interactions and stripping through ram pressure from intracluster gas may provoque a substantial gas removal.

A statistical scheme particularly suitable for the analysis of the effect of the various physical processes in the galactic evolution is the multi–phase approach studied in the last 10 years by many authors. I shortly discuss rules and variations, benefits and shortcomings of this game within the manifold of elliptical and spiral galaxies.

In Section 2, I review equilibrium and evolutionary models for the interstellar medium (ISM), presenting the aims, the physical processes considered by the authors and the consequent mathematical formulation. The main purpose of my paper is to analyze the structure more than discuss results and compare them to observations: the comparison can more profitably be found in the original literature. I will hence present the works putting emphasys on only a few aspects, relevant to my goal, with the aim to derive a few fixed points, on which different approaches and studies agree, notwithstanding the indipendent strategies.

In Section 3, I introduce a self–consistent model for the galactic evolution, in its different formulations, aimed to describe spirals and the origin of the Hubble sequence or ellipticals and in general non–isolated systems.

2 MULTI–PHASE MODELS FOR THE ISM

2.1 Equilibrium Models

The principle stated by Spitzer (1968) that the ISM phases must be kept in rough pressure equilibrium, was early used by Field, Goldsmith & Habing (1969) in their effort to justify the presence of different aggregation states in the ISM. They introduced the thermal processes of heating by low–energy cosmic rays and cooling by collisional excitation of a mixture of H, He and impurities such as C^+. Given the cosmic rays energy, they found two stable phases and evaluated their equilibrium properties: low density and high temperature ($n = 0.21\,\mathrm{cm}^{-3}, T = 9000\,\mathrm{K}$), or high density and low temperature ($n = 100\,\mathrm{cm}^{-3}, T = 18\,\mathrm{K}$).

Cox & Smith (1974) stressed the importance of a third hot component in the ISM, recognized as the medium internal to the cavity formed by a SN remnant expanding in a uniform ambient, characterized by $n < 10^{-2}\,\mathrm{cm}^{-3}, T \sim 10^6\,\mathrm{K}$ and an high filling factor ($\sim .5$). Cox & Smith spoke of a network of tunnels interesting the ISM, suggesting the view of a dynamically dominated ISM.

As a consequence, McKee & Ostriker (1977) formulated the problem of the equilibrium states of the ISM considering the possibility of three distinct phases, not in static equilibrium, but in mass and pressure balance globally. The ISM is dominated and regulated by SN explosions and the physical processes considered are: the evaporation of cloud material by hot gas, the replenishment of the high–density phase in dense shells about old SNRs, and the effects of radiative losses, both directly by radiation from the hot gas and indirectly by radiation from isothermal shocks, driven into the clouds. The McKee & Ostriker scenario provides a description of the actual situation of the ISM, considered to be in a metastable state; the authors determine the equilibrium conditions, not the detailed evolution. McKee & Ostriker recommend their model as a "self–consistent" scheme for the ISM, but this is not true; indeed, they do not consider the fundamental link between the Cold Neutral Medium (the clouds, to be explicit) with the stars formed from them. The Star Formation (SF) regulates the dynamical equilibrium of the system, through the most massive components of the stellar phase, that is the progenitors of SNae. Moreover the same nature of the particular equilibrium state can be determined by the SF processes.

2.2 Evolutionary Models

Ikeuchi and coworkers (Habe, Ikeuchi & Tanaka 1981; Ikeuchi & Tomita 1983; Ikeuchi, Habe & Tanaka 1984) faced this question: under which conditions the steady state hypothesis by Field et al. and McKee & Ostriker is realized? They produced models for the time variation of each component of the ISM by considering the mutual exchange process driven by SNae explosions. Starting from the three phase description of McKee & Ostriker, they considered progressively more complicate systems, ending with a six-phase ISM, as in Fig.1. The exchange processes considered are : (i) sweeping up of Ambient Medium (AM, a mixture of SNR and warm gas) and WG by SNR to form hot cavities (HG) and cold dense shells, broken-up into Small Clouds (SC, $M_{\rm cl} \sim 40 M_\odot$) after the death of SNR; (ii) sweeping-up of HG to form cold shells; (iii) evaporation of SC within hot cavities; (iv) formation of Molecular Clouds (MC, $M_{\rm cl} \sim 10^4 M_\odot$) from SC by encounters; (v) formation of Giant Molecular Clouds (GMC, $M_{\rm cl} \sim 3 \times 10^5 M_\odot$) from MC by gravitation; (vi) photoionization of SC surfaces by the UV flux (to WG); (vii) erosion of GMC by stars (to WG); (viii) mixing of HG in AM (after destruction of SNR shells); (ix) SF from GMC, following the suggestion by Shore (1981) and Ferrini et al. (1982) that a model for the interstellar medium must consider on the same ground the gaseous and the stellar phases to achieve self-consistency. The Star Formation Rate (SFR) introduced is proportional to the abundance of GMC via a coefficient which is the ratio of the efficiency of SF to the mean life time of a GMC. The model is not anyhow self-consistent because it does not include the restitution of processed material from the stars to the ISM.

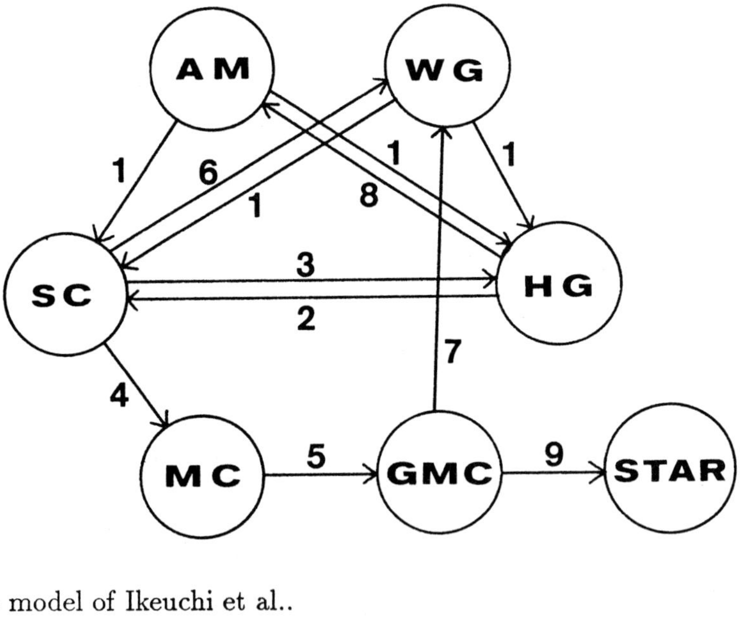

Fig.1. The model of Ikeuchi et al..

A system of coupled first order differential equations is the mathematical framework corresponding to the exchange from one phase to the other; they are master equations where the population of a certain phase is varied according to the various transition probabilities to the other components. The equations present non–linear terms, due to the intrinsic non–linearity of processes. Ikeuchi and coworkers have studied the morphology of the possible solutions in a vast parameter space, not constrained by physical limits from observations or theory; consequently, as for any system described by non–linear evolution equations they found in the space of variables stationary solutions with fixed points, but also limit cycles and chaotic behaviour. The two and three phase stationary states are reproduced respectively in the case of low SN rate and high mean gas density or high SN rate and low gas density.

Similarly, Bodifée & de Loore (1985) and Palouš (1989) studied the evolution of a three component star forming system, putting emphasys on the self–organization consequent to the non–linear dynamics: feedback mechanisms operating in the open non–equilibrium environment produce oscillations or fixed points.

Scalo & Struck–Marcell (1984, 1986), Struck–Marcell & Scalo (1984, 1987) and Vázquez & Scalo (1989) have applied an hydrodynamical approach (through kinetic equations) to the non–equilibrium behaviour of galactic gas flows, to determine the number density, mass, flow velocity and velocity dispersion of clouds in an ambient intercloud medium considering the stellar gravitational field. The source terms considered are: (i) cloud collisions, (ii) cloud formation from intercloud medium, (iii) cloud disruption due to internal SF, (iv) acceleration and formation of clouds by energetic stellar processes. Their procedure to derive from the hydrodynamic equations a set of kinetic equations relies on the argument that the macroscopic description is averaged on microscopic events. The terms corresponding to processes are introduced as rates, which have a reasonable dependence on the physical quantities and through rate coefficients.

Similarly, the chemo–dynamical models by Hensler & Burkert (1990) treat the phases exchange, introducing also a cooling depending on the metal content. The problem with these models as well as with the previous ones, is that the SF in not introduced in a self–consistent way, but *a priori*, there are no feedbacks and there is no delay in the restitution of recycled matter from stars.

Recently, Cioffi & Shull (1991) have developed numerical simulations of SN remnant shocks in a homogeneous medium, following the idea that SNae dominate the evolution of the large scale thermal and dynamical properties of the ISM. The physical processes considered are: (i) cooling by bremsstrahlung and metals, (ii) formation of clouds from the shell material, (iii) thermal conduction from the hot phase to clouds which causes their evaporation. The model contemplates two phases: diffuse gas and clouds, and suggests the necessity of a gaseous halo. They underline the lack of a criterion of star formation in their model, to make its own stars and explode SN, as an absence to make an "intelligent" totally self-consistent system.

2.3 Self–Regulation of Star Formation

Self–regulation mechanisms operate with the SF. On the scale of a single molecular cloud the regulation happens through winds from embedded low mass protostars, as discussed by Franco & Cox (1983). The heated shells can cool and gravitationally collapse to form new stars.

Franco & Shore (1984) extended the self–regulation to large scale through the energy and momentum injection from OB associations, which generate large expanding shells and second generation cloud complexes can be formed from the swept–up material by gravitational instabilities. New clouds imply new stars: this is a way to have long range self–propagating (induced) star formation.

Parravano (1988) discussed the pressure governed self–regulation processes going from warm gas to small clouds which fragment into stars. Stars controll the small clouds formation via the UV radiation; photoejection of electrons from dust grains and cosmic rays ionization fix the gas pressure to the right value for the condensation of small clouds from warm diffuse gas.

Larson (1987) considered the feedbacks – negative, cloud destruction and positive, cloud compression from diffuse gas – in the balance between SF and ISM, to put limits to the efficiency of the SF process, which is regulated by ionization and cloud destruction.

Palouš, Franco & Tenorio–Tagle (1990) studying the expansion of material expelled from OB associations in differentially rotating disks, found that molecular clouds can be formed on the tips of these elongated superstructures. Stellar associations hence induce the formation of molecular clouds, which in turn will be unstable to form new associations.

Tenorio–Tagle & Bodenheimer (1988) have summarized in a nice graphic scheme the cycle of SF (Fig.2). The structure of the ISM depends on the star formation rate, as a consequence of deposition of energy mainly from massive stars. The Star Formation is kept under controll via the self–regulation processes.

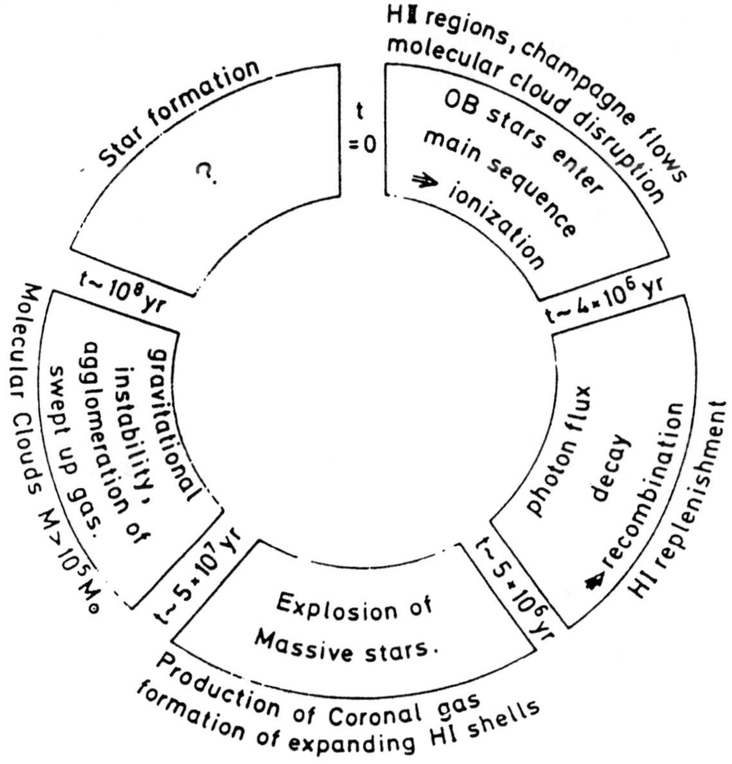

Fig.2. The star formation cycle (Tenorio–Tagle & Bodenheimer 1988).

Finally, I mention a few papers which tackled the problem of a self–propagating star formation with non–local terms. Shore (1983) and Nozakura & Ikeuchi (1984) studied the ISM evolution with diffusion terms, considering the matter flow, while Neukirch & Feitzinger (1988) used an analytical approach to the flow of SF, triggered by stars created in different places at previous times. All these authors conclude that local non–linear processes induce global spatial structures, in agreement with the discussion of Elmegreen (1987) on the relation between morphology and Star Formation, who suggested that the SF on large scale is regulated by local processes and not by the spiral density wave.

2.4 Disk–Halo Interactions through Gas

The disk region in the galaxy is not isolated, indeed there are various exchanges of matter between the disk and the halo. First, hot gas flows into the halo from the disk, cools and falls back to the disk, in several ways (cfr. Heiles this Workshop): galactic fountains (Shapiro & Field 1976; Bregman 1980; Ferrara this Workshop), chimneys

(Norman & Ikeuchi 1989), superstructures (e.g. Tenorio–Tagle & Bodenheimer 1988), photolevitation (Franco et al. 1991). Then, there is the accumulation of disk material from the halo: the disk is a secondary structure.

To describe the distribution of gas in the direction perpendicular to the galactic plane, Cox (1989) proposes the following scheme: – a thin cloud zone, $|z| \leq 100\,\mathrm{pc}$, – a thick extracloud zone, $|z| \leq 1500\,\mathrm{pc}$, with a neutral and a ionized component, – a galactic corona, with $T \sim 2 \times 10^6\,\mathrm{K}$.

2.5 Resumé

Summarizing the above discussion, we may recognize as aims, first the search for the equilibrium state conditions, then the determination of the dynamical equilibrium, through the study of the evolution in time of the ISM. Two aspects, that cannot be disregarded are the self–regulation and the large scale circulation.

As physical processes considered by the various authors, we extract: the heating and cooling processes, cloud formation from diffuse gas, cloud–cloud collisions, stellar energy injections, external couplings, disk–halo interactions via gas, spiral density wave influence on the dynamics of gaseous phases and Star Formation.

The consequent mathematical formulation has changed according to the purposes and the physical content of the models: the pressure equilibrium and the mass and energy balance equilibrium were sufficient to fix the equilibrium states, then the complex systems of non–linear evolution equations were necessary to describe the dynamics of the phases exchange.

3 A SELF–CONSISTENT MODEL FOR THE GALACTIC EVOLUTION

Since our undergraduate studies in classical mechanics, we know that the evolution of a dynamical system can be conveniently studied by knowing the Hamiltonian of the system. For a galactic subsystem the Hamiltonian would be function of the microscopic physical properties of gas and of stars, for each component (HII regions, clouds, stellar clusters,...) and this is impossible to be achieved. We may profitably refer to other fields of physics that faced the same question on how to determine the macroscopic evolution of a complex system, for which it is almost impossible to give a detailed microscopic description. This is e.g. the case of phase transitions in magnetic systems; the scaling hypothesis by Kadanoff allows to substitute to the cell Hamiltonian for a single spin, a block Hamiltonian which is the result of summing up the relevant details within a scale of n lattice constant: the long range correlations are responsible for all the singular behaviours at transition. The role of the block

Hamiltonian in our study of the galactic evolution would be played by the Hamiltonian describing the solar neighbourhood, which could be: H (stars, ISM phases; interactions with external ambient and halo, large scale dynamics,...)
This fixes the scale on which the SF processes are relevant to the description (Ferrini 1987): we do not need to know the *exact* Star Formation Rate (SFR) defined on the scale of a single cloud, nor we can use an *empirical* SFR averaged over the whole galaxy, but we need to adopt the *phenomenological* SFR defined on the solar neighbourhood via the physical processes interesting the cloud population and then fruit of the interactions of phases.

Following Pagel (1989), we recall that a good model for the evolution of the solar neighbourhood should satisfy the following constraints:
- to understand the halo–disk connection, taking into account the initial enrichment of the disk by previous star formation in the halo;
- the halo/disk mass ratio;
- the relative numbers of disk and halo stars in the solar cylinder;
- the present day gas fraction in the disk;
- the ratio of present SFR to average past SFR;
- the age–metallicity relation in the disk;
- the metallicity distribution in halo and disk (the G–dwarf problem);
- the variations in relative abundances of the most common chemical elements.

The recipe for a self–consistent model of galactic evolution must then contain several ingredients:
- the SF as result of the dynamics of phases;
- a series of physical processes concerning the interactions among ISM phases and stars: (i) cloud formation from diffuse gas; (ii) star formation from clouds, spontaneous via cloud–cloud collisions, induced via massive stars–cloud interactions; (iii) restitution from clouds to diffuse gas, due to the low efficiency of star formation; (iv) restitution of processed material from evolving stars; (v) locking–up matter in long living stars and stellar remnants;
- interactions among the structural components of galaxies: bulge, thin disk, thick disk and halo;
- delays due to the finite life–times of stars and clouds;
- the metal enrichment history and its feedback on the thermal processes;
- models for SNae type I and II formation, to evaluate in a self–consistent way their energy and elements pollution rate;
- an IMF averaged over an ensemble of clouds in the solar neighbourhood, which can be varied according to the modified thermal conditions of the cloud population;
- self–regulation;

- adaptability to different morphological situations, without changing the physics;
- possibility of treating non–isolated systems;
- non–locality.

We developed in a series of papers a class of models with the aim to satisfy these series of constraints. In the following I will shortly describe the more recent version of the models, constructed on the previous ones (Shore, Ferrini & Palla 1987; Matteucci & François 1989; Matteucci et al. 1990; Ferrini et al. 1991a). Questions connected with the non–deterministic aspects of galactic evolution have been discussed by Ferrini & Marchesoni (1984) and Ferrini, Marchesoni & Shore (1985). An explanation of the Hubble sequence in terms of different couplings between clouds due to the continuously varying strenght of the SDW has been proposed by Ferrini & Galli (1988) and Galli & Ferrini (1989).

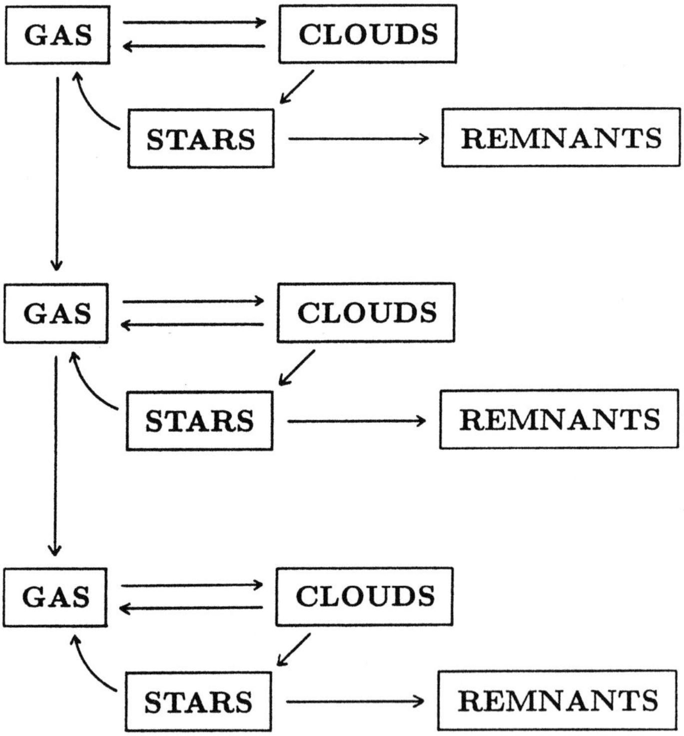

Fig.3. The multi–phase model for the three galactic zones.

3.1 Spiral Galaxies

The present version of the model (Ferrini et al. 1991b) considers three zones: thin disk, thick disk and halo, coupled via the accumulation process from the more external to the more internal component, as drawn in Fig.3. The populations are kept distinct. All the ingredients discussed in the previous paragraph have been introduced but the last two. The set of equations describing the evolution present only four parameters – the conversion from gas to clouds, the induced star formation, cloud–cloud collisions, the accumulation of gas – the same for the different regions, in principle known from observations in the solar neighbourhood and from the theory of microscopic processes. In Fig.4 I plot the resulting Star Formation history for the three components; the halo activity is concentrated in the first Gyr, the thick disk accumulates matter from the halo and looses matter to the thin disk reaching a maximum in the SFR at about 1 Gyr, the thin disk has a delayed activity and the self–regulation without lost of material allows a long lasting slowly decreasing SFR.

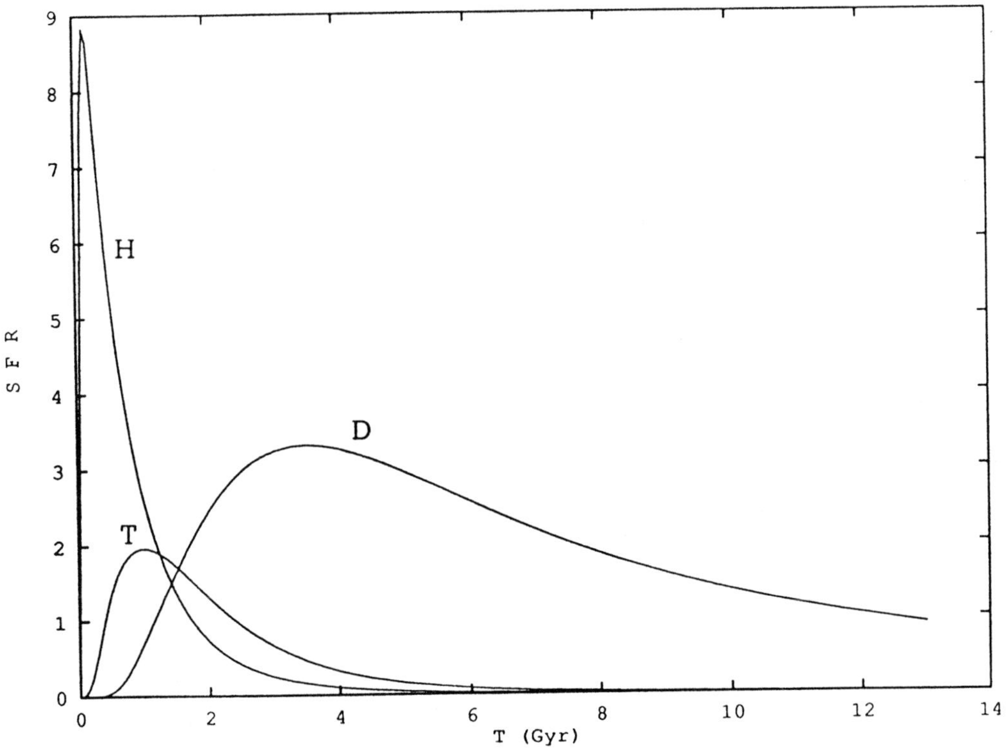

Fig.4. SFR in the three galactic zones.

The phase evolution for thick and thin disk are displayed in Fig.5. I note that the G–dwarf problem, that is the disk stellar distribution on the metal poor end, is well solved by explicitely considering the formation of the thick disk as an intermediate step between halo and thin disk.

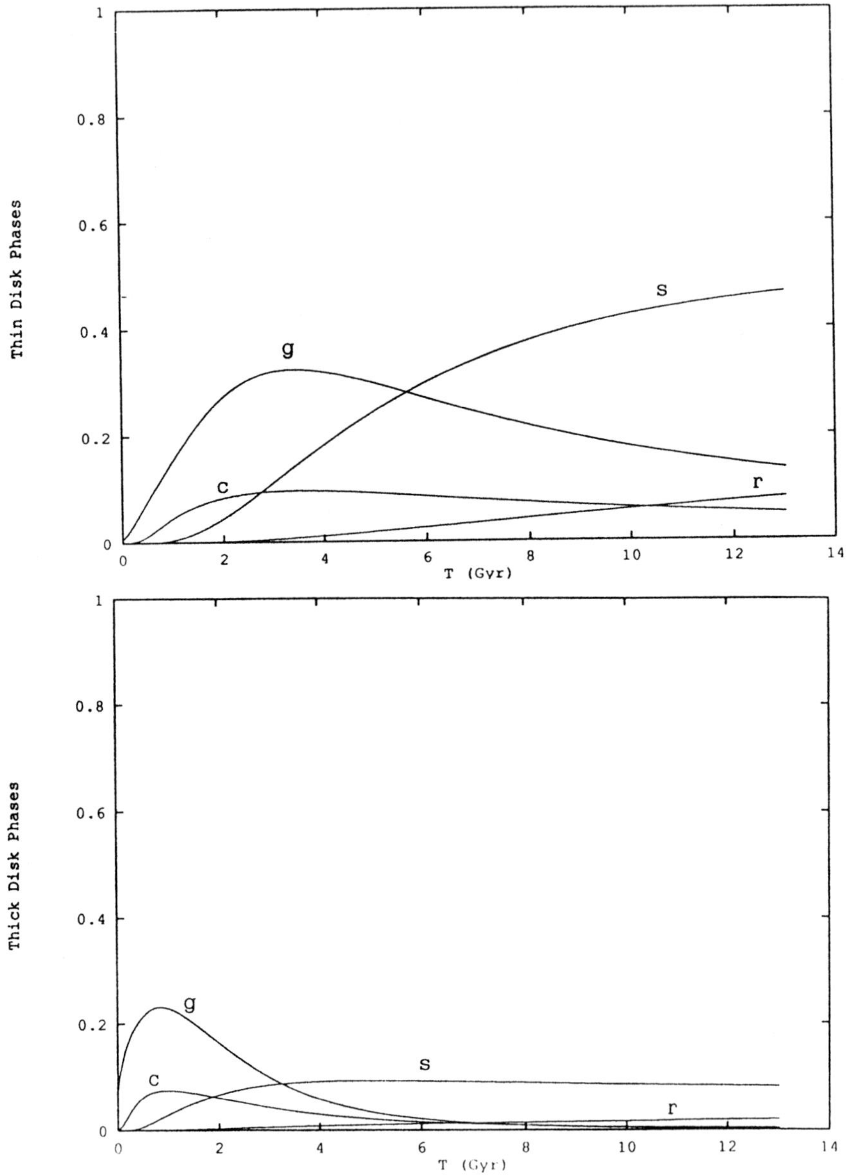

Fig.5. Phases evolution in the thin and thick disk.

3.2 Elliptical Galaxies

A two-zone model has been developed for ellipticals (Ferrini & Poggianti 1991) which considers also the dynamical action of SNae explosions on the diffuse gas, that can be removed from the galaxy when its kinetic energy overcomes the binding gravitational energy. We take into account also the lower efficiency in forming clouds from the gas, heated by the SNae energy injection. The presence of dark matter, important to the determination of the binding energy has been considered. Multiple events of outflow are present in the models, with a significant removal of mass; the enrichment hystory is steeper than for spirals, and the present abundances are well reproduced as well as a significant radial gradient.

4 CONCLUSIONS

The ISM is part of the global balance in the galaxy: interactions among components, restitution from stars, the presence of various zones, the large scale dynamics, the abundances hystory are all intervening in the ecology of the system. The multi-phase approach is an attempt at this complex evolution, based on the following points:

- interactions are more important than details;
- the common matrix of modelling is the synergetic behaviour of the system: a few variables would describe the evolution, while microphysics intervenes in the processes which determine the values of the parameters;
- evolution is dominated by collective effects, where self-regulation and propagation are relevant;
- the self-consistency is a request that cannot be forgotten in the models.

ACKNOWLEDGMENTS

I thank all the friends which whom I had the pleasure to work on these subjects in the years: D. Galli, F. Matteucci, F. Palla, C. Pardi, U. Penco, B. Poggianti and Steve Shore. A special thank to D. Boschiglio for her kind assistance. This review was written during a short visit to ESO, Garching.

REFERENCES

Bodifée, G., & de Loore, C. 1985, A&A, 142, 297

Bregman, J.N. 1980, ApJ, 236, 577

Cioffi, D.F., & Shull, J.M. 1991, ApJ, 367, 96

Cox, D.P. 1989, in IAU Colloquium 120, Structure and Dynamics of the Interstellar Medium, ed. G. Tenorio-Tagle, M. Moles & J. Melnick (Berlin: Springer-Verlag), p.500

Cox, D.P., & Smith, R.W. 1974, ApJ, 189, L105

Elmegreen, B.G. 1987, in IAU Symposium 115, Star Forming Regions, ed. M. Peimbert & J. Jugaku (Dordrecht: Reidel), p.457

Ferrini, F. 1987, in X ERMA, Evolution of Galaxies, ed. J. Palouš (Praga), p.271
Ferrini, F., & Galli, D. 1988, A&A, 195, 27
Ferrini, F., & Marchesoni, F. 1984, ApJ, 287, 17
Ferrini, F., Marchesoni, F., & Shore, S.N. 1985, Adv.Chem.Phys., LXII, 493
Ferrini, F., Marchesoni, F., & Vulpiani, A. 1982, in The First Stellar Generations, Frascati Workshop, Mem.S.A.It., 54, 217
Ferrini, F., Matteucci, F., Pardi, C., & Penco, U. 1991a, ApJ, submitted
Ferrini, F., Matteucci, F., Pardi, C., & Penco, U. 1991b, in preparation
Ferrini, F., & Poggianti, B. 1991, in preparation
Field, G.B., Goldsmith, D.W., & Habing, H.J. 1969, ApJ, 155, L149
Franco, J., & Cox, D.P. 1983, ApJ, 273, 243
Franco, J., & Shore, S.N. 1984, ApJ, 285, 813
Franco, J., Ferrini, F., Ferrara, A. & Barsella, B. 1990, ApJ, 366, 443
Galli, D., & Ferrini, F. 1989, A&A, 218, 31
Habe, A., Ikeuchi, S., & Tanaka, Y.D. 1981, PASJ, 33, 23
Hensler, G., & Burkert, A. 1990, in Chemical and Dynamical Evolution of Galaxies, ed. F. Ferrini, J. Franco & F.Matteucci (Pisa: ETS), p.539
Ikeuchi, S., Habe, A., & Tanaka, Y.D. 1984, MNRAS, 207, 909
Ikeuchi, S., & Tomita, H. 1983, PASJ, 35, 77
Larson, R.B. 1987, in Starburts and Galaxy Evolution, ed. T.X. Thuan, T. Montmerle & J. Tran Thanh Van (Gif Sur Yvette: ed. Frontières), p.467
Matteucci, F., Ferrini, F., Pardi, C., & Penco, U. 1990, in Chemical and Dynamical Evolution of Galaxies, ed. F. Ferrini, J. Franco & F.Matteucci (Pisa: ETS), p.577
Matteucci, F. & François, P. 1989, MNRAS, 239, 885
McKee, C.F., & Ostriker, J.P. 1977, ApJ, 218, L148
Neukirch, T., & Feitzinger, J.V. 1988, MNRAS, 235, 1343
Norman, C.A., & Ikeuchi, S. 1989, ApJ, 345, 372
Nozakura, T., & Ikeuchi, S. 1984, ApJ, 279, 40
Pagel, B.E.J. 1989, Rev.Mex.A.A., 18, 161
Palouš, J. 1989, in IAU Colloqium 120, Structure and Dynamics of the Interstellar Medium, ed. G. Tenorio–Tagle, M. Moles & J. Melnick (Berlin: Springer–Verlag), p.518
Palouš, J., Franco, J., & Tenorio–Tagle, G. 1990, A&A, 227, 175
Parravano, A. 1988, A&A, 205, 71
Scalo, J.M., & Struck–Marcell, C. 1984, ApJ, 276, 60
Scalo, J.M., & Struck–Marcell, C. 1986, ApJ, 301, 77
Shapiro, P.R., & Field, G.B. 1976, ApJ, 205, 762
Shore, S.N. 1981, ApJ, 249, 93
Shore, S.N. 1983, ApJ, 265, 202
Shore, S.N., Ferrini, F., & Palla, F. 1987, ApJ, 316, 663

Spitzer, L. 1968, in Diffuse Matter in Space, (New York: Interscience)
Struck–Marcell, C., & Scalo, J.M. 1984, ApJ, 277, 132
Struck–Marcell, C., & Scalo, J.M. 1987, ApJS, 64, 39
Tenorio–Tagle, G., & Bodenheimer, P. 1988, ARA&A, 26, 145
Vázquez, E.C., & Scalo, J.M. 1989, ApJ, 343, 644

The Gravitational Field Near the Sun

Gerard Gilmore

Institute of Astronomy, Cambridge, UK

ABSTRACT

Knowledge of the local Galactic gravitational potential and the fraction of it contributed by known stars and gas is a fundamental requirement for understanding the chemical and dynamical evolution of galaxies, and the energy-spatial distribution balance in the inter-stellar medium. Recent results have provided for the first time a detailed map of the local distribution of stars and gas which describes all disk matter in terms of identified components. For the first time, dynamical analyses provide evidence that there is no significant unidentified matter associated with the Galactic disk. Recent determinations of the local gravitational potential show that the fraction of all mass in the Solar neighbourhood which is in the inter-stellar medium is $\sim 30\%$, a dynamically significant value.

INTRODUCTION

Many of the most important distributions governing the evolution of the Galaxy can be determined reliably only near the Sun. Chief among these are the detailed spatial and chemical distribution of stars and the interstellar medium, and the distribution of mass in the Galactic disk. Substantial progress in determination of these distributions has been achieved recently, and is outlined here, with emphasis on the significance of these results for dynamical modelling of the Galaxy and the Solar neighbourhood.

1 THE MASS DISTRIBUTION NEAR THE SUN

A description of the amount and spatial distribution of mass near the Sun is an essential prerequisite for any study of the dynamical evolution of the solar neighbourhood. Knowledge of the amount of baryonic matter which has ever been available to the interstellar medium is clearly necessary for chemical evolutionary models, while the current spatial distribution of old stars and the fraction of matter currently in the ISM are also clearly of importance for both chemical and dynamical processes. A common factor linking all these distributions is the Galactic vertical force law, $\mathcal{K}_z(z)$. It is this force law which provides a measure of the total amount of matter near the Sun, which measures the present–day gas fraction, which allows conversion of local kinematic data to weighted contributions to a column through the Galaxy, and which identifies or precludes local dark matter as a significant source or sink term in chemical evolution.

The distribution of mass in the Galactic disk is characterized by two numbers, its local *volume* density ρ_o and its total *surface* density $\Sigma(\infty)$. The local *volume* mass density — *i.e.* the amount of mass per unit volume near the Sun, which for practical purposes is the same as the volume mass density at the Galactic plane has units of $\mathcal{M}_\odot\,\mathrm{pc}^{-3}$. The best available determination of the local mass density in *identified* material is $\sim 0.1\mathcal{M}_\odot\,\mathrm{pc}^{-3}$, with a very-poorly defined uncertainty of perhaps as much as 25% in this value. Comparison of this value with that determined from dynamical analyses is required to test for the existence of dark matter associated with the Galactic disk.

The integral surface mass density has units of $\mathcal{M}_\odot\,\mathrm{pc}^{-2}$, and is the total amount of disk mass in a column perpendicular to the Galactic plane. It is this which is required for the interpretation of rotation curves and the large-scale distribution of mass in galaxies. Recent determinations of this surface mass density lead to values of about $50\mathcal{M}_\odot\,\mathrm{pc}^{-2}$ for that fraction of the mass associated with the Galactic disk. Thus the Galactic disk provides only about 50% of the total Galactic potential at the solar galactocentric distance. The nature of the mass which generates the other half of the potential remains unknown, though it must be distributed in a Galactic halo, and not concentrated in a thin disk.

To put these values in perspective, a volume density of $0.1\mathcal{M}_\odot\,\mathrm{pc}^{-3}$ corresponds to 4.0amu/cc, or 4 hydrogen atoms per cubic centimetre. A surface mass density of $50\mathcal{M}_\odot\,\mathrm{pc}^{-2}$ corresponds to 80grams per square metre, or the surface mass density of the paper on which this is written. Roughly two-thirds of that mass is locked into stars and stellar remnants, the remainder is in gas (mostly) and dust.

1.1 Measurement of the Galactic potential

All determinations of the mass distribution in the Galactic disk (volume or surface densities) require a solution of the collisionless Boltzmann equation. In view of the inconvenience of general solutions of this equation derived from real data, in practise one utilises its vertical velocity moment, the vertical Jeans' equation:

$$\mathcal{K}_z = \frac{1}{\nu}\frac{\partial}{\partial z}(\nu\sigma_{zz}^2) + \frac{1}{R\nu}\frac{\partial}{\partial R}(R\nu\sigma_{Rz}^2) \tag{1}$$

where $\nu(R,z)$ is the space density of the stars, and $\vec{\sigma}_{ij}(R,z) = \langle v_i v_j\rangle - \langle v_i\rangle\langle v_j\rangle$ their velocity dispersion tensor.

The first term on the right hand side of equation 1 is dominant, and contains a logarithmic derivative of the stellar space density $\nu(z)$, and a derivative of the vertical velocity dispersion, σ_{zz}. Since the dominant (by mass) old disk stellar population in the solar neighbourhood is tolerably well described by an isothermal stellar population, the part of this term containing the derivative of the space density dominates the determination of $\mathcal{K}_z(z)$ near the Sun. This point is not often appreciated adequately, but means that one should determine stellar density profiles with even greater care than that required for the velocity dispersions.

The second term in the Jeans' equation 1 describes the tilt of the stellar velocity ellipsoid away from the local cylindrical–polar coordinate system in which velocity dispersions are measured. One therefore needs the R-gradients of σ_{Rz} and of ν. There are no general analytical solutions for this term, as it depends on the unknown "third integral" of the motion. Estimates of its importance may be derived by numerical integration of orbits in potentials which are thought to be realistic approximations to that of the Galaxy, and are described in detail by Kuijken & Gilmore (1989a,b).

Given a measurement of the gravitational field $\vec{\mathcal{K}}(R, z)$ in an axisymmetric galaxy, the total density ρ of gravitating matter follows from Poisson's equation:

$$\nabla \cdot \vec{\mathcal{K}} = -4\pi G\rho. \tag{2}$$

In the case of a disk galaxy we can express the R-gradient in $\nabla \cdot \vec{\mathcal{K}}$ in terms of the observed circular velocity at the Sun, v_c, or in terms of the Oort constants of Galactic rotation A and B

$$\rho = -\frac{1}{4\pi G}\left\{\frac{\partial \mathcal{K}_z}{\partial z} + 2\left(A^2 - B^2\right)\right\}. \tag{3}$$

For a disk galaxy with an approximately flat rotation curve the second term is small within a few kpc of the disk plane (Kuijken & Gilmore 1989a provide a more exact calculation; for an exactly flat rotation curve $A^2 - B^2 \equiv 0$ at $z = 0$), so we can integrate in z to obtain the total column density $\Sigma(z)$ between heights $-z$ and z relative to the disk plane $z = 0$:

$$\Sigma(z) = \int_{-|z|}^{|z|} \rho(z)dz = \frac{|\mathcal{K}_z|}{2\pi G} - \frac{(A^2 - B^2)}{\pi G}|z|. \tag{4}$$

The physical interpretation of a determination of the Galactic $\mathcal{K}_z(z)$ force law can now be seen. In effect, one measures the pressure–gravity balance of the *collisionless* stellar 'fluid'. The hydrodynamic analogy following from the description of the collisionless Boltzmann equation as the equation of stellar hydrodynamics is particularly appropriate here. The dominant first term on the right hand side of equation 1 contains a logarithmic derivative of the stellar spatial density $\nu(z)$, and the stellar velocity dispersion σ_{zz}. The spatial density term plays the role of a scale height, the velocity dispersion is analogous to a temperature, and the product $\nu\sigma_{zz}$ is a pressure.

1.2 The local volume mass density, ρ_o

Almost all determinations of ρ_o model the distribution function of the tracer stars as one or more *isothermal* components. An isothermal component by definition has constant velocity dispersion, independent of z. Then one can easily see that the velocity distribution has to be gaussian: by equation 1, if σ_{zz} is independent of z,

$$\nu(z) \propto e^{-\psi(z)/\sigma_{zz}^2}. \tag{5}$$

This method is especially simple if the tracer appears to be a single isothermal component and the analysis is restricted to stars near the plane, as then the Jeans' equation simplifies to

$$\mathcal{K}_z = \sigma_{zz}^2 \frac{\partial}{\partial z} (\ln \nu(z)) \tag{6}$$

and only the density gradient of the tracer population is needed.

The reliable deconvolution of non-isothermal populations into isothermal components is neither trivial nor robust. First, there is the question of how physically meaningful it really is to separate a stellar sample into discrete isothermal subsamples. Given continual star formation and the continuous diffusion processes thought to be active in heating the disk, this can at best be only an *ad hoc* approximation to the true distribution. A very large number of discrete isothermals is clearly necessary for a close approximation to a real galactic disk. Second, such deconvolutions are far from unique, leading to non-unique force laws. There is however no difficulty *in principle* with such deconvolutions, the difficulties lie entirely in the very large amounts of data necessary to make the deconvolution reliable.

At distance z_j, a superposition of N_{iso} isothermal components, each with velocity dispersion $\sigma^2_{zz,i}$ and density ν_i at $z = 0$ has density

$$\nu(z_j) = \sum_{i=1}^{N_{iso}} \nu_i e^{-\psi(z_j)/\sigma^2_{zz,i}}. \tag{7}$$

This makes clear that at different distances from the plane the relative densities of the components are not the same, but that those of highest velocity dispersion dominate increasingly at higher z. If we were able to specify the velocity dispersions and the spatial density normalisation at $z = 0, \nu_i(0)$ of all components, this would allow a direct solution for the potential from the density profile. However in practice we have to derive the number density of the high–velocity dispersion components at $z = 0$ from high$-z$ velocity data, as only at greater heights are these components sufficiently in evidence to allow reliable determination of their density. However, for $z \neq 0$, the relative densities of the isothermal components depend on the potential as well as the ν_i: thus we are faced with a self-consistency problem involving both the density profile and the potential far from the plane even if we wish to solve for the potential only near the plane.

1.2.1 Self Consistent Solutions for ρ_o

An almost invariable finding in early studies was that a *maximum* was found in the $\mathcal{K}_z(z)$ law a few hundred pc from the plane. Such a result is physically impossible, corresponding to a layer of negative mass. The solution to this inconvenient situation was found by Oort (1960), who imposed consistency with the Poisson equation 1 on his solution of equation 6. In effect, one assumes that uncertainties in the data rather than in the assumptions underlying the analysis are at fault, and that the theoretically reasonable solution which is most consistent with the (defective) data will produce an answer which is close to the 'true' answer. Assignment of a realistic uncertainty to the resulting answer is of course somewhat problematic in this case. More recently, Bahcall (1984a,b,c) has extended this idea, and has developed a substantial improvement in the theoretical methods with which to determine the local volume density of matter ρ_o, by deriving a joint solution of the Poisson and collisionless Boltzmann equations. This solution requires the derived $\mathcal{K}_z(z)$ law to be physically possible, which is a highly desirable and quite reasonable constraint. The analytical techniques developed by Bahcall (1984a,b,c) represent a considerable improvement over those applied previously, and for the first time allow a derivation of ρ_o which is limited by the quality of the available observational data, rather than by the approximate nature of the analysis.

In this technique, the solar neighbourhood is divided into different isothermal components, each of which responds to the potential ψ *via* its velocity dispersion:

$$\rho_i = \rho_{i,0}\, e^{-\psi/\sigma^2_{zz,i}}. \tag{8}$$

Self-consistency of the potential and the total matter density in these components then requires that Poisson's equation be satisfied, *i.e.*

$$4\pi G\rho = 4\pi G \sum_i \rho_{i,0}\, e^{-\psi/\sigma^2_{zz,i}} = \frac{d^2\psi}{dz^2}. \tag{9}$$

Here the density includes a constant 'effective halo' density, due to the halo mass and the radial gradients of the global gravitational field of the Galaxy. With the boundary conditions $\psi = \psi' = 0$ at $z = 0$, equation 9 can easily be integrated forwards to obtain $\psi(z)$. A variety of dark matter components can be added in, using the same prescription as for the visible components, and the resulting potentials calculated.

Bahcall (1984a,b) used his algorithm to reanalyse the available F dwarf and K giant high Galactic latitude data with new models which are self–consistent in the sense that the matter which generates the gravitational field itself responds to it in a manner described by the collisionless Boltzmann equation. Bahcall found that:

(i) the gravitational field due to the $0.10\mathcal{M}_\odot\,\mathrm{pc}^{-3}$ of stars and gas that are identified in the solar neighbourhood is inconsistent with the gravitational fields derived from the data;

(ii) depending on its scale height, a further 0.06–$0.14\mathcal{M}_\odot\,\mathrm{pc}^{-3}$ of unidentified matter is required. This unidentified matter is not part of a very extended halo, though that must also exist and have a local volume mass density of $\sim 0.01\mathcal{M}_\odot\,\mathrm{pc}^{-3}$ so that there is sufficient mass in the Galaxy to generate the potential required to explain the local circular velocity. Hence this result implies significant amounts of disk-like, dissipational dark matter in the solar neighbourhood.

The sensitivity of these determinations of the local volume mass density ρ_o to uncertain data lies in the modelling of the stellar velocity distribution near the Galactic plane, and in the determination of the stellar density distribution with distance from this plane. These analyses were the first in which the reliability of the answer is limited by systematic uncertainties in the data, rather than by inappropriate assumptions. The limitations of the data, and the very considerable uncertainties in the resulting deductions, are well illustrated by recent work by Bahcall, Flynn, and Gould (1991; henceforth BFG), which supersedes the results above.

BFG studied a sample of 560 K giants, towards the south Galactic pole. From this sample they isolated and removed stars with [Fe/H] ≤ -0.6, (the thick disk) with a vertical velocity dispersion measured by BFG of 46 ± 2km/s. They also isolated the most luminous and the most metal rich stars ([Fe/H]≥ 0.0, and/or $M_V \leq +0.8$), as such stars are young (their measured vertical velocity dispersion is 14 ± 2km/s) and may not be in equilibrium in the Galactic potential. [Note that the inclusion of these stars in earlier analyses was suggested by Kuijken & Gilmore (1989c) to be the cause of the high local mass density derived in 1984 by Bahcall – a result now supported by this new analysis by Bahcall *etal.*] After reduction of the contribution from the thick disk and the young disk, BFG retained a sample of 125 K giants. They then revised their absolute magnitude calibration from that adopted in Bahcall (1984b), and in an otherwise similar analysis derived a local volume mass density of $0.15 \leq \rho_o \leq 0.46 \mathcal{M}_\odot\,\mathrm{pc}^{-3}$, for a model in which the total mass is assumed distributed like the identified matter. The quoted range of solutions corresponds to one standard deviation. Thus, the local mass density derived by BFG is $\sim 1.5\sigma$ from the identified mass density, so that no significant unidentified mass is required in the Galactic disk.

The considerable remaining uncertainties in the data can be seen by noting that a change of 0.2magnitudes in their adopted K giant luminosity calibration changes the derived local volume mass density by $\sim 0.5\sigma$. 0.2 magnitudes is 1σ in the present precision of the K giant calibration, and is the systematic difference between the trigonometric parallax and the photometric absolute magnitude calibrations at present. (BFG preferred the photometric calibration.)

1.3 Determination of the disk surface mass density

Isothermal analyses of the local potential waste most of the observational distance-velocity data, by reducing the observed distribution function to the zeroth moment – the density law – and the first moment – the dispersion. We thus would prefer an analysis method which utilised all the available data.

Because high-energy stars are present at all heights above the Galactic plane, measurements of the potential very close to the plane still require knowledge of the high-energy tail of the distribution function. Therefore either the tail of the velocity distribution at low z, or the density *and* potential at high z, are required to measure the potential at low z, and hence to deduce the local volume density of matter ρ_o. The phase-space distribution function we discuss below, however, depends on the density only at points farther from the plane than the height at which data are being analysed. It is possible to capitalise on this insensitivity to the detailed shape of the potential (equivalently, the detailed mass distribution) near the plane to derive the potential at large distances from the plane from high-z data alone. Since a measurement of $\mathcal{K}_z$ at any height relates directly to the total surface density integrated to that height, this is extremely useful, allowing us to obtain meaningful determinations of the surface mass density of the Galactic disk from high-z data alone.

In most $\mathcal{K}_z$-studies, the density $\nu(z)$ is known to better precision than the velocity distribution. Instead of fixing the parameters of the latter, and then using these to model the density gradient, it is therefore preferable to work in the other direction, and predict the velocity distribution of a tracer in different model potentials, given its density. These velocity distribution models can then be compared with the observed velocity data using maximum likelihood techniques.

Given a distribution function $f_z(E_z)$ and a potential $\psi(z)$, we can calculate the density $\nu(z)$, which is just a moment of f_z:

$$\begin{aligned}\nu(z) &= \int_{-\infty}^{\infty} f_z(z, v_z) dv_z \\ &= 2\int_{\psi(z)}^{\infty} \frac{f_z(E_z)}{\sqrt{2(E_z - \psi(z))}} dE_z. \end{aligned} \tag{10}$$

Reparameterizing the z-height in terms of the potential ψ, we have

$$\nu(\psi) = 2\int_{\psi}^{\infty} \frac{f_z(E_z)}{\sqrt{2(E_z - \psi)}} dE_z. \tag{11}$$

This equation is an Abel transform, which has the well-known inversion (see *e.g.* Binney & Tremaine 1987):

$$f_z(E_z) = \frac{1}{\pi}\int_{E_z}^{\infty} \frac{-d\nu/d\psi}{\sqrt{2(\psi - E_z)}} d\psi, \tag{12}$$

so that there is a unique relation between $\nu(\psi)$ and $f_z(E_z)$. Because of this equivalence of $\nu(\psi)$ and $f_z(E_z)$, there is a triangular mathematical relationship between the three functions $\psi(z)$, $\nu(z)$ and $f_z(E_z)$: any one of them can be deduced from the other two. Abel inversions are somewhat unstable, but not as unstable as taking a direct derivative of the data.

It is important to note that equation 12 shows that $f_z(E_z)$ depends on the density only at points where the potential exceeds E_z, *i.e.* beyond the point $z = \psi^{-1}(E_z)$. It is this property which allows the derivation of $\mathcal{K}_z(z)$ at large z independently of the poorly known distribution of mass near the plane.

An analysis technique based on equation 12 has been devised by Kuijken & Gilmore (1989a,b,c;1991) for the determination of $\mathcal{K}_z(z)$, and $\Sigma(z)$. The essential feature of that analysis is that one avoids the assumption of isothermality, by instead postulating a range of potentials $\psi(z)$, and for each of them calculating $f_z(z, v_z)$ from $\nu(z)$. The range of model distribution functions was then compared to an observed distribution function of velocity–distance data using maximum likelihood, and used to select the best-fitting model potential.
Kuijken & Gilmore (1989a,b;1991) determine a dynamical total surface mass density $\Sigma_\infty = 48 \pm 9\mathcal{M}_\odot\,\mathrm{pc}^{-2}$. These same authors integrate the local observed volume mass density in stars through their derived potential, and add the directly observed mass in the inter-stellar medium, to derive an observed integral surface mass density of $\Sigma_{\infty,obs} = 48 \pm 8\mathcal{M}_\odot\,\mathrm{pc}^{-2}$. There is thus evidence that there is no significant unidentified mass associated with the Galactic disk.

2. Some Implications

The important conclusion from recent new data and analyses is clearly that there is no robust evidence for the existence of any dark mass associated with the Galactic disk. In fact, the precision of recent results, and their consistency, provides evidence that there is no significant unidentified matter associated with the Galactic disk near the Sun. There are many implications of this result, some of which are discussed by Kuijken & Gilmore (1989b). For present purposes the most important are that one can determine the local gas fraction in the disk reliably, and that one can derive the weighting factors required to convert local data to be representative of a column through the Galactic disk.

2.1 The velocity dispersion-scale height relation

One of the most direct uses of the $\mathcal{K}_z(z)$ vertical force law is the calculation of the maximum distance that a star of a given velocity in the Galactic plane will reach above the plane. Such a calculation is very uncertain at small velocities, as the potential is not well determined there. Nevertheless, it is possible to use these results to calculate the weighting factors appropriate for correction of local volume data to column integral values. Previous calculations of this correction have either assumed a harmonic potential (*i.e.*a solid–body Galactic disk), or have assumed that the inter-stellar medium is a mass-less tracer of the potential. Correct calculation (Table 1 below) shows that the harmonic assumption underestimates the importance of local halo stars by a factor of about 2.5, and local thick disk stars by a factor of about 1.75, in calculating a fair column integral distribution through the disk.

Table 1

The maximum height reached by collisionless particles with different vertical speeds at z=0. These values are derived from the local Galactic potential derived by Kuijken & Gilmore (1989b).

$v_z\,\vert_{z=0}$ (km s^{-1})	$z_{\rm max}$ (pc)	$v_z\,\vert_{z=0}$ (km s^{-1})	$z_{\rm max}$ (pc)
0	0	40.6	700
8.4	100	44.5	800
15.7	200	48.3	900
21.9	300	51.8	1000
27.3	400	58.6	1200
32.1	500	68.3	1500
36.5	600	83.2	2000

2.2 The mass in interstellar matter

It is difficult to obtain an accurate picture of the gas distribution in the Galaxy. Distances to areas of emission cannot be determined directly, but have to be derived from a kinematic model of the gas flows. This model itself has to be derived from the same data. The inhomogeneity of the distribution, coupled to distance uncertainties, makes it very difficult to obtain sensible average volume densities, although to a certain extent surface densities are easier to measure.

Excellent recent reviews of the amount and distribution of interstellar matter include this volume, Hollenbach & Thronson (1986), and Bloemen (1991). A summary is the following: From the variation of 21-cm emissivity with Galactic latitude one deduces a total surface density towards the Galactic Poles in neutral atomic hydrogen of $N_{H,\perp} = 7.4\ 10^{20}$cm^{-2}, which translates into $8\mathcal{M}_\odot\,$pc^{-2} when allowance is made for a 26% He mass fraction. About half of this is in the 'cold', half in the 'warm neutral' medium. Pulsar emission measure studies indicate the presence of a further $2\mathcal{M}_\odot\,$pc^{-2}, or $9\ 10^{19}$cm^{-2} in a hot, ionized medium, with a scale height on the order of one kpc. From tangent-point HI emission data, at Galactocentric radii 0.4–0.8R_o, Lockman (1984) fits the HI distribution with three components, of total surface density $4.6\ 10^{20}(10\mathrm{kpc}/R_o)$cm^{-2}, with no evidence for a change of scale height or surface density with radius. Inside the solar circle a larger fraction of the hydrogen is molecular than in the solar neighbourhood, though, so that it is difficult to compare the HI surface density there with the local value. Molecular hydrogen is observed by the emission of the CO molecules that occur with it. The CO emission appears to stop quite abruptly near the solar circle, so that it is difficult to define a spatial average of its surface density; a likely value is a surface density of $5.2(10\text{kpc}/R_o)\mathcal{M}_\odot\,\text{pc}^{-2}$ just inside the solar circle, with a rapid decline further out.

For the local surface density in gas, we therefore have

$$\Sigma_{\text{ISM}} = 13 \pm 3\mathcal{M}_\odot\,\text{pc}^{-2},$$

comprising $10\mathcal{M}_\odot\,\text{pc}^{-2}$ of atomic gas, and $3 \pm 3\mathcal{M}_\odot\,\text{pc}^{-2}$ (to allow for the rapidly changing radial distribution) in molecular gas. This estimate is still quite conservative, as no allowance is made for possible optical depth effects in the atomic emission. These might increase the atomic abundances by as much as 25%. The resulting gas fraction is 0.3 ± 0.1, which is dynamically significant.

At larger Galactocentric distances the gas fraction increases, and the interstellar medium increases in importance until it dominates the local disk potential. Thus, the ISM is a major dynamical constituent of the Galactic potential near the Plane at the Sun. Self-consistent treatment of the dynamics are essential for reliable analyses of the energetics.

References

Bahcall, J.N., 1984a, Astrophys. J. **276**, 156

Bahcall, J.N., 1984b, Astrophys. J. **276**, 169

Bahcall, J.N., 1984c, Astrophys. J. **287**, 926

Bahcall, J.N., Flynn, C., & Gould, A., 1991, Astrophys. J. (submitted, 2/91)

Binney, J. & Tremaine, S., 1987. *Galactic Dynamics* (Princeton U.P., Princeton)

Bloemen, H. 1991 *The Interstellar Disk-Halo Connection in Galaxies* (Kluwer, Dordrecht).

Hollenbach, D.J. & Thronson, H.A. 1986 *Interstellar Processes* (Reidel, Dordrecht).

Kuijken, K. & Gilmore, G., 1989a, Mon. Not. Roy. astr. Soc. **239** 571

Kuijken, K. & Gilmore, G., 1989b, Mon. Not. Roy. astr. Soc. **239** 605

Kuijken, K. & Gilmore, G., 1991, Astrophys. J. **367** L9

Kuijken, K. & Gilmore, G., 1989c, Mon. Not. Roy. astr. Soc. **239** 651

Lockman, J. 1984, Astrophys. J. **283** 90

Oort, J.H., 1960, Bull. Astron. Inst. Netherlands. **15** 45

Gravitational Dynamics of Galaxies

D. PFENNIGER

Geneva Observatory, CH-1290 Sauverny, Switzerland

1 INTRODUCTION

Stellar dynamics is an old central piece of our current understanding of galaxies. The main founders of this elaborate theoretical construction, namely Jeans, Eddington and Chandrasekhar, set the rules during the first half of this century at a time where very little was known about the evolution of stars and still less about the physical conditions prevailing in the interstellar medium (ISM). Nevertheless stellar dynamics, at least in its most elementary form, has kept over the years an implicit rôle in any description of galaxies, simply because these systems, as was well discerned by the founders, are to the first order ruled by collisionless gravitational physics.

However, over the years stellar dynamics has been challenged by observations of violent events occurring in active galaxies, or by the large energy exchanges between massive stars and the interstellar medium through the processes of star formation, evolution and explosion, that we call for short stellar activity. The modern view of the processes occurring in the interstellar gas, such as galactic fountains, superbubbles, etc., clashes with the classical description of galaxies as steady ensembles of stars. In the classical description, the only relevant evolution time-scale is given by Chandrasekhar's 2-body relaxation time, and the impact of the stellar energy output can be forgotten because the galaxy is supposed to be completely transparent. Stellar dynamics is also challenged in disc galaxies by the observed Wielen's (1977) diffusion, which implies that in some still unspecified way stars do not conserve well their integrals of motion over tens of rotation periods. Likewise N-body simulations of discs produce "living" models which constantly develop density fluctuations.

In this context, what part of stellar dynamics can be kept? Certainly most of the stars such as the Sun move for several tens of turns around the Galactic center without suffering close encounters (this is implied by the near circular orbit that the Earth had to maintain in order to develop life), but on the other hand, galactic potentials can hardly be considered as remaining rigid and axisymmetric over several billions of years. The corollary of this is that conservation laws, or integrals of motion, cannot be strictly applied to individual stars over such long time intervals.

In the next sections, we will start from elementary considerations about the physical state of typical quiet galaxies. Hamiltonian mechanics will show that the most solid part of stellar dynamics consists in determining the stable periodic orbits. We will discuss some successful applications of orbit theory to galaxies, and will conclude by giving some constraints on the applicability of stellar dynamics, but also of models of the ISM and chemical evolution neglecting dynamics.

2 ENERGETICS

The virial theorem tells us the following: when considering complex physical entities, a good strategy is to enumerate first the principal energies *in interactions*. For instance the Sun is essentially round because gravitational and thermal energies are in balance and dominate, say, magnetic energy on the large scale; in the virial relation the magnetic energy can be neglected without much disrupting the gravitational equilibrium. (At smaller scales, as in sunspots, the converse can be true and then magnetic forces can take over.) However considering only energies is incomplete. One has to ensure that energies are indeed interacting, i.e. to examine if the actual or virtual *powers* at hand are also important. For instance the Sun's gravitational power vanishes since the Sun is in a steady state. However the gravitational power would immediately (that is in a dynamical time, about 50 min) be available if nuclear power diverges. So the gravitational power is virtual and is estimated by the ratio of the total gravitational energy to the free-fall time. In the Sun, rest mass energy and nuclear energy largely exceed gravitational energy. Yet nuclear energy does not determine the structure of the Sun, because the nuclear energy output is strongly *regulated* to a level just compensating the surface luminosity. In comparison, the virtual gravitational power ($\approx 10^{45}\,\mathrm{erg\,s^{-1}}$) is much larger than the luminosity ($\approx 4 \cdot 10^{33}\,\mathrm{erg\,s^{-1}}$).

Can we repeat these considerations with galaxies? Nowadays, sufficient data has been accumulated on galaxies to get a clear ordering of energies and powers in typical quiescent disc galaxies, even if some particular values can be uncertain within a factor 2 to 10. This helps us to fix bounds on the applicability of current theories of galaxies such as stellar dynamics.

2.1 Energetics in the Galaxy

From the literature (e.g. Binney & Tremaine 1987), we collect, in the next Table, estimates of the averaged energy densities in the solar neighbourhood. Here the total mass density is taken to be $\rho = 0.2\,\mathrm{M_\odot\,pc^{-3}}$. A substantial fraction (≈ 0.5) of the rest mass energy ρc^2 would be released if all the mass would be transformed into neutron stars. The nuclear energy available is estimated as a small fraction (0.1) of the maximum nuclear energy produced by transforming all the hydrogen into iron $\approx 0.01\rho c^2$. The rotational velocity is taken as $V = 220\,\mathrm{km\,s^{-1}}$, the velocity dispersion $\sigma = 30\,\mathrm{km\,s^{-1}}$, and the gravitational energy (though non-local) is estimated from the virial theorem of purely gravitating systems. The ISM energies, except the chemical energy, are taken from e.g. Spitzer (1978) and Jura (1987). Chemical energy is estimated as the energy freed by transforming all the neutral H into H_2 (2.2 eV/H). The ISM energies are roughly in equipartition with similar exchange time-scales, that means that these energies are likely to be all interacting.

kind	energy density [$\mathrm{eV\,cm^{-3}}$]	time-scale [Myr]
Rest mass: ρc^2	10^{10}	$\gg 10^4$
Nuclear: $\approx 0.001\rho c^2$	10^7	$1 - 10^5$
Dynamical energies:		
Gravitational	-4200	200
Rotation: $\frac{1}{2}\rho V^2$	2000	200
Star residual motions: $\frac{3}{2}\rho\sigma^2$	100	200
ISM energies:		
gas pressure, gas turbulence, photons, cosmic rays, magnetic fields, chemistry, each:	≈ 1	$1 - 10$

So in principle rest mass energy and nuclear energy can dominate the other forms of energy, but we know that these energies dont easily interact since they need special conditions (such as high temperatures) to be released. On the other hand the ISM energies are certainly unable to play a direct rôle in shaping the galaxy at large scales, since their reservoir is much too small with respect to energies associated with dynamics. The major interacting energies are the rotational and gravitational ones. The energy stored in the random residual motions of stars is already much smaller, but is necessary and just sufficient to prevent vehement global instabilities.

2.2 Powers

Considering now the typical associated time-scales to exchange these energies, we see that the ISM can play an indirect crucial rôle by liberating nuclear energy rapidly, via star formation. The actual nuclear power is nearly equal to the Galactic bolometric luminosity. If we consider the virtual gravitational power for the luminous part of the Galaxy, i.e. where most of the stars are and where gravitation can be accounted essentially to baryonic matter, it turns out that *nuclear and gravitational powers are similar*: $\approx 0.5 - 1.0 \cdot 10^{44}\,\mathrm{erg\,s^{-1}}$. For dynamical considerations this coincidence has to be taken seriously, because if only a sizable fraction of radiation is diverted into thermal and mechanical energy, secularly it has to exercise a significant dynamical effect on the Galaxy (i.e. on the stellar region of it) by disturbing the equilibrium. Now that infrared radiation can be measured, we know that a substantial fraction (> 0.3) of the Galactic bolometric luminosity lies in the (far) infrared, which implies also that about this same fraction of stellar light has been thermalised, presumably by ionisation and absorption by dust. This is also consistent with recent claims that the luminous regions of disc galaxies are more opaque than previously thought, with typical optical depths $\tau \approx 1$ (Disney et al. 1989, Valentjin 1990, Peletier et al. 1992).

The significance of the diversion of stellar light into gas pressure is that stellar activity in normal disc galaxies could influence the mass and length scales of discs, and all the more so in starburst galaxies. This would mean that, similarly to stars, the fundamental galaxy parameters would depend weakly on the pre-collapse conditions, but directly on the intrinsic properties of matter. Further consequences are discussed in Pfenniger (1992a).

2.3 Simplifications

The previous energetics considerations show that the *equilibrium* of galaxies such as ours is dominated to first order by gravitational and rotational energies. However, the power considerations show that other energetics factors such as stellar activity limit the pure dynamical models probably to a few rotation periods, and have to be taken into account to understand how the equilibrium is reached and maintained. The traditional first order simplifications to describe the equilibrium are:

- The ISM and stellar activity are neglected, which means that energy dissipation and mass exchange are neglected.
- Stellar motions are assumed collisionless, i.e. star relaxation is neglected.
- Exact symmetries (such as time and rotational invariance) are adopted, which entails the detailed conservation of quantities such as energy and angular momentum.

At this point, we have reached classical stellar dynamics. However the problem to link self-gravity to the motion of stars is formidable, so a further step is often taken:

- Self-gravity is neglected, so collective relaxation is neglected.

This approximation is loosely substantiated by remarking that a density perturbation $\delta\rho$ changes only the second derivatives of the potential by Poisson's equation ($\delta\rho \sim \Delta\Phi$), while the orbit shapes are often determined by integrals of motion perturbed to the same order than the potential. Smooth variations of the density produce only much smaller variations of the potential. As we will see below, this approximation can sometimes be justified when considering orbits with integrals of motion, i.e. the periodic and quasi-periodic orbits, but cannot be justified when considering chaotic orbits.

With these assumptions one reaches the most elementary stellar dynamical description. A more refined description taking into account more physics can be considered as a *perturbation* of the the most elementary one. If one starts from Boltzmann's equation, where the potential Φ is the average gravitational field, and one neglects the collision term, or if one starts from Navier-Stockes' fluid equation, where one neglects the pressure gradient ∇P and the dissipative terms, one arrives at the same problem of describing the solutions of a second order differential system,

$$\ddot{\vec{x}} = -\nabla\Phi(\vec{x}, t), \qquad (1)$$

i.e. the dissipationless newtonian problem of the motion of test particles in a potential field Φ. The description of the solutions of this particular type of differential equation is the theory of orbits of classical mechanics. It benefits from an exceptionally large body of knowledge, in particular the mathematical one.

3 THEORY OF ORBITS

The main purpose of this section is to extract in an informal, non-rigorous way the most relevant aspects of modern classical mechanics and the theory of dynamical systems useful for applications in galactic dynamics.

3.1 Mechanical Systems

In many textbooks of classical mechanics, dissipative forces are mostly ignored. It doesn't need to be so and we recall the main route to follow (Goldstein 1950). Starting from d'Alembert's principle we obtain the complete Lagrange's equations for a mechanical system with N degrees of freedom along the N coordinates q_i:

$$\frac{d}{dt}\left(\frac{\partial L}{\partial \dot{q}_i}\right) - \frac{\partial L}{\partial q_i} = Q_i, \quad i = 1, \dots N, \qquad (2)$$

where the Lagrangian $L(q_i, \dot{q}_i, t) = T - U$ is the difference between the kinetic energy T and the potential energy U due to the forces which derive from a potential. The Q_i are the dissipative forces, i.e. the forces which are not a gradient of a potential. Then, the Hamiltonian $H(q_i, p_i, t)$ is defined as usual by the Legendre transformation, $p_i = \partial L/\partial \dot{q}_i$, $H(q_i, p_i, t) = \sum_i \dot{q}_i p_i - L(q_i, \dot{q}_i, t)$, $i = 1, \dots N$. The coordinates q_i and the conjugate momenta p_i form the $2N$-dimensional phase space. One derives from equ. (2) the generalised Hamilton's equations,

$$\dot{q}_i = \frac{\partial H}{\partial p_i}, \quad \dot{p}_i = -\frac{\partial H}{\partial q_i} + Q_i, \quad i = 1 \dots . N. \qquad (3)$$

The most important property distinguishing pure Hamiltonian systems from dissipative mechanical systems is the evolution of phase space volumes. If V is a volume of phase space (i.e. a compact set of initial conditions), its evolution is given by (Pfenniger & Norman 1990),

$$\frac{1}{V}\frac{dV}{dt} = \sum_{i=1}^{N} \frac{\partial Q_i}{\partial p_i}, \qquad (4)$$

i.e. a Q_i has to depend on its respective p_i to lead to phase space volume contraction.

The definition of Hamiltonian systems is that they are of the form of equ. (3) with $Q_i = 0$, i.e. they conserve their phase space volumes. This determines mostly the organisation of phase space.

In the mathematical sense, an **orbit** is the solution of equ. (3) extended over an infinite time for a particular initial condition. All the initial conditions taken along this orbit define the same orbit. The important concept is that an orbit is an *invariant subset* of phase space, i.e. all the points along an orbit form an ensemble which remains in this subset. Consequently orbits are the most elementary building blocks of phase space. In a stationary galaxy, the ensemble of stars along a given orbit forms an invariant block of the galaxy. In stellar dynamics, galaxies are therefore seen as built not of stars, but of orbits.

3.2 Phase Space Organisation of Hamiltonian Systems

Starting with Poincaré (1892) up to the recent proliferation of works on dynamical systems, the *qualitative* understanding of the solutions of general bounded Hamiltonian systems, i.e. mechanical systems without dissipative forces, is well understood. The description of the orbits, the invariant sets, is geometric, and its validity concerns all such systems.

Phase space is filled essentially by two broad classes of orbits, the *regular* and the *chaotic* ones (also called irregular, stochastic, semi-ergodic, etc.). Other orbits, such as periodic orbits, occupy a null volume of phase space. The regular orbits are well understood, while the chaotic orbits much less. In the $2N$-dimensional phase space the most frequent regular orbits have the shape of ND-tori. So in a 3D stationary galaxy potential ($N = 3$), the most frequent regular orbits are 3D-tori. 3D-tori are hard to visualise because one needs at least a 4D space to immerse them. These 3D-tori are concentric to one another "around" 2D-tori. In turn, these 2D-tori are concentric around 1D-tori, i.e. closed or periodic orbits (see Fig. 1). Often periodic orbits belong to families turning around fixed points (0D), as e.g. the one at the bottom of a potential well. 2D- and higher dimensional tori are also called quasi-periodic orbits.

The torus shape results from the combination of transverse, independent 1D oscillations. A simple oscillation (e.g. a pendulum) is an oval orbit in phase space. The surface I of this oval is called its *action*. Two independent oscillations produce a product of two ovals, a 2D-torus, etc. So a 3D-torus is characterised by 3 frequencies ω_1, ω_2, ω_3 with irrational ratios, and three actions I_1, I_2, I_3. The actions represent

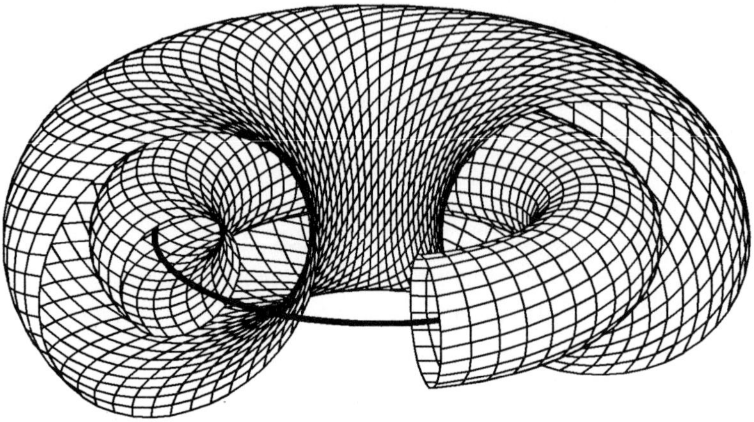

Fig. 1: *Cuts of concentric 2D-tori, represented by grids, surrounding a stable periodic orbit. Quasi-periodic orbits spiral on their torus and cover it densely over time.*

the surface in phase space spanned by each independent oscillation, they are a measure of the amplitude of the respective oscillations.

A 3D-torus degenerates into a set of 2D-tori when a ratio of two of its frequencies is rational, or one of its action vanishes. If all three frequencies are in rational ratios, or two actions vanish, then it degenerates into periodic orbits. If all three actions vanish, it is a fixed point.

In contrast to regular orbits, the Fourier spectrum of chaotic orbits is broad, so an infinity of frequencies characterises them. Instead of frequencies, chaotic orbits are described by $2N$ Liapunov exponents, which measure the average exponential divergence of neighbouring orbits in orthogonal directions of phase space. A chaotic orbit has at least one positive Liapunov exponent. The geometric dimension of chaotic orbits is poorly known, but has been shown to be fractal in specific cases.

3.3 Neighbourhood of Periodic Orbits

While the fixed points are usually exceptional, the periodic orbits are more numerous, and since regular orbits are organised around some of them, periodic orbits form the *backbone* of phase space. They summarise the motion of the surrounding quasi-periodic orbits, and are useful as an abstraction of the motion in galaxies (stars move essentially in *circular orbit* in a disc, etc.).

Periodic orbits can be *stable*, in the sense that neighbouring orbits remain close along the full length of the periodic orbits. These neighbouring orbits are precisely the higher dimensional tori, the quasi-periodic orbits described above. So, stable periodic orbits are *always* surrounded by quasi-periodic orbits. A fundamental result of the so called KAM (Kolmogorov, Arnold, Moser) theory (see e.g. Arnold 1988 for more detail), strengthened by numerical experiments, is that regular orbits are *robust* to small Hamiltonian perturbations, i.e. perturbations preserving the dissipationless nature of the problem. A small Hamiltonian perturbation essentially *deforms* most of the quasi-periodic orbits, but preserves their topological properties given by the ω_i's and I_i's. For a given perturbation amplitude, the lower the dimensionality of a regular orbit, the more robust is the orbit. Therefore, with the fixed points, *stable periodic orbits are the most robust invariant pieces of phase space.* In practical computations this is indeed verified, in the sense that even very rough galactic potentials contain already the main periodic orbits of a more refined one.

Unstable periodic orbits are usually surrounded by a region of chaotic orbits. Chaotic orbits form a "sea" in phase space, permeated by unstable periodic orbits. The precise structure of this sea is an open question yet. The main feature of chaotic and unstable periodic orbits is that they are *sensitive* to perturbations, in the sense that any perturbation amplifies the difference between the initial and the actual trajectory exponentially. If chaotic orbits are sensitive to disturbances, they are nevertheless structurally stable, in the sense that chaotic orbits remain generally chaotic after being slightly perturbed.

3.4 Resonances

Typically along a family of concentric tori the frequencies ω_i change with the actions I_i. Therefore the tori with irrational ω_i ratios must be separated by tori with rational ω_i ratios, so by tori having a lower dimensionality. The region where higher dimensional tori break into lower dimensional ones due to a rational frequency ratio defines the concept of *resonance.* The major resonances have a frequency ratio made with small integers. At a resonance, a 2D-torus family always breaks up into an integer and equal number of stable and unstable periodic orbits. This simple fact leads to the conclusion that these sets of nicely concentric tori described above are actually riddled with resonances containing stable and unstable periodic orbits, and chaotic orbits of higher order. Around each stable periodic orbit in the resonances one finds again respectively higher order tori and layers of chaotic orbits. Inevitably this decomposition must continue at smaller and smaller scales *ad infinitum.*

The map of the intersections of trajectories by a plane in phase space is called a surface of section, or Poincaré map. An example of such a map representing a typical section of phase space of an Hamiltonian system with 2 degrees of freedom is shown in Fig. 2 (left), and a chaotic layer around an unstable periodic orbit in Fig. 2 (middle).

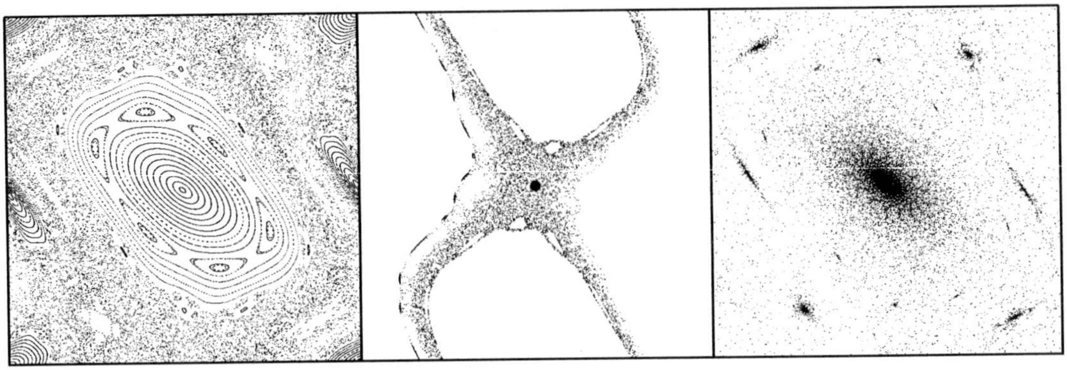

Fig. 2: Left: *A Poincaré map (the "standard map") illustrates all the typical structures of a surface of section in Hamiltonian systems, where concentric regular orbits are intermingled with chaotic ones. Inside the big central "island" of quasi-periodic orbits a resonance of order 6 can be seen.* **Middle:** *Section of a single chaotic orbit over 10'000 intersections, showing the complex structure of such orbits. The associated unstable periodic orbit (large dot) is located at the centre.* **Right:** *A small dissipative perturbation of the map at left transforms phase space radically. The orbits converge onto a few attractors (e.g. at the centre), and are modified more rapidly (as seen by the lower dot density) in the chaotic than in the regular regions.*

3.5 Dissipative Perturbations

When small dissipative forces are introduced into a Hamiltonian system, the structure of phase space is radically changed. All the trajectories then converge asymptotically onto small subsets of phase space, called attractors. These attractors are often points (at the bottom of a potential), or closed curves (limit cycles), but can also have a fractal dimension and be chaotic (strange attractors). For practical considerations, the important point to retain is that the different types of orbits of the original Hamiltonian system resist differently to a small dissipative force. In decreasing order of sensitivity, measured with the time to change significantly the orbit properties, we have first the chaotic orbits, then the 3D-tori, the 2D-tori, the periodic orbits, and finally the fixed points. In practical cases, the robustness of stable periodic orbits can typically be orders of magnitudes larger than the one of chaotic orbits (Fig. 2 right).

In conclusion, determining the major stable periodic orbits of a galaxy is a safe enterprise, as the results depend weakly on the exact mass distribution, on the presence of random fluctuations, or on the inclusion of self-gravity. Weak dissipative forces change these orbits slowly, so the time-scale over which dissipation can be neglected is much longer for a stable periodic orbit than for quasi-periodic and chaotic orbits. On the contrary, if a galaxy contains a major resonance, many chaotic orbits can be generated around it and their sensitivity to fluctuations can boost collective relaxation effects. Neglecting self-gravity is then no longer appropriate. Likewise dissipative forces have their effect amplified in the chaotic regions, i.e. around resonances.

4 APPLICATIONS OF ORBIT THEORY TO GALAXIES

Computing orbits has a long tradition in astronomy. As the applications are numerous, only a few recent advances in galactic dynamics will be discussed.

4.1 Orbit Shape Versus Galaxy Shape

About twenty years ago, the reason why galaxies adopt some particular shapes was poorly understood. Mainly since the 70's the question of the orbital content has been extensively investigated in disc and triaxial galaxies. Before undertaking the more difficult task to build self-consistent galactic models in equilibrium, i.e. mass models compatible with the orbits that their potential generates, it has proved useful to know these orbits beforehand (Schwarzschild 1979). Finding *equilibrium* models is a preliminary step before finding *stable* models.

As a simple example showing without calculation that the equilibrium of some mass distributions is unrealisable with its orbits, consider a uniformly rotating thin needle as an unsophisticated model of a rotating stellar bar. Knowing the fact that radial orbits are impossible along the needle due to the Coriolis force, the only remaining possibility is that stars are all at rest in the rotating frame, i.e. the gravitational and centrifugal forces balance exactly. The only solution, given by potential theory, is a finite uniform needle with a fixed rotation frequency. Therefore inhomogeneous rotating needles are impossible.

Another example is the orbital content of ellipticals (see a review by de Zeeuw & Franx, 1991). In non-rotating triaxial galaxy models, the orbits circling around the centre are oval and, except for some pathological exceptions, elongated perpendicularly to the galaxy major axis. Thus we know immediately that triaxial ellipticals have to be sustained by *other* orbits than these oval (loop) orbits, contrarily to discs. Ellipticals with a realistic de Vaucouleurs profile or a steep profile do contain non-rotating orbits oscillating in such a manner so as to avoid the central region (called boxlets in Miralda-Escudé & Schwarzschild 1989, Pfenniger & de Zeeuw 1989, Martinet & Udry, 1990, Udry 1991). The proportion of these boxlets in specific N-body models and their relation with morphology is presently studied by Udry (1992).

The investigations of the orbits in the bar of disc galaxy models (Athanassoula et al. 1983, Contopoulos & Papayannopoulos 1981, Pfenniger 1984, Sparke & Sellwood 1984, Pfenniger & Friedli 1991) show that the most plausible solution is that bars rotate fast, i.e. their corotation radius is close to their end. In this case the loop orbits are indeed aligned with the bar and mostly stable. Also in these barred galaxy models one finds locations of oval orbits just at the locations of the inner and outer rings observed in several SB's (Fig. 3).

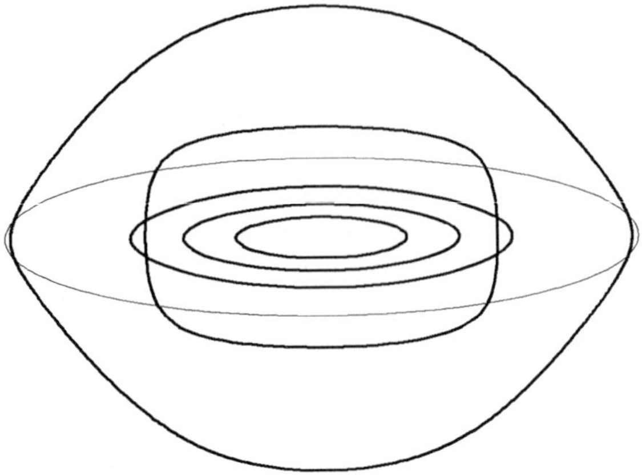

Fig. 3: *The thick curves represent main stable periodic orbits in a typical barred galaxy model (Pfenniger 1984), in the rotating frame of the bar, delineated by a thin ellipse. The corotation radius lies at the end of the bar. The inner elongated orbits have a shape compatible with the bar, while the outer more oval orbit belong to a narrow family that could sustain an inner ring.*

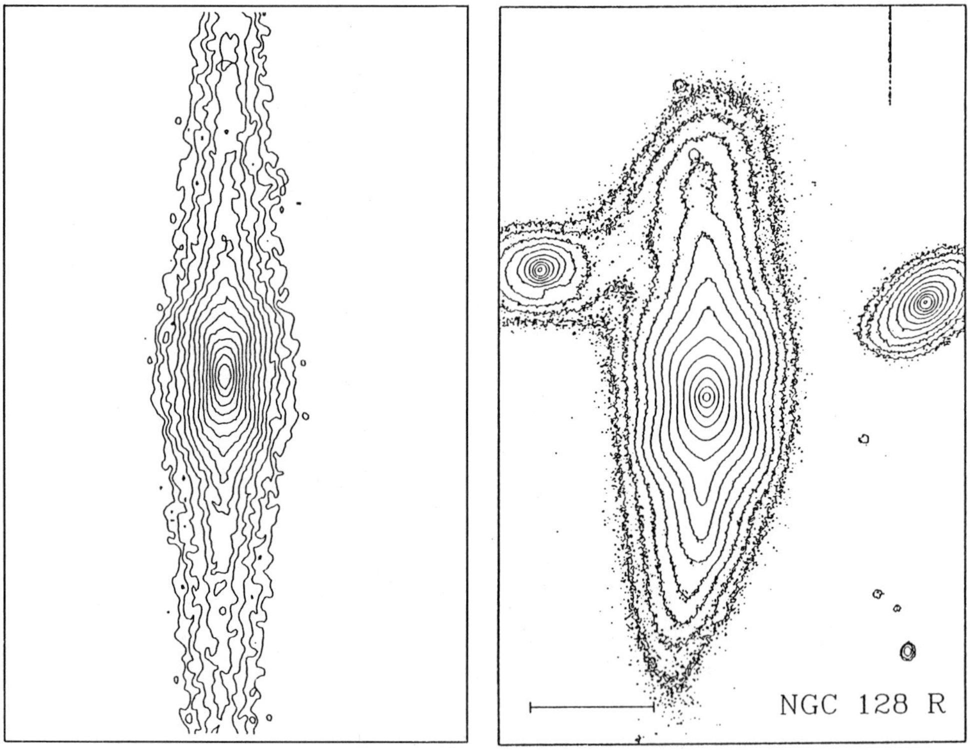

Fig. 4: *Left: A N-body bar takes spontaneously a peanut shape when seen on edge (Pfenniger & Friedli 1991). The peanut shape can be understood by the presence of a 2/2/1 family of periodic orbits. Right: The peanut-shaped galaxy N128 in the R band (Jarvis 1990). In both frames the isophotes are spaced by a half magnitude.*

4.2 Disc Heating by Bars and the Formation of Boxy Galaxies

In addition to determining the stability of periodic orbits, investigating orbits also provides the conditions of orbital instabilities. In particular, in rotating bar potentials strong resonances of the orbits sustaining the bar are found to always exist, exciting oscillations *transverse* to the galactic plane (Pfenniger 1984, 1985). These so called vertical resonances are especially strong when the bar density is inhomogeneous (Pfenniger & Norman 1990). This leads to the conclusion that a disc forming a bar is unable to prevent stars from diffusing at larger heights, up to a few kpc above the plane. So bulges should be formed at least in part by this process. Also late type galaxies forming a bar should also form a bulge after a few rotation periods, leading to a change of their Hubble type.

N-body simulations (Combes & Sanders 1981, Combes et al. 1990, Friedli & Pfenniger 1990, Raha et al. 1991) illustrate well this effect and allow to take into account the effects of self-gravity. After bar formation as predicted the plane periodic orbits sustaining the bar become unstable transversally to the plane in a specific energy range (Pfenniger & Friedli 1991). This triggers a transverse global instability that has precisely the length given by the location of the resonance. For a brief moment self-gravity allows a z-asymmetry deforming but keeping the orbital structure. When the instability saturates, the galaxy again becomes symmetric in z, as stars are trapped by 2/2/1 stable and symmetric orbits outside of the plane. This family of 3D orbits is intimately linked to the vertical resonance, as it bifurcates from it. This family of orbits gives to the model a box-peanut shape when seen edge-on, that resembles some observed edge-on galaxies (Fig. 4).

A very similar behaviour has been reported by Merritt & Hernquist (1991) and Udry (1992) in N-body simulations of eccentric *non-rotating* ellipticals, which spontaneously deform and take finally a boxy shape. Presumably, and this has to be checked, the underlying instability occurs when the long axis radial orbit is transversally unstable and generates, by a bifurcation, the previously mentioned boxlets.

4.3 Dissolution of Bars by a Strong Inner Linblad Resonance

Bars are easily produced in disc galaxies mainly because discs manage to keep their local stability parameter Q close to being critical. So large scale instabilities are easily triggered by a small disturbance, such as a small satellite. Now if bars are so easily produced, and are shown to be long-lived in N-body simulations, why are some disc galaxies still unbarred? If discs are cooling, massive halos dont help much because their dynamical effect inside the stellar region is too small.

A possible answer can be found again by considering the bar orbital structure in the galactic plane. If the density profile is steep in the central region, then an Inner Lindblad Resonance (ILR) is ineluctable. (An ILR occurs on a loop orbit when the frequency of plane oscillations across the loop orbit amounts to twice the orbital period in the rotating frame of a non-axisymmetric perturbation.) Then from the 2D studies of the orbits in bars, the stable periodic orbits inside the ILR radius are known to be always *perpendicular* to the bar major axis. Therefore if the central mass concentration is large enough, it induces an ILR at a radius comparable with the bar minor semi-axis. This must necessarily influence the orbits elongated along the bar and sustaining it, up to disrupting the bar. The prediction from orbit considerations (Hasan & Norman 1990) is that only a few percents of the galaxy mass accreted in the central part suffice to dissolve it. This prediction has been tested in 3D N-body simulations either by accreting mass at the center by a slow dissipative process (Friedli & Pfenniger 1991), or by accreting a small compact satellite by merging (Pfenniger 1992b). In both cases the bar is indeed dissolved by a small central mass and becomes a hot spheroid, strongly resembling the mildly triaxial bulge of M31 for instance. This process suggests rapid evolution phases in spirals, and could contribute to explain the statistics of spiral types: (SA : SAB : SB) $\approx$ (1 : 1 : 1).

5 CONCLUSIONS

Periodic orbits are not only the most robust part of stellar dynamics, but are also useful as elementary mental models of the dynamics. They provide a first order understanding of complex situations in galactic dynamics. The knowledge of orbits allows to discriminate between possible and impossible galaxy shapes, and allows us to predict the location of dynamical instabilities. The inclusion of self-gravity is needed to analyse in detail the development of instabilities, as instabilities produce chaotic orbits that are sensitive to density fluctuations, in particular to their own ones. Chaotic orbits are also sensitive to dissipation.

On the other hand, classical stellar dynamics in disc galaxies is limited to time-scales smaller than a few 10^8 yr, as witnessed by Wielen's diffusion, and as energetics leads to estimate. The limitation can come either from collective gravitational effects, or from ISM and stellar activity effects. Such effects should have non-negligible contributions to dynamics after a few 10^8 yr. Reciprocally, we have seen a few examples where purely dynamical processes change substantially the gravitational potential, with typical time-scales also of the order of 10^8 yr. These changes in the gravity should directly influence the ISM, star formation and consequently chemical evolution. Thus for large scale ISM modelling, dynamical effects, such as the variations of the gravitational field, can only be neglected for time-scales smaller than 10^8 yr. Over several 10^9 yr, dynamics, the ISM and stellar physics should be coupled to properly describe the evolution of disc galaxies.

References

Arnold, V.I.: 1988, *Dynamical Systems,* Vol. III, Berlin: Springer

Athanassoula, L., Bienaymé, O., Martinet, L., Pfenniger, D.: 1983, *Astron. Astrophys.* **127**, 349

Binney, J., Tremaine, S.: 1987, *Galactic Dynamics,* Princeton: Princeton Univ. Press

Combes, F., Debbasch, F., Friedli, D., Pfenniger, D.: 1990, *Astron. Astrophys.* **233,** 82

Combes, F., Sanders, R.H.: 1981, *Astron. Astrophys.* **96,** 164 p

Contopoulos, G., Papayannopoulos, Th.D.: 1980, *Astron. Astrophys.* **92,** 33

de Zeeuw, P.T., Franx, M.: 1991, *Ann. Rev. Astron. Astrophys.* (in press)

Disney, M., Davies, J., Phillips, S.: 1989, *M.N.R.A.S.* **239,** 939

Friedli, D., Pfenniger, D.: 1990, in *Bulges of Galaxies,* ESO Conference and Workshop Proceedings No. 35, eds. B.J. Jarvis, D.M. Terndrup, p. 265

Friedli, D., Pfenniger, D.: 1991, in *Dynamics of Galaxies and Molecular Clouds Distribution,* IAU Symp. No 146, eds. Casoli F., Combes, F., Dordrecht: Kluwer, p. 362

Goldstein, H.: 1950, *Classical Mechanics,* Reading: Addison-Wesley

Hasan, H., Norman, C.: 1990, *Astrophys. J.* **361,** 69

Jura, M.: 1977, in *Interstellar Processes,* eds. Hollenbach, D.J., Thronson, H.A., Dordrecht: Kluwer, p. 3

Martinet, L., Udry, S.: 1990, *Astron. Astrophys.* **235,** 69

Merritt, D., Hernquist, L.: 1991, *Astrophys. J.* **376,** 439

Miralda-Escudé, J., Schwarzschild, M.: 1989, *Astrophys. J.* **339,** 752

Peletier, R.F., Willner, S.P., Rupen, M.P.: 1992, in *The Stellar Populations of Galax-*

ies, IAU Symp. No 149, ed. B. Barbuy, (in press)

Pfenniger, D.: 1984, *Astron. Astrophys.* **134,** 384

Pfenniger, D.: 1985, *Astron. Astrophys.* **150,** 112

Pfenniger, D.: 1992ab, in *Dynamics of Disc Galaxies,* ed. B. Sundelius, (in press)

Pfenniger, D., de Zeeuw, P.T.: 1989, in *Dynamics of Dense Stellar Systems,* ed. D. Merritt, Cambridge University Press: Cambridge, p. 81

Pfenniger, D., Friedli, D.: 1991, *Astron. Astrophys.* (in press)

Pfenniger, D., Norman, C.: 1990, *Astrophys. J.* **363,** 391

Poincaré, H.: 1892, *Les méthodes nouvelles de la mécanique célèste,* Vols. 1–3, Paris: Gauthier-Villars, 1957, New-York: Dover

Raha, N., Sellwood, J.A., James, R.A., Kahn, F.D.: 1991, *Nature* **352,** 411

Schwarzschild, M.: 1979, *Astrophys. J.* **232,** 236

Sparke, L., Sellwood, J.A.: 1987, *M.N.R.A.S.* **225,** 653

Spitzer, L.: 1978, *Physical Processes in the Interstellar Medium,* New York: Wiley

Udry, S.: 1991, *Astron. Astrophys.* **245,** 99

Udry, S.: 1992, *Astron. Astrophys.* (in preparation)

Valentijn, E.A.: 1990, *Nature* **346,** 153

Wielen, R.: 1977, *Astron. Astrophys.* **60,** 263

Gas Flows in Spirals and Bars

J A Sellwood

Space Telescope Science Institute

1 A RECENT THEORY OF SPIRAL ARM FORMATION

Irregularities in the mass distribution within a galactic disc are smoothed out over time by the inexorable differential rotation and by the related process of phase mixing. It has been widely thought, therefore, that the (coarse grain) distribution function of the old stellar population should be smooth. While very efficient at smoothing non-axisymmetric irregularities, these processes are unable to smooth out axisymmetric density variations. Moreover, small-scale variations in the axisymmetric density do not merely survive, they are strongly destabilising (Lovelace & Hohlfeld, 1978; Sellwood & Kahn, 1991, hereafter SK). It now seems likely that they are responsible for much of the spiral structure which has intrigued theoreticians for so long.

Variations in the axisymmetric surface density profile of a disc cannot occur on scales smaller than the mean radial epicyclic excursions of stars. But arbitrarily narrow features can be created in the angular momentum density within the disc and are of much greater interest. In the limit of a so called cold disc (all stars on exactly circular orbits) the absence of stars over a particular narrow angular momentum range would produce a groove in the surface density also.

1.1 Groove Modes

SK used N-body simulations and local linear theory to demonstrate the strong destabilising effect of narrow grooves (and also ridges) in the angular momentum density of a simple model disc galaxy. Attention to many important details made their paper quite complicated, but the overall principle of a groove mode is easy to grasp. It is closely related to the edge mode (Toomre, 1981, 1989; Papaloizou & Savonije, 1989) which afflicts a disc having steep outer density gradient.

A groove is best thought of as two, closely-spaced, sharp edges. A small, co-rotating, non-axisymmetric perturbation near each sharp edge causes large relative variations in the surface density, as shown in Figure 1; the shading marks areas where high density material has been perturbed into regions of formerly much lower density. The corresponding perturbation to the gravitational field implies that the density maxima will attract each other in pairs, assuming them not to be precisely evenly spaced. These tangential forces perturb the angular momentum of particles in the lumps; those which are dragged forward gain angular momentum and therefore rise

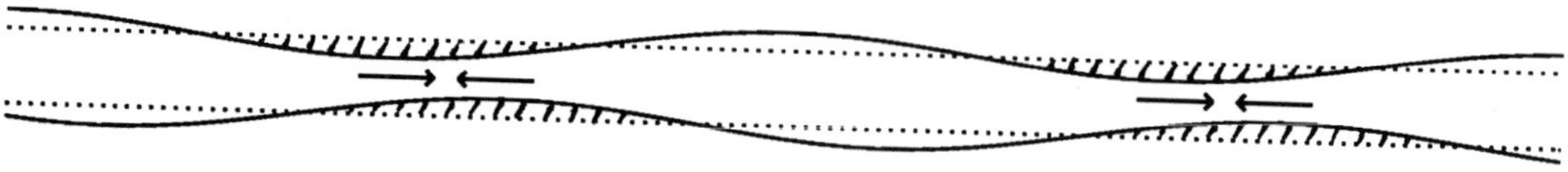

Figure 1 **A schematic illustration of the density changes caused by a sinusoidal perturbation to the two initially straight edges (dotted) of the groove. The shaded areas are regions where high density material has been moved into regions of former lower density and the arrows show the directions of the resulting gravitational disturbance forces along the groove.**

outwards in the disc, those which are retarded sink towards the centre. Provided the phase relation between the waves on the two edges is as drawn, the perturbing forces are in the right sense to cause the bulges (and the dents also) to grow – *i.e.* the groove is unstable.

The comparatively mild instability that the groove can sustain in this way, acts as a seed from which a larger-scale and altogether more serious instability develops. This is because any density enhancement within a gravitating medium attracts nearby material, and gathers a *polarisation cloud* around itself. In a stable, cool, differentially rotating disc, the polarisation response to a steady, co-rotating mass enhancement is a pronounced spiral wake (Julian & Toomre, 1966). The response of the surrounding disc to a growing non-axisymmetric perturbation along a groove is almost exactly the same, differing mainly by the fact that the spiral wake grows exponentially at the same rate as the perturbation.

In a cool massive disc, the potential perturbation from the polarisation response is much greater than that from the original disturbance on the groove, and substantially raises the growth rate of the instability – typically by a factor of five in a moderately responsive disc. [Perturbations on an isolated edge, or a ridge-like feature, are neutrally stable, and these modes are destabilised only by the polarisation response (Toomre, 1989; SK).] The overwhelming importance of the polarisation response makes it seem that the axisymmetric feature was irrelevant, but this is not so – without it, the instability would have no toe-hold from which to start (see also Papaloizou & Savonije, 1989).

Groove modes grow rapidly but saturate once the density maxima from opposite sides of the groove (Figure 1) begin to overlap. The instability therefore produces

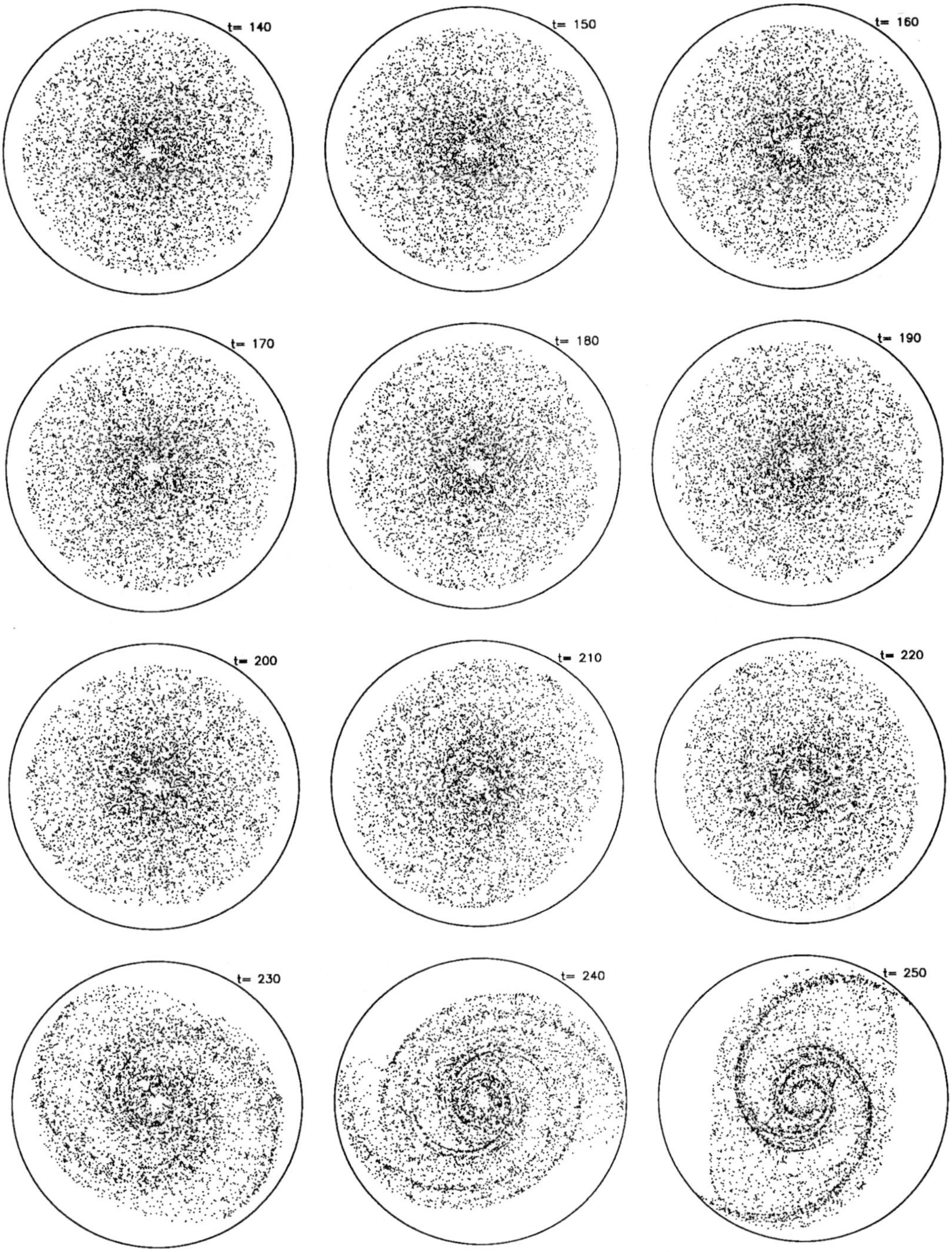

Figure 2 The later stages of evolution of a cold disc with a shallow Lorentzian groove. The model is similar to those reported by SK, but the groove is much shallower; in this case the half-width at half-depth is 0.2 and the maximum depth is 0.1. The perturbation forces were restricted to $m = 2$.

a set of mass clumps equally spaced along the groove which continue to support a strong spiral response as the clumps slowly disperse. Thus the limiting amplitude will depend mainly upon the width and depth of the groove. Those used by SK produced very intense spiral waves, strong enough to tear large holes in the disc and much stronger than those observed in galaxies. Figure 2 shows a similar simulation which has a more modest initial groove; although the spiral pattern is still very pronounced, the final amplitude is much less than in the simulations reported in SK, but may still be stronger than in real galaxies.

1.2 Origin of Grooves

Co-rotation for a groove mode lies at the groove centre, or in a warm disc, the home radius for a test particle with the angular momentum of the groove centre. The spiral response extends on both sides to the Lindblad resonances.

The large-amplitude wave produced by the mode scatters the stars which are in Lindblad resonance; those at the ILR lose angular momentum on average while at the OLR they gain (Lynden-Bell & Kalnajs, 1972). Both resonant regions are therefore de-populated, and because the resonances are quite sharp the de-populated regions are narrow and deep. The mode therefore carves new grooves in the distribution function, each of which could give rise to a new instability. A recurrent intstability cycle of this nature was observed by Sellwood & Lin (1989) in simulations of a low mass disc in nearly Keplerian rotation about a central mass. Their model, therefore, differs in many important respects from real galaxies, but a demonstration that a similar cycle of instabilities can be supported in a realistic disc galaxy model is almost complete (Sellwood, 1991).

Scattering caused by the waves raises the level of random motion of disc particles in the vicinity of the resonance. Sellwood & Lin found that only the groove carved in a previously undisturbed part of the disc caused a new mode to grow – the other Lindblad resonance lay in a region which had already been heated by two previous waves, and was unable to support a new instability. Thus the spiral "disease" progressed radially in a single direction. Once the entire disc has been heated, no further spiral waves can be sustained – unless some cooling mechanism is available (*e.g.* Sellwood & Carlberg, 1984; Carlberg & Freedman, 1986).

There still remains the question of the origin of the first groove. It is easy to think of a number of different ways in which this might have arisen; possibilities are:

1 Resonant scattering by some other type of mode, perhaps of the bar-forming or edge type.
2 The accretion of a large lump of fresh material, perhaps in the form of a satellite galaxy, or simply as part of the galaxy formation process. The large lump would be slow to disperse and would drag around a large wake and create grooves at its Lindblad resonances. It seems inconceivable that the galaxy formation process

would be so smooth and symmetric that this could be prevented.

3 The close passage of a companion galaxy which may be able to excite a moderately long-lived response in the disc.

1.3 Gas Response

It is interesting to consider how the inter-stellar medium would respond to these time-dependent spiral waves. A full calculation of the behaviour would be a major undertaking, but some preliminary indications can be obtained from the response of rings of test particles to the time varying gravitational perturbation.

Figure 3 shows the evolution of a set of such rings which were included in the calculation shown in Figure 2. Each of these massless particles was moved in the changing galactic potential, but experienced no other forces. A simple minded view of the possible gas response can perhaps be deduced if we identify each ring as a streamline in the unperturbed system. The distortion of each ring then gives some indication of how that streamline would evolve only up to the point at which the ring crosses either itself or another ring. At the positions where ring crossings occur, we might expect shocks to occur; from the moment the rings cross the distribution of these non-interacting test particles ceases to give any guide to the possible gas behaviour.

The distortions first become visible near co-rotation, since that is the point at which the mode amplitude is greatest (SK). Depending upon the phase relative to the spiral density maximum, the particles are pushed inwards or outwards and respectively run ahead or lag behind their unperturbed motion. This behaviour causes the stream lines to distort in the manner shown and eventually to wrap over themselves.

Apart from demonstrating where we might expect shocks to develop, this simple model may also give some indication of the structure of the magnetic field. Assuming the unperturbed field structure to thread the test particles along each stream line, the model indicates that we naturally expect to see field reversals in the vicinity of the spiral arms.

2 GAS FLOWS IN BARS

The original papers on gas flows in galactic discs demonstrated that a rotating oval distortion to the galactic potential could drive a spiral response in the gas disc (Sanders & Huntley, 1976) and that a stronger bar could cause offset shocks in positions similar to those of the characteristic dust lanes along bars (Sørensen *et al.*, 1976). Such calculations were thought to be reliable enough that a comparison with observed gas flows could be used to determine such otherwise unobservable properties as the M/L ratio and pattern speed of the bar (Sanders & Tubbs, 1980; Duval & Athanassoula, 1983). Despite the success of this work on matching the flow patterns within the bar (Prendergast, 1983), two problems remain outstanding which together illustrate just how little we still understand.

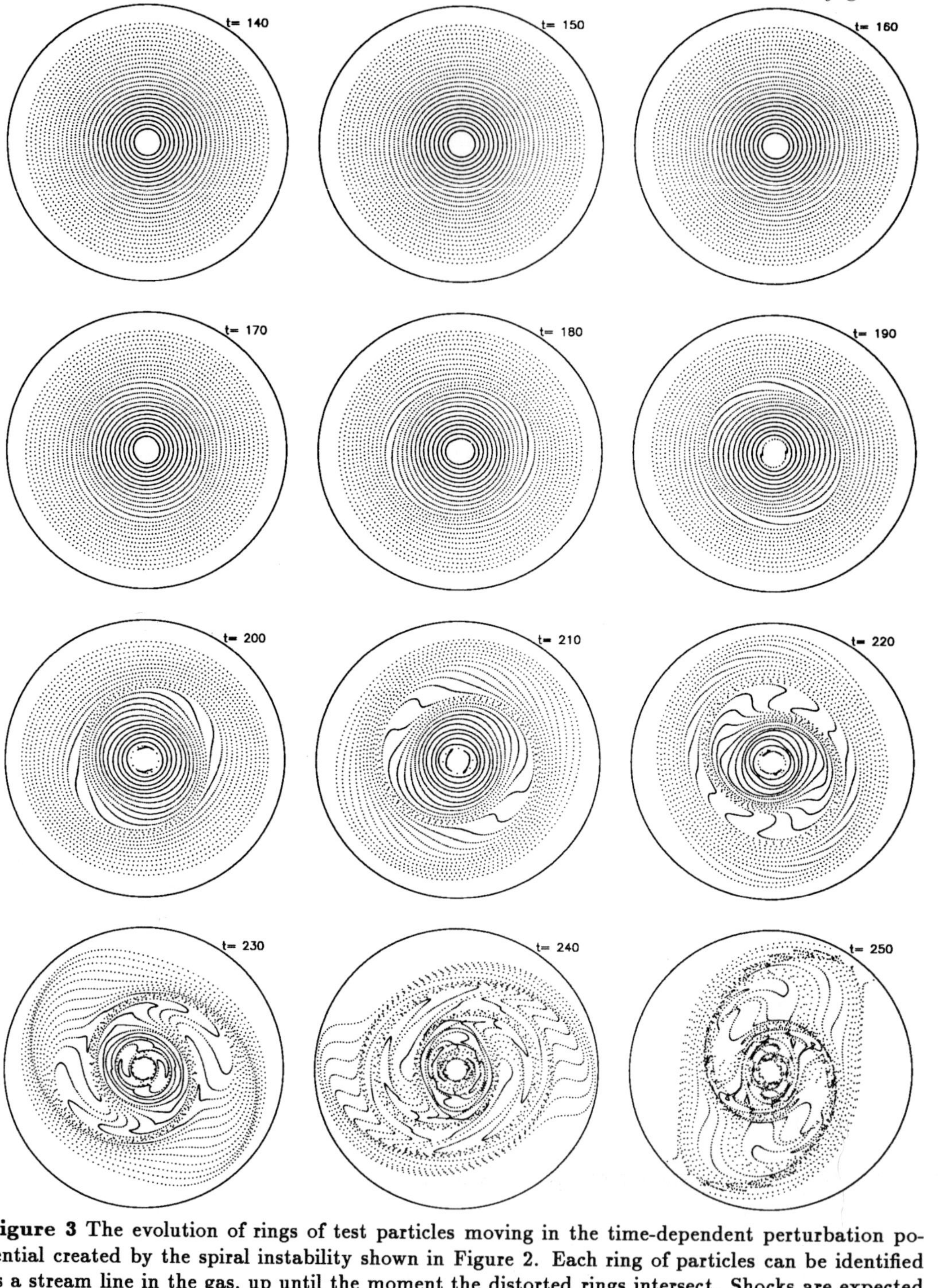

Figure 3 The evolution of rings of test particles moving in the time-dependent perturbation potential created by the spiral instability shown in Figure 2. Each ring of particles can be identified as a stream line in the gas, up until the moment the distorted rings intersect. Shocks are expected where the rings cross.

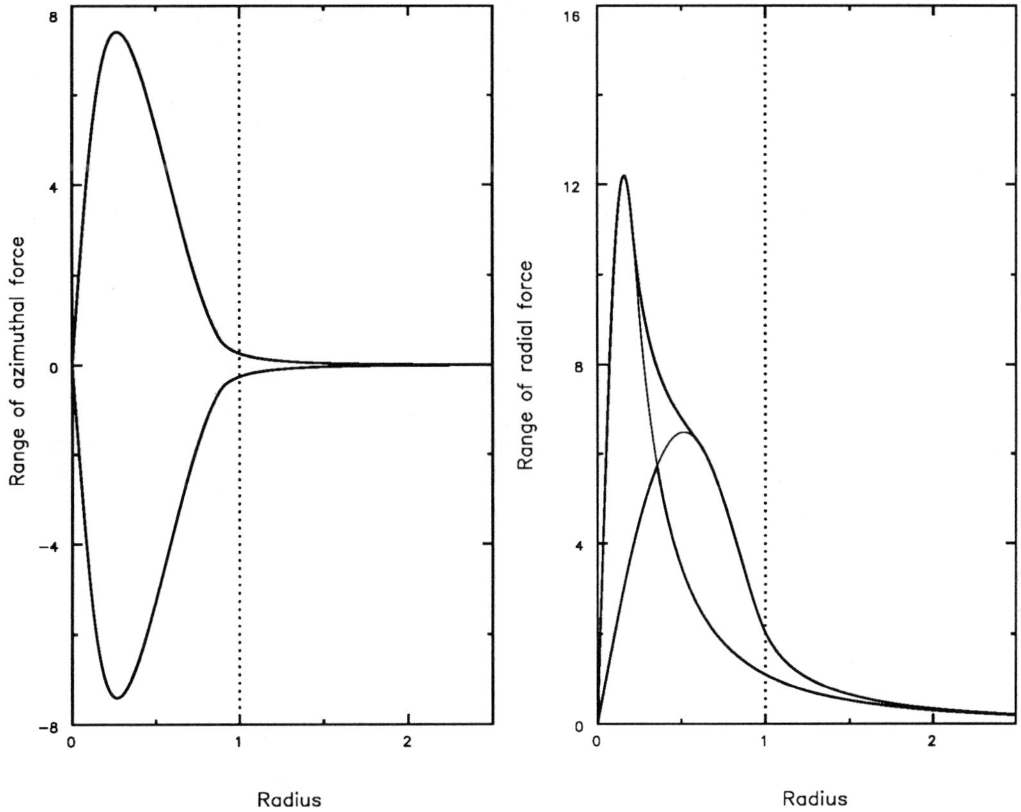

Figure 4 The radial variation of the gravitational force components in a plane containing the major axis of the prolate spheroidal Ferrers bar model having an axis ratio of 4:1. Each heavily drawn line shows the largest or smallest values of the force component around a circle at that radius and the vertical dotted line marks the position of the bar end on its major axis. The radial forces on the major and minor axes are also drawn as lighter lines in the right hand panel. Within the bar, the forces vary strongly with azimuth, but the variations die away very fast outside the bar.

2.1 Driven Spiral Arms

Several authors have commented that a bar of finite length does not drive a spiral response much beyond the end of the bar; among others, both Roberts *et al.* (1979) and England (1988) for NGC 1300 and Hunter *et al.* (1989) for NGC 3992 found that a pure bar did not produce a spiral response of the magnitude and winding seen in these galaxies. Much of the appeal of the simple bar driving model is lost if it becomes necessary to postulate *ad hoc* additional components to the driving potential in order to produce a spiral response comparable to that observed; Roberts *et al.* added an outer spiral component and the Florida group added a global oval distortion.

The principal reason for the failure of this aspect of the bar-driving model is that the quadrupole field of a bar-like mass distribution falls off too fast with distance. Figure 4 shows the peak azimuthal force and the range of the radial force exerted by a strong bar model – an inhomogeneous prolate ellipsoid of axis ratio 4:1 (Ferrers, 1877). It is clear that the forces are becoming very nearly axisymmetric only a little further from the centre than the length of the bar major axis. Since spiral driving depends, amongst other things, upon the non-axisymmetric component of the field, it is hardly surprising that the spiral amplitudes die away so fast in the models. (The non-axisymmetric forces in Sanders & Huntley's original model were a fixed fraction of the axisymmetric forces at all radii and therefore declined unrealistically slowly for a strong bar.)

2.2 Pattern Speeds

The second difficulty of the bar driven models is that the arms in galaxies often appear to lie *inside* co-rotation, whereas bars are expected to end near co-rotation. The direction of radial flow on the minor axis is the most reliable indicator of whether the spiral at that point is in- or outside co-rotation (Kalnajs, 1973), but secondary indications come from the positions of dustlanes along the arms *etc.* Galaxies for which there is evidence that a substantial length of spiral arm lies inside co-rotation include: NGC 1365 (Ondrechen & van der Hulst, 1989), NGC 1097 (Ondrechen, van der Hulst & Hummel, 1989), NGC 1300 (from the weaker evidence of dust lane morphology) and maybe also NGC 7723 (Chevalier & Furenlid, 1978).

If we cling to the assumption that the bar and spiral have the same pattern speed, these data would imply that the bars in these galaxies end well before co-rotation – an uncomfortable conclusion, because there is a significant body of largely circumstantial evidence that co-rotation should lie shortly outside the end of the bar. This has been the prejudice of theoreticians from Contopoulos (1980) onwards, but somewhat stronger evidence comes from the close resemblance noted by Sparke & Sellwood (1987) between their N-body bar, which did end near co-rotation, and the observational data on NGC 936 and from the gas flow models of Sanders & Tubbs (1980) or Duval & Athanassoula (1983) – their models best matched the observed flow *within* the bar of NGC 5383 when they placed co-rotation at the bar end.

2.3 Suggested Solution

The simplest solution which avoids *both* these problems is to drop the assumption that the spirals are driven by the bar (Sellwood & Sparke, 1988). If the spirals are an independent pattern, there is no reason to suppose that their angular rotation rate should be the same as that of the bar, avoiding the apparent dilemma of the pattern speed evidence. Moreover, if the spirals are not forced by the bar, but arise in a similar manner to those in unbarred galaxies, then the rapid fall-off with distance of the quadrupole bar field ceases to be a problem.

The conventional view that the spiral is driven by the bar stems from the *impression* that the arms generally start from the ends of the bar. A close examination of a sample of barred galaxies in any galaxy atlas should convince the reader that there are many galaxies where this is clearly untrue. Sellwood & Sparke also showed that the spirals in their N-body simulation, which rotated more slowly than the bar, gave a clear impression of starting from the bar for well over half the beat period between the two patterns – the overlap of the two decaying density (light) distributions inevitably produces an apparent continuity.

3 THE BUCKLING INSTABILITY OF BARS

Combes & Sanders (1981) were the first to remark that fully 3-D N-body models of barred galaxies can take on a peanut shape when viewed from the side. Combes *et al.* (1990) confirmed their original result in experiments employing altogether superior numerical techniques and a much wider class of barred galaxy models. These authors attribute the bar thickening normal to the plane to the instability of the planar periodic orbits, caused by a vertical ILR within the bar (Pfenniger, 1984).

3.1 Fire-Hose Instability

A similar outcome was observed by Raha *et al.* (1991), but they emphasised that the thickening was preceded by a buckling phase in which the bar bent asymmetrically out of the plane. Both their simulations behaved in a similar manner, but the buckling and disruption of the bar were more pronounced in that with the thinner disc. Figure 5 shows measurements of the mean z-height of the particles, averaged over all azimuths, as functions of both radius and time in their thicker disc simulation. The bend appears to grow exponentially up to some limiting amplitude, after which its coherence is lost as the structure quickly disrupts. In the final state, the bar regains reflection symmetry about the plane but is considerably thicker in z, especially at the ends. The disruption of the bend coincides with a reduction in the bar strength in the plane and an increase in the central density of the model.

Raha *et al.* interpret the initial bending phase in terms of the fire-hose instability, which has also recently been discovered in non-rotating bars by Merritt & Hernquist (1991). A slight departure from a precise plane requires particles moving along the length of the bar to be accelerated in the z-direction in order to stay within the bar. When the gravitational restoring force to the plane is insufficiently strong to provide the required z acceleration, the bend grows sharper, *i.e.* the bar buckles. Toomre (1966) and others (*e.g.* Fridman & Polyachenko, 1984; Araki, 1985) concluded that the random motions of the stars normal to the plane were likely to be sufficient to quench this instability in the rotationally supported disc of an unbarred galaxy. Bar formation, however, makes the orbits of stars much more eccentric and aligns their principal axes to lie close to the bar major axis, making the bar more vulnerable to buckling instabilities.

The fire-hose instability is a collective dynamical instability. Conclusive evidence that the growing asymmetry in the potential is important comes from the report by Friedli & Pfenniger (1990) that the thickening process was greatly slowed when they

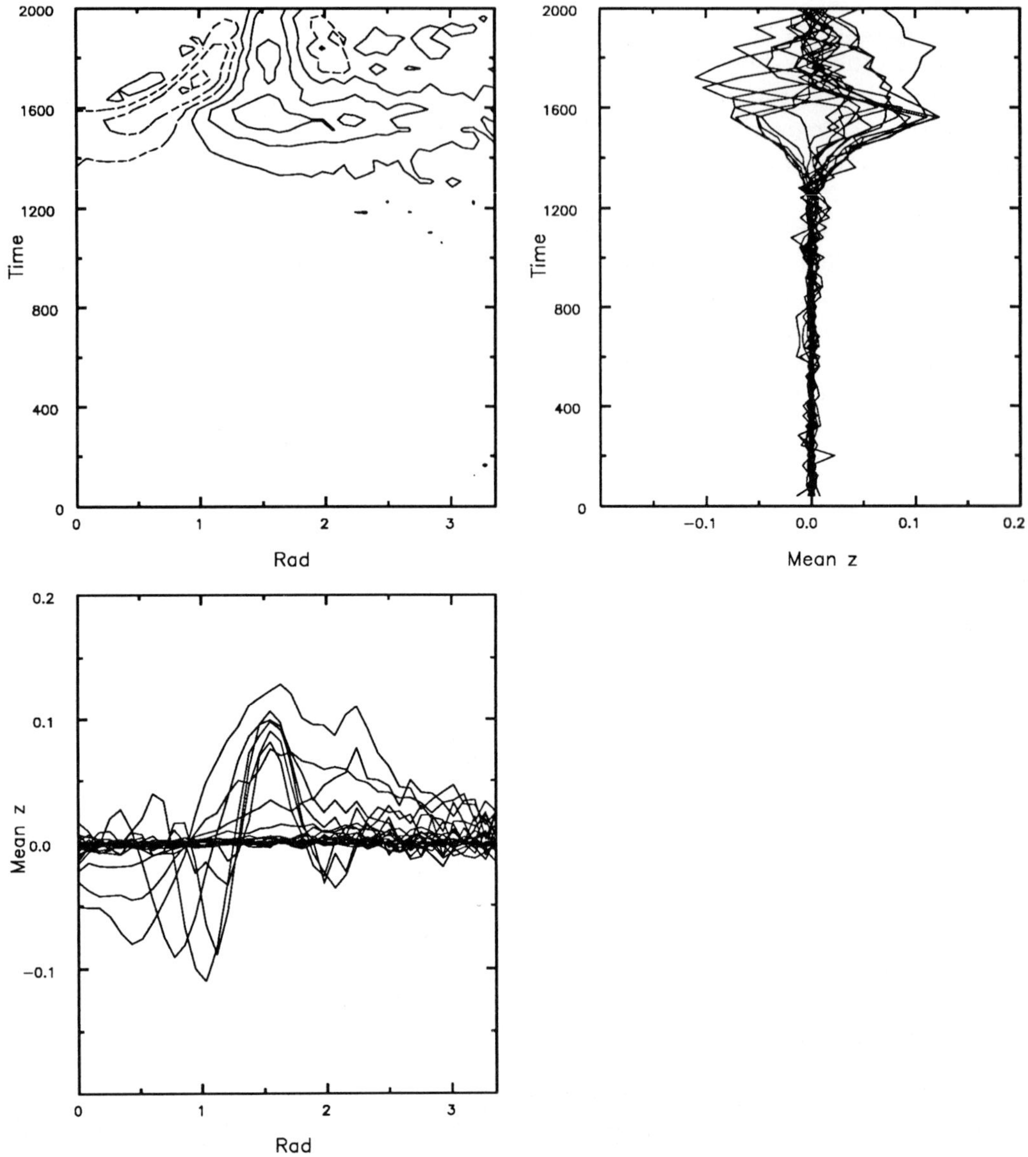

Figure 5 The top left panel shows contours of the mean z height of particles as a function of radius and time in the thick disc simulation reported by Raha *et al.* (1991); the bar in this model is not fully formed until about time 1000. Full drawn contours indicate positive (upward) displacements and downward displacements are marked by dashed contours. The other two panels show projections of this surface onto the two axes. The top right panel shows the approximate exponential growth of the displacements up to time $\sim$ 1600, and the lower left shows that the simple bend develops structure on shorter scales at late times.

imposed reflection symmetry about the galactic plane. The hypothesis by Combes *et al.*, that the peanut shape grows because the periodic orbits are unstable, therefore requires further clarification.

Combes *et al.* also drew attention to the pair of stable 2:2:1 "banana shaped" families of periodic orbits, which are bent away from the plane at both ends of the bar. It is very likely that orbits trapped about this family play an important role in supporting the final peanut shape.

3.2 Gas Flow During the Instability

As for the time-dependent spiral structure, computing the gas flow during the course of this instability would be a major undertaking and has not so far been attempted. Nevertheless, a number of considerations suggest that this would be interesting.

The flow in the mid-plane of the bar is unlikely to differ much from the type of quasi-static flow pattern referred to in §2 as the bend grows exponentially. However, as the buckling instability disrupts, the fluctuating potential pumps energy into the vertical oscillations of both the stars and gas clouds in the bar. The gas clouds must quickly dissipate their energy of random motion and resettle to the plane and, probably, also towards the galactic centre.

A further very interesting possibility is that the fire-hose instability could recur several times during the life of a galaxy. As new stars form within the gas layer in the bar, a new population of stars will be created which has large organised streaming motions along the bar, and very little random motion normal to the plane. It is not clear whether scattering processes could raise the vertical velocity dispersion of this new group of stars on a reasonable time-scale, but if the new population remains thin it must again become unstable and would stir up the gas layer again.

References

Araki, S., 1985. *PhD thesis,* MIT.

Carlberg, R. G. & Freedman, W. L., 1985. *Astrophys. J.,* **298**, 486.

Chevalier, R. A. & Furenlid, I., 1978. *Astrophys. J., ,* **225**, 67.

Combes, F. & Sanders, R. H., 1981. *Astr. Astrophys.,* **96**, 164.

Combes, F., Debbasch, F., Friedli, D. & Pfenniger, D., 1990. *Astr. Astrophys.,* **233**, 82.

Contopoulos, G. 1980. *Astr. Astrophys.,* **81**, 198.

Duval, M. F. & Athanassoula, E., 1983. *Astr. Astrophys.,* **121** 297.

England, M. N., 1989. *Astrophys. J.,* **344**, 669.

Ferrers, N. M., 1877. *Quart. J. Pure Appl. Math.*, **14**, 1.

Fridman, A. M. & Polyachenko, V. L., 1984. *Physics of Gravitating Systems*, Springer-Verlag, New York.

Friedli, D. & Pfenniger, D., 1990. In *Bulges of Galaxies*, Proceedings of ESO workshop No. 35, p. 265, eds. Jarvis, B. J. & Terndrup, D. M., ESO.

Hunter, J. H., Ball, R., Huntley, J. M., England, M. N. & Gottesman, S. T., 1988. *Astrophys. J.,* **324**, 721.

Julian, W. H. & Toomre, A., 1966. *Astrophys. J.,* **146**, 810.

Kalnajs, A. J., 1973. *Proc. Astron. Soc. Australia*, **2**, 174.
Lovelace, R. V. E. & Hohlfeld, R. G., 1978. *Astrophys. J.*, **221**, 51.
Lynden-Bell, D. & Kalnajs, A. J., 1972. *Mon. Not. R. astr. Soc.*, **157**, 1.
Merritt, D. & Hernquist, L., 1991. *Astrophys. J.*, **376**, 439-457.
Ondrechen, M. P., & van der Hulst, J. M., 1989. *Astrophys. J.*, **342**, 29-38.
Ondrechen, M. P., van der Hulst, J. M. & Hummel, E., 1989. *Astrophys. J.*, **342**, 39-48.
Papaloizou, J. C. B. & Savonije, G. J., 1989. In *Dynamics of Astrophysical Discs*, p. 103, ed. Sellwood, J. A., Cambridge University Press.
Pfenniger, D. 1984. *Astr. Astrophys.*, **134**, 373-386.
Prendergast, K. H., 1983. In *Internal Kinematics and Dynamics of Galaxies*, IAU Symposium **100**, p. 215, ed. Athanassoula, E., Reidel.
Raha, N., Sellwood, J. A., James, R. A. & Kahn, F. D., 1991. *Nature* **352**, 411-412.
Roberts, W. W., Huntley, J. M., van Albada, G. D., 1979. *Astrophys. J.*, **233**, 67.
Sanders, R. H. & Huntley, J. M., 1976. *Astrophys. J.*, **209**, 53.
Sanders, R. H. & Tubbs, A. D., 1980. *Astrophys. J.*, **235**, 803.
Sellwood, 1991. In *Dynamics of Disc Galaxies*, ed Sundelius
Sellwood, J. A. & Carlberg, R. G., 1984. *Astrophys. J.*, **282**, 61.
Sellwood, J. A. & Kahn, F. D., 1991. *Mon. Not. R. astr. Soc.*, **250**, 278 (SK).
Sellwood, J. A. & Lin, D. N. C., 1989. *Mon. Not. R. astr. Soc.*, **240**, 991.
Sellwood, J. A. & Sparke, L. S., 1988. *Mon. Not. R. astr. Soc.*, **231**, 25p.
Sørensen, S-A., Matsuda, T. & Fujimoto, M., 1976. *Astrophys. Sp. Sci.*, **43**, 491.
Sparke, L. S. & Sellwood, J. A., 1987. *Mon. Not. R. astr. Soc.*, **225**, 653.
Toomre, A., 1966. In *Geophysical Fluid Dynamics*, notes on the 1966 Summer Study Program at the Woods Hole Oceanographic Institution, ref. no. 66-46, p111.
Toomre, A., 1981. In *Structure and Evolution of Normal Galaxies*, p. 111, eds. Fall, S. M. & Lynden-Bell, D., Cambridge University Press.
Toomre, A., 1989. In *Dynamics of Astrophysical Discs*, p. 153, ed. Sellwood, J. A., Cambridge University Press.

The Cold Interstellar Gas and the Stability of Thick Galactic Discs

ALESSANDRO B. ROMEO

Tuorla Observatory, Turku University, 21500 Piikkiö, Finland

1 INTRODUCTION

The spiral structure theory relies on a number of working assumptions which allow to make the linearized system of the coupled fluid (or Vlasov) and Poisson equations more tractable. However, in some situations of astrophysical interest the validity of such assumptions may be questioned. For instance, we know that real galactic discs have finite, although small, thickness and the possibility of regarding them as infinitesimally thin depends on the relevant wavelengths of the perturbations excited. In some cases, when the underlying spiral structure has a high winding degree, finite-thickness effects should be taken into account in the stability analysis. In view of the importance that such effects may have in the *self-regulation* mechanisms which are expected to operate in galactic discs and to be at the basis of their secular heating (Romeo 1987, 1990), we have tried to evaluate them. This can be done only after that their vertical structure at equilibrium has *carefully* been investigated.

In performing this analysis, we have made use of simplified models of galactic discs in which the stars and the cold interstellar gas are treated as two different components. Although such models might be thought of as being inaccurate to describe actual galactic discs, which are known to consist of different populations of stars and gas components, they incorporate indeed the most essential features as regard their stability properties. In this context it should be noted that such a single equivalent stellar component is taken to be representative of the whole active stellar disc consisting of low-velocity dispersion stars (high-velocity dispersion stars do not participate appreciably in spiral structure), whereas the single gaseous component is taken to simulate HI regions of neutral atomic hydrogen and giant molecular clouds and complexes. Generally, even more drastically simplified galactic models are used, in which the cold interstellar gas is not taken into account. In some situations of astrophysical interest this further simplification is not justified, because the cold interstellar gas is expected to play an important or even crucial role in the stability of galactic discs due to its low turbulent velocities (cf. infinitesimally thin case discussed in Romeo 1985, 1989, 1990; Bertin and Romeo 1988).

The basis of our investigation is the stability analysis performed by Vandervoort (1970) in the case of one-component purely stellar discs. The method we have employed is in fact a straightforward extension of that developed by Vandervoort (1970) to the case in which two components are present (the consideration of a fluid component does not give rise to difficulties), but the resulting analysis is indeed much more complicated than in the one-component case. For a more detailed account of the results presented in this paper see Romeo (1991a,b).

2 ANSATZ FOR THE REDUCTION FACTORS

In contrast to the one-component case, finite-thickness corrections to the local dispersion relation do not generally reduce to a simple scaling of the surface densities of the two components. The use of two corresponding reduction factors can reasonably be justified only when the two components are not strongly coupled. For the moment we are not able to express this statement in a precise mathematical form. In most situations of astrophysical interest, such as in the case in which the two components are identified with the stars of the active disc and the cold interstellar gas, this condition is rather general and is fulfilled throughout the galactic disc, except possibly in the outermost parts where the mass density of the cold interstellar gas becomes comparable to that of low-velocity dispersion stars. It is to such cases that our analysis is devoted.

The two components are denoted by different subscripts, the stars of the active disc by "H" (HOT) and the cold interstellar gas by "C" (COLD), in order to recall that they are characterized by different velocity dispersions. When the two components are not strongly coupled (cf. the non-restrictive condition mentioned above), the corresponding reduction factors should fairly well be approximated by the **Effective ansatz**

$$\mathfrak{T}_{\mathrm{i}} \equiv \frac{\sigma_{\mathrm{red\,i}}}{\sigma_{\mathrm{i}}} \simeq \mathfrak{T}_{\mathrm{eff\,i}} \equiv \frac{1}{1+|k|z_{\mathrm{eff\,i}}} \quad [\,|k|z_{\mathrm{eff\,i}} = O(1)\,] \quad (\mathrm{i}=\mathrm{H},\mathrm{C})\,, \tag{1}$$

where we have indicated the formal ordering which we expect to be required by the underlying approximation. In this formula, k is the local radial wave-number of the perturbation and $z_{\mathrm{eff\,i}}$ are the **Effective thickness-scales** of the two components:

$$z_{\mathrm{eff\,i}} \equiv \frac{\sigma_{\mathrm{i}}}{2\rho_{0\mathrm{i}}} \quad (\mathrm{i}=\mathrm{H},\mathrm{C})\,, \tag{2}$$

σ_{i} and $\rho_{0\mathrm{i}}$ being their unperturbed surface densities and volume densities in the plane, respectively. Asymptotically, our ansatz matches the reduction factors derived by Shu (1968) when the system approaches the limiting cases in which only one component is present or two components with the same vertical velocity dispersions can be identified, apart from the obvious infinitesimally thin case.

3 LOCAL STABILITY

The qualitative discussion given in the previous section will now be quantified, by studying in the framework of a fluid approach the marginal stability curves corresponding to the ansatz introduced above. The results obtained in the infinitesimally thin case discussed in Romeo (1985, 1989, 1990), Bertin and Romeo (1988) are at the basis of this further investigation.

We have adopted the following scaling and basic parametrization (cf. infinitesimally thin case):

$$\bar{\lambda} \equiv \frac{k_{\mathrm{H}}}{|k|}, \quad \text{where} \quad k_{\mathrm{H}} \equiv \frac{\kappa^2}{2\pi G \sigma_{\mathrm{H}}}; \tag{3}$$

$$\alpha \equiv \frac{\sigma_{\mathrm{C}}}{\sigma_{\mathrm{H}}}, \quad \beta \equiv \frac{c_{\mathrm{C}}^2}{c_{\mathrm{H}}^2} \quad (0 < \alpha < +\infty, \ 0 < \beta < 1); \tag{4}$$

$$Q_{\mathrm{H}} \equiv \frac{c_{\mathrm{H}} \kappa}{\pi G \sigma_{\mathrm{H}}} \quad \textsf{(Local Stability parameter)}; \tag{5}$$

and the additional **Wave-Number parametrization**

$$\zeta_{\mathrm{H}} \equiv k_{\mathrm{H}} z_{\mathrm{eff\,H}}, \quad \beta_{z\,\mathrm{eff}} \equiv \frac{z_{\mathrm{eff\,C}}}{z_{\mathrm{eff\,H}}} \quad (0 < \beta_{z\,\mathrm{eff}} < 1), \tag{6}$$

similar to that adopted by Shu (1968). The case $\zeta_{\mathrm{H}} = 0$ represents the limit of an infinitesimally thin system, and the case $\beta_{z\,\mathrm{eff}} = 1$ represents the limit of a system in which the two components have the same scale-height (and hence the same vertical velocity dispersions). The marginal stability curve in the $(\bar{\lambda}, Q_{\mathrm{H}}^2)$ plane can be derived from its uncorrected expression, by applying the reduction factors to each dimensionless quantity involving the surface densities. A local stability criterion can be stated in analogy with the case of infinitesimally thin one-component systems. A function $\bar{Q}^2 = \bar{Q}^2(\alpha, \beta, \zeta_{\mathrm{H}}, \beta_{z\,\mathrm{eff}})$, which reduces to unity when $\alpha = 0$ and $\zeta_{\mathrm{H}} = 0$, can be defined in such a way that when $Q_{\mathrm{H}}^2 > \bar{Q}^2$ the system is locally stable at all wavelengths. For $Q_{\mathrm{H}}^2 < \bar{Q}^2$ we expect the system to be locally unstable in ranges of wavelengths defined by the marginal stability condition. Note in this context that the ratio

$$Q_{\mathrm{eff}} \equiv \frac{Q_{\mathrm{H}}}{\bar{Q}} \quad \textsf{(Effective Local Stability parameter)} \tag{7}$$

plays the role of an effective Q-parameter (cf. infinitesimally thin one-component case). Consistently with a mechanism of self-regulation, we expect

$$Q_{\mathrm{eff}}(r) = 1 \quad \textsf{(Self-Regulation constraint)} \tag{8}$$

throughout the active disc. The self-regulation constraint on the profile of the effective local stability parameter will be considered in future applications.

In presenting the results of the local stability analysis performed in this paper, we have considered the standard star-dominated and the peculiar gas-dominated regimes already investigated in the less general context of infinitesimally thin two-component models of galactic discs (Romeo 1985, 1989, 1990; Bertin and Romeo 1988). Moreover, other regimes of the relevant local parameters which are typical in the solar neighbourhood have been considered. The marginal stability curves derived according to the effective ansatz in the context of the wave-number parametrization are shown in Figure 1. As a result, when finite-thickness effects are taken into account, the system tends to be more stable. More precisely, the stabilizing role of such effects can partially counterbalance the destabilizing role of the cold interstellar gas in linear regimes. As a consequence of these competing roles, in some cases the monotonicity properties of the marginal stability curve are qualitatively different from the infinitesimally thin case. Note in this context that the profile of the effective local stability parameter and the shape of the marginal stability curve are of *primary* importance in determining the propagation properties of the spiral waves and thus the *global* stability properties of galactic discs.

REFERENCES

Bertin G. and Romeo A. B. (1988), *A. Ap.* **195**, 105.

Romeo A. B. (1985), *Tesi di Laurea*, University of Pisa and Scuola Normale Superiore, Pisa, Italy.

Romeo A. B. (1987), *M. Phil. Thesis*, SISSA, Trieste, Italy.

Romeo A. B. (1989), in *Dynamics of Astrophysical Discs*, Cambridge University Press, Cambridge, Sellwood J. A. ed., p. 209 (Manchester, UK, 1988).

Romeo A. B. (1990), *Ph. D. Thesis*, SISSA, Trieste, Italy.

Romeo A. B. (1991a), "Stability of Thick Two-Component Galactic Discs: I. Vertical Structure at Equilibrium", submitted to *M.N.R.A.S.*.

Romeo A. B. (1991b), "Stability of Thick Two-Component Galactic Discs: II. Local Finite-Thickness Effects", submitted to *M.N.R.A.S.*.

Shu F. H. (1968), *Ph. D. Thesis*, Harvard University, Cambridge, Massachusetts USA.

Vandervoort P. O. (1970), *Ap. J.* **161**, 87.

Figure 1. Two-fluid marginal stability curves in the $(\bar{\lambda}, Q_{\mathrm{H}}^2)$ plane derived evaluating the reduction factors of the two components according to the effective ansatz for some values of the local parameters ζ_{H}, $\beta_{z\,\mathrm{eff}}$, α and fixed $\beta = .01$ (top), and for values of the same local parameters corresponding to typical lower and upper bounds in the solar neighbourhood (bottom).

Figure 1

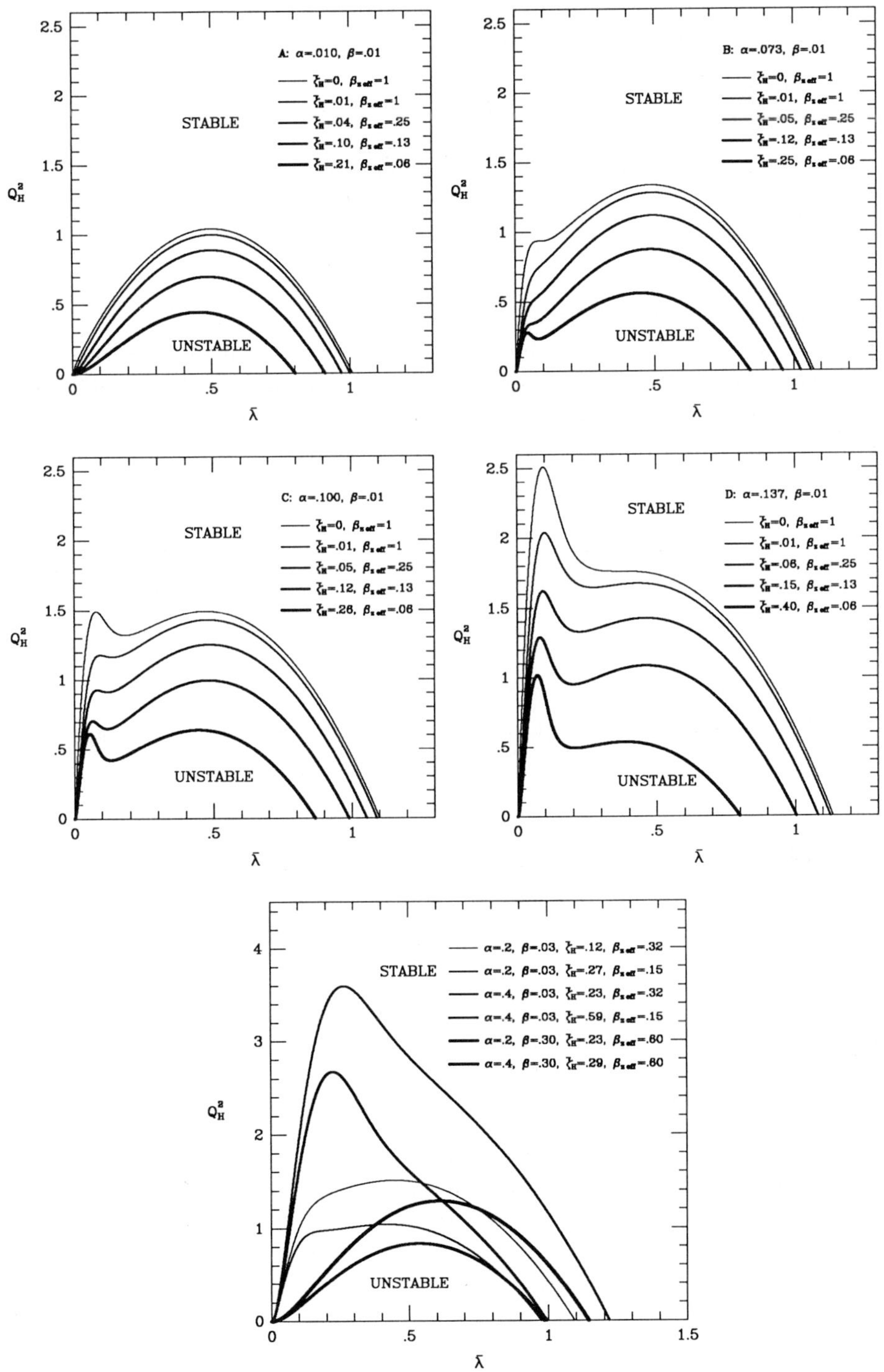
A: α=.010, β=.01
ζ_H=0, β_x eff=1
ζ_H=.01, β_x eff=1
ζ_H=.04, β_x eff=.25
ζ_H=.10, β_x eff=.13
ζ_H=.21, β_x eff=.06
STABLE
UNSTABLE
B: α=.073, β=.01
ζ_H=0, β_x eff=1
ζ_H=.01, β_x eff=1
ζ_H=.05, β_x eff=.25
ζ_H=.12, β_x eff=.13
ζ_H=.25, β_x eff=.06
STABLE
UNSTABLE
C: α=.100, β=.01
ζ_H=0, β_x eff=1
ζ_H=.01, β_x eff=1
ζ_H=.05, β_x eff=.25
ζ_H=.12, β_x eff=.13
ζ_H=.26, β_x eff=.06
STABLE
UNSTABLE
D: α=.137, β=.01
ζ_H=0, β_x eff=1
ζ_H=.01, β_x eff=1
ζ_H=.06, β_x eff=.25
ζ_H=.15, β_x eff=.13
ζ_H=.40, β_x eff=.06
STABLE
UNSTABLE
α=.2, β=.03, ζ_H=.12, β_x eff=.32
α=.2, β=.03, ζ_H=.27, β_x eff=.15
α=.4, β=.03, ζ_H=.23, β_x eff=.32
α=.4, β=.03, ζ_H=.59, β_x eff=.15
α=.2, β=.30, ζ_H=.23, β_x eff=.60
α=.4, β=.30, ζ_H=.29, β_x eff=.60
STABLE
UNSTABLE
Q_H^2
λ̄

Dynamical Systems: Transdisciplinary Mathematical Tool or Fiction ?

IVAN DVOŘÁK

Laboratory for Applied Mathematics and Bioengineering Prague Psychiatric Center Ústavní 91, 181 03 PRAGUE 8, Czechoslovakia

1. INTRODUCTION

To someone, the title of this contribution may seem provocative. It was chosen intentionally so. In our everyday scientific life we are too often dragged by the stream of facts, evidences, hypotheses and theories, perhaps we spend too little time by critical evaluation of our methods and concepts. In this contribution I want to stop for a moment and together with the reader critically reconsider one of our nowadays widely used theoretical concepts - that brought by the theory of deterministic chaos. We shall carefully weight its merits and pitfalls and we shall illustrate them by the examples of its application for analysis of possibly the most complex system ever known - that of the brain.

2. DYNAMICAL SYSTEMS AND DETERMINISTIC CHAOS

The reader of this contribution is supposed to be familiar with basic concepts of the theory of deterministic chaos. Just to unify the vocabulary and remind the way of reasoning let us summarize in this paragraph the basic definitions.

Theory of deterministic chaos is also sometimes called an ergodic theory of dynamical systems. This name or the other, it is a part of mathematics, a branch of the nonlinear dynamics. It deals with object called dynamical systems.

By a dynamical system we understand a smooth mapping $\mathbf{F}_t : \mathbf{R}^M \rightarrow \mathbf{R}^M$, where $\mathbf{R}^M$ is an M-dimensional manifold and t (time) is a parameter, that fulfills the following conditions

$$\mathbf{F}_0 = \text{identity}, \qquad \mathbf{F}_t \oplus \mathbf{F}_s = \mathbf{F}_{t+s}.$$

Dynamical systems can be either continuous or discrete. The most typical example of the continuous dynamical system is the set of (nonlinear) ordinary differential equations of the first order

$$\frac{d\mathbf{X}(t)}{dt} = \mathbf{f}(\mathbf{X}(t), \lambda), \tag{1}$$

where $\mathbf{f}$ is a smooth vector function, $\mathbf{X}$ is an m-dimensional vector from $\mathbf{R}^M$ and λ is a vector of parameters.

The most common example of the discrete dynamical systems is a set of difference equations of the first order

$$\mathbf{X}_{n+1} = \mathbf{f}(\mathbf{X}_n, \lambda). \tag{2}$$

In this equations the symbols have the same meaning as in the preceding case, only continuous dependence on time has been replaced by discrete dependence on the index n.

Definition of a dynamical system includes rather large set of different mathematical objects. In the theory of deterministic chaos we study usually much narrower set of systems that can be described by equation of type (1) or (2). Moreover we are interested only in those dynamical systems that have in some sense "complex" solution.

Manifold $\mathbf{R}^M$ is usually called the *state space*. Solution of dynamical system (1) or (2) is represented in the state space by a curve called the *state trajectory*. Complexity of the solution is then characterized by complexity of this trajectory. Usually at least fulfillment of two requirements is demanded if the trajectory should be regarded as "complex".
1) It should depend sensitively on initial conditions (slight change of initial conditions brings eventually entirely different trajectory).
2) When external conditions are kept steady and the transients have died out the state trajectory should be confined only to a limit set called an *strange attractor*, which is the subspace of the space $\mathbf{R}^M$ with fractal structure (its Hausdorff dimension is different from its topological dimension) and zero Lebesque measure. (Let us recall that strange attractor is a special case of a general attractor which should have Lebesque measure zero but need not have the fractal strucutre.)

As a result of the first condition the state trajectory of the dynamical system is unpredictable in the long run. As a result of the second condition mappings of the volume element of the state space converge for $t \to \infty$ to zero. (Let us remark that this property is characteristic only for so called *dissipative systems*. For so called conservative systems the volume element is preserved.)

If the state trajectory fulfills the listed requirements it is usually called *chaotic*. Also the solution giving rise to this trajectory is called chaotic. Dynamical system with chaotic solution is then called a *chaotic dynamical system*. Adjective *dissipative* is also added sometimes to stress this propery.

3. APPLICATIONS OF THE THEORY OF DETERMINISTIC CHAOS

Existence of chaotic systems as mathematical objects was probably known already to Poincaré by the end of the last century. Entirely new horizon have been opened, however, when chaos was detected in natural systems. E. Lorenz is always quoted as the first, who discovered chaos in physical system when he studied his model of dynamics of the Earth atmosphere (1963). Other discoveries of chaos in existing systems were to follow soon. For some time physics, and especially hydrodynamics, mastered chaos, but with surprising rapidity it get out of any control. In a decade theory of deterministic chaos has found its application in the whole spectrum of natural and social sciences starting from quantum theory and other physical or physics related

hydrodynamics	transition to turbulence	Swiney & Gollub (1981)
meteorology	weather changes climatic attractor	Lorenz (1963) Nicolis & Nicolis (1984)
astronomy	motion of planets in the solar system	Sussman and Wisdom (1988) Laskar (1989), Hartley (1990)
astrophysics	solar radiation	Kurths & Herzel (1987)
chemistry	Belousov-Zhabotinskii reaction	Roux (1984)
biochemistry	biochemical systems	Ricard &Cornish-Bowden (1984)
ecology	dynamics of biological population Canadian lynx	May (1974) Schaffer (1985)
electro-physiology	EEG analysis neuronal activity	Babloyantz (1985) Dvořák & Šiška (1986) Rapp et al. (1985)
physiology	physiological regulations cellular metabolism heartbeat rate	Mackey & Glass (1977) Rapp (1986) Glass et al. (1985)
psychophysics	response to physical stimuli	Gregson (1988)
psychophysiology	dysfunctions in pursuit eye movements	Huberman (1986)
epidemiology	incidence of measles	Schaffer & Kot (1985)
politology	international arms races	Grossman & Mayer-Kress (1989)

Table 1.: Some examples of application of the theory of deterministic chaos in natural and social sciences

sciences like astronomy, through chemical and biological sciences, to psychology, sociology and even economy and politology. Table 1., which is neither representative nor exhaustive, brings some typical examples of applications of the theory of deterministic chaos till the end of 1990.

Spreading of chaotic paradigm and the span of its applications are striking. It looks even a little suspicious. We all know that the progress of science is driven not only by rational reasoning and founded calculations but also by passion, intuition, group thinking, and sometimes also by fashion. We must therefore ask: is the theory of deterministic systems really such a powerful transdisciplinary tool? Are the chaotic systems really so common ? Or is it all fashion or even fiction ?

Our worries are not uncommon. In a broader context the problem was discussed by Ruelle (1990). In the following we are going to show that deterministic chaos theory brings very powerful and universal paradigm. In this respect widespread use of its ideas and concepts is founded and should be welcome. On the other hand, wide spreading of its mathematical methods and even numerical techniques is not founded and sometimes can be even very problematic.

4. MATHEMATICAL MODELS AND ANALYSIS OF DATA

In applications of deterministic chaos methods in natural and social sciences we can distinguish two principally different ways. They differ each other with respect to our a priori knowledge on the studied system. The first way brings applications based on the mathematical model of the studied system or process. The second brings those without such a model, where methods of deterministic chaos are used as pure data analysis techniques. We shall look at both of them in some detail.

Let us suppose we study a natural system exhibiting (supposedly) chaotic behavior. Application of deterministic chaos methods for investigation of this behavior is much easier if we know the basic laws governing the dynamics of the studied system. Starting from these laws we can derive a concrete mathematical model describing dynamics of this system. Equations of this model can be solved and values of its parameters giving rise to the chaotic solution can be determined. Obtained results can be compared with experimental data measured on the system. Results of this comparison would confirm or reject our model and in this way also our application and interpretation of the basic laws. This relatively favorable situation is encountered in meteorology (Lorenz model), hydrodynamics (Navier-Stokes equations), astronomy (Newton's gravitation law), ecology (Lotka-Volterra type equations), membranology (membrane equations) and other sciences.

Situation when the basic laws governing the behavior of the observed (supposedly) chaotic system are not known is more problematic. The only possible way how to apply the theory of deterministic chaos is to suppose some type of a mathematical model. Than we can apply some of the data analysis techniques developed in the theory (their development being usually stimulated by problems from other field of science) for the supposed type of the model. This is the case of majority of biological sciences, especially brain sciences, the case of economy, social sciences, politology etc.

Differences between the two situations are evident. While in the first case we can rely on the exact theory, in the second case we are bogged in the uncertain ground of assumptions that should be verified in each step of the data analysis. No wonder that results obtained in the first case are usually much more founded and thus reliable than results obtained in the second.

5. ANALYZING DATA BY METHODS OF CHAOTIC DYNAMICS

Situation when theoretical model is known is familiar to many a reader. We shall therefore focus our attention on the situation when the mathematical model of the studied system is not known. We shall shortly summarize the way how approaches of the theory of deterministic chaos are applied as data analysis techniques and point out pitfalls of this approach.

Let us deal with more complex case when we deal with multi-dimensional experimental time series registered on the spatio-temporal physical (biological, social) system. When the system has only time as an independent variable the situation is qualitatively simpler. These time series are supposed to be projections (smooth mappings) of the state trajectory representing the dynamics of the dynamical system supposed to underlay the dynamics of the studied system in the state space of (unknown) dimension M.

Two steps should be accomplished during the analysis. First, the state trajectory representing the dynamics of the system in the state space should be reconstructed, second the attractor of the system should be properly and simply characterized. In principle there are two ways how the trajectory representing the dynamics of the studied system can be reconstructed from the (multidimensional) data.

In the first way we use single components of the multi-dimensional time series one by one. For each we reconstruct its individual state space by the method of delayed coordinates introduced by Takens (1981) Given the i-th component of the N-dimensional series

$$X^i = \{X_1, X_2, ...X_k, X_{k+1}, ..., X_L\},$$

where L is its length, we reconstruct an M-dimensional time series in the following way

$$^{M}\mathbf{X}_{j}^{i} = [X_j, X_{j+d}, ..., X_{j+(M-1)d}],$$

where d is properly chosen time delay.

The reconstructed state space (and the state trajectory) are relevant to the site, where the component of the multidimensional signal was registered. Thus this method gives "local description".

The other methods follow from the conjecture published by Ruelle and Eckmann (1985). Components of the multidimensional vector are taken directly as components of the state space vector. This state space is naturally relevant to the whole studied system or at least to the part where the components of the signal were registered. This method gives thus "global description".

The second step in the analysis of chaotic dynamics consists usually in estimating values of *dynamical invariants* of the system. This term denotes parameters invariant with respect to smooth transformation of coordinates of the state space that are chosen to characterize quantitatively the dynamics of the studied system. There are various dynamical invariants defined for characterization of the chaotic processes (dimensions, Lyapunov characteristic exponents, K-entropy etc.). Majority of them is derived for characterization of the *probability measure* which is generated by the reconstructed state trajectory. These dynamical invariants are also called *complexity measures*. Complexity measures usually classify processes generated by the studied dynamical systems according to their positions on a scale between the constant (the most simple) process and the white noise (the most complex process). For completeness let us remark that sometimes parameters that need not be invariant with respect to coordinate transformation are also called the complexity measures.

Reconstruction of the state space is complicated by the fact that we do not know a priori its dimension M. Let us suppose the system dynamics is N-dimensional, that means it can be properly represented in a natural state space with finite dimension N (which we just do not know). Choosing M - the dimension of reconstructed state space - smaller than N brings loss of information. If M is greater than N, the representation of the studied dynamics may be faithful. There is mathematical theorem (Takens, 1981) proving that if M equals 2N+1 or more, faithfulness of the representation is ensured. Reconstructed state space then represents an embedding in the sense of Whithey and it is therefore homeomorphic to the natural state space. M should be therefore chosen large enough to ensure this property (M is usually called the *embedding dimension*). In practice, proper M is found by repeating the procedure of reconstructing the state space and estimating the value of some complexity measure for increasing values of M. It is supposed that when proper value of M is reached, the estimated value of the complexity measure should saturate on some level. This level is then declared as the value of this complexity measure for the given signal.

In context of EEG studies it is the complexity measure called *correlation exponent* that is the most popular. It is the scaling exponent of the Renyi H_r entropy of order two which is defined as

$$H_R(r) = \mathrm{Log} \sum_i p_i^2(r),$$

where p_i is the probability of visiting i-th cube of the r-grid covering the state space. It is supposed that when r - the scale of the grid - goes to zero, H_R scales as

$$H_R(r) \simeq r^\nu,$$

where ν is just the correlation exponent.

Correlation exponent has convenient properties. For constant processes the correlation exponent has value zero while, for the white noise it equals to M (the embedding dimension) and with increasing embedding dimension tends to infinity. For deterministic processes the value of correlation exponent is finite, for chaotic processes it even need not be integer. The higher the value of the correlation exponent the less determined the process. Common interpretation of the correlation exponent states that rounded up to nearest integer it gives the number of independent variables that are fully sufficient for characterization of the studied process. This interpretation enlightens the meaning of complexity measured by the correlation exponent: the more complex process needs more independent variables for its full characterization.

Grassberger and Procaccia (1983, 1984) published an efficient algorithm (GPA) for estimating the correlation exponent from reconstructed state trajectory. This algorithm has been later improved in order to compensate various sources of bias by Dvořák & Klaschka (1990). It is this algorithm that is used in EEG studies described in the following.

6. DETERMINISTIC CHAOS IN THE BRAIN

Processes taking place within the brain bear highly irregular, noise-like character. In spite of this they reflect highly coordinated activity demonstrated by purposeful behavior. Since even the most complex processes are on the microscopic level realized by combination of many and many various chemical and physical processes, which can be described by deterministic laws, it is reasonable to put forward a hypothesis that these processes are deterministic. Regarding their evidently irregular character, on the other hand, if they should be regarded as deterministic, they can be regarded only as deterministic chaotic processes. How can this hypothesis be tested and possibly verified ?

7. ELECTROENCEPHALOGRAPH - EEG

The most natural way of testing the hypothesis is to use some "objective" information provided by measuring of some physical signal on the brain. The most suitable for this purpose is the electroencephalogram - EEG, which is routinely registered not only in research laboratories but also in clinics and hospitals. EEG represents changes of electric potential registered by flat electrodes on the surface of the scalp. As electrical phenomena play leading role in the brain activity, these measurements are a powerful source of information on the brain processes taking place in the upper layer of the cortex.

Irregular character of EEG reflects very complex nature of the brain processes (see Fig. 1). Amplitude of the signal is between 3 and 100 μV, its frequency range (bearing the physiologically meaningful information) is 0.5 - 40 Hz. EEG is usually registered from 4 to 32 (or even 64) registration leads placed according to one of generally consented registration schemes called "montages" (so called 10 - 20 Jasper system is the most widespread) and sampled with sampling frequency 128 - 256 Hz.

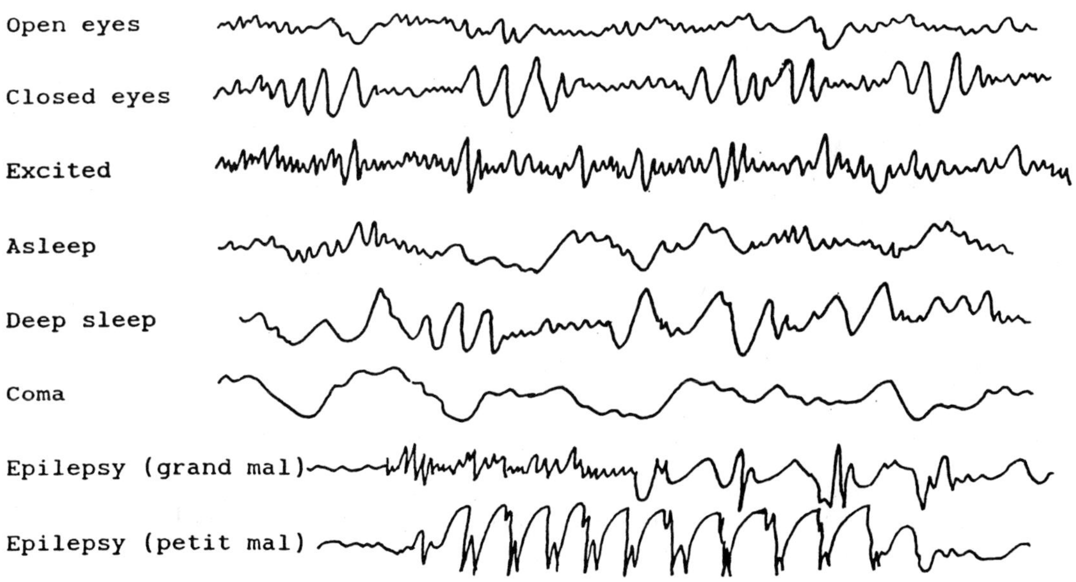

Fig.1 EEG for various states of the brain

Precision of registration is usually 8 - 14 bits. Since length of the registered signal may vary from 1 minute to 8 hours, numerical calculations must deal with 4 - 32 dimensional vectors with $10^4 - 10^7$ samples in each component.

8. RESULTS OF ESTIMATES OF THE CORRELATION EXPONENT OF EEG

Since 1985 many papers were published documenting that correlation exponent of EEG is finite. This is a strong argument supporting the basic hypothesis that EEG can be regarded as a deterministic chaotic process. Usually correlation exponent from the single channel EEG, or the *local correlation exponent* was calculated (c.f. Babloyantz, 1985, Rapp, 1986, Dvořák & Šiška, 1986). It bears information on determinacy of the local dynamics of the brain processes taking place under the registration site. Later also the *global correlation exponent* of EEG for more channels together (c.f. Destexhe et al., 1990, Dvořák 1990), which characterizes coordination of processes taking places in different parts of the cortex, was calculated.

As the discussed methodology is new, it suffers by many children diseases. It has been therefore very difficult to compare results obtained in different laboratories as they have applied slightly different numerical algorithms, different registration techniques and different numerical parameters of calculations. When more data had been accumulated, however, some general conclusion could be drawn (c.f. Dvořák & Holden, 1991).

Value of local correlation exponent for healthy subjects usually varies between 4 and 8. It is lower for states with closed eyes and higher for states with open eyes. In case of epilepsy this values falls under 3 or even 2, on the contrary in case of anesthesia it grows to 9. Global correlation exponent is around 5 or 6 for healthy subjects and also falls down in the case of epilepsy. Value of the correlation exponent changes under impact of psychotropic drugs. Some results obtained in Prague Psychiatric Center indicate that value of the global correlation exponent falls down under influence of ethanol. In cooperation with University of Zürich we were able to prove this effect statistically for large doses of nootropic drug piracetam. On the other hand discrimination between mental states seems to be very difficult because of large inter-individual and inter-individual variances of EEG. So, at present by calculating correlation exponent of EEG can be detected brain pathology with clear biological causes (epilepsy) and impact of large doses of some psychotropic drugs. This documents applicability of this approach for brain diagnostics.

Possible application of the studied concept and methodology for diagnostics of the brain processes is very encouraging. We must not forget, however, that we have based our analyses on purely phenomenological approach. When we stress not only *description* of the brain processes (which is the core of diagnostics of brain function) but also *understanding* of their nature a lot of questions arises. We must naturally ask : what are we measuring when we find the value of the correlation exponent of a given EEG in (say) temporal area of the scalp for a relaxed subject with closed eyes to be 5.2 ? What is the relation between local and global correlation exponents ? What is the physiological meening of six or nine variables that are sufficient for full characterization of EEG ? To grasp how complicated anwers to these questions may be must return to the very beginning and reconsider the physical and physiological nature of the brain.

9. LORENZ ATMOSPHERE AND THE BRAIN ARE DIFFERENT

Development of methods of chaotic dynamics was primarily stimulated by problems of turbulence. Thus implicitly, though not always consciously, turbulent fluid was used as a mental model for the physical realization of studied mathematical objects. Naturally, methods and techniques developed within the deterministic paradigm reflect this state of affairs.

There is profound difference between turbulent fluid and the brain. Turbulent fluid is composed of many identical elements that interact in a uniform way. Globally turbulent fluid is isotropic if not homogeneous. Interactions of the elements are only of a short range and there are no time delays in the system. All this results in a global coordination of flow of fluid particles in such a way that - though the system is spatio-temporal - the dynamics of the system can be characterized by the state trajectory in the finite-dimensional state space.

The brain - on the other hand - is composed of many elements - neurons - of many various kinds. Their interactions are far from being uniform, they are of a short as well as of a long range. Interactions on the long range bring into the system - due to finite velocity of the action potentials - significant time delays. Moreover, the brain is highly nonuniform and anisotropic. Shortly, it is hard to imagine objects more different by its material composition than the brain and the turbulent fluid.

It is only natural to doubt that parameters developed with description of turbulent fluid in mind will provide also valid description of a system qualitatively more complex. Of course, formal application of methods discussed gives results, but their interpretation is unclear. As was pointed out in the preceding paragraph, we do not know what our results - values of correlation exponent in various states of the brain, for instance - really mean.

The situation described is direct consequence of application of deterministic chaos methods and techniques as purely data analyzing techniques, that is without proper mathematical model. Let us stress once more that this type of application does not precludes valuable results important for improvement of brain diagnostics. For *understanding* of the brain processes, however, elaboration of the theoretical model seems to be inevitable. This model should describe the control structure of the dynamics of the studied brain processes and enable us in this way to interpret properly the parameters describing qualitatively this dynamics.

10. CONCLUSIONS

Irregular, noise-like processes can be found in many a branch of science. Often they are generated by systems governed by deterministic laws. In principle all such processes can be approached as chaotic processes and analyzed accordingly. In this respect transdisciplinarity of chaotic dynamic approach is natural.

Analysis of application of methods of deterministic chaos for studying natural and/or social systems, however, leads to differentiation between applications based on a mathematical model and those without such a model, that are limited only to formal analysis of experimental data. When the appropriate mathematical model of the dynamics of the system in question ensures necessary inside into its functioning, obtained results may have high heuristic value.

The second approach was extensively illustrated by analysis of dynamics of the brain processes studied via EEG. It is shown that though this analysis gives interesting results that may be valuable for improvement of neurologic and psychiatric diagnostics, for better understanding the brain processes a mathematical model describing the causal structure of control of their dynamics processes is necessary.

Returning to our opening question we can conclude that deterministic chaos is surely challenging and highly stimulating paradigm. It is good thing to care, however, in which situation the paradigm is applied. When the approach results only in a more or less formal analysis of data the results should be accepted with care. Especially when mathematical techniques developed with specific branch of science in mind are applied directly in another branch of science the real value of obtained results may be doubtful.

LITERATURE

Babloyantz,A., Nicolis,C., Salazar,M.: Phys. Lett. A 83 (1985), 152.
Destexhe, A., Sepulchre, J.A., Babloyantz,A.: Phys.Lett. A 132 (1988), 101.
Dvořák, I. & Šiška, J.: Phys. Lett. A 118 (1986), 63.

Dvořák, I. & Klaschka, J.: Phys. Lett. A 143 (1990)
Dvořák, I.: Phys. Lett. A 151 (1990), 225.
Dvořák, I. & Holden A.V.(eds.): Mathematical Approaches to Brain Functioning Diagnostics, Manch. Univ. Press, Manchester (1991).
Glass, L., Shrier, A., Bélair, J.: In: Chaos, A.V.Holden (ed.), Manch. Univ. Press., Manchester, (1986), 237.
Grassberger, P. & Procaccia, I.: Phys. Rev. A 28 (1983), 2591.
Grassberger, P. & Procaccia, I.: Physica 13 D (1984), 34.
Gregson, R.: Nonlinear Psychophysical Systems, Lawrence Erlbaum Assoc., Hillsdale (1988).
Grossman, S. & Mayer-Kress, G.: Nature 337 (1989), 701.
Hartley, K.: Astronomy, May (1990), 34.
Huberman, B.A.: Proc. of conference Biological Dynamics and Theoretical Medicine, Bethesda (1986).
Kurths, J. & Herzel, H.: Physica 25 D (1987), 165.
Laskar, J.: Nature 338 (1989), 237.
Lorenz, E.N.: J.Atmos. Sci 20 (1983), 130.
May, R.M.: Science 186 (1974), 645.
Mackey, M.C & Glass, L.: Science 197 (1977), 287.
Nicolis, C. & Nicolis, G.: Nature 311 (1984), 529.
Rapp, P.E., Zimmerman, I.D., Albano, A.M., deGuzman, G.C., Greenbaum, N.N.: Phys. Lett. A 110 (1985), 335.
Rapp, P.E.: In: Chaos, A.V.Holden (ed.), Manch. Univ. Press, Manchester (1986).
Ricard, J. & Cornish-Bowden, A.: Dynamics of Biochemical Systems, Plenum Press, New York (1984).
Roux, J.C.: In: Singularities and Dynamical Systems, S.N.Pnevmatikos (ed.), Elsevier, Amsterdam (1985), 345.
Ruelle, D.: Proc. R. Soc. London, A427 (1990), 241.
Ruelle, D. & Eckmann, J.P.: Rev. Mod. Phys. 53 (1985), 617.
Schaffer, W.M.: Ecology 66 (1985), 93.
Schaffer, W.M. & Kot M.: J. theor. Biol. 112 (1985), 403.
Sussman, G.J. & Wisdom, J.: Science 241 (1988) 433.
Swiney, H.L. & Gollub, J.P.: Hydrodynamic Instabilites and the Transitions to Turbulence, Springer, Berlin (1981).
Takens, F.: In: Dynamical Systems and Turbulence, Lect. Notes. in Math. 898, Warwick (1981), 336.

DYNAMICS OF HI-GAS IN THE PECULIAR GALAXY NGC 1808

B. KORIBALSKI [1], M. DAHLEM [2]

[1]MPI FÜR RADIOASTRONOMIE, 5300 BONN 1, GERMANY

[2]HAMBURGER STERNWARTE, 2050 HAMBURG 80, GERMANY

1. INTRODUCTION

HI line studies are one of the most powerful techniques to investigate galaxy dynamics. Here we present recently obtained *VLA* spectral line data of the peculiar starburst galaxy **NGC 1808**. This beautiful spiral galaxy (type Sbc) is mostly known because of its prominent *dust filaments* perpendicular to the central disk, indicating ongoing disk-halo interactions, which are reminiscent of those in M82. Our **HI** observations reveal more peculiarities as for example a *warp*, indicated by the shape of the HI intensity distribution and a strongly falling rotation curve, and a ***broad absorption line*** against the central continuum source, whose origin will be subject to further investigations.

2. OBSERVATIONS

NGC 1808 was observed with the *C/D* and *B/C* configurations of the *V*ery *L*arge *A*rray during 1988 June and 1990 October. Both data sets were reduced with the AIPS package using mostly standard procedures and will later be combined in order to obtain high resolution images, which are not affected by missing spacing problems. The *C/D*-array data, which have a resolution of 45"x65"x20.7 km/s, are published by Dahlem et al. (1991). Therefore in the following we will concentrate on the new *B/C*-array data. Owing to the improved spectral capabilities at the *VLA* we could obtain *B/C*-array data with both high angular resolution *and* high velocity resolution (15"x21"x5.2 km/s). Combination of two intermediate frequency (IF) bands resulted in 102 "good" channels out of which 74 contain HI line emission. First results of our **B/C–array** observations are presented in the next chapter.

3. RESULTS

Neutral hydrogen (HI) from the southern galaxy NGC 1808 has been detected in both *emission* and *absorption* over nearly the same velocity range ($\triangle V = 385$ km/s). While the HI emission is distributed over an even larger area than the optical emission, the absorption, which is seen only against the central continuum source (Saikia et al., 1990), remains unresolved. The map of the integrated HI line intensity (Fig.1) shows three pronounced features: * A near-circular HI envelope, which is associated with the outer spiral arms, * strong line emission from the central disk, showing an asymetric S-shape, and * a "hole" at the position of the nucleus.

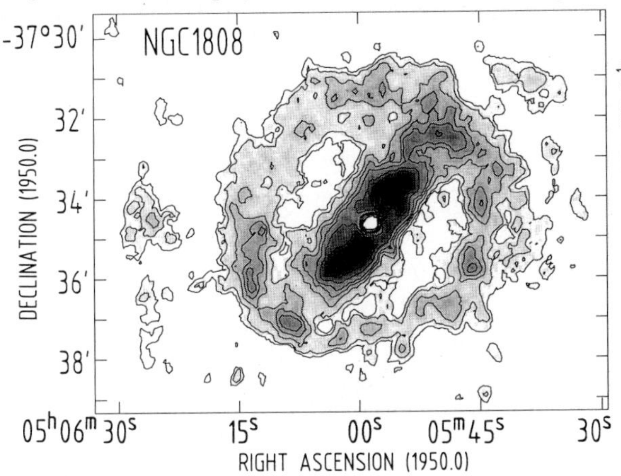

Fig.1–HI intensity distribution of the galaxy NGC 1808. Contour levels are 25, 50, 100, 150, 200, 300, 400, 500, 600 Jy/beam m/s.

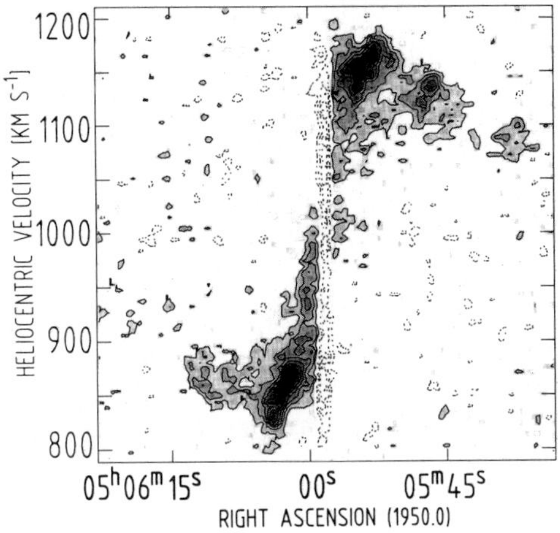

Fig.2–pv-diagram along the major axis of NGC 1808. Contour levels go from -5.2 to 7.8 mJy (step 1.3 mJy).

If we assume the general shape of the galaxy to be the same in the inner and outer regions, then the observed projection of the galaxy NGC 1808 clearly indicates a *warp* of the outer spiral arms against the inner disk. This interpretation is supported by the steady decrease of the "rotation curve" (Fig.2). The position/velocity diagram in Figure 2 is taken at a position angle of 310° and has a width of 15". Another remarkable feature in the *pv-* diagram is *HI absorption* (indicated by dashed lines) over a very broad velocity range, seen as a "hole" in the HI intensity distribution (Fig.1). Similar absorption lines have been observed in NGC 4945 (Whiteoak et al., 1990) and NGC 3079 (Irwin et al., 1991). Possible explanations for the kinematics of HI gas near the nucleus are *a)* a small fast rotating ring or *b)* individual clouds distributed randomly, moving towards and away from the nucleus.

REFERENCES

Dahlem, M., Koribalski, B., Mebold, U.: 1991, *Astron. Astrophys.* (subm.)

Irwin, J., and Seaquist, E.R.: 1991 *Astroph. J.* **371**, 111

Saikia, D.J., Unger, S.W., Pedlar, A., Yates, G.J., Axon, D.J., Wolstencroft, R.D., Taylor, K., Gyldenkerne, K.: 1990, *Mon. Not. Roy. astr. Soc.* **245**, 397

Whiteoak, J.B., and Wilson, W.E.: 1990, *Mon. Not. R. astr. Soc.* **245**, 665

ABUNDANCE GRADIENTS IN ELLIPTICALS

F. MATTEUCCI[1] and M. STIAVELLI [2]

1) Max-Planck Institut für Astrophysik, K.Schwarzschild Str.1, D-8046 Garching, Germany

2) European Southern Observatory, K. Schwarzschild Str.2, D-8046 Garching, Germany

ABSTRACT

A three-dimensional collisionless code is used to simulate the formation of elliptical galaxies with a timescale for star formation comparable to the collapse time. A one-zone chemical evolution scheme is used to compute the evolution of the abundances of several chemical elements. The initial "clumpy" system is made of gas and, in some cases, of warm dark matter. In the latter case, we show that abundance gradients are produced. For a galaxy of $10^{11}M_\odot$ of gas and $3\cdot 10^{11}M_\odot$ of warm dark matter, the predicted gradients of Fe and Mg, inside an effective radius, are: $\frac{\Delta[Fe/H]}{\Delta logR} = -0.11$ and $\frac{\Delta[Mg/H]}{\Delta logR} = -0.05$. In the absence of warm dark matter a greater amount of dissipation is required to produce abundance gradients. The mechanism proposed here for the formation of abundance gradients should be relevant for bright and extended objects where there is evidence for the presence of dark matter.

Scientific rationale

While dissipationless collapse can naturally provide a physical justification for the luminosity profile of ellipticals (see van Albada, 1982), it cannot reproduce their observed color and metallicity gradients. On the other hand, the dissipative scenario (Larson, 1975 and references therein) can naturally explain metallicity gradients, but it requires a fine tuning of parameters as well as "ad hoc" assumptions on the amount of dissipation, in order to reproduce the observed luminosity profile. In this paper, we investigate the collapse of a system initially made of gas and warm dark matter, where the star formation time scale is of the order of the collapse time. We obtain a collapse which, from the dynamical point of view, maintains the characteristics of the dissipationless collapse, while introducing an "effective dissipation" sufficient to produce abundance gradients.

Model description

The simulations were carried out with a code based on a spherical harmonics expansion of density and potential truncated at the order $l = m = 6$ inclusive. The number of particles was $2\cdot 10^4$. Several initial conditions were explored. In all cases the initial velocity dispersion of gas and dark matter was identical.

Gaseous particles are converted into stellar particles following the law:

$$d\rho/dt = \nu\rho_{gas}$$

where the timescale for star formation is $\nu^{-1} \simeq 0.1 Gyr$. To the stellar particles, when formed, is assigned a velocity distribution, with a dispersion σ_* ten times smaller than that of the gas particles. The chemical composition of stars at their birth, e.g. that of the gas out of which they formed, is computed by means of a detailed chemical evolution model for elliptical galaxies as described by Matteucci and Tornambè (1987) (see also Stiavelli and Matteucci, 1991 for a more complete description of the chemo-dynamical model). We assume that elements ejected from stars are instantaneously mixed (but not instantaneously produced), so that the chemical composition of the gaseous medium is homogenous at any time. As a consequence of this assumption, our predicted metallicity gradients represent a lower limit for gradients in ellipticals.

Model results

Our major results can be summarized as follows :

- abundance gradients are produced, provided that the star formation rate timescale is not shorter than the collapse rate timescale;
- the iron abundance gradient is found to be steeper than that of Mg This conclusion depends only on the assumed stellar nucleosynthesis;
- dark matter can help in producing abundance gradients by slowing down the collapse, provided that its velocity dispersion is high enough;
- models with realistic gradients are characterized by a relatively small anisotropy in their velocity dispersion;
- isochromes appear to be flatter than isophotes, and both curves appear to be boxy. The flattening is due to velocity anisotropy rather than rotation.

Our conclusions are that objects characterized by photometric and kinematics properties as well as abundance gradients similar to the observed ones, can be reproduced in a "delayed" collapse with a small amount of dissipation. Given the fact that there are indications for the presence of dark matter only for bright and extended ellipticals (Saglia et al., 1991), the proposed mechanism for abundance gradient formation could be relevant for these objects. For fainter and more compact elliptical galaxies, where no evidence of dark matter has been found (e.g. NGC5077, Bertola et al., 1991), a more dissipative model than that presented here would be required to account for abundance gradients.

References

Bertola, F., Bettoni, D., Danziger, I.J.,Sadler, E., Sparke, L.,de Zeeuw, T., 1991, *Astrophys. J.*, in press

Larson, R.B., 1975, *Mon. Not. R. astr. Soc.*, **173**, 671

Matteucci, F., Tornambè, A., 1987, *Astron. Astrophys.* **185**, 51

Saglia, R.P., Bertin, G., Stiavelli, M., 1991, *Astrophys. J.*, in press

Stiavelli, M., Matteucci, F., 1991, *Astrophys. J. Letters*, in press

van Albada, T.S., 1982, *Mon. Not. R. astr. Soc*, **201**, 939

Polar-Ring Galaxies and the Possibility of their Formation in Triaxial Potential

IGOR G. KOLESNIK

Main Astronomical Observatory, Ukrainian Academy of Sciences 252127 Kiev, Goloseevo, USSR

Polar-ring galaxies have a very unusual and amazing appearance. A most conspicuous feature in these galaxies is connected with a large portion of matter that rotates in the plane perpendicular to the rotation of a galactic body. The first galaxy that was understood to have a polar ring was NGC 2685 (Schechter and Gunn, 1978). From that time some other similar objects were discovered and studied: A0136-0801, NGC 4650A, and ESO 415-626, by Schweizer, Whitmore, and Rubin (1983), Whitmore, McElroy, and Schweizer (1987), van Gorcom, Schechter, and Kristian (1987); and UGC 7576 and II Zw 73, by Schechter et al. (1984). The most complete list and atlas of polar-ring galaxies and similar objects is that given by Whitmore et al. (1990).

On the basis of studies carried out earlier we can formulate the properties of polar-ring galaxies as follows:

1. In most cases, the cores belong to S0 galaxies. Only one polar-ring galaxy, AM 2020-504, is known to have a well pronounced polar ring; its core is an elliptical galaxy (Whitmore, McElroy and Schweizer 1987b). But many elliptical galaxies show dust lanes aligned along the minor axis or the major axis (Bertola 1987). These can be considered as indirect indications on the existence of ring structures around elliptical galaxies. In this case, the polar rings when compared to S0 galaxies are thinner and more compact and are located closer to the core.

According to Haud (1988), the high-velocity, high-latitude HI clouds in our Galaxy can be organized into a polar-ring structure. It is interesting to search for the existence of these type structures in spiral galaxies, also inspite of difficulties in distinguishing them unambiguously from other structural and kinematic features in these objects.

2. In all cases, the cores have a red colour but the rings have a blue colour. This shows that the polar rings are comparatively young structures, where star forming activity can continue.

3. The HI emission is connected with polar rings. The 21-cm line profile has the two-horned shape that is typical of rotating disk galaxies. The masses of neutral hydrogen $M_{HI} = \ 0.01\ M$, where M is the total mass. Thus, the polar rings are massive selfgravitating structures.

How common the polar-ring galaxies are is not yet well known. But there are some peculiarities in galaxies that can be explained successfully by the existence of polar-ring structures.

Hoag's object contains the first class of such galaxies and Hoag-type galaxies (Schweizer et al. 1987, Wakamatsu 1990). They consist of a surprisingly round core inside a circular ring. The list of Hoag-type galaxies contains 13 objects (Wakamatsu 1990). We believe that the nature and properties of this type of objects can be explained in the frame of a model that consists of a triaxial galaxy surrounded by an elliptical polar ring (Berczik and Kolesnik 1991a). At the appropriate inclination, the central body and the ring can be seen as round images simultaneously. If all Hoag-type galaxies are objects with polar rings then the number of known polar-ring galaxies is more than doubled.

Our model of a polar-ring galaxy can be relied upon to explain the case of counter-rotation of outer parts with respect to the inner stellar body. A remarkable object of this type is the elliptical galaxy IC 2006 (Schweizer, van Gorkom, and Sceitzer 1989). As was shown by Berczik and Kolesnik (1991b), the properties of IC 2006 can be reproduced in the model of a triaxial elliptical galaxy with a circular polar ring. Other cases of counter-rotation in elliptical galaxies (Bender 1988) and in dust-lane ellipticals (Bertola, Buson, and Zeilinger 1988) can be considered in the same model.

Therefore the systems with polar-rings must be widespread among galaxies. As a rule, when the problem of the origin of ring structures in galaxies is discussed, the merging or the close interactions of galaxies with a mass exchange have to be taken into consideration. But in many cases there is no sign of a strong interaction in polar-ring galaxies. So it is interesting to note that the internal processes, not only the accidental events, can lead to polar ring formation.

A model of successively embedded triaxial ellipsoids is proposed here. The most common shape of a protogalaxy is a triaxial figure (Fig. 1). In such figures, two types of regular motions dominate (Schwarzschild 1979): Those in the plane perpendicular to the minor axis, and those in the plane perpendicular to the major axis. Around the major axis, gas rotates in the internal, most dense, regions of a protogalaxy. Motions around the minor axis are related to the external, relatively low density, envelope. When the conditions for galaxy formation are created inside such a body, the dense internal regions contract ra pidly and create an object that rotates in the plane perpendicular to rotation of the outer layers. Under appropriate conditions, a polar ring can be formed. This is a qualitative picture of polar-ring galaxy formation.

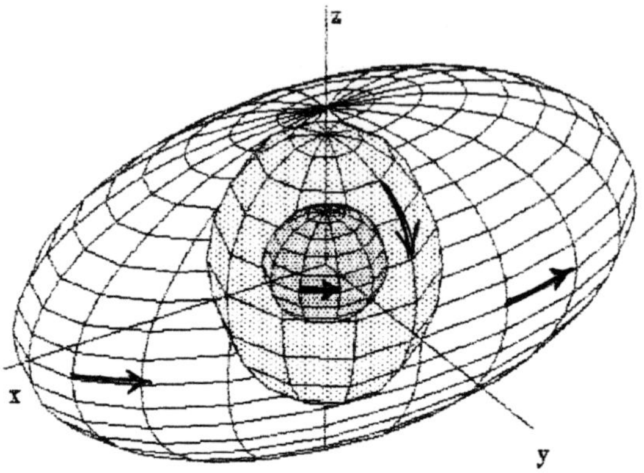

Fig.1

It is easy to explain properties of the wide class of ring galaxies, from the spindle-like type, such as UGC 7576, NGC 4650A, A0136-0801, to the optically prolate galaxies with rings around the major axis, such as ESO 415-G26, by different orientations of such systems (Fig. 2). Minor axis dust-lane galaxies are probably of the same nature.

Evidently, the properties of the forming system depend strongly on the relative mass and momentum in the core and in the envelope.

In some cases, if the internal body is essentially triaxial, the core in its turn can be divided into the compact object and the ring. The whole system can be looked upon as a compact near-spherical core inside of the two mutually perpendicular rings. There is known one galaxy, ESO 474-626, that has this type of structure (Whitmore et al. 1990).

Therefore, on the basis of internal dynamics and thermal evolution of protogalactic gas in triaxial potentials, a wide variety of forms and types of galaxies can be explained.

We think it is probable that polar-ring galaxies have retained signs of a distant history in galaxy formation and our model makes it possible to study these processes hidden in time.

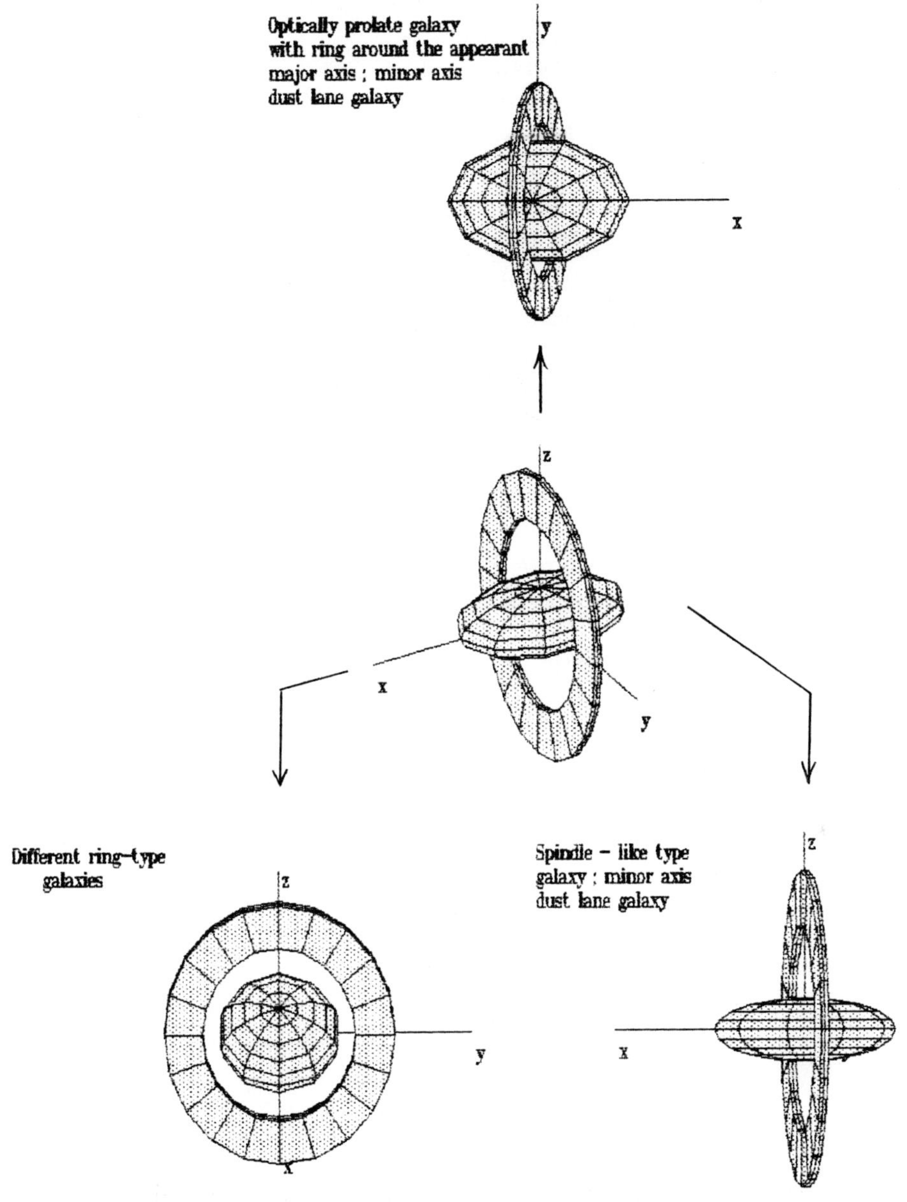

Fig.2 The polar ring galaxy in space and its images in different lines of sight.

REFERENCES

Bender, R., 1988, Astron. Astrophys. 202, L5
Berczik, P., Kolesnik, I.G., 1991a, this volume
Berczik, P., Kolesnik, I.G., 1991b, this volume
Bertola, F., 1987, in "Structure and Dynamics of Elliptical Galaxies", IAU Symp. 127, ed. T.de Zeeuw, Reidel, p.135
Bertola, F., Buson, L.M., Zeilinger, W.W., 1988, Nature 335, 705
Haud, U., 1988, Astron. Astrophys. 198, 125
Schechter, P.L., Gunn, J.E., 1978, AJ 83,1360
Schechter, P.L., Sancisi, R., van Woerden, H., Lynds, C.R., 1984, MNRAS 208, 111
Schwarzschild, M., 1979, ApJ 232, 236
Schweizer, F., Whitmore, B.C., Rubin, W.C., 1983, AJ 88,909
Schweizer, F., Ford, W.K., Jr., Jerdrzejewsky, R., Giovanelli,R., 1987, ApJ 320, 454
Schweizer, F., van Gorkom, J.H., Sceitzer, P., 1989, ApJ 338, 770
van Gorkom, J.H., Schechter, P.L., Kristian, J., 1987, ApJ 314, 457
Wakamatsu, K., 1990, ApJ 348, 448
Whitmore, B.C., McElroy, D.B., Schweizer, F., 1987a, ApJ 314, 439
Whitmore, B.C., Lucas, R.A., McElroy, D.B., Steiman-Cameron, T.Y., Sackett, P.D., Olling, R.P., 1990, AJ 100, 1489
Whitmore, B.C., McElroy, D.B., Schweizer, F., 1987b, in "Structure and Dynam ics of Elliptical Galaxies", IAU Symp. 127, ed. T. de Zeeuw, Reidel, p.413

Triaxial Galaxy Models with Stäckel Potentials

C. HUNTER[1] & P.T. DE ZEEUW[2]

[1] Florida State University, Tallahassee, FL 32306–3027, U.S.A.
[2] Sterrewacht, Postbus 9513, 2300 RA Leiden

We have constructed self–consistent distribution functions for non–rotating triaxial galaxy models with Stäckel potentials. All stellar orbits are regular and belong to four distinct families: boxes (B), short–axis tubes (S), inner long–axis tubes (I), and outer long–axis tubes (O). Stars in any tube orbit have a definite sense of rotation about their axis, though this may be in either direction. Hence the models are generally endowed with angular momentum about both long and short axes, and the total angular momentum lies in the plane defined by these axes. Stars in box orbits have no sense of rotation and do not contribute to any mean–streaming.

Self–consistent triaxial galaxy models can be built in many ways because of the existence of the four orbit families, each of which penetrate to almost all regions of space. We have restricted each of the three families of tube orbits to be infinitesimally thin and with no radial epicyclic motion. This has allowed us to obtain explicit expressions for their phase–space distribution functions, and then their densities and velocity moments throughout space. The box orbits are of unrestricted form, and their distribution function satisfies an integral equation of Volterra type which cannot be solved explicitly. Even with the considerable restriction to thin tube orbits, our models are not unique, and there is a choice of allocation between the two kinds of long–axis tubes. Once this choice has been made, no other arbitrariness remains. Our analysis confirms Statler's (1987) numerical result that orbits of all four families are needed to build a self–consistent model.

Our models have been constructed for figures whose equidensity surfaces are similar ellipsoids with semi–axes $a \geq b \geq c$. They depend upon a single parameter k, and the larger k is, the more inner long–axis tubes and the fewer short–axis tubes there are. Generally, there are also more boxes and fewer outer long–axis tubes. Mass fractions for models with $c/a = 0.5$ and for the two extreme values of the parameter k are shown in the left panel of Figure 1. The abscissa is the triaxiality parameter $T = (1 - b^2/a^2)/(1 - c^2/a^2)$. The right panel shows the maximum values that the components of total angular momentum about the long x–axis and short z–axis would have if all the tube orbits circulate in the same sense. The minimum value

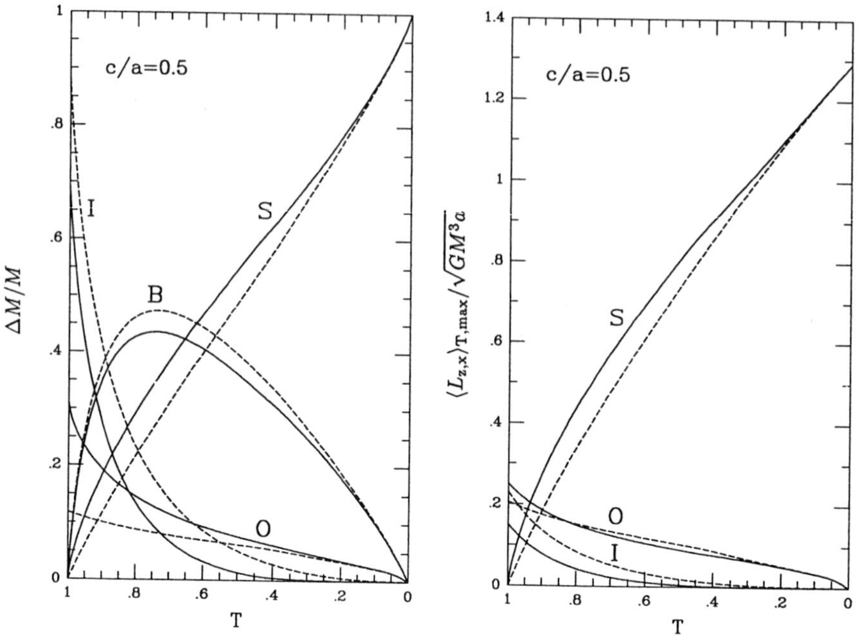

Fig. 1. Mass fractions (left) and angular momenta (right) as a function of T for c/a=0.5. Solid and dashed curves are for minimum and maximum values of k respectively.

of k gives the largest angular momentum about the short axis, and also the least angular momentum of inner long–axis tubes. But the total angular momentum of outer long–axis tubes does not depend simply on their number, and does not change monotonically with k. Outer tubes carry more angular momentum than inner tubes, but the amount does vary considerably with their form. It is largest for outer tubes which are circular, and the largest angular momentum about the long axis is achieved with a heavy weighting of nearly circular outer tubes.

Figure 1 shows that rapid changes occur away from the prolate limit of $T = 1$ ($b = c$). Boxes replace inner long–axis tubes and short–axis tubes replace outer long–axis tubes, which has a substantial effect on angular momentum. Inner long–axis tubes are relatively insignificant except when T is near 1. When their stars cross the plane perpendicular to the long axis, they are moving faster than any material on closed elliptical orbits in this plane. They are then at the throat of their orbit, and their rotation about the long axis is largest. A central region of rapid rotation along the true minor axis is likely to be an indication of a significant population of inner long–axis tubes, and hence of near–prolate shape. A full account of this work has been submitted for publication in the Astrophysical Journal.

Reference

Statler, T. 1987, *Astrophys. J.*, **321**, 113.

Hoag's Object as a Polar-Ring Galaxy

PETER BERCZIK AND IGOR G. KOLESNIK

Main Astronomical Observatory, Ukrainian Academy of Sciences, 252127 Kiev, Goloseevo, USSR

Hoag's object has an amazingly regular image that consists of a round core inside of a circular ring. Its nature has not yet been explained. Schweizer et al. (1987) recently studied properties of Hoag's object. At a surface brightness of $g_0 = 24.5\ mag\ arcsec^{-1}$, the core radius lies at $r_c = 12.8\ kpc$, the ring's inner edge at $r_1 = 16\ kpc$, and the ring's outer edge at $r_2 = 30\ kpc$. The ellipticity of the core is very small, $\epsilon \approx 0.03$. The major-axis position angle is $(P.A.)_c = 19^\circ$. The ellipticity of the ring is also very small, $\epsilon \approx 0.07$. Its major axis points approximately at $(P.A.)_r = 91^\circ$.

The ring has been detected in HI emission. The 21-cm line profile has the two-horned shape typical of rotating disk galaxies. The distance between the horns of the line is about $200\ km\ s^{-1}$. The mass of neutral hydrogen is $M_{HI} \approx 1.8 \times 10^{10} M_\odot$ (for $H_0 = 50\ km\ s^{-1} Mpc^{-1}$). The estimated total mass within $r = 32\ kpc$ is about $M_{tot} \approx 7 \times 10^{11} M_\odot$, which corresponds to $M/L_B \approx 14(M/L_B)_\odot$. This ratio is relatively large in comparison with $M/L_B \approx 6$, 4 and 2 for Sa, Sb, and Sc galaxies, respectively.

As far as its properties are concerned, the core is a normal S0 galaxy. The ring has a blue color and shows star formation activity. The list of Hoag-type galaxies contains 13 objects (Wakamatsu 1990).

The first problem that must be solved concerns the surprisingly round form of Hoag's object. It is practically impossible to see that all Hoag's type galaxies precisely from their pole. Among S0 galaxies there are objects with polar rings (Whitmore et al. 1990).

We surmise that Hoag's object also is an S0 galaxy with a polar ring, but in this case the polar ring definitely has an elliptical form. In such a case, under appropriate orientation, such a polar-ring galaxy can look like a Hoag object (Fig. 1).

A model giving good results is one containing a core as a triaxial body with axial ratio $a : b = 3 : 2$ and a polar ring with axial ratio $B : C = 3 : 4$, situated in the plane (y,z) perpendicular to the major axes of the galaxy. If we incline it at an angle $\alpha = 42^\circ$ around the East-West line and at an angle $\beta = 3^\circ$ the around South-North line, the model gives an image that resembles the one of Hoag's object. By rotation of the model around the line of sight at an appropriate position angle, kinematics of the model and of the observed object agree (Fig. 2). This model shows that the rotational velocity of the ring is $V \approx 150\ km\ s^{-1}$, which is about two times smaller than that estimated by Schweizer et al. (1987). This yields the total mass $M_{tot} \approx 1.8 \times 10^{11} M_\odot$ and the mass-to-light ratio $M/L_B \approx 3.5\ (M/L_B)_\odot$, which is consistent with the M/L_B for galaxies.

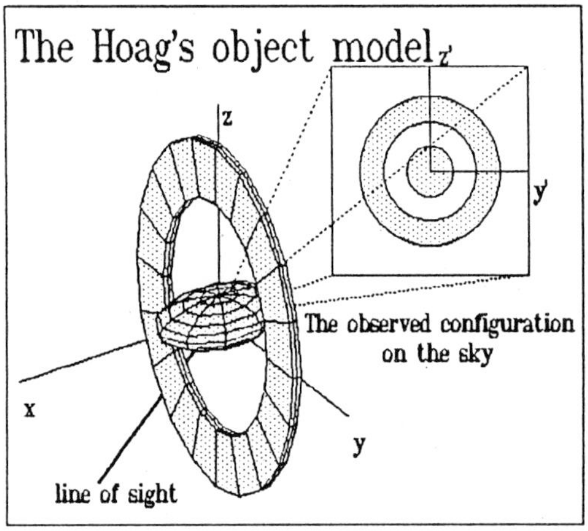

Fig.1

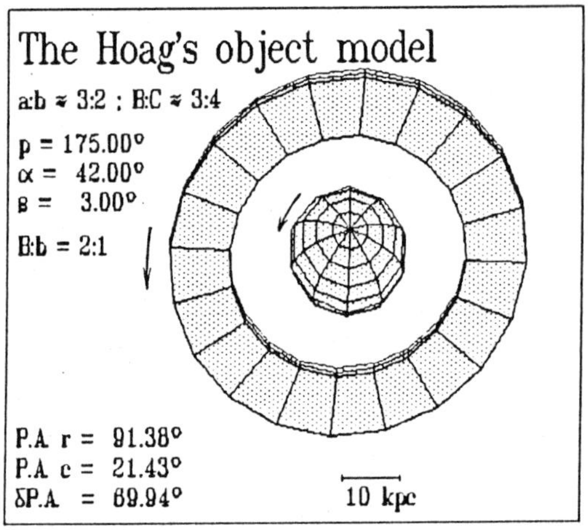

Fig.2

It can be concluded that Hoag's object and other similar objects can be co nsidered as S0 or early-type barred galaxies surrounded by elliptical polar rings. This significantly increases the number of objects known as polar-ring galaxies.

REFERENCES

Schweizer, F., Ford, W.K., Jedrzejewski, R., Giovanelly, R., 1987, ApJ, 320, 454
Wakamatsu, K., 1990, ApJ, 348, 448
Whitmore, B.C., McElroy, D.B., Schweizer, F., 1987, ApJ, 314, 439

IC 2006 as an Elliptical Galaxy with a Polar Ring

PETER BERCZIK AND IGOR G. KOLESNIK

Main Astronomical Observatory, Ukrainian Academy of Sciences, 252127 Kiev, Goloseevo, USSR

A regular ring of luminous matter around the E1 galaxy IC2006 was discovered by Schweizer (1987). Soon after that Schweizer, van Gorkom, and Seitzer (1989) detected HI emission from this ring. The galaxy may be specified by the following parameters: semimajor axis at a surface brightness $B = 25\ mag\ arcsec^{-1}$, $a_{25} = 8\ kpc$; ellipticity $\epsilon \approx 0.2$, and $(P.A.)_c = 34^\circ$. The radius of the ring is $R_r = 19\ kpc$. The ring is very thin, $\delta R/R < 0.17$. The position angle of its apparent ellipticity is $(P.A.)_r = 38^\circ$. The ring contains $4.8 \times 10^8 M_\odot$ of HI. The most interesting aspect o f the kinematics of IC 2006 is that the gas rotates in the direction opposite to that of the star component. The apparent rotational velocity of the stars is $V_{star} \approx 65\ km\ s^{-1}$, that of the HI ring is $V_{gas} = 125\ km\ s^{-1}$. From thes e data, the mass of the galaxy within the ring radius is about $M_r = 1.8 \times 10^8 M_\odot$, and $M/L_B \approx 16.3(M/L_B)_\odot$.

To explain the counter-rotation of the stars and the gas we assume that the ring of IC 2006 occupies a polar orbit (Fig. 1). There are some other cases of elliptical galaxies with polar rings. A prominent example is AM 2020-504 (Whitmore, McElroy, and Schweizer 1987).

The observed properties of IC 2006 can be explained by a model that contains, as the inner body, a prolate ellipsoid with axial ratio $a : b = 5 : 4$ and has the circular polar ring with $B = C \approx 2.5\ b$ (Fig. 2). In Fig. 2, the angle α corresponds to the rotation of the model around the East-West line, the angle β gives the rotation around the South-Nor th line, and the positional angle p corresponds to the rotation around the line of sight. In the model considered here, the ring velocity is

$$V_{ring} = \frac{V_{gas}}{\sin\alpha} \approx 165\ km\ s^{-1}$$

This gives

$$M_r \approx 1.1 \times 10^{11} M_\odot, \qquad \text{and} \qquad M/L_B \approx 10(M/L_B)_\odot$$

which is only two times larger than the value $M/L_B \approx 5\ (M/L_B)_\odot$ estimated directly from the brightness distribution of the galaxy.

The proposed model of a polar-ring galaxy thus explains the properties of IC 2006 and describes the counter-rotation of the ring and the central body. We speculate that the interesting cases of elliptical galaxies having cores rotating in the opposite direction from the envelopes (as described by Bender (1988)) can be explained in terms of a model in which the angular momentum of the inner and outer parts have mutually perpendicular orientations.

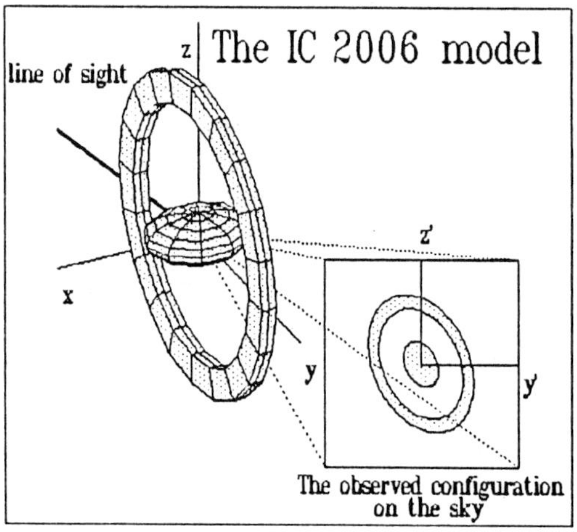

Fig.1

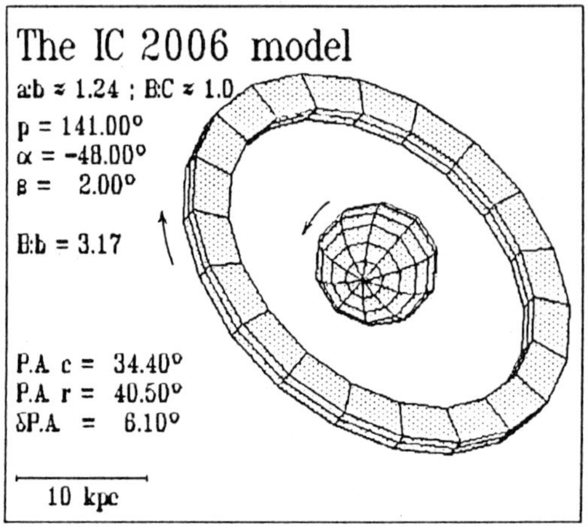

Fig.2

We believe that polar-ring formation is connected with the fundamental properties and with the evolution of the protogalaxy. This possibility was considered by Kolesnik (1991) for the case of a protogalaxy consisting of successively imbedded triaxial ellipsoids. In this case the internal motions can generate the polar-ring structures. Polar-ring galaxies may be are the key objects relevant to problems concerning the origin of galaxies of different types.

REFERENCES

Bender, R., 1988, Astron. Astrophys., 202, L5

Kolesnik, I.G., 1991, this volume.

Schweizer, F., 1987, IAU Symp., 127, Proc.

van Gorkom, J.H., and Seitzer, F., 1989, ApJ, 338, 770

Whitmore, B.C., McElroy, D.B., Schweizer, F., 1987, IAU Symp., 127, Proc.

Part IV

INTERACTING GALAXIES AND STARBURSTS

INTERSTELLAR GAS IN COLLIDING GALAXIES

I.F. MIRABEL
Section d'Astrophysique, C.E. Saclay, 91191 Gif/Yvette, France.

ABSTRACT Galaxy-galaxy collisions cause massive concentrations of neutral gas which trigger powerful starbursts and nuclear activity. Observational studies of the atomic and molecular gas in luminous infrared galaxies, Seyfert galaxies, radio galaxies, and quasars, show that gas accretion toward the central regions takes place in all these four categories of active extragalactic objects.

TIDAL INTERACTIONS AND INTERSTELLAR GAS

The idea that the environment, and in particular, tidal interactions play an important role in triggering the nuclear activities observed in the optical, radio, and infrared wavelength bands, has now strong observational support. Statistical studies show that close companion galaxies enhance star formation (MacKenty 1989), and that Seyfert galaxies are more likely to have close companion galaxies than field galaxies (Dahari, 1984). Morphological features that are characteristic of violent galaxy interactions have been identified in nearby quasars (i.e. Stockton, 1990), as well as in samples of powerful radio galaxies (i.e. Heckman et al. 1986). Furthermore, NTT images by Melnick and Mirabel (1990) have clearly shown that **all** nearby ultraluminous infrared galaxies ($L_{ir} \geq 10^{12}$ $L_\odot$) are colliding galaxies with critical separations of $\leq$ 10 kpc, confirming that ultraluminous infrared galaxies are in fact advanced mergers (Sanders et al., 1988a).

The evolution of the interstellar medium in merging galaxies must play a crucial role in the production of circumnuclear starbursts and the stimulation of compact sources at the centers of galaxies. Hernquist (1989) has shown that the fuel for both types of activities is likely to be in the form of cold interstellar gas. To generate luminosities in the range of 10^{11}-10^{13} $L_\odot$ during 10^8-10^9 years, 10^7-10^{10} $M_\odot$ masses of interstellar gas are required (Shlosman, 1990). To assemble in the nuclear regions such large quantities of gas in $\leq 10^8$ years, a mechanism that reduces the angular momentum by one to two orders of magnitude is needed. Barnes and Hernquist (1991) have shown that due to the dissipative nature of the gas, tidal forces between galaxies provoke the inward motion of the disk gas, triggering starbursts and nuclear activity in the central regions.

The occurence of circumnuclear enhancements of massive star formation (Starbursts) and the presence of high energy compact sources at the centers of galaxies (AGNs) are often concurrent phenomena. This does not imply necessarily, as has been assumed, a causal connection between Starbursts and AGNs. It is possible that these phenomena appear associated simply because both are consequences of a common cause, namely, the sudden supply of fuel into the nuclear regions by galaxy-galaxy interactions.

INFLOW OF COLD GAS

Massive Nuclear Concentrations of Neutral Gas

The first evidences for high density clouds of neutral gas in the nuclear regions of Seyfert and infrared luminous galaxies were obtained from λ21 cm observations of atomic hydrogen (Dickey, 1982; Mirabel, 1982). From the HI absorption spectra against the nuclear continuum sources present in this type of galaxies, column densities of 10^{21}-10^{23} atoms cm^{-2} are derived. The absorption arises in clouds located in the nuclear regions, which have turbulent motions with velocities in the range of 200-1000 km s^{-1}. The observations of OH megamasers at λ18 cm (i.e. Baan et al. 1986; Martin et al. 1989) reveal similar high turbulent motions of molecular gas in the inner 500 parsecs of these galaxies.

A major step forward in the study of the molecular gas content in active galaxies was made with the recent developements in millimeter technology. Strong CO(1$\rightarrow$0) emission is detected from luminous infrared galaxies (Sanders and Mirabel, 1985), powerful radio galaxies (Mirabel, Sanders and Kazès, 1989), and a few nearby quasars (e.g. Sanders et al. 1988b). Interferometric observations of the CO(1$\rightarrow$0) emission show that $\geq$ 50% of the emission comes from the nuclear regions. Scoville et al. (1991) derive masses of 10^9-10^{10} $M_\odot$ inside a few hundred parsecs, which yields mean densities of 10^3-10^4 cm^{-3}.

CS observations by Solomon, Radford and Downs (1990), and HCN and H_2CO^+ observations by Radford et al. (1991) confirmed the presence of high density molecular gas in some active galaxies. Supergiant molecular clouds may reside at the centers of some Seyferts and infrared luminous galaxies.

Kinematic evidences for infalling gas

The high concentrations of interstellar matter at the center of mergers must be a consequence of rapid inward motions of the disk gas during collisions. To concentrate $\geq$ 50% of the overall interstellar gas in the nuclear region, gas from distances $\geq$ 5 kpc must be driven into the center in $\leq 10^8$ years. This implies infalling velocities that are $\geq$ 50 km s^{-1}.

If the systemic velocity of galaxies is known, observations at λ21 cm of atomic hydrogen can be a sensitive probe for the kinematics of cold gas along the line of sight to active nuclei. Even at high redshifts, HI gas can be detected in absorption against strong radio continuum sources. Since the HI absorptions may be broad with typical half-widths greater than 100 km s^{-1}, and the optical redshifts may have large errors, the analysis is of statistical nature.

From statistical differences between the HI and optical redshifts in Seyfert galaxies, Mirabel and Wilson (1984) had tentatively favored infall of cold gas. Furthermore, Hutching, Gower, and Prince (1987), and Condon, Hutching, and Gower (1986) found similar differences between the 21 cm and optical redshifts of Seyferts 1s and nearby quasars.

A striking statistical difference between the optical redshifts and the redshifts of the HI absorptions was latter found by Mirabel and Sanders (1988) for a large sample of luminous infrared galaxies. Although the discrepancy between the radio and optical velocities is usually smaller than the velocity width of the HI absorptions, there is a clear statistical trend for the radio redshifts to be greater than the optical redshifts. Among 18 galaxies with HI absorption, there are 15 with $V_{HIabs} \geq V_{opt}$, and only 3 with $V_{HIabs} \leq V_{opt}$. The mean value of V_{HIabs} - V_{opt} is + 90 km s^{-1}.

It is interesting that the HI absorptions detected in radio galaxies, which are mostly elliptical galaxies, show the same trend (van Gorkom et al. 1989; Mirabel, 1990). In 9 of the 10 cases known up to date, the HI absorption is redshifted with respect to the optical velocities. Of particular interest are the HI narrow absorption features with velocities of 400 km s^{-1} and 2,700 km s^{-1} in the reference frames of the giant elliptical galaxies 4C 31.04 and NGC 1275. The estimated infall rates in radio galaxies are adequate to fuel the nuclear radio power.

Infall of HI or outflow of line-emitting gas?

What is the systemic velocity of active galaxies? How is it measured?. These questions must be answered before interpretations of the statistical discrepancy between the optical velocities and radio velocities of active galaxies.

The optical redshifts of the sample galaxies used for the statistical comparisons are usually taken from redshift surveys in studies of the large scale structures in the Universe. These redshifts are usually determined from the emission lines seen in low dispersion spectra. On the other hand, from high dispersion spectra of active nuclei it is known that the emission lines are usually asymmetric with extended blue wings (Heckman et al. 1981). In this context, the statistical discrepancy between the radio and optical redshifts could be due to optical line-emitting gas moving radially, probably outward. If such outwardly moving line-emitting gas is mixed with dust, the attenuation of emission from the far side leads to an observed optical redshift below systemic. If this is the case, the discrepancies between radio and optical redshifts could be considered as statistical evidence for large-scale winds in active nuclei.

Although some of the difference between the optical and radio redshifts could be accounted for by radial motions of the optical line-emitting gas, some fraction of this difference may actually be due to infall of HI toward the central source. One way out of this dilemma will be the determination of the systemic velocity of galaxies using stellar absorption lines, rather than interstellar gas emitting lines. One possibility is the use of the near-IR CaII triplet absorption lines. Terlevich et al. (1990) find these lines in absorption. However, in Seyfert 1s and quasars Persson (1988) has found these lines in emission. We are currently exploring this possibility.

REFERENCES

Baan, W.A., van Gorkom, J., Schmelz, J.T., and Mirabel, I.F. 1986, *Ap. J.*, **313**, 102.
Barnes, J.E. and Hernquist, L.E. 1991, *Ap. J. (Letters)*, **370**, L65
Condon, J.J., Hutchings, J.B., and Gower, A.N. 1985, *A.J.* **90**, 1642.
Dahari, O. 1984, *A.J.* **89**, 966.
Dickey, J.M. 1982, *Ap. J.*, **263**, 87.
Heckman, T.M. et al. 1986, *Ap. J.*, **311**, 526.
Heckman, T.M. et al. 1881, *Ap. J.*, **247**, 403.
Hernquist, L. 1989, Nature **340**, 687.
Martin, J.M. et al. 1989, *Acadèmie de Sciences*, **308**, 287.
Hutchings, J.B., Gower, A.C., Price, R. 1987, *A.J.*, **92**, 6.
MacKenty, J. W. 1989, *Ap. J.*, **343**, 125.
Mirabel, I.F. 1982, *Ap. J.*, **260**, 75.
Mirabel, I.F. and Wilson, A.S. 1984, *Ap. J.*, **277**, 92.
Mirabel, I.F. and Sanders, D.B. 1988, *Ap. J.*, **335**, 104.
Mirabel, I.F., Sanders, D.B., and Kazès, I. 1989, *Ap. J. (Letters)*, **340**, L9.
Melnick, J. and Mirabel, I.F. 1990, *Astr. Ap.*, **231**, L19.
Persson, S.E. 1988, *Ap. J.*, 330, 751.
Radford, et al. 1991, *Dynamics of Galaxies and Molecular Clouds Distribution*, ed. F. Combes, in press.
Terlevich, E., Díaz, A.I., and Terlevich, R. 1990, *M.N.R.A.S*, 242, 271.
Shlosman, I. 1990, *NASA Conference Publication 3098*, eds. J.W. Sulentic and W.C. Keel, 689.
Scoville, et al. 1991, *Ap. J. (Letters)*, **366**, L1.
Sanders, D.B. et al. 1988a, *Ap. J.*, **325**, 74.
Sanders, D.B. et al. 1988b, *Ap. J. (Letters)*, **335**, L1.
Solomon, P.M., Radford, S.J.E., and Downs, D. 1990, *Ap. J. (Letters)*, **348**, L53.
Stockton, A. 1990, *Dynamics and Interactions of Galaxies*, ed. R. Wielen, Springer-Verlag, 440.

"HOT" MODEL OF GALAXY FORMATION: EVOLUTION OF A CLOUDY PROTOGALAXY INTERACTING WITH A GALACTIC SUPERWIND

A.A. SUCHKOV AND V.G. BERMAN

Rostov State University, Rostov-on-Don, USSR

ABSTRACT

We argue that evidence gained both from galactic and extragalactic studies suggests that a period of major star formation in giant galaxies is preceded by an evolutionary phase in which a strong galactic wind driven by the initial burst of star formation enriches the gaseous protogalaxy with metals and heats it up, so that the latter reverts from contraction to expansion. The result is the ejection of enriched material from the outer part of the protogalaxy into intergalactic space, while the inner part, after a delay of about one to a few Gyr, finally contracts and cools down to form the galactic major stellar component (the "hot" model of galaxy formation).

In this paper a specific mechanism is presented which produces a hot protogalaxy according to which an early galactic wind is imparting energy and momentum into a collapsing protogalaxy whose mass is contained mainly in clouds, and only a small portion is in the intercloud gas that provides confinement of pressure for the nearly equal mass and iron abundance in cluster giant galaxies and the intracluster gas provided that the observationally plausible input parameters for giant galaxies and early galactic winds are adopted.

We also predict the formation of long-lived X-ray coronae with characteristics similar to those observed around giant ellipticals.

1 ACTIVE PHASES OF GALACTIC EVOLUTION

There are currently a number of observational facts which seem to provide the most significant clues for the formation and early evolution of galaxies. Among them the following are of primary interest: (i) the discreteness of the multicomponent structure of the Galaxy; (ii) violent bursts of star formation observed or traced in the past histories of many galaxies; (iii) the enormous mass of hot gas with nearly solar iron abundance in clusters of galaxies. We have incorporated these facts into an evolutionary scenario which we have called the "hot" model of galaxy formation (Berman and Suchkov, 1991).

The starting point for the "hot" model was the discovery of discrete metallicity groups in globular clusters (Marsakov and Suchkov, 1976), which led to a hypothesis that the Galaxy experienced two or three distinct enrichment events associated with corresponding bursts of star formation. The bursts were supposed to produce most of the metals observed.

The key idea was that a large energy release at a burst would quickly inhibit further star formation, so that the metals produced in a particular burst would enter into stars only after a long delay, in the next period (burst) of star formation. This would make the resulting distribution of metallicity in the galactic population discontinuous; this also would produce discreteness in kinematical and geometrical parameters, since the dynamical evolution of the gaseous component of the Galaxy between the sequential epochs of star formation would change both its kinematics and spatial configuration. Thus, the very existence of two or three distinct galactic subsystems could be attributed to the existence of two or three "active phases" in the history of the Galaxy (Suchkov, 1981).

The term "active phase" emphasizes that a burst of star formation not only produces stars and metals but also strongly affects the star formation itself, due to the enormous release of energy. Specifically, a burst may ultimately inhibit further star formation by heating up and destroying the interstellar gas clouds (Suchkov and Shchekinov, 1979) or by developing a strong galactic wind that heats the ambient protogalactic gas to such very high temperatures that not only the star formation but even the overall protogalaxy contraction turns out to be impossible for a long time (Suchkov, 1981; Berman and Suchkov, 1991).

In this way the model based on the concept of active phases predicts age discontinuities between galactic subsystems as well as chemical, kinematical, and geometrical discreteness in galactic stellar populations.

2 "HOT" MODEL OF GALAXY FORMATION

Within the framework of the "hot" model of galaxy formation, a gaseous protogalaxy, whose mass is contained initially mainly in dense cold clouds, becomes "hot" as it interacts with a strong galactic superwind (see Heckman et al., 1987) resulting from a burst of star formation. The model has three major ingredients: a cloudy protogalaxy, a galactic wind, and a hot protogalaxy.

The requirement for a galactic wind emerges from the necessity to find, first, an energy source and, second, a source of metal enrichment to make a hot, metal-rich protogalaxy. Observationally, there are at least two empirical facts that require the existence of early galactic winds. The first fact concerns the mass-metallicity relation: the decrease of galactic heavy element abundance with decreasing galactic mass is commonly interpreted as being due to mass loss in young galaxies following supernova-driven galactic winds that develop as a result of strong bursts of star formation. The second fact concerns the large mass and high metal abundance of the intracluster hot gas in clusters of galaxies: these are attributed to early metal-rich galactic winds that carry the heavy elements synthesized in cluster galaxies out into the intergalactic medium.

Hot protogalaxies are needed mainly to account for the age gaps between the galactic components and corresponding discontinuities in their metallicity, kinematics, and geometrical configuration. The most natural way to stop star formation for a few Gyr and produce an age gap like the one between the disk and the halo of the Milky Way is to heat up the gaseous remainder of a galaxy to temperatures in excess of that corresponding to the virial equilibrium; otherwise, as argued by Suchkov and Shchekinov (1979), the lull in star formation cannot last more than a few hundreds Myr.

It is also relevant to the production of hot protogalaxies that the ma ss and metal abundance of the hot intracluster gas are of the same order as in cluster giant galaxies. A logical suggestion is that the giant galaxies have lost about half of their initial baryonic mass, enriched with metals, while the other half has been retained to produce the major stellar component of metallicity which is nearly the same as that of the lost material. Since the mass of the wind can hardly be larger than about a tenth of the baryonic mass of a giant galaxy, the bulk of the ejected material must have been initially contained in the massive remainder of the protogalactic cloud.

It is also necessary to our considerations that the protogalaxy be cloudy. In the case of a homogeneous circumgalactic gas, a galactic wind drives a shock though this gas, while the wind material remains confined by the gas to the inner part of the galaxy. If the shock expels some fraction of the gas into the intergalactic space, it will be the unenriched gas since all metals are contained in the wind material. In this case there would be no metal-rich intracluster gas. Alternatively, if the shock is strong enough to eject all of the envelope and let the enriched wind material escape into the intracluster gas, the parent galaxy will be left with no material to subsequently form metal-rich stellar components. Thus either option turns out to be unsatisfactory. One has to think of a way in which wind material was carried to the outskirts of a protogalaxy, more or less uniformly imparting energy to it and enriching its mass with metals, with a substantial fraction of this mass to be retained in the galaxy. The problem can be solved by invoking a cloudy protogalaxy. The idea is that a wind can blow between the clouds, thus transporting through the protogalaxy energy, momentum, and metals outwards. On its way through the protogalaxy, the wind collides with the clouds that gain energy from it. As a result, temperature and pressure of the clouds will increase and eventually dispel material within the tenuous wind, completely mixing up with it. The outcome will be the formation of a hot protogalaxy, more or less uniformly enriched by the absorbed wind.

3 WHY DO GALAXIES HAVE GALACTIC SIZES?

Unlike many other theories of galaxy formation, the "hot" model predicts the size of a galaxy: this is a powerful test for the model. After formation and violent expansion of a hot protogalaxy, its outer envelope escapes from the system and contributes to the hot, metal-rich intergalactic gas. The inner gas, after being gravitationally decelerated, slowly begins to contract as it gradually cools off. At this stage, the gas is in quasi-hydrostatic equilibrium maintained by the balance of radiative cooling and gravitational heating (as the gas is compressed by the gravitational field). This implies that the cooling time,

$$t_{cool} \sim \frac{3kT}{2\Lambda n}$$

where $\Lambda(T)$ is the cooling function and T and n are the characteristic gas temperature and density respectively, is larger than the dynamical time scale. If the latter is determined mainly by the gravity of the massive dark halo whose mass distribution is $M_a(r) \propto r$ (e.g. as indicated by flat rotation curves in outer parts of spiral galaxies), then its value is

$$t_{dyn} \sim r \left(\frac{GM_a(r)}{r} \right)^{-\frac{1}{2}}$$

As the gas contracts and the mean density increases, the cooling time decreases and finally becomes smaller than the dynamical time scale. At this stage, the bulk of the gas experiences runaway cooling which is accompanied by thermal instability. As a result, the hot protogalaxy decays into dense, cold clouds embedded in the hot, tenuous intercloud gas, and the further dynamical behaviour is essentially the free fall of clouds in the galactic gravity field (with the intercloud medium contracting quasi-hydrostatically and producing new clouds). The clouds form a dissipationless collapsing system which is known to be capable of shrinking by no more than by a factor of two. This means that the onset of runaway cooling and thermal instability fixes the size of the forthcoming stellar system: this size is about half of that which the gaseous protogalaxy has at the onset of the thermal instability (to be more accurate, a dissipationless systems attains equilibrium in the gravity field $\varphi \propto \ln r$ as it contracts by a factor of about $e^{-\frac{1}{2}}$).

Taking into account that the gas temperature in a hydrostatic equilibrium is close to

$$T \sim T_{eq} = \frac{G\mu m_H M_a(r)}{3kr}$$

we can derive the size of the protogalactic cloud at the brink of the runaway cooling by equating the two time scales, $t_{cool} = t_{dyn}$, to obtain the final size of the stellar system as

$$r_s = M_s^{-\frac{1}{2}} \left(\frac{\Lambda}{2\pi e}\right)^{\frac{1}{2}} (kT_{eq})^{-\frac{3}{4}} (3\mu m_H)^{-\frac{1}{4}}.$$

Here we assume that the bulk of the gaseous mass becomes stars whose total mass is M_s. For the range of equilibrium temperature $10^6 - 10^7 K$, which is typical for giant galaxies, the temperature dependance of the cooling function $\Lambda(T)$ is close to $\Lambda \propto T^{-\frac{1}{2}}$, with $\Lambda = 10^{-23} erg\ cm^3\ s^{-1}$ at $T = 10^7\ K$ (for the sola r-abundance gas). Subsitution into the formula for r_s together with $\mu = 0.6$, yields (Suchkov, 1988)

$$r_s \sim 10 \left(\frac{M_s}{10^{11} M_\odot}\right)^{1/2} \left(\frac{M_a}{10^{12} M_\odot}\right)^{-1} \left(\frac{r}{10kpc}\right) kpc$$

This order-of-magnitude estimate fits in perfectly with the observed sizes of giant ellipticals and of halos of giant spirals.

The predicted size depends not only on the stellar mass of the galaxy but also on the mass of the dark halo and its distribution, i.e. on the M_a/r ratio. This fact allows us to make one more prediction. Identifying the hydrostatic temperature, $T_{eq} \propto M_a/r_a$, with the temperature of the hot X-ray corona of a giant galaxy (cf. Forman et al., 1985), and adopting the mass-to-light radio $M/L \propto L^{1/4}$ (Faber et al., 1986), we obtain from the formula for r_s the relationship (Suchkov, 1989):

$$\Sigma_e \propto T_x^2$$

where Σ_e is the mean surface brightness of the galaxy. Thus the "hot" model predicts that the optical surface brightness of ellipticals and perhaps of halos of spiral galaxies is proportional to the square of the galactic corona temperature.

4 EVOLUTION OF A PROTOGALAXY INTERACTIN WITH A GALACTIC WIND

A more quantitative analysis confirms the above order-of-magnitude estimate for characteristic galactic sizes. Berman and Suchkov (1991) examined a series of model protogal axies and numerically integrated time-dependent, spherically-symmetric, gas-dynamic equations describing the evolution of an early galactic wind interacting with a surrounding extended envelope of dense clouds and tenious intercloud gas. The "basic" model was specified as follows. The cloudy gas of mass $M_{cl} = 10^{11} M_\odot$ and the intercloud gas of mass $M_{int} = 10^{10} M_\odot$ are distributed between the inner radius $r_w = 5\ kpc$ and the outher radius $r_B = 100\ kpc$; the dark halo of mass $2\times10^{12} M_\odot$ is confined to 20 *kpc*. At $r = r_w$ the boundary conditions for the galactic wind are set: its temperature is $T_w^0 = 5 \times 10^7 K$ and its speed is $u_w^0 = 1.5 \times 10^3 km\ s^{-1}$; the wind blows over the span of time $t_w^0 = 10^8\ yr$ at a rate $\dot{M}_w = 10^2 M_\odot yr^{-1}$; the flux of the wind thermal energy is $\dot{E}_w = 1.4 \times 10^{44} erg\ s^{-1}$.

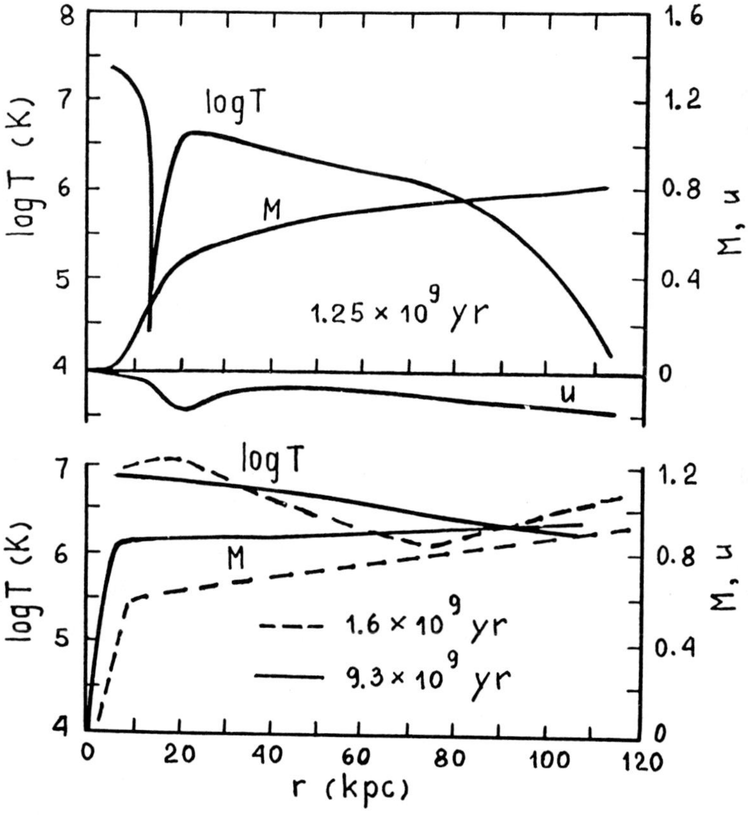

Fig.1. Mass, temperature, and velocity profiles of the protogalactic gas.

The clouds interacting with the hot material of the wind and shocked intercloud gas disperse on the time scale $t_{cl} = 2.5 \times 10^8\ yr$, mixing up with this matter (to allow for this interaction, the corresponding gas-dynamic equations for the wind and intercloud gas material are coupled via the source terms with the equation for the clouds).

The cooling function $\Lambda(T)$ is from Dalgarno and McCray (1972), and Raymond et al. (1976) for the solar-abundance gas. The mean molecular weight μ is taken to be 0.6.

Figure 1 shows the dynamical behavior of the "basic" model described above, showing for three instances the radial profiles of mass, temperature, and velocity of the galactic wind and of the intercloud protogalactic gas clouds. A particularly striking moment is shown in the upper panel of Figure 1. At a distance of about 13 kpc from the center, the gas density reaches a value at which the radiative energy loss overcomes the heating rate due to compression; the result is runaway cooling and compression of the corresponding gas layer, thus providing the first portion of material available to become stars. This moment may be identified with the onset of star formation.

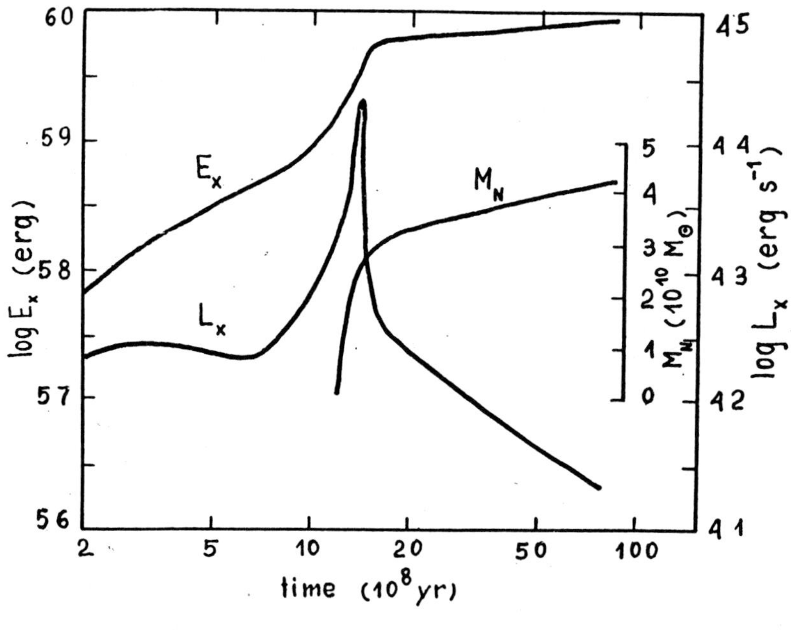

Fig.2. The evolution of the X-ray luminosity L_x and emitted energy E_x, and the formation of the cold nucleus in the hot protogalaxy model.

After that, the inner layers, $r < 10 - 13$ *kpc* , remain hot and compressed for a while; the next layers that undergo rapid cooling are those coming outwards as they reach the same distance of about 10 to 13 kpc. The total mass cooled over this distance

range amounts to $1.5 \times 10^{10} M_{\odot}$. After that the region of runaway cooling shifts to the inner layers, beginning with those nearest to the center. They cool consecutively over a distance range from ~ 300 *pc* to ~ 5 *kpc* and add another $1.5 \times 10^{10} M_{\odot}$ to the cooled mass in the center of the protogalaxy. The bulk of this mass is formed over a rather short time interval: in $3 \times 10^{8} yr$ the mass of the cold central gas aproaches $3 \times 10^{10} M_{\odot}$. After that the rate at which
this mass grows sharply drops, and another $1.5 \times 10^{10} M_{\odot}$ are gained over $\sim 5 \times 10^{9}$ *yr* (see Fig.2). In fact, the system becomes quasi-stationary: as seen in the lower panel of Figure 1, the variation of all parameters over many Gyr is quite small, with velocities close to zero (not plotted in the lower panel).

At this stage, the system may be regarded as being composed of two distinct components with a sharp boundary between them. One component is the dense, cold nucleus containing the bulk of the retained gaseous mass ($M_N = 4.6 \times 10^{10} M_{\odot}$). As mention ed above, this mass is assumed to convert into the major stellar component of the galaxy. The other component is the extended, tenuous, hot envelope which is identified with the X-ray emitting corona ($M_{cor} = 3 \times 10^{9} M_{\odot}$). The gas in the outer part of the system is assumed to be the ejected material which forms the hot, metal-rich intergalactic gas in clusters of galaxies ($M_{ej} = 4.7 \times 10^{10} M_{\odot}$). The dynamical situation at this late stage can be described in terms of quasi-hydrostatic cooling flows as they are defined, e.g., in Fabian (1988).

If we assume that the mass and the size of the cold nucleus of the hot protogalaxy are relevant to the characteristics of the galactic stellar component to come, we must then recognize that the above results, taken at face value, provide an answer to the question why galaxies, at least normal ellipticals, and the halos of spirals, have galactic sizes. Indeed, the typical size of a giant elliptical is about 5 kpc - just as is obtained for half a diameter of the cold nucleus. One can easily see that the emerging size agrees quite well with the above analytical estimate for r_s.

5 SUMMARY

The "hot" model of galaxy formation explains quantitatively, in a selfconsistent way, why giant galaxies have galactic sizes, why they are constituted of distinct (discrete) components, why the mass and iron abundance of the intracluster gas and cluster giant galaxies are comparable, and why giant elliptical galaxies do not have coronae whose mass is 0.1-0.01 of the galactic mass.

REFERENCES

Berman, V.G. and Suchkov, A.A. 1991, Astrophys. Space Sci., in press

Dalgarno, A. and McCray, R.A. 1972, Ann. Rev. Astron. Astrophys. 10, 375
Faber, S.M., Dressler, A.,Davies, R.L., Burstein, D., Lynden-Bell, D., Terlevich, R., and Wegner, G. 1986, in: Nearly Normal Galaxies, ed. S.M. Faber, Springer-Verlag,p.175
Fabian, A.C. 1988, in: Cooling Flows in Clusters and Galaxies, ed. A.C. Fabian, Kluwer Academic Publishers, Dordrecht, Holland, p. 315
Forman, W., Jones, C., and Tucker, W. 1985, Astrophys J. 293, 102
Heckman, T.M., Armus, L., and Miley, G.K. 1987, Astron. J. 93, 276
Raymond, J.C.,Cos, D.P., and Smith, B.W. 1976, Astrophys. J. 204, 290
Suchkov, A.A. 1981, Astrophys. Space Sci. 77, 3
Suchkov, A.A. 1988, Soviet Astron. 32, 1
Suchkov, A.A. 1989, Soviet Astron. Letters, 15, No.2
Suchkov, A.A. and Shchekinov, Yu.A. 1979, Soviet Astron. 23, 26

Starbursts Stimulated by Thermal Conductivity and Heating Radiation

YURI SHCHEKINOV

Institute of Physics, University of Rostov, Russia

1 INTRODUCTION

The ensemble of starburst galaxies contains quite a number of isolated systems (see review, Scalo, 1987). This means that starburst phenomena in such galaxies are connected with their internal properties. It is quite reasonable to assume that there exists a positive feedback mechanism in these galaxies which ensures a burst-like increase of star formation (SF).

Self-propagating SF stimulated by shock waves from active stars (Elmegreen and Lada, 1977, Seiden et al., 1982) provides such a positive feedback, but the characteristic spreading velocity of the SF process is about $3 - 10$ kms^{-1}. However, starburst galaxies show enhanced SF over a large scale of about $1 - 3$ kpc in a comparatively short time $\sim 10^7 - 10^8$ yr, so that the characteristic velocity of the SF process may be up to 100 kms^{-1}.

In such cases, it is necessary to assume the existence of long-range forces providing coherent behaviour of SF on large scales. There can be no doubt that gravitational forces are a plausible source of coherency of star-gaseous systems, although in nearly virial systems the large-scale gravitational potential changes with characteristic times larger than the free-fall times.

The purpose of this paper is to show the existence of nongravitational forces which are able to provide positive feedback on large scales.

The action of these forces is connected with implosion (or squeezing) of interstellar clouds embedded in hot coronal gas or exposed to heating radiation. It is known that heating of the external layers of a cloud by thermal conductivity leads to its evaporation. But actually the pressure excess arising at the edge of conductively heated cloud simultaneously causes the inflow of adjoining gas towards its center. Quite similar dynamics take place when an optically thick cloud is exposed to heating radiation: the excess pressure in the absorbing layer drives both inflow and outflow of the cloud material (Kovalenko and Shchekinov, 1991).

This effect provides a fairly substantial decrease in the critical mass M_{cr} for the clouds to be subjected to self-gravitating contraction, and it can lead to an increase of the SF process due to the enlargement of the mass interval of clouds which are able to be transformed into stars.

A detailed quantitative description of the nongravitational stimulation of starbursts will be published elsewhere (Cammerer and Shchekinov, 1991, Shchekinov, 1991).

2 CONDUCTIVELY-DRIVEN STARBURSTS

The principal idea of the starburst mechanism connected with the implosion of interstellar clouds by thermal conductivity is that the ensemble of clouds occasionally becomes surrounded by hot coronal gas covering large scales. If the SF rate is assumed to be proportional to the total number of clouds with individual masses larger than the critical mass:

$$SF\ rate = At_c^{-1} \int_{M_{cr}}^{\infty} \varphi(M) dM \qquad (1)$$

where $\varphi(M)$ is cloud mass spectrum. $M_{cr} = M_{cr}(T_{hot})$ is the critical mass depending on the temperature of hot coronal gas and t_c is the compression time of a cloud, then such an event leads to enhanced SF when the more numerous population of clouds is covered by hot gas.

The possibility for the cloud ensemble to be surrounded by hot coronal gas on the large scales of about 1 - 3 kpc diameter, depends on the intersections of SN remnants. One such possibility is described by Ikeuchi and Yoshioka (1989), who show that interactions between SN remnants lead to coherent oscillations of the filling factor f_h of hot gas with a peak value of the filling factor about unity: in a volume $V = (2\ kpc)^3$ the characteristic period of such oscillations is about 10 Myr if the porosity of SNRs equals Q = 0.4. If this region contains a sufficient amount of cloud gas during the period of increased filling factor of hot gas, it leads to the enhanced SF results.

In the model of Ikeuchi and Yoshioka, SN explosions were distributed homogeneously in a given volume. Another limiting case corresponds to the clustered SN explosions (Heiles, 1990). In such a case, the filling factor of hot gas on large scale is a random function of time, and there exists a finite probability for the system with quite a moderate value of SN rate to have a high-level crossing of f_h.

Due to this high-level crossing the interstellar gas turns into the state with pervasive coronal gas. The duration of this new state depends on many factors, such as size of the region covered by such global fluctuations, evaporation and dispersion of cloud gas, and the ability of SN explosions to maintain a hot coronal phase. In any case, this time is not less than $10^7 yr$, and it is sufficient to induce enhanced SF due to conductively driven implosion, since the characteristic time of the cloud's implosion is substantially less than its hydrodynamical time ($\leq$ 3 Myr, Kovalenko and Shchekinov, 1991). The intensity of SF stimulated in such a way can be quite high because of a dramatic decrease of M_{cr} for the clouds embedded in hot coronal gas, when compared to the Jeans mass.

The probability for a star-gaseous system to have a high-level crossing of porosity factor q is given by the formula:

$$P(q) \simeq \tau^{-1} exp(-\frac{(q-Q)^2}{Q}\frac{V}{V_o})(1 + O(\frac{1}{\sqrt{\nu\tau}})), \qquad (2)$$

for $\nu\tau \gg 1$. Here Q is the mean value of porosity for a given SN rate and ambient gas density, V; the volume of the system, V_o is the maximum volume of the remnant of cumulative explosion of SNs, ν is the SN rate per year in the volume V, and τ is the life-time of the cavity.

Thus, for star-gaseous systems with parameters similar to those of our Galaxy: $\frac{\nu}{V} \sim 10^{-13}\ pc^{-3}yr^{-1}$ mean SN rate, $n_a = 1cm^{-3}$, $p_a = 10^4 Kcm^{-3}$ ambient density and pressure, so that for $\tau \sim 10Myr, V_o \sim 10^8 pc$, $E = 10^{53} erg$ (for cumulative explosion) equation (2) gives the probability for the transition from $Q \sim 0.3$ (near the percolation threshold for hot coronal gas) to $q \sim 0.8$ (corresponding to fairly developed percolation of hot gas), namely $P(q) \sim 10^{-11} - 10^{-12} yr^{-1}$ for $V = 1kpc^3$. This means that in a galaxy with a quite moderate value of the SN rate and a mean porosity factor of 3 - 30, such events covering the scales of 1 kpc can occur during a Hubble time. But for a galaxy with a mean density of interstellar gas about $0.1 - 0.3cm^{-3}$, this estimate increases by one or two orders of magnitude.

3 RADIATIVELY DRIVEN STARBURTS

Another possibility for stimulating starbursts by an energy input from massive stars is connected with squeezing interstellar clouds by heat ing ($\lambda > 912$Å) radiation.

If at the initial time t $= 0$ a group of massive stars is born at any place $\overline{r}$, then the excess of UV radiation generated by these stars,

$$\delta U_\nu \sim \frac{\delta s\ e^{-\tau(\overline{x},\overline{r})}}{|\overline{x} - \overline{r}|^2}$$

induces additional SF in the volume with a radius of $|\overline{x}-\overline{r}| < (\sigma_o n)^{-1}$ due to implosion of interstellar clouds by radiation. Here σ_o is the extinction cross-section, n is the mean interstellar density, $\tau(x,r)$ is the optical depth along the $(x-r)$-vector, and δs is the number of massive stars born.

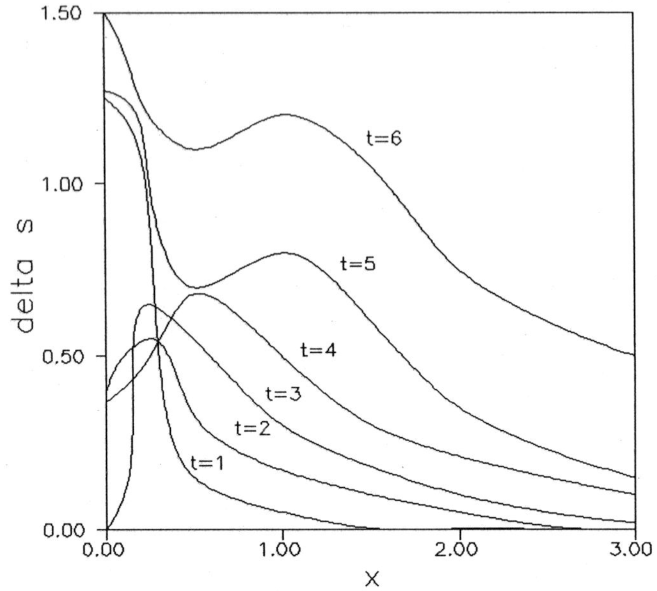

Fig.1. Dynamics of star formation by heating radiation. The initial perturbation of stars is $\delta s = 1$ in a cell $\Delta x = 0.25$ near $x = 0$. Curves are marked by the time from the beginning in units of $t_l = 3\ 10^7$ yr - lifetime of 7 $M_\odot$.

After the time needed for imploded interstellar clouds to become main-sequence stars, they produce a new portion of UV radiation and excite the formation of a new generation of stars. This leads to the penetration of the SF process through the system.

The characteristic velocity of SF-penetration is determined by the ratio of the extinction length $(\sigma_o n)^{-1}$ to the time delay from the beginning of the cloud's implosion to the formation of a star Δt :

$$v_{SF} \sim \frac{1}{\sigma_o n \Delta t}$$

For $(\sigma_o n)^{-1} \sim 300pc$ and $\Delta t \sim 3-10$ Myr, $v_{SF} \sim 30-100 kms^{-1}$; so that this process has kinematic characteristics similar to those of starbursts.

A numerical example of such spreading SF for plane-parallel symmetry is plotted in Fig. 1 (Cammerer and Shchekinov, 1991). Here the SF rate is assumed to be determined by equation (1) with the critical mass depending on the UV-flux and cloud mass spectrum $\varphi(M) \sim M^{-3}$ clouds per mass interval; the initial excess of massive stars relative to the unperturbed homogeneous stars' number in unit volume was assumed to equal $\delta s = 1$ in a cell $\Delta x = 0.25$; the unit of time is $t_l = 30 Myr$, with a main-sequence life-time of a 7 $M_\odot$ star, and the unit of length is extinction length $(\sigma_0 n)^{-1}$.

4 CONCLUSIONS

The above considerations demonstrate that the input of heat to interstellar clouds by UV radiation with $\lambda > 912$ Å(or cosmic rays) and thermal conductivity can stimulate SF and provide a positive feedback in star-forming systems.

The heating radiation can cause positive feedback in systems with optically thick clouds, unlike the situation in systems with optically thin clouds, where heating radiation plays a role of negative feedback (Parravano, 1988).

The heating radiation can initiate SF on a large scale in star-forming systems with optically thick interstellar clouds, and it can be a factor providing coherent behaviour of SF in star-forming systems. Thermal conductivity can stimulate burst-like enhanced SF in systems with a randomly increased SN rate and filling factor of hot gas.

REFERENCES

Cammerer, M., and Shchekinov, Yu., 1991, Astron. and Astrophys., submited
Elmegreen, B., and Lada, C., 1977, Astrophys.J., 214, 725
Heiles, C., 1990, Astrophys.J.,
Ikeuchi, S., and Yoshioka, S., 1989, Comments Astrophys., 13, 117
Kovalenko, I., and Shchekinov, Yu., 1991, this volume, and Astron. and Astrophys. Transactions, submited
Paravanno, A., 1988, Astron. and Astrophys., 205, 71
Scalo, J., 1987, Publ. Astron. Inst. Czechosl. Acad. Sci, 69, 101
Seiden, P.E., Schulmann, L.S., and Feitzinger, J.V., 1982, Astrophys.J., 253, 91
Shchekinov, Yu., 1991, in preparation

Star Formation Activity in the Interacting Galaxy NGC2976

N. BROUILLET [(1)], P. NOTNI [(2)], W. BRONKALLA [(2)], A. BAUDRY [(3)], C. HENKEL [(1)]

[(1)]Max-Planck-Institut für Radioastronomie, Auf dem Hügel 69, D-5300 Bonn 1, Germany
[(2)]Zentralinstitut für Astrophysik, Sternwarte Babelsberg, O-1591 Potsdam, Germany
[(3)]Observatoire de Bordeaux, B.P. 89, F-33270 Floirac, France

The M81 group of galaxies

NGC2976, classified as a Sc pec with $M \simeq 3 \times 10^9 M_\odot$, is one of the main members of the central M81 group of galaxies. M82, NGC3077 and NGC2976 share a common property: they are linked to M81 by HI bridges (Appleton et al., 1981). In M82 and NGC3077, this gravitational interaction leads to strong optical and HI perturbations (Brouillet et al., 1991). We investigate in this work the strength of the interaction and its role on the star formation activity in NGC2976.

The star population of NGC2976

UBV photometry (Bronkalla and Notni, 1990, 1991) reveals two different populations: an old disk population ($\geq 2 \times 10^9$ yrs) forming the smooth outer exponential disk of stars, and a younger disk population which is superposed onto the inner disk and clumped in many blue knots of age about 2×10^7 yrs, mostly prominent also in Hα (Hodge and Kennicutt, 1983).

The gas content of NGC2976

We have mapped the CO emission of NGC2976 with the IRAM 30-m radiotelescope in the 2-1 and 1-0 lines (spatial resolution of 23" and 12" respectively). The distribution of the molecular gas is limited to the starburst region with sharp edges. It presents two large prominent complexes in the surroundings of the youngest blue knots, located near the turn-over points of the rotation curve. We find some correlation of the maxima of CO emission with faint regions, probably strong extinction (Bronkalla and Notni, 1991), near the most luminous knots (see Figure 1). The CO emission agrees well in distribution and velocity with the HI emission mapped at Westerbork (L. Tacconi, private communication). Whereas the blue and Hα emissions show a more or less uniformly clumped disk, a spiral structure and a faint bar are indicated in the CO and HI distributions.

The rotation curve deduced from our CO observations shows a rigid rotation. It is in agreement with the Hα data (Carozzi-Meyssonnier, 1980) and confirms that all these components really trace the disk of NGC2976 in which, moreover, no sign of a bulge is seen. Using the conversion factor of Bloemen et al. (1986) and assuming a distance of 3.25 Mpc, we find from our CO data a molecular mass of $M_{H2} \simeq 3.5 \times 10^7$ $M_\odot$. A rough estimate of the HI mass is about 6×10^7 $M_\odot \simeq 0.01$ M_{total}. About 1% of the gas mass is currently involved in the bright blue knots.

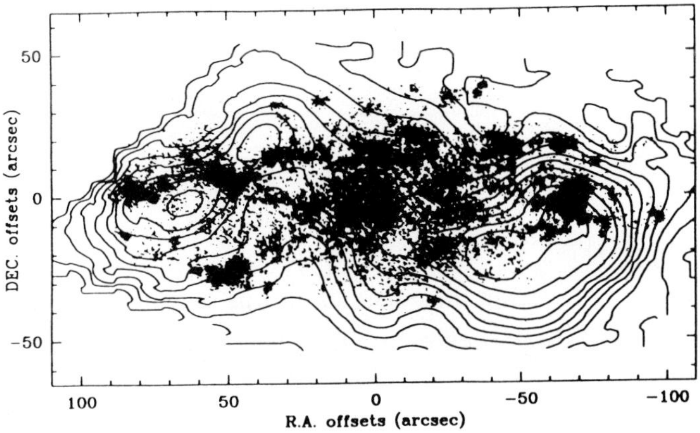

Fig. 1. Comparison of the CO(1-0) integrated emission (contour map) with the UV map of NGC2976. The contour levels are from 0.5 to 5.0 K.km/s with 0.5 K.km/s increment.The (0,0) position corresponds to: $\alpha_{1950}= 09^h43^m10.2^s$, $\delta_{1950}= 68°08'54''$.

Origin of the star forming activity

No strong spiral structure is present, and NGC2976 is similar to a dwarf galaxy with an extended weak starburst in the inner part of the galaxy. The HI gas in NGC2976 is clearly related to the HI bridge towards M81. However it is unlikely that the star formation activity is triggered directly by an earlier close encounter with M81 like for M82 and NGC3077, or from an outflow of gas as it would require too much energy from the galaxy (NGC2976 doesn't show any sign of strong perturbation in the star distribution and the gas velocity field). An infall of intergalactic gas, whose velocity is close to the systemic velocity of NGC2976, is thus more likely to have induced the star forming activity.

References

Appleton, P.N., Davies, R.D., Stephenson, R.J, 1981, MNRAS 195, 327

Bloemen, J.B., Strong, A.W., Blitz, L., Cohen, R.S., Dame, T.M., Grabelsky, D.A., Hermsen, W., Lebrun, F., Mayer-Hasselwander, H.A. and Thaddeus, P. 1986, Astron. Astrophys. 154, 25

Bronkalla, W. and Notni, P. 1990, Astronomische Gesellschaft, Abstract Series 5, 85

Bronkalla, W. and Notni, P. 1991, Astron. Nachr., submitted

Brouillet, N., Baudry, A., Combes, F., Kaufman, M. and Bash, F. 1991, Astron. Astrophys. 242, 35

Carozzi-Meyssonnier, N., 1980, Astron. Astrophys. 92, 189

Hodge, P.W. and Kennicutt, R.C. 1983, PAPS Doc. ANJOA88-296-300

Merging in Groups of Galaxies

Gastão B. Lima Neto and Françoise Combes

Observatoire de Meudon - DEMIRM — 92190 - Meudon - France

1 INTRODUCTION

The brightest galaxies found in clusters and groups of galaxies are usually classified as 'cD' galaxies. They are very extended objects and present some peculiar characteristics:

- a small dispersion of absolute magnitude within a given metric radius,
- large envelopes reaching out ~1 Mpc,
- double or multiple nuclei.

For their masses ($\sim 10^{13}$ $M_{\odot}$), they play a fundamental role in the dynamics of the cluster. The interpretations of these features are still debated (Hausman and Ostriker 1978; Merrit 1984; Ulson *et al.* 1990). The 'Cannibalism' during the evolution of a cluster or group of galaxies may play a fundamental role in the origin of the first ranked galaxies. In this work, we follow the evolution of a group of galaxies through N-body numerical simulation. In this way, physical phenomena like dynamical friction, merging, tidal stripping and deformation, are treated with no *ad hoc* hypothesis.

2 THE MODEL

We have followed the time evolution of groups of 15 disc galaxies (which can also be seen as the central region of rich clusters) through a self-consistent, three dimensional 'Tree-code' (Hernquist 1988). The simulations start with the galaxies distributed randomly in a sphere with a Plummer density profile. The galaxies velocities are chosen to result in a virialised cluster. Each galaxy is modeled with three components: a central bulge, a disc and a halo, each one following the potential-density pair given by Nagai and Miyamoto (1976). Supposing the galaxy in a steady state and that the velocity ellipsoid has its principal axes always aligned with the coordinate directions of the system, then the Jeans equations can be solved to obtain the velocity dispersions within the galaxy. Taking $\langle v_R \rangle = 0$, $\langle v_z \rangle = 0$ and $\langle v_\theta \rangle = (\langle v_\theta^2 \rangle - \langle v_R^2 \rangle)^{1/2}$ we are able to 'build' our galaxies.

3 RESULTS

Figures 1.a - 1.b show snapshots of the group in X-Y projection. The simulations show that shortly after the experiment starts, a first merger occurs and the remnant of this merger will start to cannibalise the other galaxies. We observe, then, the multiple nuclei feature in this object. Since our experiments start with all particles bound to the galaxies, the intracluster medium is quickly enhanced by particles that are stripped from their parent galaxies due to close encounters, mergers and tidal disruptions. Even the boundest particles, the ones that were in the bulges, contribute to the formation of this medium. The first ranked object is surounded by an envelope composed by this intra-cluster matter.

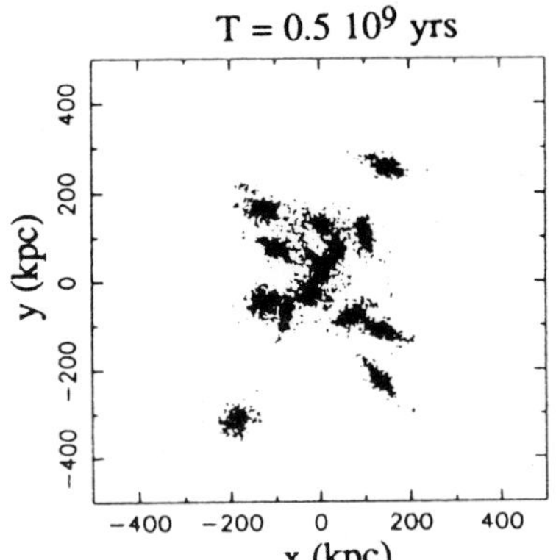

Figure 1.a

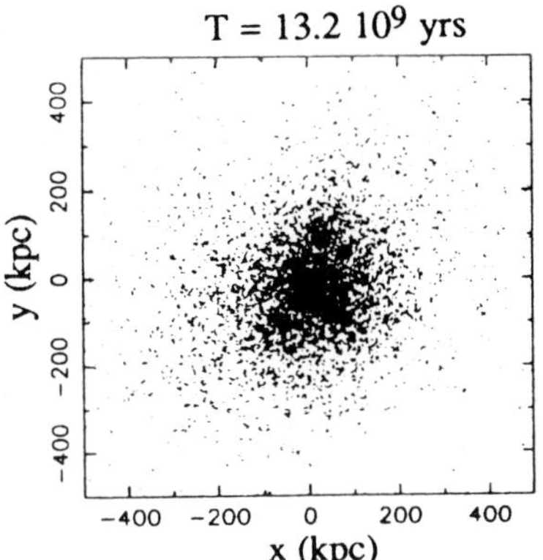

Figure 1.b

Figure 2 shows the surface density of each component at the end of the experiment. The continuous lines are fits of de Vaucouleurs' profile ($R^{1/4}$). Note that the halo particles seem to follow a power law. Figure 3 shows the three-dimensional velocity dispertion as a function of radial distance. Halo particles are hotter on the envelope.

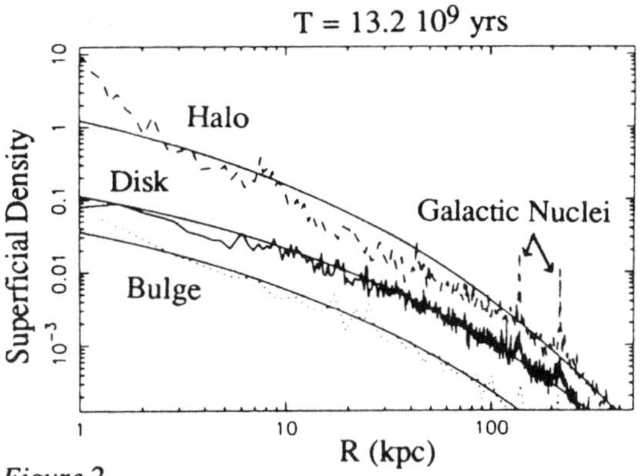

Figure 2

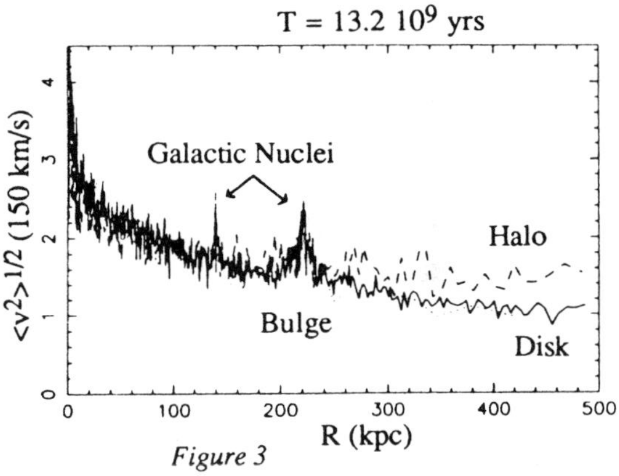

Figure 3

The evolution of $\varepsilon \equiv \frac{\text{orbital kinetic energy}}{\text{internal kinetic energy}}$ with time suggests a behaviour $(d\varepsilon/dt) \propto -\varepsilon$.

No disk galaxy survives through the evolution of the group. In view of this result we can suggest that the observed groups that have a spiral first ranked galaxy should be either dynamically young or the result of chance projection on the sky.

REFERENCES

Hausman, M.A. and Ostriker, J.P. 1978, Ap. J. **224**, 320
Hernquist, L. 1988, Comp. Phys. Comm. **48**, 107
Merrit, D. 1984, Ap. J. Letters **280**, L5
Nagai, R. and Miyamoto, M. 1976, Publ. Astron. Soc. Japan **28**, 1
Ulson, J.M., Boughn, S.P., Kuhn, J.R. 1990, Science **250**,539

Part V

CONCLUDING REMARKS

Star Formation in Galaxies—Some Thoughts on This Workshop*

IVAN R. KING

University of California, Berkeley, U.S.A.

At the close of a conference like this one, it seems appropriate to do more than just summarize. We should consider the broad setting that we are in. We are honored to be the members of the first Workshop organized by the Center for Theoretical Study. That Center is dedicated to the interplay between various fields of intellectual endeavor, and I shall try to do what I can to look broadly at our efforts here, while at the same time trying to summarize what we have done and heard.

After a spectacle that takes place in Shakespeare's last play, *The Tempest*, the chief character, Prospero, addresses the others, saying,

> Our revels now are ended.
>
> The Tempest, IV, I, 148

The end of a conference always reminds me of that line. But Prospero then goes on to say,

> These our actors,
> As I foretold you, were all spirits and
> Are melted into air, into thin air;
> And like the baseless fabric of this vision,
> The cloud-capp'd towers, the gorgeous palaces,
> The solemn temples, the great globe itself,
> Yea, all which it inherit, shall dissolve,
> And, like this insubstantial pageant faded,
> Leave not a rack behind.
>
> The Tempest, IV, I, 148-156

There is indeed some of that feeling at the end of a conference, but the underlying metaphor doesn't fit at all. It *won't* all dissolve and disappear, and we will in fact go home enriched by the scenes that have been played here. I can't say I haven't heard a paper or two that I wish would go away; but for the most part, what we have heard will remain solidly standing, and will be strengthened before much time has passed at all.

* Like many "Concluding Remarks," this written version represents not what the speaker actually said, but rather what he wishes he had said.

These concluding summaries are a chore, though. You've got to attend all the papers and listen to them, and stay awake; and you can't relax till the meeting is completely over. The question of staying awake brings me to my first point: I stayed awake nearly all the time, which any of my colleagues will tell you is quite unusual; so *there's* an objective measure of how interesting this meeting was. It was especially so for me, because I've learned something about many subjects from which I've been rather remote. For example, this is the first time in my life that I've been expected to get up and talk in a learned way about the interstellar medium.

Musings like this help me to realize why an astronomical topic, and this one in particular, has been such a good choice for the First Workshop of the Center for Theoretical Study. Astronomy is, in its own way, a microcosm of the interdisciplinary needs that beset the whole intellectual world. We work, usually, in our own separate compartments; and it is too rarely that we try to join them together. That is what we have done at this meeting, where we have ranged from atomic processes in the interstellar medium to interactions between galaxies. But more of that when I get to the actual astronomy.

Just a few more remarks about the Center that is our host. It strikes me that the last word of its name is in the singular form—not "Studies," but "Study." This implies that it's all part of a single thing. I applaud the aims of the Center, and I wish its Director, Ivan Havel, all luck in developing it. (I don't need to wish him strength; it's clear that he already has that.) But to this beautiful Czechoslovak nation that is hosting us, I wish luck *and* strength. They will *need* strength to rebuild after the many decades of difficulties that they have survived so nobly. And I cannot fail to wish to all our Soviet friends the luck and the strength that *they* will need to pass though this difficult period of *their* history and to realize the tremendous potential that is inherent in their nations.

But now to the Workshop itself. As I mentioned, we have combined here a wide range of astronomical topics. We have heard about supernova shells, about HII regions, about giant molecular clouds, about processes and regions of star formation, about the general structure of the interstellar medium, about its overall distribution in galaxies, about the stellar populations that have formed from it, about the dynamical processes that mould the disks and especially their interstellar component, and finally about collisions between galaxies and their star-forming results, both in galaxy tails and in active galactic nuclei. Quite a mouthful, isn't it?—or rather, I should say, quite a head-full.

In spite of this wide variety of topics, our central unifying theme has been star formation. Yes, the workshop is entitled, "Evolution of Interstellar Matter and Dynamics of Galaxies"; but what we have been talking about, again and again and throughout, is star formation. (Mind you, I have no quarrel with the title; if this Workshop had been thoughtlessly entitled "Star Formation," we galaxy people would have stayed home, as would many others too, and it would have been an overly specialized meeting.) But putting aside the question of titles, star formation is a totally appropriate topic, because it is one of the most active and rapidly moving areas of present-day astronomy.

During the past few decades we have learned a great deal about stellar evolution and about the structure of galaxies. Even though the detailed processes of star formation are still an enigma, we are at grips with physical problems that are clearly relevant. An increasing wealth of facts is emerging, and from these facts will surely come increased understanding.

Now for our meeting itself. I will not summarize but rather categorize. What topics did we touch on, and how do they relate to each other? (In doing so, I will try to touch on all reviews and contributed papers, but I will make no attempt to cover the poster papers; I apologize to the presenters of poster papers, but one can do only so much.) In covering the papers I think that it will make some sense to go from the smallest scale of processes to the largest.

On the smallest spatial scale, we had a number of papers that dealt with the expanding region of a supernova shell. A good starting point is the review by Yorke, which covered many of the difficult points that come up in an elaborately planned simulation. He also presented results that showed the effect of changing various physical assumptions. This review was nicely complemented by the one from Nelson, who went in detail through the calculations that have been made about colliding clouds. We should also be grateful to him for the memorable admission that "It is conceivable that in not too many years simulations will be realistic."

Bruce Elmegreen's review was an interesting contrast; he concentrated completely on the physical processes that can cause a small region to contract into a protostar. He considered particularly the role of a local magnetic field. My impression was that you're damned if you do and damned if you don't; compression across the field causes collapse for one reason, compression along the field for a different reason.

We also had a number of papers presenting particular calculations of the behavior of an expanding region. Cantó presented a picture of mixing layers at the place where the expanding gas impinges on the surrounding interstellar medium. Franco had a somewhat similar paper, in which lateral motions in his 2-D calculation destroyed the neatness of the shell. (I had occasion to comment that his results did not look as we expect a theory to; instead of neat theoretical curves he had something that looked instead like a real supernova shell!)

There were two papers in which the authors had, for lack of adequate computing power, chosen to confine their attention to the narrow shell of interaction. Palouš had chosen to try a realistic matching of the HI in a particular region, including Galactic gravitational forces. He found that he could get a good fit only with particular assumptions about the nature of the hydrogen layers, and more than one of the observers remarked that this theory had predictive capabilities that might be quite valuable. Silich, in a somewhat similar calculation, had chosen instead to treat a more general problem; his results will presumably apply to a wider range of situations, but at the cost of the specific predictions that Palouš is able to make.

And, even though it was not a paper on phenomena in a small region, I should mention here the discussion by Gilmore of the force perpendicular to the Galactic plane, which is needed for the papers that I just mentioned. More importantly, however, this force is our way of detecting any otherwise unknown mass, and Gilmore argued cogently that there is no reason to suppose the existence of any. (Incidentally, this is a field that I do know something about. Although Gilmore is quite right that the basic physics is trivially simple, he modestly neglected to inform you that the statistical problems of analyzing the data are quite horrid, and that he and his former student Kuijken have developed and carried out heroic measures for handling them.)

Now I have to mention two papers whose implications left me rather puzzled. The papers were fine, but they suggest conclusions that I find worrisome. Bykov called attention to the possibility that magnetic-field irregularities within an expanding shell give rise to low-energy cosmic rays, which then have an important effect on the pressure and the ionization of the whole region. Dorfi, by contrast, emphasized that high-energy cosmic rays are generally presumed to originate in supernovae, so that the surrounding shell may have its energy balance strongly influenced by these particles. The perplexity with which these papers leave me is: if all the papers previously mentioned omit the effect of cosmic rays, are they partially or seriously wrong?

Another set of important papers dealt with a somewhat larger scale: that of a giant molecular cloud. Blitz's interesting review emphasized the ubiquity of star formation in GMC's (with one GMC standing out as a provocative exception). He also dwelt on the importance of studying GMC's of different ages, so as to see how they develop with time. Along with a wealth of information, he left us with several perplexing questions. Why are the rotations of GMC's generally retrograde? Why do their velocities tend to be independent of mass? Why do the masses of the clumps in them have a distribution that so often follows the same power law? Does the maintenance of a steady population of GMC's require a Galactic infall of gas?

From this and other presentations we seemed to be getting a picture of a layered structure, in which central star formation has created an H II region that is surrounded by a dense cloud of CO, surrounded in turn by a large quantity of H I.

Burton's discussion dealt with the association between the IRAS cirrus and neighboring H I, which generally seems to have anomalous velocities. These latter may in turn be associated with disturbances in the normal-velocity H I. He gave us detailed descriptions of these connections in a number of regions.

By contrast Junkes had looked for IRAS point sources, near supernova remnants. He had numerous successes, indicative of vigorous star formation in these regions.

Fiedler presented an interesting new way of probing the ISM: rapid variations in QSO brightnesses, for which he made a convincing case that the cause was irregularities in interstellar refraction. With this technique the turbulent structure of the ISM can be observed from A.U. scale on up.

Now we pass to a larger scale: the thickness of the Galactic disk. (We should note again here Gilmore's paper on the density and the z-force, although its relevance was to other papers that dealt with smaller scales.)

Heiles regaled us with tales of chimneys, supershells, and worms. A chimney is formed when supernovae blow a hole completely out of the Galactic disk; above and below the chimney hover the two ends of the supershell that created it, evidenced by a systematic pattern of high velocities in H I. These supershells probably create the so-called Reynolds layer. A worm is the somewhat twisted cylinder that surrounds a chimney, a region where there is a noticeable regularity in the velocity structure. If a supershell is powerful enough to form a chimney, its energy goes into the halo; if not, the energy remains in the disk. In either case the horizontal resistance of the interstellar medium will squeeze the supershell into a worm. Heiles also gave a picture of leakage of UV radiation from an H II region, leading to the formation of the warm interstellar medium and the diffuse recombination lines.

(But I can't forgive Heiles for including in his title the worst pun perpetrated at this meeting.)

Very much in this vein, Kunze presented a calculation of the formation of a chimney with its associated supershell, as a result of multiple supernova events in a stellar association.

A different phenomenon is the infall of the high-velocity clouds. For this, Gómez, using polarization measures to probe the magnetic field, drew us a picture of clouds falling down the bends in the magnetic field that could be produced by a Parker instability.

Onward and outward. The next scale is a galactic one, with discussions of the distribution of clouds, and of stellar populations and chemical abundances, on the scale of a whole galaxy.

Dickey took on the task of comparing the interstellar content of M31 with that of the Milky Way. M31 has much less molecular gas, largely on account of the large central hole in its distribution. Although its content of H I had been thought to be less that of the Milky Way, a recent increase in the quantity of observational data indicates that the H I content of M31 is probably similar to ours. Dickey also discussed the separation of the warm from the cool H I, by using an interplay of H I emission and absorption observations; and he went at length into the significance of observed antenna temperatures, in terms of how much warm hydrogen is in the line of sight to the nearest cool cloud. I admit that I didn't follow every single thing that he said; so I can't claim the free beer that he offered to anyone who did. However, I would gladly keep *him* in beer while he explained it again to me.

Combes presented a quite interesting simulation of the behavior of the interstellar gas in M51, with clouds growing by collision and coalescence, while GMC's disperse and inject energy into the ISM. As well as I can remember, she had only one free parameter, the pattern speed of the spiral structure; yet she succeeded in getting a good fit to M51.

We heard from Bystrova about the H I mapping that is going on at Pulkovo, including details around dark nebulae and at least one supershell.

Suchkov presented his "hot model" of galaxy formation. The theory is a quantitative one, but qualitatively it appears that the most important feature is that a major burst of star formation inhibits any other star formation for some length of time. Since the metals created by the stars of one burst can enter only into the next burst, the theory predicts a multi-modal distribution of abundances, which one can identify with the halo, the thick disk, and the thin disk.

Finally I want to take special note of the review of the development of stellar populations and mass functions by Ferrini. I can't possibly summarize it; I can only say that he reviewed it thoroughly. There is an unfortunate tendency for some invited reviewers to take the extended time allotted merely as a license to talk for a longer time about their own work. In this case, however, we can thank Dr. Ferrini for avoiding any such temptation and putting a large amount of effort into preparing a well-balanced review.

Still on a galaxy scale, but now looking at a galaxy as a whole, we had a number of papers that dealt with whole disks. Debbie Elmegreen's paper was based on quite novel methods of image processing and display, combined with applications of density-wave theory. She really presented two topics: the display and fitting of spiral patterns and the extraction of the n-fold symmetries of a galaxy. Her fit of a density wave to M81 was impressive; nevertheless there is one thing about it that disturbs me. Her spiral structure ran from the outer Lindblad resonance inward to the circle of 4:1 resonance, and the arms were strongest near corotation. Yet it had seemed a high point of density-wave theory when Roberts *et al.* (1975) fitted galaxies by assuming that corotation was marked by the outermost H II region, and during the seventies Jim Mark (*e.g.*, Mark 1976) reflected things off corotation. Maybe it's just my own lack of understanding, but I wish density-wave theory would make up its mind about some things that seem pretty fundamental.

The other part of the paper I found very interesting. I can't claim to remember all the details of how she extracted the symmetries, but it was fascinating to see galaxies emerge with threefold and fourfold symmetries.

I was sorry that she did not include flocculent galaxies in her discussion, but Bruce Elmegreen told me afterward that he had discussed all types together in another review (Elmegreen 1991). Even so, I would have liked to hear about them here.

Marochnik dealt with the formation of shocks in the passage through a density wave. After referring to quite a number of the earlier papers on the subject, he then discussed his own work, which dealt with the conditions that are needed to produce a shock in the neighborhood of a slowly moving wave.

Yuan discussed the role of rings interior to a grand-design structure; he interpreted them as tightly wound spirals. He emphasized the role of Q in determining where waves can exist.

We also had two papers on the stability of disks. First, Sellwood, using simulations, dealt with a number of types of instability. He showed examples of how a small perturbation can produce a strong spiral instability. He also discussed bar instabilities, showing bars forming and disappearing.

Then Pfenniger emphasized the importance of studying orbits, which he called the building-blocks of galaxies, and he showed some of the fascinating ways in which structures can be built. He concentrated particularly on bars, and he left me for the first time with the feeling that we are beginning to understand these configurations.

The largest scale of all was that of interactions between galaxies. Mirabel emphasized the importance of the gas in interacting galaxies—how a large part of the gas ends up near the center and creates starbursts or fuels active galactic nuclei. I wonder how many of us remember the original suggestion by Baade and Minkowski, back in 1952, that active objects such as Cygnus A were really collisions between galaxies. For many of the intervening decades this idea was completely out of favor, but it looks as if it has now swung full circle again. Mirabel described a sample of super-luminous IRAS galaxies, none of them very close to us. Most of the ones that he and Melnick had studied turned out to be interacting pairs. My experience with Hubble Space Telescope makes me wish immediately for HST images, to reveal in more detail the morphology of these provocative objects. Mirabel also described the distribution of the hydrogen in interacting systems; a large amount of it is in extended tails—one wonders where it eventually goes. Perhaps the answer is in the final study to which he referred, in which there was a strong tendency toward infall of the gas.

Shchekinov concentrated on the theory of starbursts, in which velocities occur that are much higher than those associated with density-wave shocks. His theory called on both gas pressure and radiation pressure as initiating forces.

Also on a galactic scale, Kolesnik described the rather bizarre forms that rings around galaxies can take, emphasizing the role of triaxial potentials in their formation and maintenance.

I have saved for last the bold presentation by Dvořák, who described efforts at what he called a "transfer of technology," using concepts developed in chaos theory (one of whose origin points was in astronomy) but applying them to the workings of the brain. Electroencephalograms apparently produce what can be considered to be a long time-series of a high but somewhat poorly known dimensionality, and their behavior may turn out to be chaotic, in the mathematical sense. Beyond this I cannot go, knowing little about either chaos theory or the nature of the brain; but I was impressed with the boldness of this approach to one of the most challenging of human problems. The advice of Socrates was, "Know thyself," and it may be that at some point in the future we will succeed.

In fact, there is a lot in our culture that applies to our astronomical endeavors. I always think of our science as Faustian, in Goethe's sense. Faust's pact with the Devil gives him power as long as he is still unsatisfied; if he once says, about a pleasing moment, "Verweile doch, du bist so schön! (Oh, linger, you are so beautiful!)," the contract is at an end and the Devil will take his soul. Now there is something downright pleasing about contemplating a good result, but I think that we are all basically Faustian, and are never satisfied to stop.

But I also think of Plato's metaphor of the cave, where we are never allowed to stare at reality, but only at the flickering shadows that it throws on the wall. That's what our always imperfect observations are, but from them we somehow try to construct something that has in it truth, and perhaps even beauty.

Well, our revels really are ended. It has been a good week, and I am indeed tempted to say, "Verweile doch, du bist so schön!" Remember, after all, that the reward of Goethe's Faust was that the Devil never did get him. A host of angels swooped down, proclaiming,

> Wer immer strebend sich bemüht,
> Den können wir erlösen.

("The one who never ceases to strive, him we can save.") May we all be saved!

REFERENCES

Elmegreen, B.G. 1991, in *Dynamics of Galaxies and Their Molecular Cloud Distributions*, eds. F. Combes and F. Casoli (Dordrecht: Kluwer), p. 113.
Mark, J.W.-K., 1976, *Ap. J.*, **205**, 363.
Roberts, W.W., Roberts, M.S., and Shu, F.H. 1975, *Ap. J.*, **196**, 381.